Studies in Systems, Decision and Control

Volume 46

About this Series

The series "Studies in Systems, Decision and Control" (SSDC) covers both new developments and advances, as well as the state of the art, in the various areas of broadly perceived systems, decision making and control- quickly, up to date and with a high quality. The intent is to cover the theory, applications, and perspectives on the state of the art and future developments relevant to systems, decision making, control, complex processes and related areas, as embedded in the fields of engineering, computer science, physics, economics, social and life sciences, as well as the paradigms and methodologies behind them. The series contains monographs, textbooks, lecture notes and edited volumes in systems, decision making and control spanning the areas of Cyber-Physical Systems, Autonomous Systems, Sensor Networks, Control Systems, Energy Systems, Automotive Systems, Biological Systems, Vehicular Networking and Connected Vehicles, Aerospace Systems, Automation, Manufacturing, Smart Grids, Nonlinear Systems, Power Systems, Robotics, Social Systems, Economic Systems and other. Of particular value to both the contributors and the readership are the short publication timeframe and the world-wide distribution and exposure which enable both a wide and rapid dissemination of research output.

More information about this series at http://www.springer.com/series/13304

Anish Deb · Srimanti Roychoudhury
Gautam Sarkar

Analysis and Identification of Time-Invariant Systems, Time-Varying Systems, and Multi-Delay Systems using Orthogonal Hybrid Functions

Theory and Algorithms with MATLAB®

Anish Deb
Department of Applied Physics
University of Calcutta
Kolkata
India

Gautam Sarkar
Department of Applied Physics
University of Calcutta
Kolkata
India

Srimanti Roychoudhury
Department of Electrical Engineering
Budge Budge Institute of Technology
Kolkata
India

ISSN 2198-4182 ISSN 2198-4190 (electronic)
Studies in Systems, Decision and Control
ISBN 978-3-319-26682-4 ISBN 978-3-319-26684-8 (eBook)
DOI 10.1007/978-3-319-26684-8

Library of Congress Control Number: 2015957098

Printed on acid-free paper

This Springer imprint is published by SpringerNature
The registered company is Springer International Publishing AG Switzerland

To my grandson Aryarup Basak who reads only in class four, but wants to be a writer

—Anish Deb

To my mother Dipshikha Roychoudhury, my sister Sayanti Roychoudhury and to my father Sujit Roychoudhury whose affection and loving presence I always feel in my heart

—Srimanti Roychoudhury

To my wife Sati Sarkar and my beloved daughter Paimi Sarkar

—Gautam Sarkar

Preface

The book deals with a new set of orthogonal functions, termed as 'hybrid functions' (HFs). This new set is a combination of 'sample-and-hold functions' (SHFs) and 'triangular functions' (TFs), which are also orthogonal as well. The set of hybrid functions are apt to approximate functions in a piecewise linear manner. From this starting point, the presented analysis takes off and explores many aspects of control system analysis and identification.

Application of non-sinusoidal piecewise constant orthogonal functions was initiated by Walsh functions, which was introduced by J.L. Walsh in 1922 in a conference and a year later was published in the American Journal of Mathematics. The very look of the Walsh function set was very different from the set of sine–cosine functions in the basic sense that it did not contain any curved lines at all!

But Walsh function set was not the first of its kind. Its forerunner was the set of Haar functions, proposed in 1910, which belonged to the same class of piecewise constant orthogonal functions. However, the Haar function set could not make a very significant stir for many decades. But with the advent of wavelet analysis in the 1960s, wider cross section of researchers came to take notice of the Haar function set, now known to be the first ever wavelet function.

For more than four decades, the Walsh function set remained dormant by way of its applications. It became attractive to a few researchers only during the mid-1960s. But in the next 10–15 years, the Walsh function set found its application in many areas of electrical engineering such as communication, solution of differential as well as integral equations, control system analysis, control system identification, and in other various fields. But from the beginning of the 1980s, the spotlight shifted to block pulse functions (BPFs). The BPF set was also orthogonal and piecewise constant. Further, it was related to Walsh functions and Haar functions by similarity transformation. This function set was the most fundamental and simplest of all piecewise constant basis functions (PCBFs). So it is no wonder that the BPF set has been enjoying moderate popularity till date.

In the last decade of the twentieth century and in the first decade of the twenty-first century, few other function sets were introduced in the literature by

Anish Deb and his co-researchers. These are the sample-and-hold function set (1998) and the triangular function set (2003).

In 2010, Anish Deb and his co-workers invented and introduced yet another new set of piecewise linear orthogonal hybrid functions (HFs). This new set could approximate square integrable time functions of Lebesgue measure in a piecewise linear manner, and it used the *samples* of the function as *expansion coefficients*, without using the traditional integration formula employed for orthogonal function-based expansions. Compared to Walsh, block pulse function, and other PCBF-based approximations, this was the main advantage of the HF set because it reduced the computational burden appreciably. Moreover, HF-based approximation incurred much less mean integral square error (MISE) as compared to BPF and other PCBF-based approximations.

In the preliminary chapters, the following topics have been discussed in detail with suitable numerical examples:

(i) properties of hybrid function (HF) and its operational rules,
(ii) function approximation and error estimates,
(iii) integration and differentiation using HF domain operational matrices,
(iv) one-shot operational matrices for integration,
(v) solution of linear differential equations, and
(vi) convolution of time functions.

In later parts of the book, in general, analysis and synthesis of many linear continuous time control systems, which include time-invariant systems, time-varying systems, and multi-delay systems, of homogeneous as well as non-homogeneous types, are discussed. And what attractive results the HF domain technique yielded!

In later chapters, the discussed topics are as follows:

(i) time-invariant and time-varying system analysis via state-space approach,
(ii) multi-delay system analysis via state-space approach,
(iii) time-invariant system analysis using the method of convolution,
(iv) time-invariant and time-varying system identification in state-space environment,
(v) time-invariant system identification using 'deconvolution,' and
(vi) parameter estimation of transfer function from impulse response data.

All the topics are supported with relevant numerical examples. And to make the book user-friendly, many MATLAB programs are appended at the end of the book.

Now, about the hybrid functions and the three authors. The first author started working on this function set in 2005, and the third author was associated with him constantly. The second author got interested in the hybrid function set and joined the other two authors in 2010. Then, from 2012, after publication of a few works on hybrid functions, the first author dreamt about a whole book on hybrid functions and the other two authors strongly supported the dream and joined the mission. Then, all of them toiled and toiled to make the dream come true. It was a great feeling to work with HF, and though toil was the major component during the past few years, it never was able to overtake the academic enjoyment of the authors.

Finally, the authors acknowledge the support of the Department of Applied Physics, University of Calcutta, and the second author acknowledges the support of her institute Budge Budge Institute of Technology, Kolkata, India, during preparation of this book. Also, the support of Dr. Amitava Biswas, Associate Professor, Department of Electrical Engineering, Academy of Technology, Hooghly, India, is gratefully acknowledged.

Kolkata
September 2015

Anish Deb
Srimanti Roychoudhury
Gautam Sarkar

Contents

About the Authors

Anish Deb (b.1951) obtained his B.Tech. (1974), M.Tech. (1976), and Ph.D. (Tech.) (1990) degrees from the Department of Applied Physics, University of Calcutta. Presently, he is a professor (1998) in the Department of Applied Physics, University of Calcutta. His research interest includes automatic control in general and application of 'alternative' orthogonal functions in systems and control. He has published more than 70 research papers in different national and international journals and conference proceedings. He is the principal author of the books 'Triangular Orthogonal Functions for the Analysis of Continuous Time Systems' published by Elsevier (India) in 2007 and Anthem Press (UK) in 2011, and 'Power Electronic Systems: Walsh Analysis with MATLAB' published by CRC Press (USA) in 2014.

Srimanti Roychoudhury (b.1984) did her B.Tech. (2006) from Jalpaiguri Government Engineering College, under West Bengal University of Technology and M.Tech. (2010) from the Department of Applied Physics, University of Calcutta. Presently, she is an assistant professor (from 2010) in the Department of Electrical Engineering, Budge Budge Institute of Technology under West Bengal University of Technology and pursuing her Ph.D. Her research area includes application of 'alternative' orthogonal functions in different areas of systems and control. She has published four research papers in different national and international journals.

Gautam Sarkar (b.1953) received his B.Tech. (1975), M.Tech. (1977), and Ph.D. (Tech) (1991) degrees from the Department of Applied Physics, University of Calcutta. He is presently in the chair of Labanyamoyee Das Professor. His area of research includes automatic control, fuzzy systems, smart grids, and application of piecewise constant basis functions in systems and control. He has published more than 50 research papers in different national and international journals and conference proceedings. He is the co-author of the book 'Triangular Orthogonal Functions for the Analysis of Continuous Time Systems' published by Elsevier (India) in 2007 and Anthem Press (UK) in 2011.

About the Authors

Principal Symbols

δ_{pq}	Kronecker delta
φ_n	$(n + 1)$th component function of a Walsh function set
$\mathbf{\Phi}$	Walsh function vector
ψ_n	$(n + 1)$th component function of a block pulse function set
$\mathbf{\Psi}_{(m)}$	Block pulse function vector of dimension m, having m component functions
μ_i	A point in the ith interval
μ_{max}	Maximum value of μ_i
AMP error	Average of Mod of Percentage error
$\mathbf{A}$	System matrix in state-space model
$\mathbf{A}(t)$	Time-varying system matrix in state-space model
$\mathbf{B}$	Input vector in state-space model
$\mathbf{B}(t)$	Time-varying input vector in state-space model
$\mathbf{C}$	Output matrix in state-space model
$\mathbf{C}(t)$	Time-varying output matrix in state-space model
$\mathbf{D}$	Direct transmission matrix in state-space model
$\mathbf{D}(t)$	Time-varying direct transmission matrix in state-space model
$\mathbf{D}_{G1}$	A diagonal matrix of order m whose entries are the elements of the vector **G1**
$\mathbf{D}_{G2}$	A diagonal matrix of order m whose entries are the elements of the vector **G2**
$\mathbf{D}_{R1}$	A diagonal matrix of order m whose entries are the elements of the vector **R1**
$\mathbf{D}_{R2}$	A diagonal matrix of order m whose entries are the elements of the vector **R2**
$\mathbf{D}_S$	Differentiation matrix for the sample-and-hold function component
$\mathbf{D}_T$	Differentiation matrix for the triangular function component
$f(t)$	Time function
$f(t - \tau)$	Function $f(t)$ delayed by τ seconds

$\dot{f}_{\max}$	Maximum value of first-order derivative of $f(t)$
$\dot{f}(\mu_i)$	First order derivative of $f(t)$ in the $(i + 1)$th interval at the point μ_i
$\bar{f}(t)$	Function approximated in hybrid function domain
$\hat{f}(t)$	Reconstructed function
h	Sampling period
$H_i(t)$	$(i + 1)$th component function of a hybrid function set
$\mathbf{H}_{(m)}(t)$	Hybrid function set of dimension m, having m component functions
LTI	Linear time invariant
LTV	Linear time varying
m	Number of subintervals considered in a time period T
m_i	Slope of the reconstructed function in the $(i + 1)$th interval
MISE	Mean integral square error
$P_i(t)$	$(i + 1)$th Legendre polynomial
P1ss	Sample-and-hold part of the first-order integration operational matrices for integration of the SHF component
P1st	Triangular part of the first-order integration operational matrices for integration of the SHF component
P1ts	Sample-and-hold part of the first-order integration operational matrices for integration of the TF component
P1tt	Triangular part of the first-order integration operational matrices for integration of the TF component
$\mathbf{P}_{(m)}$	Operational matrix for integration in block pulse function domain of dimension m
Pnss	Sample-and-hold part of nth-order integration operational matrices for integration of the SHF component
Pnst	Triangular part of nth-order integration operational matrices for integration of the SHF component
Pnts	Sample-and-hold part of nth-order integration operational matrices for integration of the TF component
Pntt	Triangular part of nth-order integration operational matrices for integration of the TF component
$S_i(t)$	$(i + 1)$th component of a sample-and-hold function set
$\mathbf{S}_{(m)}(t)$	An orthogonal function set of dimension m, sample-and-hold function set of order m, having m component functions
t	Time in seconds
T	Time period
$\mathbf{T1}_{(m)}(t)$	Left-handed triangular function vector of dimension m, having m component functions
$T1_i(t)$	$(i + 1)$th component function of a left-handed triangular function vector
$\mathbf{T2}_{(m)}(t)$	Right-handed triangular function vector of dimension m, having m component functions

$\mathbf{T}_{(m)}(t)$	Same as $\mathbf{T2}_{(m)}(t)$, but renamed as *triangular function vector* of dimension m, having m component functions
$T2_i(t)$	$(i + 1)$th component function of a right-handed triangular function vector
$T_i(t)$	Same as $T2_i(t)$, but renamed as $(i + 1)$th *component function of a triangular function vector*

Chapter 1
Non-sinusoidal Orthogonal Functions in Systems and Control

Abstract This chapter discusses different types of non-sinusoidal orthogonal functions such as Haar functions, Walsh functions, block pulse functions, sample-and-hold functions, triangular functions, non-optimal block pulse functions and a few others. It also discusses briefly the application of Walsh, block pulse and triangular functions, three major members of the non-sinusoidal orthogonal function family, in the area of systems and control. Finally, this chapter proposes a new set of orthogonal functions named 'Hybrid Function' (HF). At the end of the chapter, more than hundred useful references are given.

Orthogonal properties [1] of familiar sine-cosine functions have been known for more than two centuries; but the use of such functions to solve complex analytical problems was initiated by the work of the famous mathematician Baron Jean-Baptiste-Joseph Fourier [2]. Fourier introduced the idea that an arbitrary function, even the one defined by different equations in adjacent segments of its range, could nevertheless be represented by a single analytic expression. Although this idea encountered resistance at the time, it proved to be central to many later developments in mathematics, science, and engineering.

In many areas of electrical engineering the basis for any analysis is a system of sine-cosine functions. This is mainly due to the desirable properties of frequency domain representation of a large class of functions encountered in engineering design and also immense popularity of sinusoidal voltage in most engineering applications.

In the fields of circuit analysis, control theory, communication, and the analysis of stochastic problems, examples are found extensively where the completeness and orthogonal properties of such a system of sine-cosine functions lead to attractive solutions. But with the application of digital techniques in these areas, awareness for other more general complete systems of orthogonal functions has developed. This "new" class of functions, though not possessing some of the desirable properties of sine-cosine functions, has other advantages to be useful in many applications in the context of digital technology. Many members of this class of orthogonal functions are piecewise constant binary valued, and therefore indicate their possible suitability and applicability in the analysis and synthesis of systems leading to piecewise constant solutions.

A. Deb et al., *Analysis and Identification of Time-Invariant Systems, Time-Varying Systems, and Multi-Delay Systems using Orthogonal Hybrid Functions*,
Studies in Systems, Decision and Control 46, DOI 10.1007/978-3-319-26684-8_1

1.1 Orthogonal Functions and Their Properties

Any continuous time function can be synthesized completely to a tolerable degree of accuracy by using a set of orthogonal functions. For such accurate representation of a time function, the orthogonal set should be complete [1].

Let a time function $f(t)$, defined over a time interval [0, T), be represented by an orthogonal function set $\mathbf{S}_{(n)}(t)$. Then

$$f(t) = \sum_{j=0}^{\infty} c_j s_j(t) \tag{1.1}$$

where, c_j is the coefficient or weight connected to the (j + 1)th member of the orthogonal set.

The members of the function set $\mathbf{S}_{(n)}(t)$ are said to be orthogonal in the interval $0 \le t \le T$ if for any positive integral values of p and q, we have

$$\int_0^T s_p(t) s_q(t)\, \mathrm{d}t = \delta_{pq} \text{ (a constant)} \tag{1.2}$$

where, δ_{pq} is the Kronecker delta and

$$\delta_{pq} = \begin{cases} 0 & \text{for } p \neq q \\ \text{constant} & \text{for } p = q \end{cases}$$

When $\delta_{pq} = 1$, the set is said to be orthonormal. An orthogonal set is said to be complete or closed if no function can be found which is normal to each member of the defined set, satisfying Eq. (1.2).

Since, only a finite number of terms of the series $\mathbf{S}_{(n)}(t)$ can be considered for practical realization of any time function $f(t)$, right-hand side (RHS) of Eq. (1.1) has to be truncated and we write

$$f(t) \approx \sum_{j=0}^{N} c_j s_j(t) \tag{1.3}$$

where N is an integer. A point to remember is, N has to be large enough to come up with a solution of the problem with the desired accuracy.

When N is appreciably large, the accuracy of representation is good enough for all practical purposes. Also, it is necessary to choose the coefficients c_j's in such a manner that the mean integral square error (MISE) [3] is minimized. The MISE is defined as

$$\text{MISE} \triangleq \frac{1}{T}\int_0^T \left[f(t) - \sum_{j=0}^{N} c_j s_j(t)\right]^2 \mathrm{d}t \tag{1.4}$$

and its minimization is achieved by making

$$c_j = \frac{1}{T}\int_0^T f(t)s_j(t)\,\mathrm{d}t \tag{1.5}$$

For a complete orthogonal function set, the MISE in Eq. (1.4) decreases monotonically to zero as N tends to infinity.

1.2 Different Types of Non-sinusoidal Orthogonal Functions

For more than four decades different piecewise constant basis functions (PCBF) have been employed to solve problems in different fields of engineering including the area of control theory.

1.2.1 Haar Functions

In 1910, Hungarian mathematician Haar [4] proposed a complete set of piecewise constant binary-valued orthogonal functions that are shown in Fig. 1.1. In fact, Haar functions have three possible states 0 and $\pm A$ where A is a function of $\sqrt{2}$. Thus, the amplitude of the component functions varies with their place in the series.

The component functions of the Haar function set have both scaling and shifting properties. These properties are a necessity for any wavelet [5]. That is why it is now recognized as the first known wavelet basis and at the same time, it is the simplest possible wavelet.

An m-set of Haar functions may be defined mathematically in the semi-open interval $t \in [0, 1)$ as given below.

The first member of the set is

$$\text{har}(0, 0, t) = 1, \quad t \in [0, 1)$$

while the general term for other members is given by

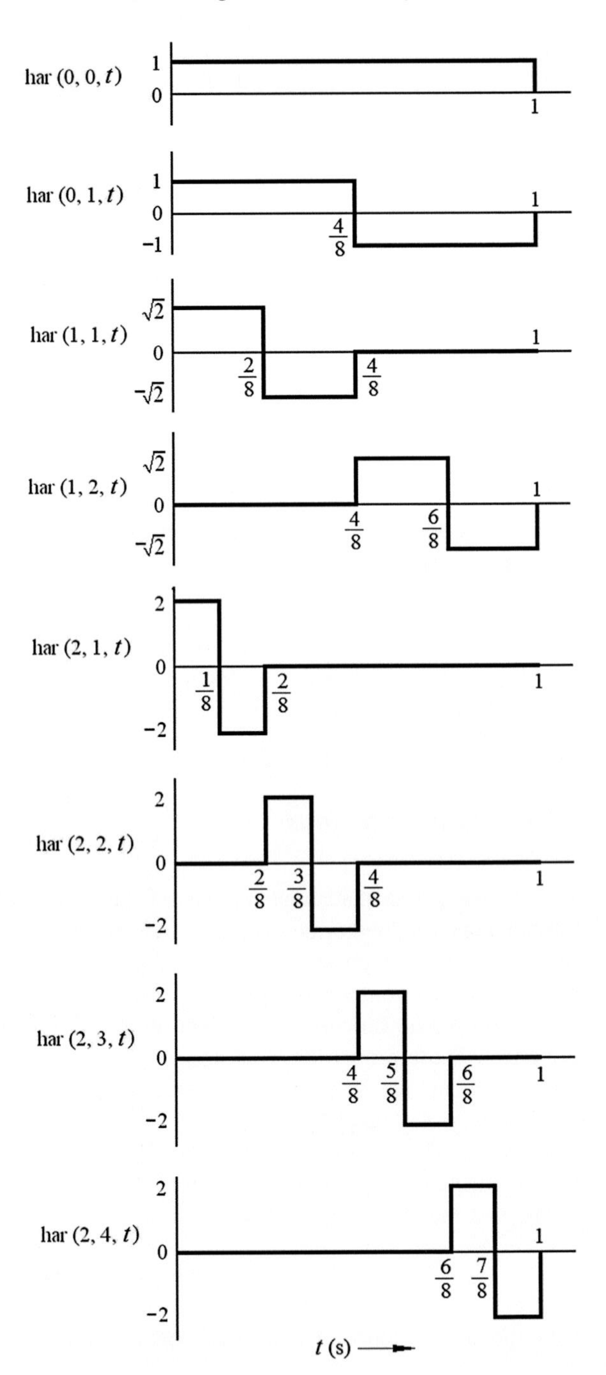

Fig. 1.1 A set of Haar functions

$$\mathrm{har}(j,n,t)=\begin{cases}2^{j/2}, & (n-1)/2^j\leq t<(n-\frac{1}{2})/2^j\\ -2^{j/2}, & (n-\frac{1}{2})/2^j\leq t<n/2^j\\ 0, & \text{elsewhere}\end{cases}\qquad(1.6)$$

where, j, n and m are integers governed by the relations $0\leq j<\log_2(m)$, $1\leq n\leq 2^j$. The number of members in the set is of the form $m = 2^k$, k being a positive integer. Following Eq. (1.6), the members of the set of Haar functions can be obtained in a sequential manner. In Fig. 1.1, k is taken to be 3, thus giving $m = 8$.

Haar's set is such that the formal expansion of a given continuous function in terms of these new functions converges uniformly to the given function.

1.2.2 Rademacher Functions

In 1922, inspired by Haar, German mathematician H. Rademacher presented another set of two-valued orthonormal functions [6] that are shown in Fig. 1.2. The set of Rademacher functions is orthonormal but incomplete. As seen from Fig. 1.2, the function rad(n, t) of the set is given by a square wave of unit amplitude and 2^{n-1} cycles in the semi-open interval [0, 1). The first member of the set rad(0, t) has a constant value of unity throughout the interval.

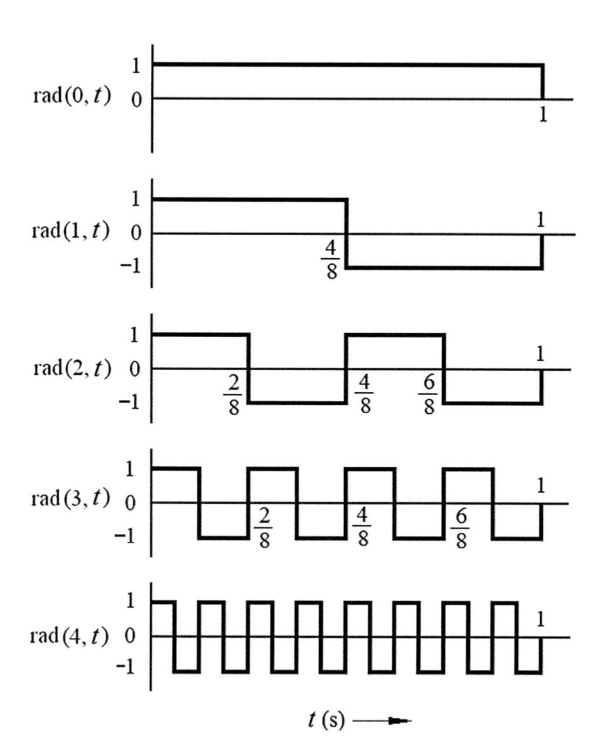

Fig. 1.2 A set of Rademacher functions

1.2.3 Walsh Functions

After the Rademacher functions were introduced in 1922, around the same time, American mathematician J.L. Walsh independently proposed yet another binary-valued complete set of normal orthogonal functions Φ, later named Walsh functions [3, 7], that is shown in Fig. 1.3.

As indicated by Walsh, there are many possible orthogonal function sets of this kind and several researchers, in later years, have suggested orthogonal sets [8–10] formed with the help of combinations of the well-known piecewise constant orthogonal functions.

In his original paper Walsh pointed out that, "… Haar's set is, however, merely one of an infinity of sets which can be constructed of functions of this same character." While proposing his new set of orthonormal functions Φ, Walsh wrote

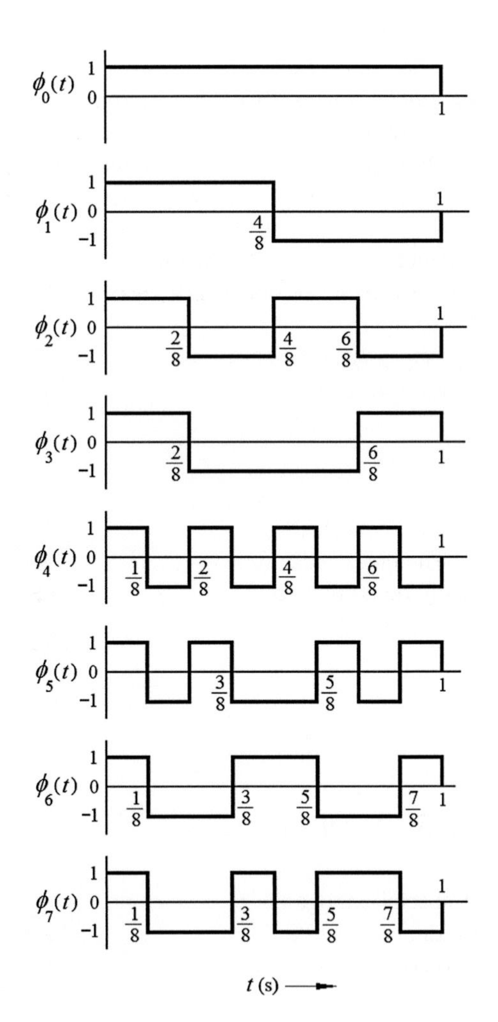

Fig. 1.3 A set of Walsh functions arranged in dyadic order

"… each function φ takes only the values +1 and −1, except at a finite number of points of discontinuity, where it takes the value zero."

However, the Rademacher functions were found to be a true subset of the Walsh function set. The Walsh function set possesses the following properties all of which are not shared by other orthogonal functions belonging to the same class. These are:

(i) Its members are all two-valued functions,
(ii) It is a complete orthonormal set,
(iii) It has striking similarity with the sine-cosine functions, primarily with respect to their zero-crossing patterns.

1.2.4 Block Pulse Functions (BPF)

During the 19th century, voltage and current pulses, such as Morse code signals, were generated by mechanical switches, amplified by relays and finally detected by different magneto-mechanical devices. These pulses are nothing but block pulses—the most important function set used for communication.

However, until the 80s of the last century, the set of block pulses received less attention from the mathematicians as well as application engineers possibly due to their apparent incompleteness. But disjoint and orthogonal properties of such a function set were well known.

A set of block pulse functions [11–13] in the semi-open interval $t \in [0, T)$ is shown in Fig. 1.4.

An m-set block pulse function is defined as

$$\psi_i(t) = \begin{cases} 1 & \text{for } ih \le t < (i+1)h \\ 0 & \text{elsewhere} \end{cases}$$

where, $i = 0, 1, 2, \ldots, (m - 1)$ and $h = \frac{T}{m}$.

The block pulse function set is a complete [14] orthogonal function set and can easily be normalized by defining the component functions in the interval [0, T) as

$$\psi_i(t) = \begin{cases} \frac{1}{\sqrt{h}} & \text{for } ih \le t < (i+1)h \\ 0 & \text{elsewhere} \end{cases}$$

1.2.5 Slant Functions

A special orthogonal function set, known as the slant function set, was introduced by Enomoto and Shibata [15] for image transmission analysis. These functions are also applied successfully to image processing problems [16, 17].

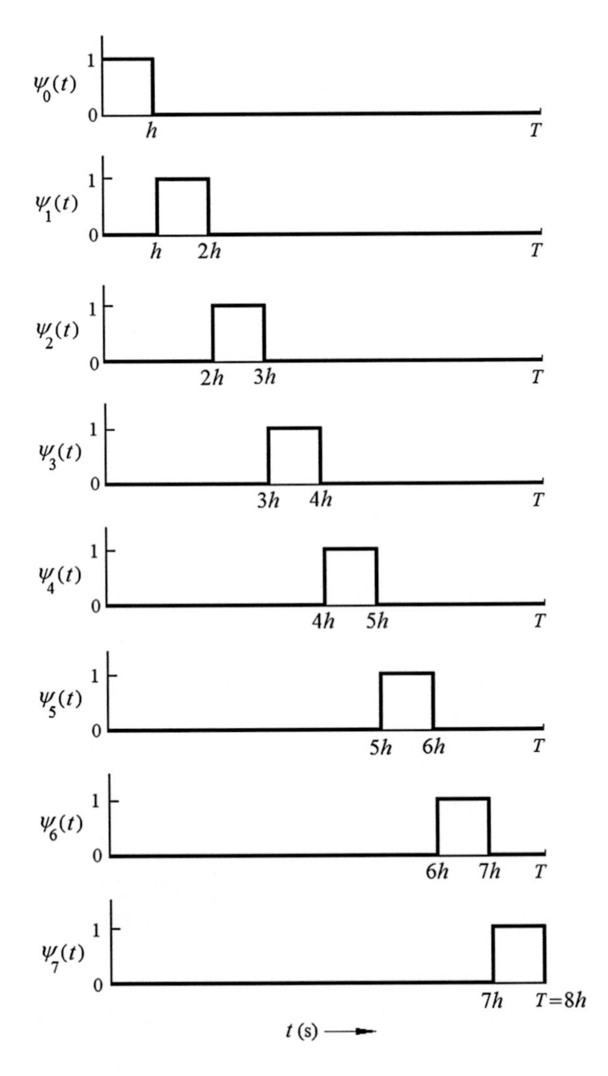

Fig. 1.4 A set of block pulse functions

Slant functions have a finite but a large number of possible states as can be seen from Fig. 1.5. The superiority of the slant function set lies in its transform characteristics, which permit a compaction of the image energy to only a few transformed samples. Thus, the efficiency of image data transmission in this form is improved.

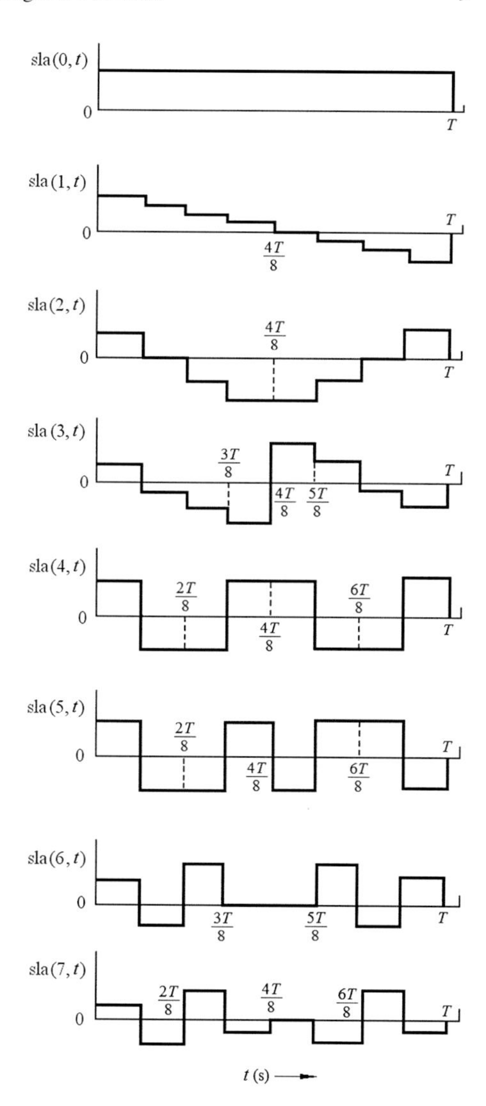

Fig. 1.5 A set of slant functions

1.2.6 Delayed Unit Step Functions (DUSF)

Delayed unit step functions, shown in Fig. 1.6, were suggested by Hwang [18] in 1983. Though not of much use due to its dependency on BPFs, shown by Deb et al. [13], it deserves to be included in the record of piecewise constant basis functions as a new variant. The $(i + 1)$th member of this function set is defined as

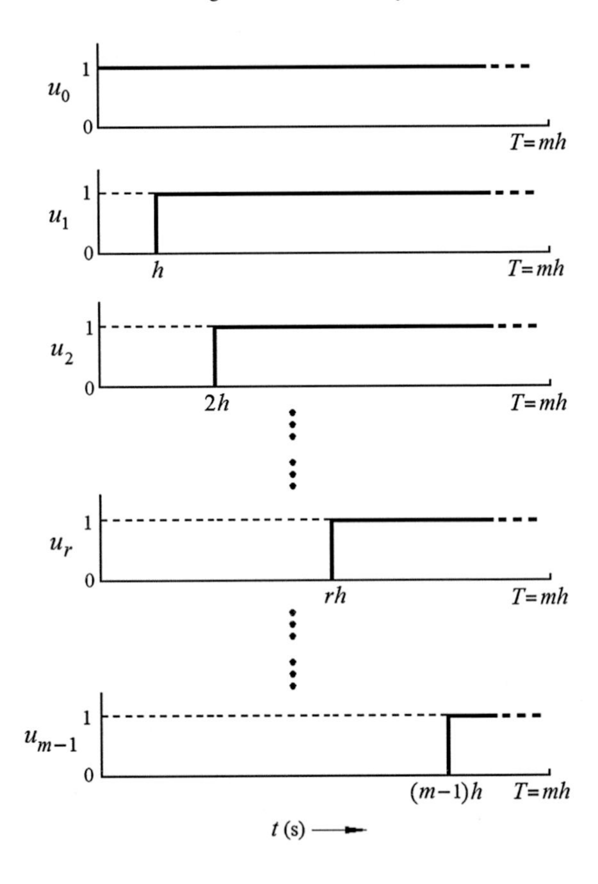

Fig. 1.6 A set of DUSF for m-component functions

$$u_i(t) = \begin{cases} 1 & t \geq ih \\ 0 & t < ih \end{cases}$$

where, $i = 0, 1, 2, \ldots, (m-1)$.

1.2.7 *General Hybrid Orthogonal Functions (GHOF)*

So far the discussion centered on different types of orthogonal functions having a piecewise constant nature. The major departure from this class was the formulation of general hybrid orthogonal functions (GHOF) introduced by Patra and Rao [19, 20]. While sine-cosine functions or orthogonal polynomials can represent a continuous function quite nicely, these functions/polynomials become unsatisfactory for approximating functions with discontinuities, jumps or dead time. For representation of such functions, undoubtedly piecewise constant orthogonal functions such as Walsh or block pulse functions, can be used more advantageously. But with functions having both continuous nature as well as a number of discontinuities in the time interval of interest, it is quite clear that none of the orthogonal functions/polynomials of continuous nature is suitable for approximating the function with a reasonable degree of accuracy. Also, piecewise constant orthogonal functions are not suitable for the job either.

Hence, to meet the combined features of continuity and discontinuity in such situations, the framework of GHOF was proposed and applied by Patra and Rao and it seemed to be more appropriate. The system of GHOF formed a hybrid basis which was both flexible and general.

However, the main disadvantage of GHOF seems to be it requires a priori knowledge about the nature as well as discontinuities of the function, which are to be matched with the segment boundaries of the system of GHOF comprised of different types of orthogonal function sets chosen for the analysis. This also requires a complex algorithm for better results.

1.2.8 Variants of Block Pulse Functions

In 1995, a pulse-width modulated version of the block pulse function set was presented by Deb et al. [21, 22] where, the pulse-width of the component functions of the BPF set was gradually increased (or, decreased) depending upon the nature of the square integrable function to be handled.

In 1998, a further variant of the BPF set was proposed by Deb et al. [23] where, the set was called sample-and-hold function (SHF) set and the same was utilized for the analysis of sampled data systems with zero-order hold.

1.2.9 Sample-and-Hold Functions (SHF)

Any square integrable function $f(t)$ may be represented by a sample-and-hold function set [23] in the semi-open interval $[0, T)$ by considering the $(i + 1)$th member of the set to be

$$f_i(t) = f(ih), \quad i = 0, 1, 2, \ldots, (m - 1)$$

where, h is the sampling period $(=T/m)$, $f_i(t)$ is the amplitude of the function $f(t)$ at time ih and $f(ih)$ is the first term of the Taylor series expansion of the function $f(t)$ around the point $t = ih$, because, for a zero order hold (ZOH) the amplitude of the function $f(t)$ at $t = ih$ is held constant for the duration h.

A set of SHF, comprised of m component functions, is defined as

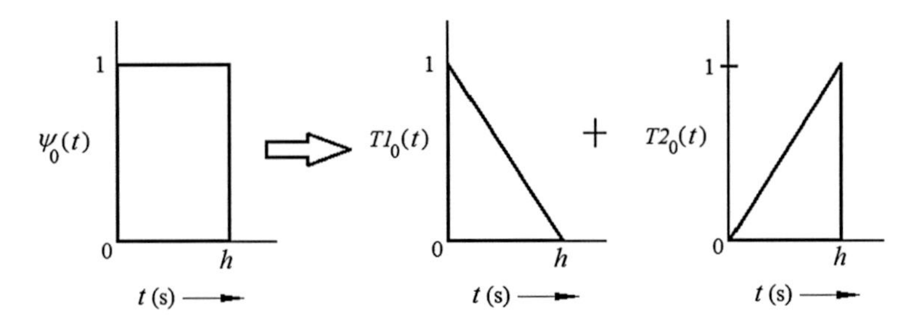

Fig. 1.7 Dissection of the first member of a BPF set into two triangular functions

$$s_i(t) = \begin{cases} 1 & \text{for } ih \leq t < (i+1)h \\ 0 & \text{elsewhere} \end{cases} \tag{1.7}$$

where, $i = 0, 1, 2, \ldots, (m - 1)$.

The basis functions of the SHF set are look-likes of the members of the BPF set shown in Fig. 1.4. Only the method of computation of the coefficients differs in respective cases. That is, the expansion coefficients in SHF domain do not depend upon the traditional integration Formula (1.5).

1.2.10 Triangular Functions (TF)

A rectangular shaped block pulse function can be dissected along its two diagonals to generate two triangular functions [24–26]. That is, when we add two component triangular functions, we get back the original block pulse function. This dissection process is shown in Fig. 1.7, where the first member $\psi_0(t)$ is resolved into two component triangular functions $T1_0(t)$ and $T2_0(t)$.

From a set of block pulse function, $\mathbf{\Psi}_{(m)}(t)$, we can generate two sets of orthogonal triangular functions (TF), namely $\mathbf{T1}_{(m)}(t)$ and $\mathbf{T2}_{(m)}(t)$ such that

$$\mathbf{\Psi}_{(m)}(t) = \mathbf{T1}_{(m)}(t) + \mathbf{T2}_{(m)}(t)$$

These two TF sets are complementary to each other. For convenience, we call $\mathbf{T1}_{(m)}(t)$ the left handed triangular function (LHTF) vector and $\mathbf{T2}_{(m)}(t)$ the right handed triangular function (RHTF) vector. Figure 1.8a, b show the orthogonal triangular function sets, $\mathbf{T1}_{(m)}(t)$ and $\mathbf{T2}_{(m)}(t)$, where m has been chosen arbitrarily as 8. For triangular function domain expansion of a time function, the coefficients are computed from function samples only [26], and they do not need any assistance from Eq. (1.5).

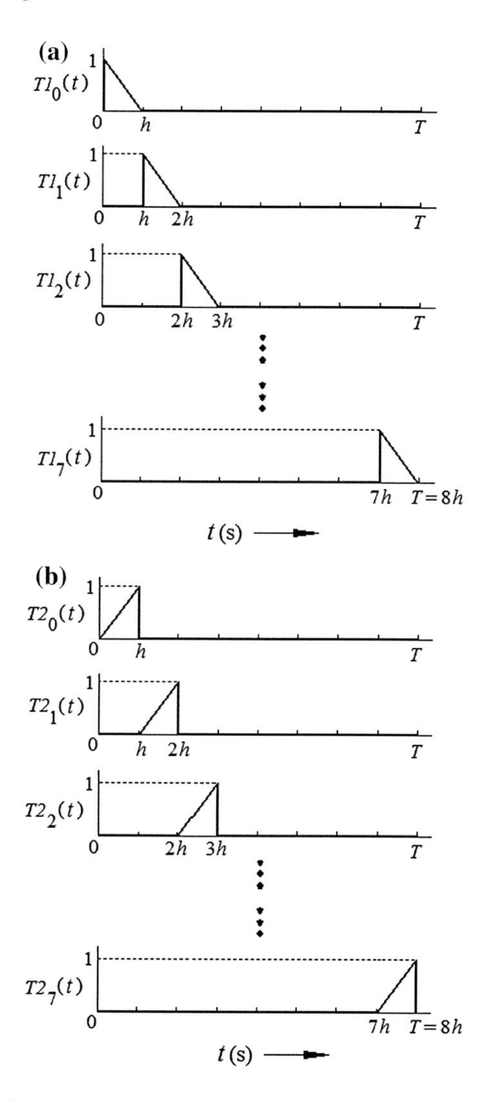

Fig. 1.8 **a** A set of eight LHTF $\mathbf{T1}_{(8)}(t)$. **b** A set of eight RHTF $\mathbf{T2}_{(8)}(t)$

1.2.11 Non-optimal Block Pulse Functions (NOBPF)

The 'non-optimal' method of block pulse function coefficient computation has been suggested by Deb et al. [27] which employs trapezoidal [28] integration instead of exact integration. The approach is 'new' in the sense that the BPF expansion coefficients of a locally square integrable function have been determined in a more 'convenient' manner.

This 'non-optimal' expansion procedure for computation of coefficients uses the trapezoidal rule for integration where only the samples of the function to be approximated are needed in any particular time intervals to represent the function in NOBPF [27] domain and thus reduces the computational burden.

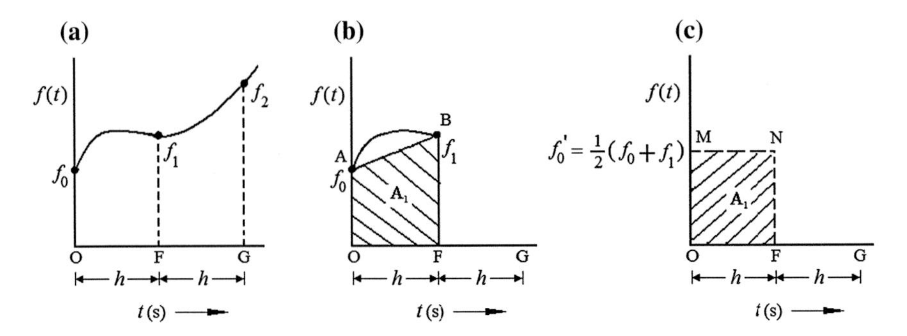

Fig. 1.9 Function approximation in non-optimal block pulse function (NOBPF) domain where area preserving transformation is employed

Let us employ the well-known trapezoidal rule for integration to compute the non-optimal block pulse function coefficients of a time function $f(t)$. Calling these coefficients f_i''s, we get

$$f_i' \approx \frac{\frac{1}{2}[f(ih) + f\{(i+1)h\}] \cdot h}{h} = \frac{[f(ih) + f\{(i+1)h\}]}{2} \tag{1.8}$$

f_i''s are 'non-optimal' coefficients computed approximately from Eq. (1.8). It is observed that f_i''s are in effect, the average values of two consecutive samples of the function $f(t)$, and this is again a significant deviation from the traditional Formula (1.5).

The process of function approximation in non-optimal block pulse function (NOBPF) domain is shown in Fig. 1.9.

A time function $f(t)$ can be approximated in NOBPF domain as

$$f(t) \approx \begin{bmatrix} f_0' & f_1' & f_2' & \cdots & f_i' & \cdots & f_{(m-1)}' \end{bmatrix} \boldsymbol{\Psi}_{(m)}'(t) = F'^{\mathrm{T}} \boldsymbol{\Psi}_{(m)}'(t) \tag{1.9}$$

It is evident that $f(t)$ in (1.9) will not be approximated with guaranteed mean integral square error (MISE).

Figure 1.10 shows a time scale history of all the functions discussed above.

1.3 Walsh Functions in Systems and Control

Among all the orthogonal functions outlined earlier, Walsh function based analysis first became attractive to the researchers from 1962 onwards [7, 29–31]. The reason for such success was mainly due to its binary nature. One immediate advantage is the task of analog multiplication. To multiply any signal by a Walsh function, the problem reduces to an appropriate sequence of sign changes, which makes this usually difficult operation both simple and potentially accurate [29]. However, in

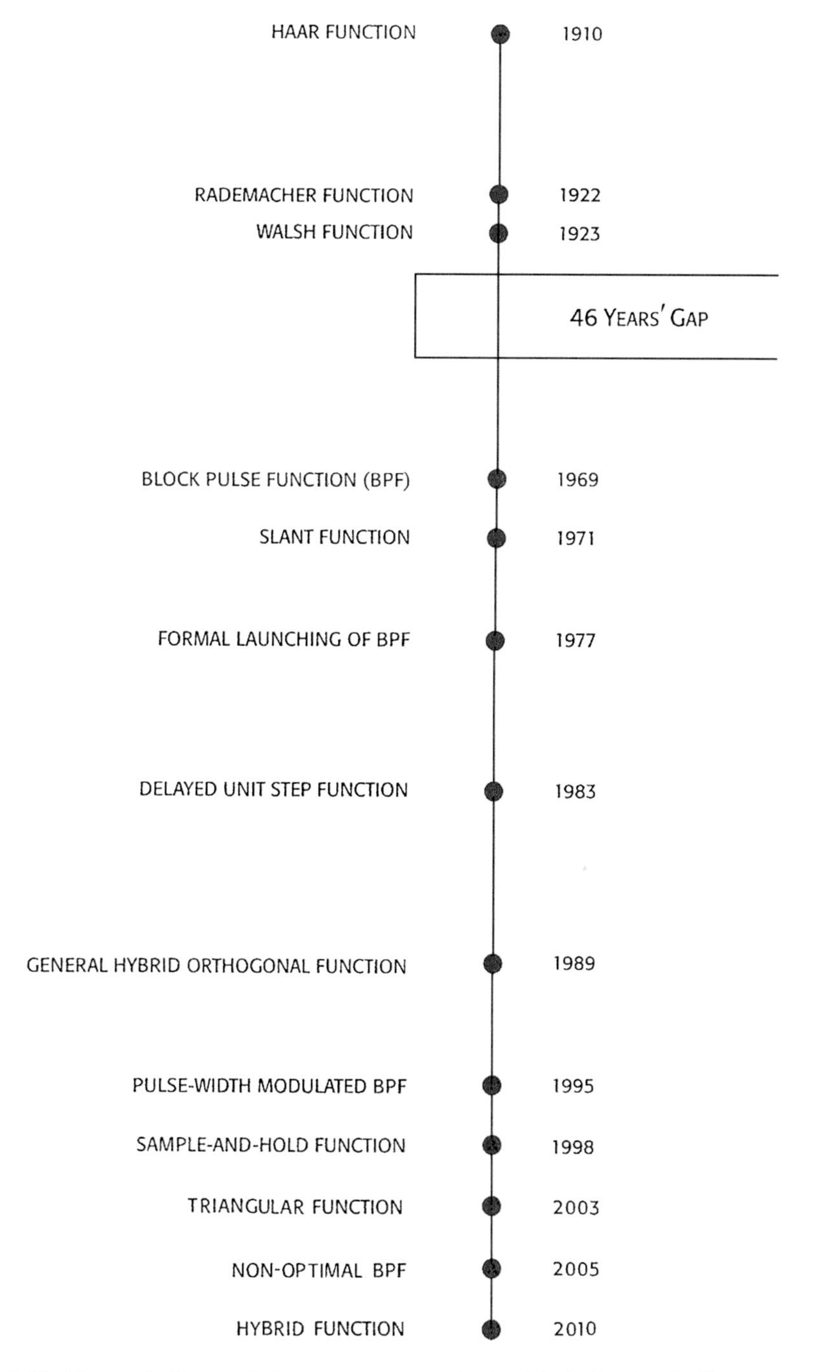

Fig. 1.10 Time scale history of piecewise constant and related basis function family

system analysis, Walsh functions were employed during the early 1970s. As a consequence, the advantages of Walsh analysis were unraveled to the workers in the field compared to the use of conventional sine-cosine functions. Ultimately, the mathematical groundwork of the Walsh analysis became strong to lure interested researchers to try every new application based upon this function set.

In 1973, it was Corrington [32] who proposed a new technique for solving linear as well as nonlinear differential and integral equations with the help of Walsh functions. In 1975, important technical papers relating Walsh functions to the field of systems and control were published. New ideas were proposed by Rao [33–41] and Chen [42–47]. Other notable workers were Le Van et al. [48], Tzafestas [49], Chen [50–53], Mahapatra [54], Paraskevopoulos [55], Moulden [56], Deb and Datta [57–59], Lewis [60], Marszalek [61], Dai and Sinha [62], Deb et al. [63–68], and others.

The first positive step for the development of the Walsh domain analysis was the formulation of the operational matrix for integration. This was done independently by Rao [33], Chen [42], and Le Van et al. [48]. Le Van sensed that since the integral operator matrix had an inverse, the inverse must be the differential operator in the Walsh domain. However, he could not represent the general form of the operator matrix that was done by Chen [42, 43]. Interestingly, the operational matrix for integration was first presented by Corrington [32] in the form of a table. But he failed to recognize the potentiality of the table as a matrix.

This was first pointed out by Chen and he presented Walsh domain analysis with the operational matrices for integration as well as differentiation:

(i) to solve the problems of linear systems by state space model [42];
(ii) to design piecewise constant gains for optimal control [43];
(iii) to solve optimal control problem [44];
(iv) in variational problems [45];
(v) for time domain synthesis [46];
(vi) for fractional calculus as applied to distributed systems [47].

Rao used Walsh functions for:

(i) system identification [33];
(ii) optimal control of time-delay systems [35];
(iii) identification of time-lag systems [36];
(iv) transfer function matrix identification [38] and piecewise linear system identification [39];
(v) parameter estimation [40];
(vi) solving functional differential equations and related problems [41].

Rao first formulated the operational matrices for stretch and delay [41]. He proposed a new technique for extension of computation beyond the limit of initial normal interval with the help of “single term Walsh series” approach [37], and estimated the error due to the use of different operational matrices [40]. Rao and Tzafestas indicated the potentiality of Walsh and related functions in the area of systems and control in a review paper [69]. Tzafestas [70] assessed the role of

Walsh functions in signal and system analysis and design, in a rich collection of papers.

W.L. Chen defined a "shift Walsh matrix" for solving delay-differential equations [51], and used Walsh functions for parameter estimation of bilinear systems [50] as well as for the analysis of multi-delay systems [53]. Paraskevopoulos determined the transfer functions of a single input single output (SISO) system from its impulse response data with the help of Walsh functions and a fast Walsh algorithm [55]. Tzafestas applied Walsh series approach for lumped and distributed system identification [49]. Mahapatra used Walsh functions for solving matrix Riccati equation arising in optimal control studies of linear diffusion equations [54]. Moulden's work was concerned with the application of Walsh spectral analysis of ordinary differential equations in a very formal manner [56]. Deb applied Walsh functions to analyze power-electronic systems [57, 58], Deb and Datta was the first to define the Walsh Operational Transfer Function (WOTF) for the analysis of linear SISO systems [57, 58, 63–65]. Deb was the first to notice the oscillatory behavior in the Walsh domain analysis of first-order systems [68].

1.4 Block Pulse Functions in Systems and Control

The earliest work concerning completeness and suitability of BPF for use in place of Walsh functions, is a small technical note of Rao and Srinivasan [71]. Later Kwong and Chen [14], Chen and Lee [72], and Sloss and Blyth [73] discussed convergence properties of BPF series and the BPF solution of a linear time invariant system.

Sannuti's paper [74] on the analysis and synthesis of dynamical systems in state space was a significant step toward BPF applications. Shieh et al. [75] dealt with the same problems. The doctoral dissertation of Srinivasan [76] contained several applications of BPF to a variety of problems. Rao and Srinivasan proposed methods of analysis and synthesis for delay systems [77] where an operational matrix for delay via BPF was proposed.

Chen and Jeng [78] considered systems with piecewise constant delays. BPF's are also used to invert Laplace transforms [79–82] numerically. Differential equations, related to the dynamics of current collection mechanism of electric locomotives, contain terms with a stretched argument. Such equations have been treated in Ref. [83] using BPF. Chen [84] also dealt with scaled systems. BPFs have been used in obtaining discrete-time approximations of continuous-time systems. Shieh et al. [75] and recently Sinha et al. [85] gave some interesting results in this connection. The BPF method of discretization has been compared with other techniques employing bilinear transformation, state transition matrix, etc.

The higher powers of the operational matrix for integration accomplished the task of repeated integration. However, the use of higher powers led to accumulation of error at each stage of integration. This has been recognized by Rao and Palanisamy who gave one shot operational matrices for repeated integration via

BPF and Walsh functions. Wang [86] deals with the same aspect suggesting improvements in operational matrices for fractional and operational calculus. Palanisamy reveals certain interesting aspects of the operational matrix for integration [87]. Optimal controls for time-varying systems have also been worked out [88]. Kawaji [89] gave an analysis of linear systems with observers.

Replacement of Walsh function by block pulse took place in system identification algorithms for computational advantage. Shih and Chia [90] used BPF in identifying delay systems. Jan and Wong [91] and Cheng and Hsu [92] identified bilinear models. Multidimensional BPFs have been proposed by Rao and Srinivasan [93]. These were used in solving partial differential equations. Nath and Lee [94] discussed multidimensional extensions of block pulse with applications.

Identification of nonlinear distributed systems and linear feedback systems via block pulse functions were done by Hsu and Cheng [95] and Kwong and Chen [96]. Palanisamy and Bhattacharya also used block pulse functions in system identification [97] and in analyzing stiff systems [98]. Solution of multipoint boundary value problems and integral equations were obtained using a set of BPF [99, 100]. In parameter estimation of bilinear systems Cheng and Hsu [101] applied block pulse functions. Still many more applications of block pulse functions remain to be mentioned.

Thus block pulse function continued to reign over other piecewise constant orthogonal functions with its simple but powerful attributes. But numerical instability is observed when deconvolution operation [102] in BPF domain is executed for system identification. Also, oscillations where observed [68] for system analysis in BPF domain.

1.5 Triangular Functions (TF) in Systems and Control

It was Corrington [32] who initiated application of Walsh functions in the area of systems and control by solving differential and integral equations with this new set. Only four years later, Chen et al. [47] took up the trail and came up with formal representation of block pulse functions and partially used them in conjunction with Walsh functions for solving problems related to distributed systems. Two years earlier, operational matrices for integration and differentiation in Walsh domain were proposed by Chen et al. [46]. It was also independently presented by Rao and Sivakumar [33] and Le Van et al. [48]. Such operational matrices played a vital role in the analysis and synthesis of control systems.

With the onset of the 80s, the block pulse function set proved to be more elegant, simple and computationally attractive compared to Walsh functions. Thus, it enjoyed immense popularity for more than a decade. Later variants of block pulse functions were employed successfully by Deb et al. [22, 23]. In the last decade, a new set of functions, namely, triangular function set were introduced by Deb et al. [24–26]. Gradually, these new sets of functions attracted many researchers [103–106].

1.6 A New Set of Orthogonal Hybrid Functions (HF): A Combination of Sample-and-Hold Functions (SHF) and Triangular Functions (TF)

In the present work, a new set of piecewise orthogonal hybrid function (HF), derived from sample-and-hold [23] functions and triangular functions [24–26], is proposed. This new set of functions is used for the analysis and synthesis of various types of linear continuous time non-homogeneous as well as homogeneous control systems, namely, time-invariant systems, time varying systems and multi-delay systems. The main advantage of the hybrid function set is it works with function samples. Also, it provides the time solutions in a piecewise linear manner in two parts: sample-and-hold function component and the triangular function component. And if we leave out the TF component of the solution, we are left with the sample-and-hold function component which is sometimes needed for analyzing digital control systems and related research.

Function approximation in traditional BPF domain requires many integration operations. That is, computation of each coefficient means performing one numerical integration, thus requiring more time as well as memory space increasing the computational burden. But, this 'new' HF domain approach is more suited to handle complicated functions because it works with function samples as coefficients. Also, for identification of control systems, block pulse domain technique gives rise to numerical instability [102], while the HF domain approach does not. Thus, the HF domain approach seems to be more efficient in many ways than traditional approaches, the main reason being this set works with samples only.

With this rich background, it seems worthwhile to explore this field.

References

1. Sansone, G.: Orthogonal Functions. Interscience Publishers Inc., New York (1959)
2. Fourier, J.B.: Thèorie analytique de la Chaleur, 1828. English edition: The analytic theory of heat, 1878, Reprinted by Dover Pub. Co, New York (1955)
3. Beauchamp, K.G.: Walsh functions and their applications. Academic Press, London (1975)
4. Haar, Alfred: Zur theorie der orthogonalen funktionen systeme. Math. Annalen **69**, 331–371 (1910)
5. Mix Dwight, F., Olejniczak, K.J.: Elements of Wavelets for Engineers and Scientists. Wiley Interscience (2003)
6. Rademacher, H.: Einige sätze von allegemeinen orthogonal funktionen. Math. Annalen **87**, 122–138 (1922)
7. Walsh, J.L.: A closed set of normal orthogonal functions. Am. J. Math. **45**, 5–24 (1923)
8. Harmuth, H.F.: Transmission of information by orthogonal functions, 2nd edn. Springer, Berlin (1972)
9. Fino, B.J., Algazi, V.R.: Slant-Haar transform. Proc. IEEE **62**, 653–654 (1974)
10. Huang, D.M.: Walsh-Hadamard-Haar hybrid transform. In: IEEE Proceedings of 5th International Conference on Pattern Recognition, pp. 180–182 (1980)

11. Wu, T.T., Chen, C.F., Tsay, Y.T.: Walsh operational matrices for fractional calculus and their application to distributed systems. In: IEEE Symposium on Circuits and Systems, Munich, Germany, April, 1976
12. Jiang, J.H., Schaufelberger, W.: Block Pulse Functions and Their Application in Control System. LNCIS, vol. 179. Springer, Berlin (1992)
13. Deb, Anish, Sarkar, Gautam, Sen, S.K.: Block pulse functions, the most fundamental of all piecewise constant basis functions. Int. J. Syst. Sci. **25**(2), 351–363 (1994)
14. Kwong, C.P., Chen, C.F.: The convergence properties of block pulse series. Int. J. Syst. Sci. **12**(6), 745–751 (1981)
15. Enomoto, H., Shibata, K.: Orthogonal transform coding system for television signals. In: Proceedings of the 1971 Symposium on Application of Walsh Functions, Washington DC, USA, pp. 11–17 (1971)
16. Pratt, W.K., Welch, L.R., Chen, W.: Slant transform for image coding. In: Proceedings of the 1972 Symposium on Application of Walsh Functions, Washington DC, USA, pp. 229–234, March 1972
17. Pratt, W.K.: Digital Image Processing. Wiley, New York (1978)
18. Hwang, Chyi: Solution of functional differential equation via delayed unit step functions. Int. J. Syst. Sci. **14**(9), 1065–1073 (1983)
19. Patra, A., Rao, G.P.: General hybrid orthogonal functions—a new tool for the analysis of power-electronic systems. IEEE Trans. Ind. Electron. **36**(3), 413–424 (1989)
20. Patra, A., Rao, G.P.: General Hybrid Orthogonal Functions and Their Applications in Systems and Control. Springer, Berlin (1996)
21. Deb, Anish, Sarkar, Gautam, Sen, S.K.: Linearly pulse-width modulated blockpulse functions and their application to linear SISO feedback control system identification. Proc. IEE, Part D, Control Theory Appl. **142**(1), 44–50 (1995)
22. Deb, Anish, Sarkar, Gautam, Sen, S.K.: A new set of pulse width modulated generalised block pulse functions (PWM-GBPF) and their application to cross/auto-correlation of time varying functions. Int. J. Syst. Sci. **26**(1), 65–89 (1995)
23. Deb, Anish, Sarkar, Gautam, Bhattacharjee, Manabrata, Sen, S.K.: A new set of piecewise constant orthogonal functions for the analysis of linear SISO systems with sample-and-hold. J. Franklin Inst. **335B**(2), 333–358 (1998)
24. Deb, Anish, Sarkar, Gautam, Dasgupta, Anindita: A complementary pair of orthogonal triangular function sets and its application to the analysis of SISO control systems. J. Inst. Eng. (India) **84**(December), 120–129 (2003)
25. Deb, Anish, Dasgupta, Anindita, Sarkar, Gautam: A complementary pair of orthogonal triangular function sets and its application to the analysis of dynamic systems. J. Franklin Inst. **343**(1), 1–26 (2006)
26. Deb, Anish, Sarkar, Gautam, Sengupta, Anindita: Triangular Orthogonal Functions for the Analysis of Continuous Time Systems. Anthem Press, London (2011)
27. Deb, A., Sarkar, G., Mandal, P., Biswas, A., Sengupta, A.: Optimal block pulse function (OBPF) vs. Non-optimal block Pulse function (NOBPF). In: Proceedings of International Conference of IEE (PEITSICON) 2005, Kolkata, 28–29 Jan 2005, pp. 195–199
28. Riley, K.F., Hobson, M.P., Bence, S.J.: Mathematical Methods for Physics and Engineering, 2nd edn. Cambridge University Press, UK (2004)
29. Beauchamp, K.G.: Applications of Walsh and Related Functions with An Introduction to Sequency Theory. Academic Press, London (1984)
30. Maqusi, M.: Applied Walsh Analysis. Heyden, London (1981)
31. Hammond, J.L., Johnson, R.S.: A review of orthogonal square wave functions and their application to linear networks. J. Franklin Inst. **273**, 211–225 (1962)
32. Corrington, M.S.: Solution of differential and integral equations with Walsh functions. IEEE Trans. Circuit Theory **CT-20**(5), 470–476 (1973)
33. Rao, G.P., Sivakumar, L.: System identification via Walsh functions. Proc. IEE **122**(10), 1160–1161 (1975)

34. Rao, G.P., Palanisamy, K.R.: A new operational matrix for delay via Walsh functions and some aspects of its algebra and applications. In: 5th National Systems Conference, NSC-78, PAU Ludhiana (India), Sept 1978, pp. 60–61
35. Rao, G.P., Palanisamy, K.R.: Optimal control of time-delay systems via Walsh functions. In: 9th IFIP Conference on Optimisation Techniques, Polish Academy of Sciences, System Research Institute, Poland, Sept 1979
36. Rao, G.P., Sivakumar, L.: Identification of time-lag systems via Walsh functions. IEEE Trans. Autom. Control **AC-24**(5), 806–808 (1979)
37. Rao, G.P., Palanisamy, K.R., Srinivasan, T.: Extension of computation beyond the limit of initial normal interval in Walsh series analysis of dynamical systems. IEEE Trans. Autom. Control **AC-25**(2), 317–319 (1980)
38. Rao, G.P., Sivakumar, L.: Transfer function matrix identification in MIMO systems via Walsh functions. Proc. IEEE **69**(4), 465–466 (1981)
39. Rao, G.P., Sivakumar, L.: Piecewise linear system identification via Walsh functions. Int. J. Syst. Sci. **13**(5), 525–530 (1982)
40. Rao, G.P.: Piecewise Constant Orthogonal Functions and Their Application in Systems and Control. LNC1S, vol. 55. Springer, Berlin (1983)
41. Rao, G.P., Palanisamy, K.R.: L Walsh stretch matrices and functional differential equations. IEEE Trans. Autom. Control **AC-21**(1), 272–276 (1982)
42. Chen, C.F., Hsiao, C.H.: A state space approach to Walsh series solution of linear systems. Int. J. Syst. Sci. **6**(9), 833–858 (1975)
43. Chen, C.F., Hsiao, C.H.: Design of piecewise constant gains for optimal control via Walsh functions. IEEE Trans. Autom. Control **AC-20**(5), 596–603 (1975)
44. Chen, C.F., Hsiao, C.H.: Walsh series analysis in optimal control. Int. J. Control **21**(6), 881–897 (1975)
45. Chen, C.F., Hsiao, C.H.: A Walsh series direct method for solving variational problems. J. Franklin Inst. **300**(4), 265–280 (1975)
46. Chen, C.F., Hsiao, C.H.: Time domain synthesis via Walsh functions. IEE Proc. **122**(5), 565–570 (1975)
47. Chen, C.F., Tsay, Y.T., Wu, T.T.: Walsh operational matrices for fractional calculus and their application to distributed systems. J. Franklin Inst. **303**(3), 267–284 (1977)
48. Le Van, T., Tam, L.D.C., Van Houtte, N.: On direct algebraic solutions of linear differential equations using Walsh transforms. IEEE Trans. Circuits Syst. **CAS-22**(5), 419–422 (1975)
49. Tzafestas, S.G.: Walsh series approach to lumped and distributed system identification. J. Franklin Inst. **305**(4), 199–220 (1978)
50. Chen, W.L., Shih, Y.P.: Parameter estimation of bilinear systems via Walsh functions. J. Franklin Inst. **305**(5), 249–257 (1978)
51. Chen, W.L., Shih, Y.P.: Shift Walsh matrix and delay differential equations. IEEE Trans. Autom. Control **AC-23**(6), 1023–1028 (1978)
52. Chen, W.L., Lee, C.L.: Walsh series expansions of composite functions and its applications to linear systems. Int. J. Syst. Sci. **13**(2), 219–226 (1982)
53. Chen, W.L.: Walsh series analysis of multi-delay systems. J. Franklin Inst. **303**(4), 207–217 (1982)
54. Mahapatra, G.B.: Solution of optimal control problem of linear diffusion equation via Walsh functions. IEEE Trans. Autom. Control **AC-25**(2), 319–321 (1980)
55. Paraskevopoulos, P.N., Varoufakis, S.J.: Transfer function determination from impulse response via Walsh functions. Int. J. Circuit Theory Appl. **8**(1), 85–89 (1980)
56. Moulden, T.H., Scott, M.A.: Walsh spectral analysis for ordinary differential equations: Part 1—Initial value problem. IEEE Trans. Circuits Syst. **CAS-35**(6), 742–745 (1988)
57. Deb, Anish, Datta, A.K.: Time response of pulse-fed SISO systems using Walsh operational matrices. Adv. Model. Simul. **8**(2), 30–37 (1987)
58. Deb, A.: On Walsh domain analysis of power-electronic systems. Ph.D. (Tech.) dissertation, University of Calcutta (1989)

59. Deb, Anish, Sen, S.K., Datta, A.K.: Walsh functions and their applications: a review. IETE Tech. Rev. **9**(3), 238–252 (1992)
60. Lewis, F.L., Mertzios, B.G., Vachtsevanos, G., Christodoulou, M.A.: Analysis of bilinear systems using Walsh functions. IEEE Trans. Autom. Control **35**(1), 119–123 (1990)
61. Marszalek, W.: Orthogonal functions analysis of singular systems with impulsive responses. Proc. IEE, Part D, Control Theory Appl. **137**(2), 84–86 (1990)
62. Dai, H., Sinha, N.K.: Robust coefficient estimation of Walsh functions. Proc. IEE, Part D, Control Theory Appl. **137**(6), 357–363 (1990)
63. Deb, Anish, Datta, A.K.: Analysis of continuously variable pulse-width modulated system via Walsh functions. Int. J. Syst. Sci. **23**(2), 151–166 (1992)
64. Deb, Anish, Datta, A.K.: Analysis of pulse-fed power electronic circuits using Walsh functions. Int. J. Electron. **62**(3), 449–459 (1987)
65. Deb, A., Sarkar, G., Sen, S.K., Datta, A.K.: A new method of analysis of chopper-fed DC series motor using Walsh function. In: Proceedings of 4th European Conference on Power Electronics and Applications (EPE '91), Horence, Italy, 1991
66. Deb, A., Datta, A.K.: On Walsh/block pulse domain analysis of power electronic circuits Part 1. Continuously phase-controlled rectifier. Int. J. Electron. **79**(6), 861–883 (1995)
67. Deb, A., Datta, A.K.: On Walsh/block pulse domain analysis of power electronic circuits Part 2. Continuously pulse-width modulated inverter. Int. J. Electron. **79**(6), 885–895 (1995)
68. Deb, A., Fountain, D.W.: A note on oscillations in Walsh domain analysis of first order systems. IEEE Trans. Circuits Syst. **CAS-38**(8), 945–948 (1991)
69. Rao, G.P., Tzafestas, S.G.: A decade of piecewise constant orthogonal functions in systems and control. Math. Comput. Simul. **27**(5 and 6), 389–407 (1985)
70. Tzafestas, S.G. (ed.): Walsh functions in signal and systems analysis and design. Van Nostrand Reinhold Co., New York (1985)
71. Rao, G.P., Srinivasan, T.: Remarks on "Author's reply" to "Comments on design of piecewise constant gains for optimal control via Walsh functions". IEEE Trans. Autom. Control **AC-23**(4), 762–763 (1978)
72. Chen, W.L., Lee, C.L.: On the convergence of the block-pulse series solution of a linear time-invariant system. Int. J. Syst. Sci. **13**(5), 491–498 (1982)
73. Sloss, G., Blyth, W.F.: A priori error estimates for Corrington's Walsh function method. J. Franklin Inst. **331B**(3), 273–283 (1994)
74. Sannuti, P.: Analysis and synthesis of dynamic systems via block pulse functions. Proc. IEE **124**(6), 569–571 (1977)
75. Shieh, L.A., Yates, R.E., Navarro, J.M.: Representation of continuous time state equations by discrete-time state equations. IEEE Trans. Signal Man Cybern. **SMC-8**(6), 485–492 (1978)
76. Srinivasan, T.: Analysis of dynamical systems via block-pulse functions. Ph. D. dissertation, Department of Electrical Engineering, I.I.T., Kharagpur, India (1979)
77. Rao, G.P., Srinivasan, T.: Analysis and synthesis of dynamic systems containing time delays via block-pulse functions. Proc. IEE **125**(9), 1064–1068 (1978)
78. Chen, W.L., Jeng, B.S.: Analysis of piecewise constant delay systems via block-pulse functions. Int. J. Syst. Sci. **12**(5), 625–633 (1981)
79. Hwang, C., Guo, T.Y., Shih, Y.P.: Numerical inversion of multidimensional Laplace transforms via block-pulse function. Proc. IEE, Part. D, Control Theory Appl. **130**(5), 250–254 (1983)
80. Marszalek, W.: Two-dimensional inverse Laplace transform via block-pulse functions method. Int. J. Syst. Sci. **14**, 1311–1317 (1983)
81. Jiang, Z.H.: New approximation method for inverse Laplace transforms using block-pulse functions. Int. J. Syst. Sci. **18**(10), 1873–1888 (1987)
82. Shieh, L.A., Yates, R.E.: Solving inverse Laplace transform, linear and nonlinear state equations using block-pulse functions. Comput. Electr. Eng. **6**, 3–17 (1979)
83. Rao, G.P., Srinivasan, T.: An optimal method of solving differential equations characterizing the dynamic of a current collection system for an electric locomotive. J. Inst. Math. Appl. **25** (4), 329–342 (1980)

84. Chen, W.L.: Block-pulse series analysis of scaled systems. Int. J. Syst. Sci. **12**(7), 885–891 (1981)
85. Sinha, N.K., Zhou, Q.: Discrete-time approximation of multivariable continuous-time systems. Proc. IEE, Part D, Control Theory Appl. **130**(3), 103–110 (1983)
86. Wang, C.-H.: On the generalisation of block pulse operational matrices for fractional and operational calculus. J. Franklin Inst. **315**(2), 91–102 (1983)
87. Palanisamy, K.R.: A note on block-pulse function operational matrix for integration. Int. J. Syst. Sci. **14**(11), 1287–1290 (1983)
88. Hsu, N.-S., Cheng, B.: Analysis and optimal control of time-varying linear systems via block-pulse functions. Int. J. Control **33**(6), 1107–1122 (1981)
89. Kawaji, S.: Block-pulse series analysis of linear systems incorporating observers. Int. J. Control **37**(5), 1113–1120 (1983)
90. Shih, Y.P., Chia, W.K.: Parameter estimation of delay systems via block-pulse functions. J. Dyn. Syst. Measur. Control **102**(3), 159–162 (1980)
91. Jan, Y.G., Wong, K.M.: Bilinear system identification by block pulse functions. J. Franklin Inst. **512**(5), 349–359 (1981)
92. Cheng, B., Hsu, N.-S.: Analysis and parameter estimation of bilinear systems via block-pulse functions. Int. J. Control **36**(1), 53–65 (1982)
93. Rao, G.P., Srinivasan, T.: Multidimensional block pulse functions and their use in the study of distributed parameter systems. Int. J. Syst. Sci. **11**(6), 689–708 (1980)
94. Nath, A.K., Lee, T.T.: On the multidimensional extension of block-pulse functions and their applications. Int. J. Syst. Sci. **14**(2), 201–208 (1983)
95. Hsu, N.-S., Cheng, B.: Identification of nonlinear distributed systems via block pulse functions. Int. J. Control **36**(2), 281–291 (1982)
96. Kwong, C.P., Chen, C.F.: Linear feedback systems identification via block pulse functions. Int. J. Syst. Sci. **12**(5), 635–642 (1981)
97. Palanisamy, K.R., Bhattacharya, D.K.: System identification via block-pulse functions. Int. J. Syst. Sci. **12**(5), 643–647 (1981)
98. Palanisamy, K.R., Bhattacharya, D.K.: Analysis of stiff systems via single step method of block-pulse functions. Int. J. Syst. Sci. **13**(9), 961–968 (1982)
99. Kalat, J., Paraskevopoulos, P.N.: Solution of multipoint boundary value problems via block pulse functions. J. Franklin Inst. **324**(1), 73–81 (1987)
100. Kung, F.C., Chen, S.Y.: Solution of integral equations using a set of block pulse functions. J. Franklin Inst. **306**(4), 283–291 (1978)
101. Cheng, B., Hsu, N.S.: Analysis and parameter estimation of bilinear systems via block pulse functions. Int. J. Control **36**(1), 53–65 (1982)
102. Deb, A., Sarkar, G., Biswas, A., Mandal, P.: Numerical instability of deconvolution operation via block pulse functions. J. Franklin Inst. **345**, 319–327 (2008)
103. Hoseini, S.M., Soleimani, K.: Analysis of time delay systems via new triangular functions. J. Appl. Math. **5**(19) (2010)
104. Babolian, Z.M., Saeed, H.-V.: Numerical solution of nonlinear Volterra-Fredholm integro-differential equations via direct method using triangular functions. Comput. Math. Appl. **58**(2), 239–247 (2009)
105. Han, Z., Li, S., Cao, Q.: Triangular orthogonal functions for nonlinear constrained optimal control problems. Res. J. Appl. Sci. Eng. Technol. **4**(12), 1822–1827 (2012)
106. Almasieh, H., Roodaki, M.: Triangular functions method for the solution of Fredholm integral equations system. Ain Shams Eng. J. **3**(4), 411–416 (2012)

Chapter 2
The Hybrid Function (HF) and Its Properties

Abstract Starting with a brief review of block pulse functions (BPF), sample-and-hold functions (SHF) and triangular functions (TF), this chapter presents the genesis of hybrid functions (HF) mathematically. Then different elementary properties and operational rules of HF are discussed. The chapter ends with a qualitative comparison of BPF, SHF, TF and HF.

In this chapter, we propose a new set of orthogonal functions [1]. The function set is named 'hybrid function (HF)'. This set is a combination of the sample-and-hold function (SHF) set [2] and a right handed triangular function (RHTF) set [3, 4]. This new function set is different from piecewise constant orthogonal functions [5] and approximates square integrable time functions of Lebesgue measure in a piecewise linear manner.

In the following, we discuss different properties of the proposed hybrid function (HF) set. That is, its elementary properties and the operational rules like, addition, subtraction, multiplication and division in HF domain are discussed.

2.1 Brief Review of Block Pulse Functions (BPF) [6]

Referring to Fig. 1.4 and definition of block pulse functions given in Sect. 1.2.4, a square integrable time function $f(t)$ of Lebesgue measure may be expanded into an m-term BPF series in $t \in [0,\ T)$ as

$$\begin{aligned} f(t) &\approx \sum_{i=0}^{m-1} f_i \psi_i(t) = \begin{bmatrix} f_0 & f_1 & f_2 & \cdots & f_i & \cdots & f_{(m-1)} \end{bmatrix} \boldsymbol{\Psi}_{(m)}(t) \quad \text{for} \quad i = 0,\ 1,\ 2, \ldots, (m-1) \\ &\triangleq \mathrm{F}_{(m)}^{\mathrm{T}} \boldsymbol{\Psi}_{(m)}(t) \end{aligned} \tag{2.1}$$

A. Deb et al., *Analysis and Identification of Time-Invariant Systems, Time-Varying Systems, and Multi-Delay Systems using Orthogonal Hybrid Functions*,
Studies in Systems, Decision and Control 46, DOI 10.1007/978-3-319-26684-8_2

where, $[\cdots]^{\mathrm{T}}$ denotes transpose and the $(i + 1)$th BPF coefficient f_i is given by

$$f_i = \frac{1}{h} \int_{ih}^{(i+1)h} f(t)\psi_i(t)\mathrm{d}t \tag{2.2}$$

where, $h = \frac{T}{m}$s.

Coefficients evaluated via Eq. (2.2) always ensures minimum mean integral square error (MISE) [6] with respect to function approximation. Thus, the coefficients f_i's may be termed 'optimal'.

If Eq. (2.2) is computed via trapezoidal rule, the coefficients will slightly deviate from the f_i's of (2.2) due to inexact integration. However, such 'approximate' computation leading to 'approximate' coefficients f_i''s has the advantage of working with function samples only. Thus

$$f_i = \frac{1}{h} \int_{ih}^{(i+1)h} f(t)\psi_i(t)\mathrm{d}t \approx \frac{[f(ih) + f\{(i+1)h\}]}{2} = f_i' \tag{2.3}$$

We call such BPF expansion a 'non-optimal' one and any analysis based upon this technique may be called 'non-optimal' BPF (NOBPF) analysis.

2.2 Brief Review of Sample-and-Hold Functions (SHF) [2]

Any square integrable function $f(t)$ may be represented by a sample-and-hold function set in the semi-open interval $[0, T)$ by considering

$$f_i(t) = f(ih), \quad i = 0, 1, 2, \ldots, (m-1)$$

where, h is the sampling period ($=T/m$), $f_i(t)$ is the amplitude of the function $f(t)$ at time ih and $f(ih)$ is the first term of the Taylor series expansion of the function $f(t)$ around the point $t = ih$, because, for a zero order hold (ZOH) the amplitude of the function $f(t)$ at $t = ih$ is held constant for the duration h.

SHFs are similar to BPFs in many aspects. The $(i + 1)$th member of an SHF set, comprised of m component functions, is defined as

$$S_i(t) = \begin{cases} 1 & \text{for } ih \le t < (i+1)h \\ 0 & \text{elsewhere} \end{cases} \tag{2.4}$$

where, $i = 0, 1, 2, \ldots, (m-1)$.

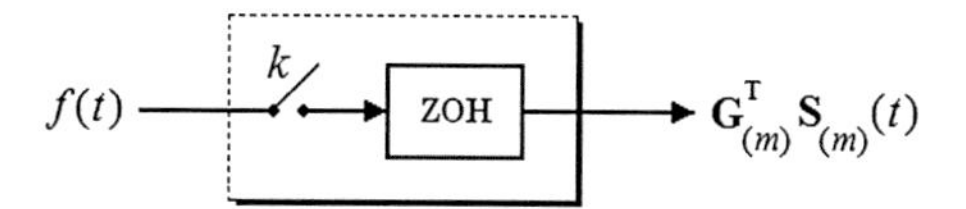

Fig. 2.1 A sample-and-hold device

A square integrable time function *f*(*t*) of Lebesgue measure may be expanded into an *m*-term SHF series in $t \in [0,\ T)$ as

$$f(t) \approx \sum_{i=0}^{(m-1)} g_i S_i(t) = \begin{bmatrix} g_0 & g_1 & g_2 & \cdots & g_i & \cdots & g_{(m-1)} \end{bmatrix} S_{(m)}(t) \quad \text{for} \quad i = 0, 1, 2, \ldots, (m-1)$$
$$\triangleq G_{(m)}^{\mathrm{T}} S_{(m)}(t) \tag{2.5}$$

where $[\cdots]^{\mathrm{T}}$ denotes transpose and $g_i = f(ih)$, the $(i+1)$th sample of the function *f* (*t*). In fact, g_i's are the samples of the function *f*(*t*) with the sampling period *h*.

Considering the nature of the SHF set, which is a look alike of the BPF set, it is easy to conclude that this set is orthogonal as well as complete in $t \in [0,\ T)$. However, the special property of the SHF is revealed by using the sample-and-hold concept in deriving the required operational matrices. If a time signal *f*(*t*) is fed to a sample-and-hold device as shown in Fig. 2.1, the output of the device approximates *f*(*t*) as per Eq. (2.5).

2.3 Brief Review of Triangular Functions (TF) [3, 4]

From a set of block pulse function, $\boldsymbol{\Psi}_{(m)}(t)$, we can generate two sets of orthogonal triangular functions (TF) [3, 4], namely $\mathbf{T1}_{(m)}(t)$ and $\mathbf{T2}_{(m)}(t)$ such that

$$\boldsymbol{\Psi}_{(m)}(t) = \mathbf{T1}_{(m)}(t) + \mathbf{T2}_{(m)}(t) \tag{2.6}$$

Figure 1.8a, b show the orthogonal triangular function sets, $\mathbf{T1}_{(m)}(t)$ and $\mathbf{T2}_{(m)}(t)$, where *m* has been chosen arbitrarily as 8. These two TF sets are complementary to each other. For convenience, we call $\mathbf{T1}_{(m)}(t)$ the left-handed triangular function (LHTF) vector and $\mathbf{T2}_{(m)}(t)$ the right handed triangular function (RHTF) vector.

Using the component functions, we could express the *m*-set triangular function vectors as

$$\mathbf{T1}_{(m)}(t) \triangleq \begin{bmatrix} T1_0(t) & T1_1(t) & T1_2(t) & \cdots & T1_i(t) & \cdots & T1_{m-1}(t) \end{bmatrix}^{\mathrm{T}}$$
$$\mathbf{T2}_{(m)}(t) \triangleq \begin{bmatrix} T2_0(t) & T2_1(t) & T2_2(t) & \cdots & T2_i(t) & \cdots & T2_{m-1}(t) \end{bmatrix}^{\mathrm{T}}$$

where $[\cdots]^{\mathrm{T}}$ denotes transpose.

The $(i + 1)$th component of the LHTF vector $\mathbf{T1}_{(m)}(t)$ is defined as

$$T1_i(t) = \begin{cases} 1 - (t - ih)/h, & \text{for } ih \le t < (i+1)h \\ 0, & \text{elsewhere} \end{cases} \tag{2.7}$$

and the $(i + 1)$th component of the RHTF vector $\mathbf{T2}_{(m)}(t)$ is defined as

$$T2_i(t) = \begin{cases} (t - ih)/h, & \text{for } ih \le t < (i+1)h \\ 0, & \text{elsewhere} \end{cases} \tag{2.8}$$

where $i = 0, 1, 2, \ldots, (m - 1)$.

A square integrable time function $f(t)$ of Lebesgue measure may be expanded into an m-term TF series in $t \in [0, T)$ as

$$\begin{aligned} f(t) &\approx [c_0 \quad c_1 \quad c_2 \quad \cdots \quad c_{m-1}]\mathbf{T1}_{(m)}(t) + [d_0 \quad d_1 \quad d_2 \quad \cdots \quad d_{m-1}]\mathbf{T2}_{(m)}(t) \\ &\triangleq \mathbf{C}^{\mathrm{T}}\mathbf{T1}_{(m)}(t) + \mathbf{D}^{\mathrm{T}}\mathbf{T2}_{(m)}(t) \end{aligned} \tag{2.9}$$

The constant coefficients $c_i's$ and $d_i's$ in Eq. (2.9) are given by

$$c_i \triangleq f(ih) \quad \text{and} \quad d_i \triangleq f[(i+1)h] \tag{2.10}$$

and the relation $c_{i+1} = d_i$ holds between $c_i's$ and $d_i's$.

2.4 Hybrid Function (HF): A Combination of SHF and TF

We can use a set of sample-and-hold functions and the right handed triangular function set to form a new function set, which we name a '*Hybrid Function set*'. To define a hybrid function (HF) set, we express the $(i + 1)$th member $H_i(t)$ of the m-set hybrid function $\mathbf{H}_{(m)}(t)$ as

$$H_i(t) = a_i S_i(t) + b_i T2_i(t)$$

where, $i = 0, 1, 2, \ldots, (m - 1)$, a_i and b_i are scaling constants, $0 \le t < T$, S_i and $T2_i$ are the $(i + 1)$th component sample-and-hold function and right handed triangular function.

For convenience, in the following, we write $\mathbf{T}$ instead of $\mathbf{T2}$. The above equation can now be expressed as

$$H_i(t) = a_i S_i(t) + b_i T_i(t) \tag{2.11}$$

Let us now illustrate how a function $f(t)$ is represented via a set of hybrid functions.

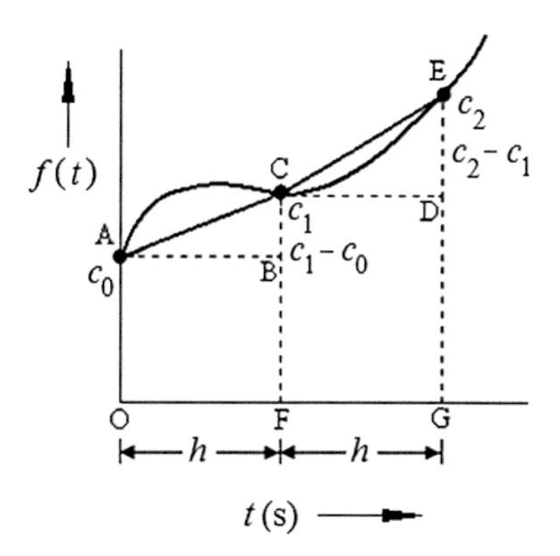

Fig. 2.2 Function approximation via hybrid functions (*HF*) domain

In Fig. 2.2, the function *f*(*t*) is sampled at three equidistant points (sampling interval *h*) A, C and E respectively with corresponding sample values c_0, c_1 and c_2. Now, *f*(*t*) can be approximated in a piecewise linear manner by the two straight lines AC and CE, which are the sides of two adjacent trapeziums. The trapezium ACFO may be considered to be a combination of the SHF block ABFO, and the triangular block ACB. Similar is the case for the second trapezium CEGF.

Hence, for the first trapezium, the hybrid function representation may be written as a combination of SHF and TF as

$$H_0(t) = c_0 S_0(t) + (c_1 - c_0) T_0(t)$$

Then the function *f*(*t*) may be represented in an interval $t \in [0, 2h)$ as

$$\begin{aligned} f(t) &= H_0(t) + H_1(t) \\ &= \{c_0 S_0(t) + (c_1 - c_0) T_0(t)\} + \{c_1 S_1(t) + (c_2 - c_1) T_1(t)\} \\ &= \{c_0 S_0(t) + c_1 S_1(t)\} + \{(c_1 - c_0) T_0(t) + (c_2 - c_1) T_1(t)\} \\ &\triangleq \mathbf{C}^{\mathrm{T}} \mathbf{S}_{(2)}(t) + \mathbf{D}^{\mathrm{T}} \mathbf{T}_{(2)}(t) \end{aligned}$$

where, $[c_0 \quad c_1] = \mathbf{C}^{\mathrm{T}}$ and $[(c_1 - c_0) \quad (c_2 - c_1)] = \mathbf{D}^{\mathrm{T}}$.

Generalizing this, we can extend the concept for an *m*-component function set as

$$f(t) \triangleq \mathbf{C}^{\mathrm{T}} \mathbf{S}_{(m)}(t) + \mathbf{D}^{\mathrm{T}} \mathbf{T}_{(m)}(t) \tag{2.12}$$

where, $\mathbf{C}^{\mathrm{T}} = [c_0 \quad c_1 \quad \cdots \quad c_{m-1}]$ and $\mathbf{D}^{\mathrm{T}} = [(c_1 - c_0) \quad (c_2 - c_1) \quad \cdots \quad (c_m - c_{m-1})]$

The radical difference between block pulse domain representation and hybrid function domain representation of a function is, the BPF representation and subsequent analysis always provides us with a staircase solution, while the HF domain technique provides us with piecewise linear results.

It is noted that, the SHF coefficients are simply the samples at the sampling instants, while the TF coefficients are the differences between two consecutive samples, e.g., $(c_i - c_{i-1})$, i being a positive integer. An added advantage of HF domain representation is, by dropping the TF domain components, we are left with only the SHF domain representation. This is sometimes convenient in function analysis, especially, for digital control systems or sample-and-hold systems.

2.5 Elementary Properties of Hybrid Functions [7]

In solving certain problems of control engineering, the advantages of using the hybrid function technique are their easy operations and satisfactory approximations. These advantages are due to the distinct properties of hybrid functions. The elementary properties are as follows.

2.5.1 *Disjointedness*

Hybrid function (HF) set is a combination of the sample-and-hold function (SHF) set and the triangular function (RHTF) set.

The sample-and-hold functions are disjoint with each other in the interval $t \in [0, T)$. This property can be formulated as

$$S_i S_j = \begin{cases} 0 & \text{where } i \neq j \\ S_i & \text{where } i = j \end{cases} \tag{2.13}$$

where i, j = 0, 1, 2,..., $(m - 1)$ and the argument (t) is dropped for simplicity. Henceforth, for both sample-and-hold and triangular functions, the argument indicating time dependency will be discarded.

The property of Eq. (2.13) can directly be obtained from the definition of sample-and-hold functions.

Triangular functions (TF) are derived from the block pulse function set. A block pulse function can be dissected along its two diagonals to generate two triangular functions. That is, when we add two component triangular functions, we get back the original block pulse function. This dissection process has been shown in Fig. 1.7.

Since, the component block pulse functions of a set are mutually disjoint, the triangular functions of the LHTF set are disjoint, and so are the functions of the RHTF set.

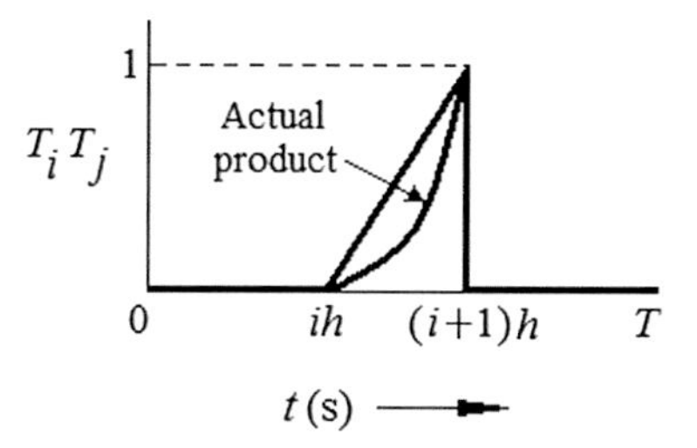

Fig. 2.3 The product $(T_i T_j)$ and its *triangular* function representation

Hence, the product of two right handed triangular functions T_i and T_j in the semi-open interval $t \in [0, T)$ is

$$T_i T_j = 0 \quad \text{where } i \neq j \text{ and } i, j = 0, 1, 2, \ldots, (m-1)$$

when $i = j$, the product at the sample points, namely ih and $(i+1)h$ are 0 and 1, respectively. Since we are concerned only with the triangular function representation of the product, shown in Fig. 2.3, the result is T_i only. Thus [3]

$$T_i T_j \approx T_i \quad \text{where } i = j$$

This property can be formulated as

$$T_i T_j = \begin{cases} 0 & \text{where } i \neq j \\ T_i & \text{where } i = j \end{cases} \tag{2.14}$$

This property can directly be obtained from the definition of triangular functions.

2.5.2 *Orthogonality [1]*

The sample-and-hold functions of a SHF set are orthogonal with each other in the interval $t \in [0, T)$:

$$\int_0^T S_i S_j \mathrm{d}t = \begin{cases} 0 & \text{where } i \neq j \\ \delta_{ij} & \text{where } i = j \end{cases} \tag{2.15}$$

where, $i, j = 0, 1, 2, \ldots, (m-1)$. This property can directly be obtained from the disjointedness of sample-and-hold functions.

Similarly, the triangular functions of a TF set are orthogonal with each other in the interval $t \in [0, T)$:

$$\int_0^T T_i T_j \mathrm{d}t = \begin{cases} 0 & \text{where } i \neq j \\ \delta_{ij} & \text{where } i = j \end{cases} \tag{2.16}$$

where, $i, j = 0, 1, 2, \ldots, (m-1)$.

Now, the hybrid function is orthogonal because each of the SHF and TF sets are orthogonal.

2.5.3 Completeness

Like the block pulse functions, the sample-and-hold function set is also complete when i approaches infinity. This means that we have:

$$\int_0^T f^2(t)dt = \sum_{i=0}^{\infty} f_i^2 \|S_i\|^2 \tag{2.17}$$

for any real bounded function $f(t)$ which is square integrable in the interval $t \in [0, T)$.

Here, the expression:

$$\|S_i(t)\| = \left[\int_0^T S_i^2 \mathrm{d}t\right]^{1/2} \tag{2.18}$$

is the norm of S_i.

From a set of block pulse function, we can generate two sets of orthogonal triangular functions (TF) [3], namely $\mathbf{T1}_{(m)}$ and $\mathbf{T2}_{(m)}$. The triangular function sets are complete when i approaches infinity. This means that we have:

$$\int_0^T f^2(t)\mathrm{d}t = \sum_{i=0}^{\infty} \|[c_i T1_i + d_i T2_i]\|^2 \tag{2.19}$$

for any real bounded function $f(t)$ which is square integrable in the interval $t \in [0, T)$.

Here, the expression:

$$\|[c_i T1_i + d_i T2_i]\| = \left[\int_0^T \left[c_i T1_i + d_i T2_i^2 \mathrm{d}t\right] \right]^{1/2} \tag{2.20}$$

is the norm of $[c_i T1_i + d_i T2_i]$.

The hybrid function set is also complete when i approaches infinity. This means that we have:

$$\begin{aligned} &\int_0^T f^2(t)dt = \sum_{i=0}^{\infty} \|f_i S_i + d_i T2_i\|^2 \\ \text{or} \quad &\int_0^T f^2(t)dt = \sum_{i=0}^{\infty} \|f_i S_i + d_i T_i\|^2 \end{aligned} \tag{2.21}$$

for any real bounded function $f(t)$ which is square integrable in the interval $t \in$ [0,T) and piecewise linear approximation.

The completeness of hybrid functions guarantees that an arbitrarily small mean square error can be obtained for a real bounded function, which has only a finite number of discontinuous points in the interval $t \in [0, T)$, by increasing the number of terms in the sample-and-hold function series and the triangular function series.

2.6 Elementary Operational Rules

2.6.1 Addition of Two Functions

For addition of two time functions $f(t)$ and $g(t)$, the following cases are considered.

Let, $a(t) = f(t) + g(t)$, where, $a(t)$ is the resulting function.

We call the samples of each functions $f(t)$ and $g(t)$ in $t \in [0, T)$ $f_0, f_1, f_2, \ldots f_i, \ldots, f_m$ and $g_0, g_1, g_2, \ldots g_i, \ldots g_m$ respectively, as shown in Fig. 2.4.

(a) ***The continuous functions are expanded in HF domain separately and then added***

The functions $f(t)$ and $g(t)$ can be expanded directly into hybrid functions series and then added to find the resultant function in HF domain. The HF domain expanded form of $a(t)$ may be called $\bar{a}(t)$. It should be noted that $\bar{a}(t)$ is piecewise linear in nature. Thus

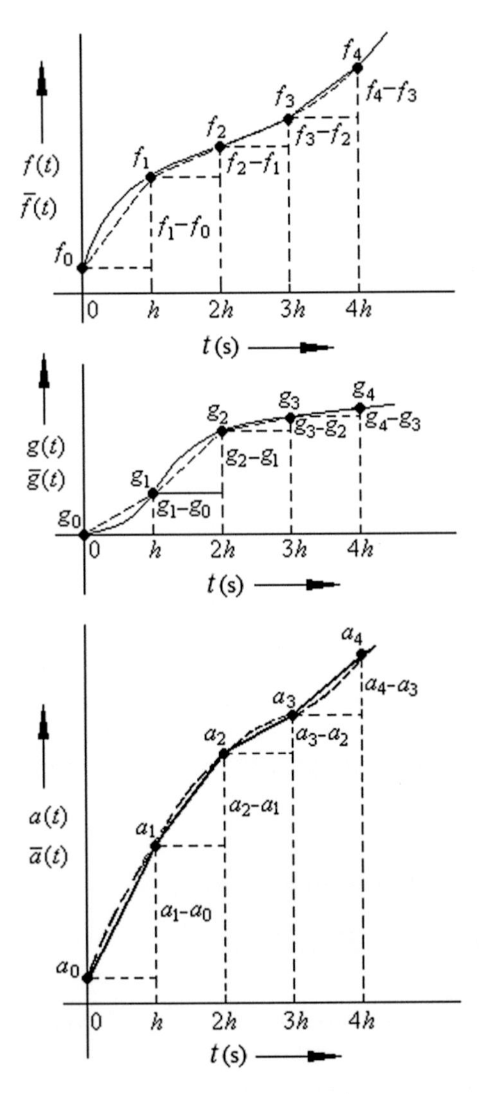

Fig. 2.4 Two time functions $f(t)$ and $g(t)$ expressed in HF domain and their sum $\bar{a}(t)$ in HF domain

$$
\begin{aligned}
a(t) = f(t) + g(t) \approx & \left[f_0 \quad f_1 \quad f_2 \quad \cdots \quad f_i \quad \cdots \quad f_{m-1}\right] \mathbf{S}_{(m)} \\
& + \left[f_0' \quad f_1' \quad f_2' \quad \cdots \quad f_i' \quad \cdots \quad f_{m-1}'\right] \mathbf{T}_{(m)} \\
& + \left[g_0 \quad g_1 \quad g_2 \quad \cdots \quad g_i \quad \cdots \quad g_{m-1}\right] \mathbf{S}_{(m)} \\
& + \left[g_0' \quad g_1' \quad g_2' \quad \cdots \quad g_i' \quad \cdots \quad g_{m-1}'\right] \mathbf{T}_{(m)} \\
= & \left[\sum_{i=0}^{m-1} f_i S_i + \sum_{i=0}^{m-1} f_i' T_i\right] + \left[\sum_{i=0}^{m-1} g_i S_i + \sum_{i=0}^{m-1} g_i' T_i\right] \\
\triangleq & \left[\mathbf{F}_{\mathrm{S}}^{\mathrm{T}} \mathbf{S}_{(m)} + \mathbf{F}_{\mathrm{T}}^{\mathrm{T}} \mathbf{T}_{(m)}\right] + \left[\mathbf{G}_{\mathrm{S}}^{\mathrm{T}} \mathbf{S}_{(m)} + \mathbf{G}_{\mathrm{T}}^{\mathrm{T}} \mathbf{T}_{(m)}\right]
\end{aligned}
\tag{2.22}
$$

where,

$$
\begin{aligned}
\mathbf{F}_{\mathrm{S}}^{\mathrm{T}} & \triangleq \left[f_0 \quad f_1 \quad f_2 \quad \cdots \quad f_i \quad \cdots \quad f_{m-1}\right] \\
\mathbf{F}_{\mathrm{T}}^{\mathrm{T}} & \triangleq \left[f_0' \quad f_1' \quad f_2' \quad \cdots \quad f_i' \quad \cdots \quad f_{m-1}'\right] \\
f_i' & = (f_{i+1} - f_i) \\
\mathbf{G}_{\mathrm{S}}^{\mathrm{T}} & \triangleq \left[g_0 \quad g_1 \quad g_2 \quad \cdots \quad g_i \quad \cdots \quad g_{m-1}\right] \\
\mathbf{G}_{\mathrm{T}}^{\mathrm{T}} & \triangleq \left[g_0' \quad g_1' \quad g_2' \quad \cdots \quad g_i' \quad \cdots \quad g_{m-1}'\right] \\
\text{and} \quad g_i' & = (g_{i+1} - g_i)
\end{aligned}
$$

Now,

$$
\begin{aligned}
\bar{a}(t) \triangleq & \left[\mathbf{F}_{\mathrm{S}}^{\mathrm{T}} + \mathbf{G}_{\mathrm{S}}^{\mathrm{T}}\right] \mathbf{S}_{(m)} + \left[\mathbf{F}_{\mathrm{T}}^{\mathrm{T}} + \mathbf{G}_{\mathrm{T}}^{\mathrm{T}}\right] \mathbf{T}_{(m)} \\
= & \left[(f_0 + g_0) \quad (f_1 + g_1) \quad \cdots \quad (f_i + g_i) \quad \cdots \quad (f_{m-1} + g_{m-1})\right] \mathbf{S}_{(m)} \\
& + \left[(f_0' + g_0') \quad (f_1' + g_1') \quad \cdots \quad (f_i' + g_i') \quad \cdots \quad (f_{m-1}' + g_{m-1}')\right] \mathbf{T}_{(m)} \\
\triangleq & \mathbf{A}_{\mathrm{S}}^{\mathrm{T}} \mathbf{S}_{(m)} + \mathbf{A}_{\mathrm{T}}^{\mathrm{T}} \mathbf{T}_{(m)}
\end{aligned}
\tag{2.23}
$$

where, $\mathbf{A}_{\mathrm{S}}^{\mathrm{T}} = \left[\mathbf{F}_{\mathrm{S}}^{\mathrm{T}} + \mathbf{G}_{\mathrm{S}}^{\mathrm{T}}\right]$ and $\mathbf{A}_{\mathrm{T}}^{\mathrm{T}} = \left[\mathbf{F}_{\mathrm{T}}^{\mathrm{T}} + \mathbf{G}_{\mathrm{T}}^{\mathrm{T}}\right]$

$$
\begin{aligned}
\mathbf{A}_{\mathrm{S}}^{\mathrm{T}} & \triangleq \left[a_0 \quad a_1 \quad a_2 \quad \cdots \quad a_i \quad \cdots \quad a_{m-1}\right] \\
\text{and} \quad \mathbf{A}_{\mathrm{T}}^{\mathrm{T}} & \triangleq \left[(a_1 - a_0) \quad (a_2 - a_1) \quad (a_3 - a_2) \quad \cdots \quad (a_i - a_{i-1}) \quad \cdots \quad (a_m - a_{m-1})\right]
\end{aligned}
$$

Equation (2.23) shows that the hybrid function coefficients of the sum $a(t)$ are the sums of the hybrid functions coefficients of the individual functions $f(t)$ and $g(t)$, in each subinterval. This is shown in Fig. 2.4.

(b) ***The continuous functions f(t) and g(t) are first added and then the resulting function $a(t) = f(t) + g(t)$ is expressed in HF domain***

In this case, the resulting continuous function $a(t) = f(t) + g(t)$ is expanded in HF domain as

$$a(t) \approx \bar{a}(t) = [a_0 \quad a_1 \quad a_2 \quad \cdots \quad a_i \quad \cdots \quad a_{m-1}]\,\mathbf{S}_{(m)} \\ + [(a_1 - a_0) \quad (a_2 - a_1) \quad \cdots \quad (a_i - a_{i-1}) \quad \cdots \quad (a_m - a_{m-1})]\,\mathbf{T}_{(m)} \tag{2.24}$$

where $a_0, a_1, a_2, \ldots, a_i, \ldots, a_m$ are the samples of $a(t)$ or $\bar{a}(t)$ at time instants $0, h, 2h, \ldots, ih, \ldots, mh$.

It is evident from Fig. 2.4 that at sampling instants the function values i.e., samples will be added as before. Hence,

$$\begin{aligned} \bar{a}(t) &= \bar{f}(t) + \bar{g}(t) \\ &= [(f_0 + g_0) \quad (f_1 + g_1) \quad \cdots \quad (f_i + g_i) \quad \cdots \quad (f_{m-1} + g_{m-1})]\,\mathbf{S}_{(m)} \\ &\quad + \left[(f_0' + g_0') \quad (f_1' + g_1') \quad \cdots \quad (f_i' + g_i') \quad \cdots \quad (f_{m-1}' + g_{m-1}')\right]\mathbf{T}_{(m)} \end{aligned}$$

That is

$$\bar{a}(t) \triangleq \mathbf{A}_{\mathrm{S}}^{\mathrm{T}} \mathbf{S}_{(m)} + \mathbf{A}_{\mathrm{T}}^{\mathrm{T}} \mathbf{T}_{(m)} \tag{2.25}$$

From Eqs. (2.23) and (2.25), it is seen that both the results of addition are identical.

Considering two functions $f(t) = 1 - \exp(-t)$ and $g(t) = \exp(-t)$, the result of addition of these two functions, using their individual coefficients, is shown in Fig. 2.5.

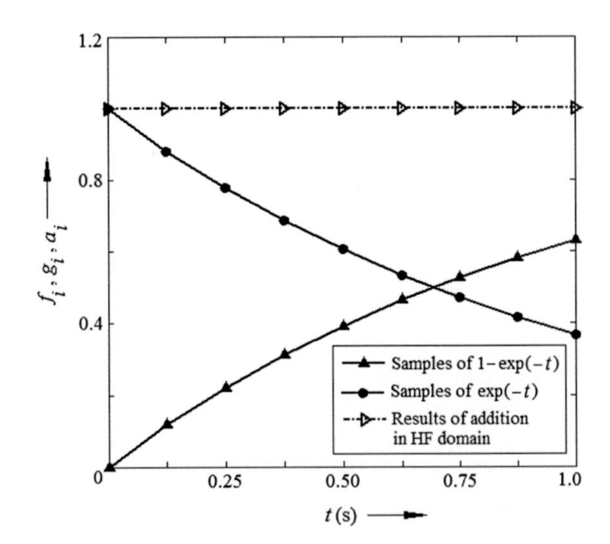

Fig. 2.5 Hybrid function expansion of $f(t) = 1 - \exp(-t)$ and $g(t) = \exp(-t)$ and the result of their addition $\bar{f}(t) + \bar{g}(t) = \bar{a}(t)$ in HF domain with $m = 8$ and $T = 1$ s. Due to high degree of accuracy, the piecewise *linear curves* look like *continuous curves* (vide Appendix B, Program no. 1)

2.6.2 Subtraction of Two Functions

For subtraction of two time functions $f(t)$ and $g(t)$, the following cases are considered.

Let, $s(t) = f(t) - g(t)$, where, $s(t)$ is the resulting function of subtraction.

As before, the samples of individual functions $f(t)$ and $g(t)$ in time domain are f_0, $f_1, f_2, \ldots, f_i, \ldots, f_m$ and $g_0, g_1, g_2, \ldots, g_i, \ldots, g_m$ respectively.

The functions $f(t)$ and $g(t)$ are expressed in HF domain. Then we subtract the HF domain expanded functions to obtain $s(t)$ in HF domain. The HF domain expanded form of $s(t)$ may be called $\bar{s}(t)$.

(a) ***The continuous functions are expanded in HF domain separately and then subtracted***

The difference of two time functions $f(t)$ and $g(t)$ can be expressed via HF domain as

$$\begin{aligned} s(t) = f(t) - g(t) &\approx [f_0 \quad f_1 \quad f_2 \quad \cdots \quad f_i \quad \cdots \quad f_{m-1}]\,\mathbf{S}_{(m)} \\ &+ [f'_0 \quad f'_1 \quad f'_2 \quad \cdots \quad f'_i \quad \cdots \quad f'_{m-1}]\,\mathbf{T}_{(m)} \\ &- [g_0 \quad g_1 \quad g_2 \quad \cdots \quad g_i \quad \cdots \quad g_{m-1}]\,\mathbf{S}_{(m)} \\ &- [g'_0 \quad g'_1 \quad g'_2 \quad \cdots \quad g'_i \quad \cdots \quad g'_{m-1}]\,\mathbf{T}_{(m)} \\ &\triangleq \left[\mathbf{F}_S^T\mathbf{S}_{(m)} + \mathbf{F}_T^T\mathbf{T}_{(m)}\right] - \left[\mathbf{G}_S^T\mathbf{S}_{(m)} + \mathbf{G}_T^T\mathbf{T}_{(m)}\right] \end{aligned} \tag{2.26}$$

$$\begin{aligned} \text{Hence, } \bar{s}(t) &\triangleq \left[\mathbf{F}_S^T - \mathbf{G}_S^T\right]\mathbf{S}_{(m)} + \left[\mathbf{F}_T^T - \mathbf{G}_T^T\right]\mathbf{T}_{(m)} \\ &= [f_0 - g_0 \quad f_1 - g_1 \quad \cdots \quad f_i - g_i \quad \cdots \quad f_{m-1} - g_{m-1}]\,\mathbf{S}_{(m)} \\ &\quad + [f'_0 - g'_0 \quad f'_1 - g'_1 \quad \cdots \quad f'_i - g'_i \quad \cdots \quad f'_{m-1} - g'_{m-1}]\,\mathbf{T}_{(m)} \\ &\triangleq \mathbf{S}_S^T\mathbf{S}_{(m)} + \mathbf{S}_T^T\mathbf{T}_{(m)} \end{aligned} \tag{2.27}$$

where,

$$\mathbf{S}_S^T \triangleq \left[\mathbf{F}_S^T - \mathbf{G}_S^T\right] \triangleq [s_0 \quad s_1 \quad s_2 \quad \cdots \quad s_i \cdots \quad s_{m-1}]$$

and,

$$\mathbf{S}_T^T = \left[\mathbf{F}_T^T - \mathbf{G}_T^T\right] \triangleq [(s_1 - s_0) \quad (s_2 - s_1) \quad (s_3 - s_2) \quad \cdots \quad (s_i - s_{i-1}) \quad \cdots \quad (s_m - s_{m-1})]$$

Equation (2.27) shows that the hybrid function coefficients of the difference of two functions $s(t)$ is the differences of the hybrid functions coefficients of the individual functions at each sampling points. This is shown in Fig. 2.6.

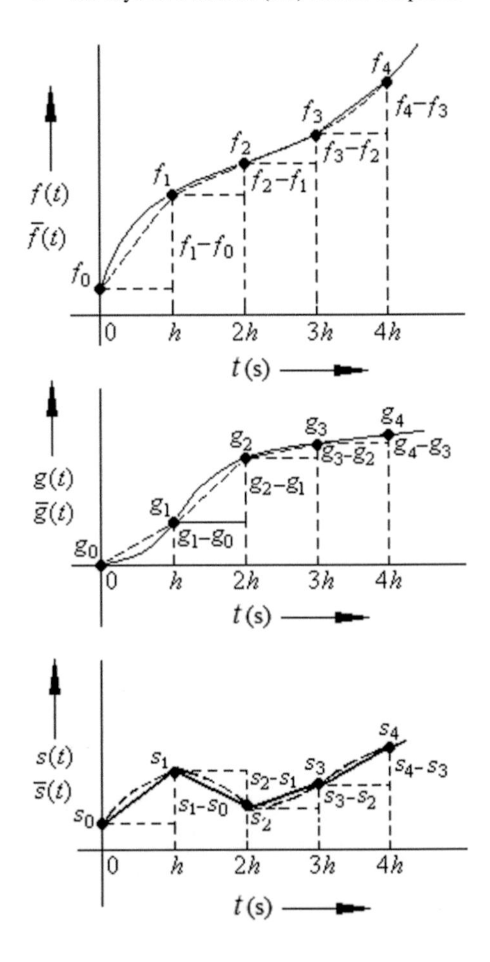

Fig. 2.6 Two time functions $f(t)$ and $g(t)$ expressed via HF domain and their difference $\bar{s}(t)$ in HF domain

(b) ***The continuous functions f(t) and g(t) are first subtracted and then the resulting function s(t)=f(t)−g(t) is expanded* via *HF domain***
In this case, the resulting continuous function $s(t) = f(t) - g(t)$ is expanded in HF domain as

$$\begin{aligned}\bar{s}(t) = & \begin{bmatrix} s_0 & s_1 & s_2 & \cdots & s_i & \cdots & s_{m-1} \end{bmatrix} \mathbf{S}_{(m)} \\ & + \begin{bmatrix} (s_1 - s_0) & (s_2 - s_1) & \cdots & (s_i - s_{i-1}) & \cdots & (s_m - s_{m-1}) \end{bmatrix} \mathbf{T}_{(m)}\end{aligned}$$

Coefficients of the resulting function $s(t)$ are found directly by subtracting the corresponding coefficients of both the time functions at same sampling instants. From Fig. 2.6, we can write the resultant function $s(t)$ are as follows

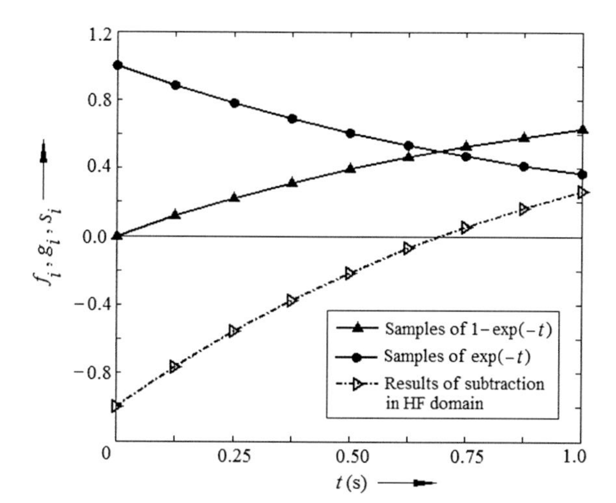

Fig. 2.7 Hybrid function expansion of $f(t) = 1 - \exp(-t)$ and $g(t) = \exp(-t)$ and the result of their subtraction $\bar{f}(t) - \bar{g}(t) = \bar{s}(t)$ in HF domain, with $m = 8$ and $T = 1$ s. Due to high degree of accuracy, the piecewise *linear curves* look like *continuous curves*

$$\begin{aligned}\bar{s}(t) &= \bar{f}(t) - \bar{g}(t) \\ &= \left[(f_0 - g_0) \quad (f_1 - g_1) \quad \cdots \quad (f_i - g_i) \quad \cdots \quad (f_{m-1} - g_{m-1})\right] \mathbf{S}_{(m)} \\ &\quad + \left[(f_0' - g_0') \quad (f_1' - g_1') \quad \cdots \quad (f_i' - g_i') \quad \cdots \quad (f_{m-1}' - g_{m-1}')\right] \mathbf{T}_{(m)}\end{aligned}$$

$$\bar{s}(t) \triangleq \mathbf{S}_\mathrm{S}^\mathrm{T}\mathbf{S}_{(m)} + \mathbf{S}_\mathrm{T}^\mathrm{T}\mathbf{T}_{(m)} \tag{2.28}$$

Thus, the result of Eq. (2.28) is same as expression (2.27).

Considering two functions $f(t) = 1 - \exp(-t)$ and $g(t) = \exp(-t)$, the result of subtraction of these two functions in HF domain is shown in Fig. 2.7.

2.6.3 Multiplication of Two Functions

We consider multiplication of two functions $r(t)$ and $g(t)$ in HF domain.

Let, $r(t) \times g(t) = m(t)$, where, $m(t)$ is the result of multiplication.

Let the samples of the individual time functions $r(t)$ and $g(t)$ be $r_0, r_1, r_2, \ldots, r_i, \ldots, r_m$ and $g_0, g_1, g_2, \ldots, g_i, \ldots, g_m$ respectively at the sampling instants $0, h, 2h, \ldots, ih, \ldots. mh$. Now the time functions are expanded in HF domain, as shown in Fig. 2.8.

(a) ***The functions r(t) and g(t) are expanded separately in HF domain and then multiplied***

For functions $r(t)$ and $g(t)$, we expand them via hybrid function series and then multiply. The resulting output function $m(t)$ can be expressed as

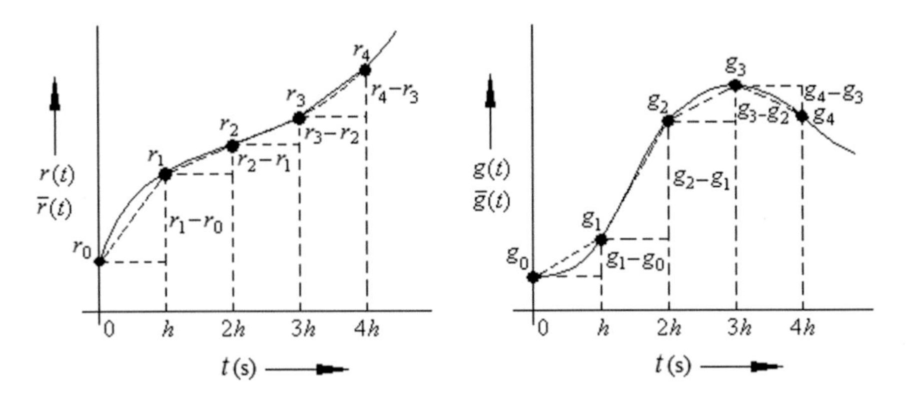

Fig. 2.8 Two time functions $r(t)$ and $g(t)$ are expanded in HF domain

$$\text{Let } m(t) = r(t) \times g(t)$$

and,

$$\begin{aligned} r(t) \approx \bar{r}(t) = & [r_0 \quad r_1 \quad r_2 \quad \cdots \quad r_i \quad \cdots \quad r_{m-1}]\,\mathbf{S}_{(m)} \\ & + [r'_0 \quad r'_1 \quad r'_2 \quad \cdots \quad r'_i \quad \cdots \quad r'_{m-1}]\,\mathbf{T}_{(m)} \end{aligned}$$

Also,

$$\begin{aligned} g(t) \approx \bar{g}(t) = & [g_0 \quad g_1 \quad g_2 \quad \cdots \quad g_i \quad \cdots \quad g_{m-1}]\,\mathbf{S}_{(m)} \\ & + [g'_0 \quad g'_1 \quad g'_2 \quad \cdots \quad g'_i \quad \cdots \quad g'_{m-1}]\,\mathbf{T}_{(m)} \end{aligned}$$

Then

$$\begin{aligned} \bar{r}(t) \times \bar{g}(t) = & \left[\sum_{i=0}^{m-1} r_i S_i + \sum_{i=0}^{m-1} r'_i T_i\right] \times \left[\sum_{i=0}^{m-1} g_i S_i + \sum_{i=0}^{m-1} g'_i T_i\right] \\ \triangleq & \left[\mathbf{R}_\mathrm{S}^\mathrm{T}\mathbf{S}_{(m)} + \mathbf{R}_\mathrm{T}^\mathrm{T}\mathbf{T}_{(m)}\right] \times \left[\mathbf{G}_\mathrm{S}^\mathrm{T}\mathbf{S}_{(m)} + \mathbf{G}_\mathrm{T}^\mathrm{T}\mathbf{T}_{(m)}\right] \end{aligned} \tag{2.29}$$

where,

$$\begin{aligned} \mathbf{R}_\mathrm{S}^\mathrm{T} & \triangleq [r_0 \quad r_1 \quad r_2 \quad \cdots \quad r_i \quad \cdots \quad r_{m-1}] \\ \mathbf{R}_\mathrm{T}^\mathrm{T} & \triangleq [r'_0 \quad r'_1 \quad r'_2 \quad \cdots \quad r'_i \quad \cdots \quad r'_{m-1}] \\ \mathbf{G}_\mathrm{S}^\mathrm{T} & \triangleq [g_0 \quad g_1 \quad g_2 \quad \cdots \quad g_i \quad \cdots \quad g_{m-1}] \\ \text{and } \mathbf{G}_\mathrm{T}^\mathrm{T} & \triangleq [g'_0 \quad g'_1 \quad g'_2 \quad \cdots \quad g'_i \quad \cdots \quad g'_{m-1}] \end{aligned}$$

In Eq. (2.29), there are three types of products involved, related to sample-and-hold functions and triangular functions. Results of these three

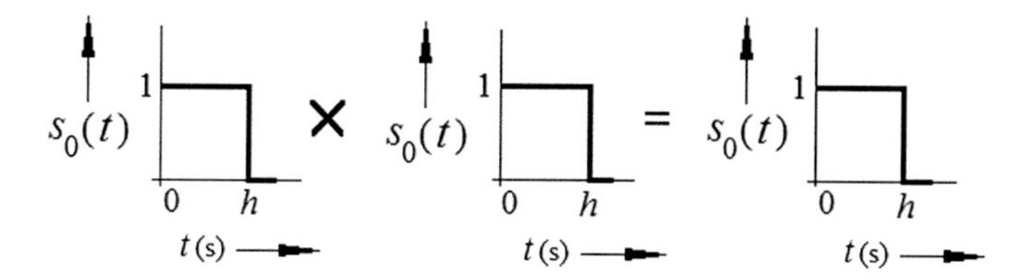

Fig. 2.9 Multiplication of the first members of two SHF sets

types of products of the basis component functions are studied in the following:

(i) **(Sample-and-hold function) × (Sample-and-hold function)**
Two sample-and-hold functions of an SHF set are mutually disjoint. Thus, the product rule is

$$S_iS_j = \begin{cases} 0 & \text{where } i \neq j \\ S_i & \text{where } i = j \end{cases}$$

This also holds for multiplication of two sample-and-hold functions belonging to two different but equivalent SHF sets. That is, both the sets having the same h and T.
For example, the product of the first members of two such SHF sets in HF domain is shown in Fig. 2.9.

(ii) **(Triangular function) × (Sample-and-hold function)**
If a triangular function of a TF set is multiplied with a sample-and-hold function of an SHF set, both the sets having the same h and T, and matched along the time scale, then the product rule is

$$T_iS_j = \begin{cases} 0 & \text{where } i \neq j \\ T_i & \text{where } i = j \end{cases}$$

For example, the product of the first members of two such TF and SHF sets, in HF domain, is shown in Fig. 2.10.

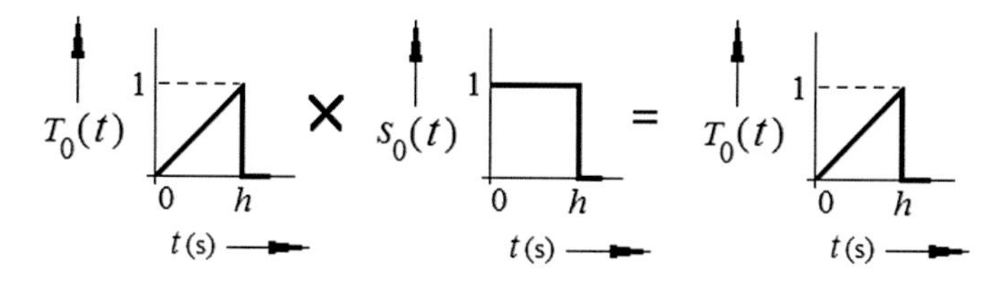

Fig. 2.10 Multiplication of the first member of TF set and the first member of SHF set

(iii) **(Triangular function) × (Triangular function)**
Two triangular functions of a TF set are mutually disjoint. Thus, the product rule is

$$T_iT_j = \begin{cases} 0 & \text{where } i \neq j \\ T_i & \text{where } i = j \end{cases}$$

This also holds for multiplication of two triangular functions belonging to two different TF sets, both having the same h and T, and matched along the time scale.
The product of the first members of two such TF sets, in HF domain, is shown in Fig. 2.11.
The result of multiplication of the first components of two equivalent triangular function sets, is converted to HF domain. The sample-and-hold function component of the product being zero, as seen from Fig. 2.11, the multiplication result is represented by only a triangular function component.
Following above rules, the result of multiplication of the two functions $r(t)$ and $g(t)$ are expressed in HF domain for $m = 4$ and $T = 1$ s as follows

$$\begin{aligned}\bar{m}(t) &\triangleq [r_0g_0 \quad r_1g_1 \quad r_2g_2 \quad r_3g_3]\,\mathbf{S}_{(m)} \\ &\quad + [(r_1g_1 - r_0g_0) \quad (r_2g_2 - r_1g_1) \quad (r_3g_3 - r_2g_2) \quad (r_4g_4 - r_3g_3)]\,\mathbf{T}_{(m)}\end{aligned}$$

The generalized result of multiplication of two time functions $r(t)$ and $g(t)$ are expressed in HF domain as follows :

$$\begin{aligned}\bar{m}(t) &\triangleq [r_0g_0 \quad r_1g_1 \quad \cdots \quad r_ig_i \quad \cdots \quad r_{m-1}g_{m-1}]\,S_{(m)} \\ &\quad + [(r_1g_1 - r_0g_0) \quad (r_2g_2 - r_1g_1) \quad \cdots \quad (r_ig_i - r_{i-1}g_{i-1}) \quad \cdots \\ &\quad (r_mg_m - r_{m-1}g_{m-1})]\,T_{(m)}\end{aligned} \tag{2.30}$$

$$\begin{aligned}\text{Now, } \bar{m}(t) &\triangleq \mathbf{R}_S^T\mathbf{D}_{G1}\mathbf{S}_{(m)} + \left[\mathbf{R}'^T_S\mathbf{D}_{G2} - \mathbf{R}_S^T\mathbf{D}_{G1}\right]\mathbf{T}_{(m)} \\ &\triangleq \mathbf{M}_S^T\mathbf{S}_{(m)} + \mathbf{M}_T^T\mathbf{T}_{(m)}\end{aligned} \tag{2.31}$$

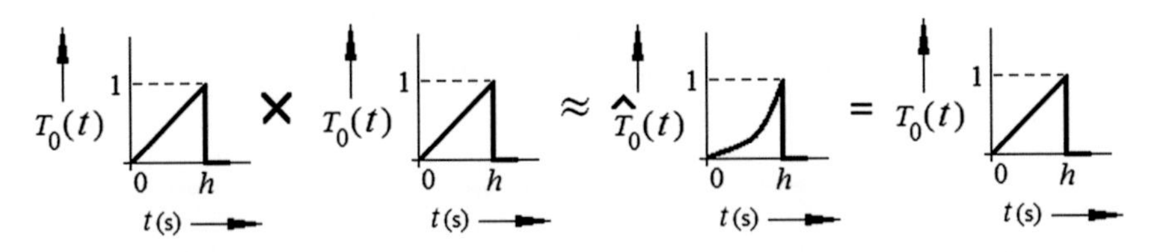

Fig. 2.11 Multiplication of the first members of two TF sets. The actual product and its HF domain equivalent are also shown

where, $\mathbf{M}_S^T \triangleq \mathbf{R}_S^T \mathbf{D}_{G1}$ and $\mathbf{M}_T^T \triangleq \left[\mathbf{R'}_S^T \mathbf{D}_{G2} - \mathbf{R}_S^T \mathbf{D}_{G1}\right]$

$$\begin{aligned}
\mathbf{R}_S^T &\triangleq [r_0 \quad r_1 \quad r_2 \quad \cdots \quad r_i \quad \cdots \quad r_{m-1}] \\
\mathbf{R'}_S^T &\triangleq [r_1 \quad r_2 \quad r_3 \quad \cdots \quad r_{i+1} \quad \cdots \quad r_m] \\
\mathbf{M}_S^T &\triangleq [m_0 \quad m_1 \quad m_2 \quad \cdots \quad m_i \quad \cdots \quad m_{m-1}] \\
\mathbf{M}_T^T &\triangleq [(m_1 - m_0) \quad (m_2 - m_1) \quad (m_3 - m_2) \quad \cdots \quad (m_i - m_{i-1}) \quad \cdots \quad (m_m - m_{m-1})]
\end{aligned}$$

and $\mathbf{D}_{G1}$ denotes a diagonal matrix whose entries are the m elements of the vector **G1**, i.e. $g_0, g_1, g_2, \ldots, g_{m-1}$ and $\mathbf{D}_{G2}$ denotes a diagonal matrix whose entries are the m elements of the vector **G2**, i.e. $g_1, g_2, g_3, \ldots, g_m$.

$$\begin{aligned}
\mathbf{D}_{G1} &= \operatorname{diag}(g_0, g_1, g_2, \ldots, g_{m-1}) \\
\text{and} \quad \mathbf{D}_{G2} &= \operatorname{diag}(g_1, g_2, g_3, \ldots, g_m)
\end{aligned}$$

(b) ***The functions r(t) and g(t) are first multiplied and then the resulting function m(t) is expanded* via *HF domain***

In this case, the resulting continuous function $m(t) = r(t) \times g(t)$ is expressed in HF domain as

$$\begin{aligned}
m(t) &= r(t) \times g(t) \approx \bar{r}(t) \times \bar{g}(t) \\
&= [m_0 \quad m_1 \quad m_2 \quad \cdots \quad m_i \quad \cdots \quad m_{m-1}]\, \mathbf{S}_{(m)} \\
&\quad + [(m_1 - m_0) \quad (m_2 - m_1) \quad \cdots \quad (m_i - m_{i-1}) \quad \cdots \quad (m_m - m_{m-1})]\, \mathbf{T}_{(m)}
\end{aligned}$$

where $m_0, m_1, m_2, \ldots, m_i, \ldots, m_m$ are the samples of $m(t)$ at time instants $0, h, 2h, \ldots, ih, \ldots, mh$.

At sampling instants the function values i.e., samples, will be multiplied as before. Hence,

$$\begin{aligned}
\bar{m}(t) &= [r_0 g_0 \quad r_1 g_1 \quad r_2 g_2 \quad \cdots \quad r_i g_i \quad \cdots \quad r_{m-1} g_{m-1}]\, \mathbf{S}_{(m)} \\
&\quad + [(r_1 g_1 - r_0 g_0) \quad (r_2 g_2 - r_1 g_1) \quad \cdots \quad (r_i g_i - r_{i-1} g_{i-1}) \quad \cdots \quad (r_m g_m - r_{m-1} g_{m-1})]\, \mathbf{T}_{(m)}
\end{aligned}$$

That is

$$\bar{m}(t) \triangleq \mathbf{M}_S^T \mathbf{S}_{(m)} + \mathbf{M}_T^T \mathbf{T}_{(m)} \tag{2.32}$$

From Eqs. (2.31) and (2.32), it is seen that both the results of multiplication are identical.

Considering two functions $r(t) = 1 - \exp(-t)$ and $g(t) = \exp(-t)$, the result of multiplication of these two functions, using their individual coefficients, is shown in Fig. 2.12.

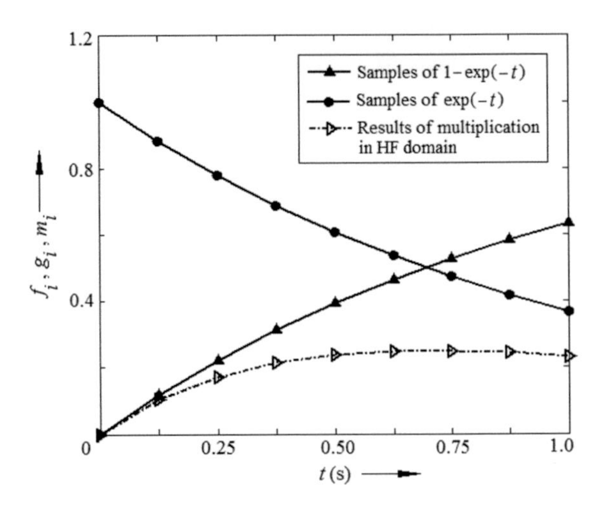

Fig. 2.12 Hybrid function expansion of $r(t) = 1 - \exp(-t)$ and $g(t) = \exp(-t)$ and the result of their multiplication $\bar{m}(t)$ in HF domain with $m = 8$ and $T = 1$ s. Due to high degree of accuracy, the piecewise *linear curves* look like *continuous curves*

2.6.4 Division of Two Functions

We consider division of two nonzero time functions $r(t)$ and $y(t)$ in HF domain.

Let, $d(t) = \frac{y(t)}{r(t)}$, where, $d(t)$ is the resulting continuous function after division. The HF domain expanded form of $d(t)$ may be called $\bar{d}(t)$.

Thus, we have,

$$y(t) = r(t) \times d(t) \tag{2.33}$$

The m number of samples of the functions $r(t)$ and $y(t)$ with sampling period h are $r_0, r_1, r_2, \ldots, r_i, \ldots r_m$ and $y_0, y_1, y_2, \ldots, y_i, \ldots y_m$ respectively. After division, let the resulting function $d(t)$ have the samples $d_0, d_1, d_2, \ldots, d_i, \ldots d_m$ with the same sampling period h.

(a) ***The functions r(t) and y(t) are expanded in HF domain and then the division operation is executed***

Using Eqs. (2.30), (2.33) can be written as follows

$$\begin{aligned} y(t) &= r(t) \times d(t) \\ &\approx \begin{bmatrix} r_0 d_0 & r_1 d_1 & r_2 d_2 & \cdots & r_i d_i & \cdots & r_{m-1} d_{m-1} \end{bmatrix} \mathbf{S}_{(m)} \\ &\quad + \left[(r_1 d_1 - r_0 d_0) \quad (r_2 d_2 - r_1 d_1) \quad \cdots \quad (r_i d_i - r_{i-1} d_{i-1}) \right. \\ &\quad \left. \cdots \quad (r_m d_m - r_{m-1} d_{m-1}) \right] \mathbf{T}_{(m)} \end{aligned} \tag{2.34}$$

Again, the time function $y(t)$ is expressed in HF domain as

$$y(t) \approx [y_0 \quad y_1 \quad y_2 \quad \cdots \quad y_i \quad \cdots \quad y_{m-1}]\,\mathbf{S}_{(m)} \\ + [(y_1 - y_0) \quad (y_2 - y_1) \quad \cdots \quad (y_i - y_{i-1}) \quad \cdots \quad (y_m - y_{m-1})]\,\mathbf{T}_{(m)} \tag{2.35}$$

Comparing the coefficients of the Eqs. (2.34) and (2.35), we have $y_0 = r_0 d_0$, $y_1 = r_1 d_1, \ldots y_i = r_i d_i \ldots$ and so on.
Thus, $d_0 = \frac{y_0}{r_0}$, $d_1 = \frac{y_1}{r_1}, \ldots d_i = \frac{y_i}{r_i} \ldots$ and so on.
Now, the above results can be expressed as follows

$$[d_0 \quad d_1 \quad d_2 \quad \cdots \quad d_i \quad \cdots \quad d_{m-1}]\,\mathbf{S}_{(m)} \\ + [(d_1 - d_0) \quad (d_2 - d_1) \quad \cdots \quad (d_i - d_{i-1}) \quad \cdots \quad (d_m - d_{m-1})]\,\mathbf{T}_{(m)} \\ = \left[\frac{y_0}{r_0} \quad \frac{y_1}{r_1} \quad \cdots \quad \frac{y_i}{r_i} \quad \cdots \quad \frac{y_{m-1}}{r_{m-1}}\right]\mathbf{S}_{(m)} \\ + \left[\left(\frac{y_1}{r_1} - \frac{y_0}{r_0}\right) \quad \left(\frac{y_2}{r_2} - \frac{y_1}{r_1}\right) \quad \cdots \quad \left(\frac{y_i}{r_i} - \frac{y_{i-1}}{r_{i-1}}\right) \quad \cdots \quad \left(\frac{y_m}{r_m} - \frac{y_{m-1}}{r_{m-1}}\right)\right]\mathbf{T}_{(m)}$$

Now, the resulting function $\bar{d}(t)$ of the division is expressed as

$$\bar{d}(t) = \sum_{i=0}^{m-1}\left(\frac{y_i}{r_i}\right)S_i + \sum_{i=0}^{m-1}\left(\frac{y_{i+1}}{r_{i+1}} - \frac{y_i}{r_i}\right)T_i$$

$$\text{or} \quad \bar{d}(t) \triangleq \mathbf{Y}_S^T\mathbf{D}_{R1}^{-1}\mathbf{S}_{(m)} + \left[\mathbf{Y'}_S^T\mathbf{D}_{R2}^{-1} - \mathbf{Y}_S^T\mathbf{D}_{R1}^{-1}\right]\mathbf{T}_{(m)} \\ \triangleq \mathbf{D}_S^T\mathbf{S}_{(m)} + \mathbf{D}_T^T\mathbf{T}_{(m)} \tag{2.36}$$

where,

$$\mathbf{Y}_S^T \triangleq [y_0 \quad y_1 \quad y_2 \quad \cdots \quad y_i \quad \cdots \quad y_{m-1}] \\ \mathbf{Y'}_S^T \triangleq [y_1 \quad y_2 \quad y_3 \quad \cdots \quad y_{i+1} \quad \cdots \quad y_m] \\ \mathbf{D}_S^T = \mathbf{Y}_S^T\mathbf{D}_{R1}^{-1} \triangleq [d_0 \quad d_1 \quad d_2 \quad \cdots \quad d_i \quad \cdots \quad d_{m-1}] \\ \mathbf{D}_T^T = \left[\mathbf{Y'}_S^T\mathbf{D}_{R2}^{-1} - \mathbf{Y}_S^T\mathbf{D}_{R1}^{-1}\right] \triangleq [(d_1 - d_0) \quad (d_2 - d_1) \quad \cdots \quad (d_i - d_{i-1}) \quad \cdots \quad (d_m - d_{m-1})]$$

and $\mathbf{D}_{R1}$ denotes a diagonal matrix whose entries are the m elements of the vector **R1**, i.e. $r_0, r_1, r_2, \ldots, r_{m-1}$, and $\mathbf{D}_{R2}$ denotes another diagonal matrix whose entries are the m elements of the vector **R2**, i.e. $r_1, r_2, r_3, \ldots, r_m$. That is

$$\mathbf{D}_{R1} = \mathrm{diag}(r_0, r_1, r_2, \ldots, r_{m-1}) \\ \text{and} \quad \mathbf{D}_{R2} = \mathrm{diag}(r_1, r_2, r_3, \ldots, r_m)$$

$$\mathbf{D}_{\mathrm{R1}}^{-1} = \mathrm{diag}\left(\frac{1}{r_0}, \frac{1}{r_1}, \frac{1}{r_2}, \ldots, \frac{1}{r_{m-1}}\right)$$

$$\text{and} \quad \mathbf{D}_{\mathrm{R2}}^{-1} = \mathrm{diag}\left(\frac{1}{r_1}, \frac{1}{r_2}, \frac{1}{r_3}, \ldots, \frac{1}{r_m}\right)$$

(b) ***The functions y(t) is first divided by r(t) and then the resulting function d(t) is expanded in HF domain***

In this case, the samples of the resulting continuous function $d(t) = \frac{y(t)}{r(t)}$ will be the results of division of the corresponding samples of the functions $y(t)$ and $r(t)$. That is

$$d_i = \frac{y_i}{r_i}$$

Hence,

$$\begin{aligned} d(t) &= \frac{y(t)}{r(t)} \approx \bar{d}(t) = \frac{\bar{y}(t)}{\bar{r}(t)} \\ &= [d_0 \quad d_1 \quad d_2 \quad \cdots \quad d_i \quad \cdots \quad d_{m-1}]\, \mathbf{S}_{(m)} \\ &\quad + [(d_1 - d_0) \quad (d_2 - d_1) \quad \cdots \quad (d_i - d_{i-1}) \quad \cdots \quad (d_m - d_{m-1})]\, \mathbf{T}_{(m)} \end{aligned}$$

where, $\bar{d}(t)$ is the HF domain representation of $d(t)$ and $d_0, d_1, d_2, \ldots, d_i, \ldots, d_m$ are the samples of $d(t)$ at time instants $0, h, 2h, \ldots, ih, \ldots mh$.

So, at sampling instants, the sample value of $d(t)$ will be division as discussed before.

Hence,

$$\begin{aligned} \bar{d}(t) &= \left[\frac{y_0}{r_0} \quad \frac{y_1}{r_1} \quad \frac{y_2}{r_2} \quad \cdots \quad \frac{y_i}{r_i} \quad \cdots \quad \frac{y_{m-1}}{r_{m-1}}\right] \mathbf{S}_{(m)} \\ &\quad + \left[\left(\frac{y_1}{r_1} - \frac{y_0}{r_0}\right) \quad \left(\frac{y_2}{r_2} - \frac{y_1}{r_1}\right) \quad \cdots \quad \left(\frac{y_i}{r_i} - \frac{y_{i-1}}{r_{i-1}}\right) \quad \cdots \quad \left(\frac{y_m}{r_m} - \frac{y_{m-1}}{r_{m-1}}\right)\right] \mathbf{T}_{(m)} \end{aligned}$$

That is

$$\bar{d}(t) \triangleq \mathbf{D}_{\mathrm{S}}^{\mathrm{T}} \mathbf{S}_{(m)} + \mathbf{D}_{\mathrm{T}}^{\mathrm{T}} \mathbf{T}_{(m)} \tag{2.37}$$

From Eqs. (2.36) and (2.37), it is seen that both the results of division are identical.

Considering two functions $y(t) = 1 - \exp(-t)$ and $r(t) = \exp(-t)$, the result of their division $d(t) = \frac{y(t)}{r(t)} = \exp(t) - 1$, using their individual coefficients (i.e., samples), is shown in Fig. 2.13.

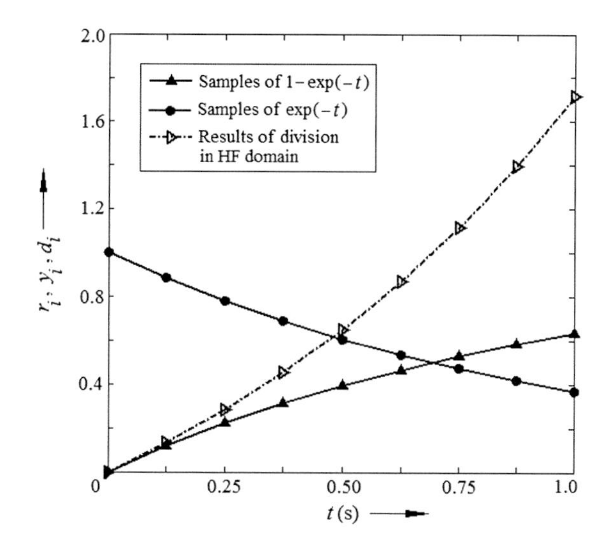

Fig. 2.13 Hybrid function expansion of $r(t) = \exp(-t)$ and $y(t) = 1 - \exp(-t)$ and the result of their division $\bar{d}(t)$ in HF domain with $m = 8$ and $T = 1$ s. Due to high degree of accuracy, the piecewise *linear curves* look like *continuous curves* (vide Appendix B, Program no. 2)

2.7 Qualitative Comparison of BPF, SHF, TF and HF

The basic properties of BPF, SHF, TF and HF are tabulated in Table 2.1 above to provide a qualitative appraisal.

2.8 Conclusion

A new set of orthogonal hybrid function (HF) has been proposed for function approximation, and its subsequent application to control system analysis and identification. The set of hybrid functions is formed using the set of sample-and-hold functions and the set of triangular functions.

The hybrid function set works with function samples, and this makes it more convenient for use. That is, the expansion coefficients of SHF and TF components of the HF set are simply the samples of the function to be approximated indicating a non-optimal approach. Thus, like traditional orthogonal function sets, the HF set does not use the well known integration formula for coefficient computation. This presents a faster algorithm, makes the mathematics less involved, and also, reduces the computation time.

The comparison of the basic qualitative properties of the hybrid function set with different related orthogonal functions have been presented in Table 2.1.

In the following chapters, it will be shown that the hybrid function set is not only suitable for function approximation, but it can efficiently integrate time functions as well. Furthermore, it is a strong tool for various applications in the area of control theory.

Table 2.1 Qualitative comparison of BPF, SHF, TF and HF

Property	BPF	SHF	TF	HF
Piecewise constant	Yes	Yes	No (piecewise linear)	No (piecewise linear)
Orthogonal	Yes	Yes	Yes	Yes
Finite	Yes	Yes	Yes	Yes
Disjoint	Yes	Yes	Yes	Yes
Orthonormal	Can easily be normalized	Can easily be normalized	Can easily be normalized	Can easily be normalized
Implementation	Easily implementable	Easily implementable	Implementation is relatively complex	Easily implementable
Coefficient determination of $f(t)$	Involves integration of $f(t)$ and scaling	Needs only samples of $f(t)$	Needs only samples of $f(t)$	Needs only samples of $f(t)$
Accuracy of analysis	Staircase solution having more error than TF	Staircase solution having less error than BPF for sample-and-hold systems	Piecewise linear solution having less error than BPF	Provides two part (SHF and TF) piecewise linear solution having less error than BPF as well as SHF. Staircase solution in SHF mode is available as a 'by-product' by setting the TF part of the solution equal to zero. This gives an edge to HF analysis over TF analysis

References

1. Sansone, G.: Orthogonal functions. Interscience, New York (1959)
2. Deb, A., Sarkar, G., Bhattacharjee, M., Sen, S.K.: A new set of piecewise constant orthogonal functions for the analysis of linear SISO systems with sample-and-hold. J. Franklin Inst. **335B** (2), 333–358 (1998)
3. Deb, A., Sarkar, G., Senupta, A.: Triangular orthogonal functions for the analysis of continuous time systems. Anthem Press, London (2011)
4. Deb, A., Dasgupta, A., Sarkar, G.: A complementary pair of orthogonal triangular function sets and its application to the analysis of dynamic systems. J. Franklin Inst. **343**(1), 1–26 (2006)
5. Rao, G.P.: Piecewise constant orthogonal functions and their applications in systems and control, LNC1S, vol. 55. Springer, Berlin (1983)
6. Jiang, J.H., Schaufelberger, W.: Block pulse functions and their application in control system, LNCIS, vol. 179. Springer, Berlin (1992)
7. Biswas, A.: Analysis and synthesis of continuous control systems using a set of orthogonal hybrid functions. Doctoral dissertation, University of Calcutta (2015)

Chapter 3
Function Approximation via Hybrid Functions

Abstract In this chapter, square integrable time functions of Lebesgue measure are approximated via hybrid functions and such approximations are compared with similar approximations using BPF and Legendre polynomials. For handling discontinuous functions, a modified method of approximation is suggested in hybrid function domain. This modified approach, named $\mathrm{HF_m}$ approach, seems to be more accurate than the conventional HF domain technique, termed as $\mathrm{HF_c}$ approach. The mean integral square errors (MISE) for both the approximations are computed and compared. Finally, error estimates for the SHF domain approximation and TF domain approximation are derived. The chapter contains many tables and graphs along with six illustrative examples.

In this chapter, similar to block pulse function [1, 2] domain approximation, we use the complementary hybrid function (HF) set, combination of the sample-and-hold function (SHF) set [3] and the triangular function (TF) set [4–6], for function approximation. Earlier, we presented the principle for the proposed hybrid function domain expansion where the expansion coefficients were the sample values of the function to be approximated. The hybrid function set may now be utilized for approximating square integrable functions in a piecewise linear manner.

The HF set obeys the conditions of orthogonality because it approximates functions using the linear combination of two orthogonal function sets, namely SHF set and TF set. For each of the orthogonal function sets, the members of the set satisfy the criteria of completeness, and hence, this complementary orthogonal set is a complete orthogonal function set.

3.1 Function Approximation via Block Pulse Functions (BPF)

A square integrable time function $f(t)$ of Lebesgue measure [7] may be expanded into an m-term BPF series in $t \in [0, T)$ as

A. Deb et al., *Analysis and Identification of Time-Invariant Systems, Time-Varying Systems, and Multi-Delay Systems using Orthogonal Hybrid Functions*,
Studies in Systems, Decision and Control 46, DOI 10.1007/978-3-319-26684-8_3

$$
\begin{aligned}
f(t) &\approx \sum_{i=0}^{m-1} f_i \psi_i(t) \quad \text{for } i = 0, 1, 2, \ldots, (m-1) \\
&= [f_0 \quad f_1 \quad f_2 \quad \ldots \quad f_i \quad \ldots \quad f_{m-1}] \mathbf{\Psi}_{(m)}(t) \\
&\triangleq \mathbf{F}_{(m)}^{\mathrm{T}} \mathbf{\Psi}_{(m)}(t)
\end{aligned}
\tag{3.1}
$$

where, $[\cdots]^{\mathrm{T}}$ denotes transpose and $f_0, f_1, f_2, \ldots, f_i, \ldots f_{(m-1)}$ are the coefficients of block pulse function expansion. The $(i + 1)$th BPF coefficient f_i is given by

$$
f_i = \frac{1}{h} \int_{ih}^{(i+1)h} f(t)\psi_i(t)\mathrm{d}t \tag{3.2}
$$

where, $h = \frac{T}{m}$s.

The coefficients f_is are determined in such a way that the integral square error [ISE] [7] is minimized.

3.1.1 Numerical Examples

At first we determine the coefficients of a time function $f(t)$ in BPF domain using Eq. (3.2). We consider the following two examples:

Example 3.1 Let us expand the function $f_1(t) = t$ in block pulse function domain taking $m = 8$ and $T = 1$ s. Following the method mentioned above, the result is

$$
\begin{aligned}
f_1(t) \approx [&0.06250000 \quad 0.18750000 \quad 0.31250000 \quad 0.43750000 \\
&0.56250000 \quad 0.68750000 \quad 0.81250000 \quad 0.93750000] \mathbf{\Psi}_{(8)}(t)
\end{aligned}
\tag{3.3}
$$

Figure 3.1 shows the original function along with its BPF approximation.

Example 3.2 Now we take up the function $f_2(t) = \sin(\pi t)$ and express it in block pulse function domain for $m = 8$ and $T = 1$ s. The result is

$$
\begin{aligned}
f_2(t) \approx [&0.19383918 \quad 0.55200729 \quad 0.82613728 \quad 0.97449537 \\
&0.97449537 \quad 0.82613728 \quad 0.55200729 \quad 0.19383918] \mathbf{\Psi}_{(8)}(t)
\end{aligned}
\tag{3.4}
$$

Figure 3.2 shows the original function along with its BPF domain approximation.

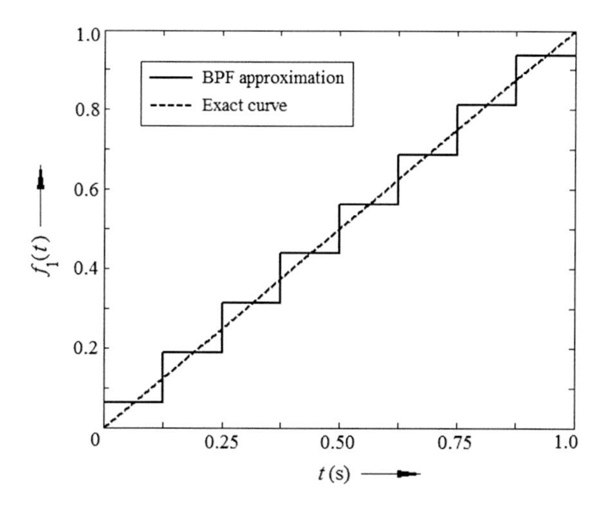

Fig. 3.1 Exact curve for $f_1(t) = t$ and its block pulse function approximation for $m = 8$ and $T = 1$ s

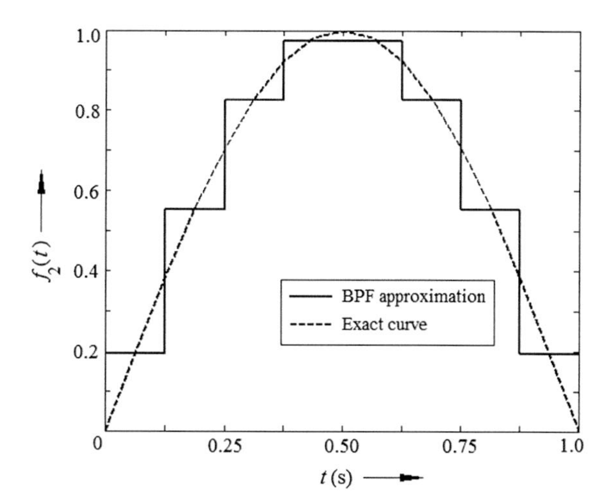

Fig. 3.2 Exact curve for $f_2(t) = \sin(\pi t)$ and its block pulse function approximation for $m = 8$ and $T = 1$ s (vide Appendix B, Program no. 3)

3.2 Function Approximation via Hybrid Functions (HF) [8, 9]

Consider a function $f(t)$ in an interval $t \in [0, T)$. If we consider $(m + 1)$ equidistant samples $f_0, f_1, f_2, \ldots, f_i, \ldots, f_m$ of the function with a sampling period h (i.e., $T = mh$), $f(t)$ can be expressed as per Eq. (2.12) as discussed in Sect. 2.4.

$$
\begin{aligned}
f(t) \approx & \left[f_0 \quad f_1 \quad f_2 \quad \cdots \quad f_i \quad \cdots \quad f_{(m-1)}\right]\mathbf{S}_{(m)} \\
& + \left[(f_1 - f_0) \quad (f_2 - f_1) \quad (f_3 - f_2) \quad \cdots \quad (f_i - f_{i-1}) \quad \cdots \quad (f_m - f_{(m-1)})\right]\mathbf{T}_{(m)} \\
& = \sum_{i=0}^{m-1} f_i S_i + \sum_{i=0}^{m-1} (f_{i+1} - f_i) T_i \\
& \triangleq \mathbf{F}_S^T \mathbf{S}_{(m)} + \mathbf{F}_T^T \mathbf{T}_{(m)}
\end{aligned}
\tag{3.5}
$$

3.3 Algorithm of Function Approximation via HF

The algorithm of function approximation via hybrid function domain is explained below in Fig. 3.3.

3.3.1 Numerical Examples

A few functions are now approximated in hybrid function domain using Eq. (3.5). We consider the following two examples:

Example 3.3 Let us expand the function $f_1(t) = t$ in hybrid function domain taking $m = 8$ and $T = 1$ s. Following the method presented above, the result is

$$
\begin{aligned}
f_1(t) \approx & [0.00000000 \quad 0.12500000 \quad 0.25000000 \quad 0.375000000 \\
& 0.50000000 \quad 0.62500000 \quad 0.75000000 \quad 0.87500000]\mathbf{S}_{(8)} \\
& + [0.12500000 \quad 0.12500000 \quad 0.12500000 \quad 0.12500000 \\
& 0.12500000 \quad 0.12500000 \quad 0.12500000 \quad 0.12500000]\mathbf{T}_{(8)}
\end{aligned}
\tag{3.6}
$$

Figure 3.4 shows the plot of the function $f_1(t) = t$ and its hybrid function approximation, as per Eqs. (3.5) and (3.6). It is observed that the exact curve entirely matches with its HF domain approximation, as is expected for a ramp function.

Hence, it is apparent that hybrid function domain representation can generally give the exact presentation of a piecewise linear function.

Example 3.4 Now we take up the function $f_2(t) = \sin(\pi t)$ and express it via hybrid functions for $m = 8$ and $T = 1$ s. The result is presented below:

$$
\begin{aligned}
f_2(t) \approx & [0.00000000 \quad 0.38268343 \quad 0.70710678 \quad 0.92387953 \\
& 1.00000000 \quad 0.92387953 \quad 0.70710678 \quad 0.38268343]\mathbf{S}_{(8)} \\
& + [0.38268343 \quad 0.32442335 \quad 0.21677275 \quad 0.07612047 \\
& -0.07612047 \quad -0.21677275 \quad -0.32442335 \quad -0.38268343]\mathbf{T}_{(8)}
\end{aligned}
\tag{3.7}
$$

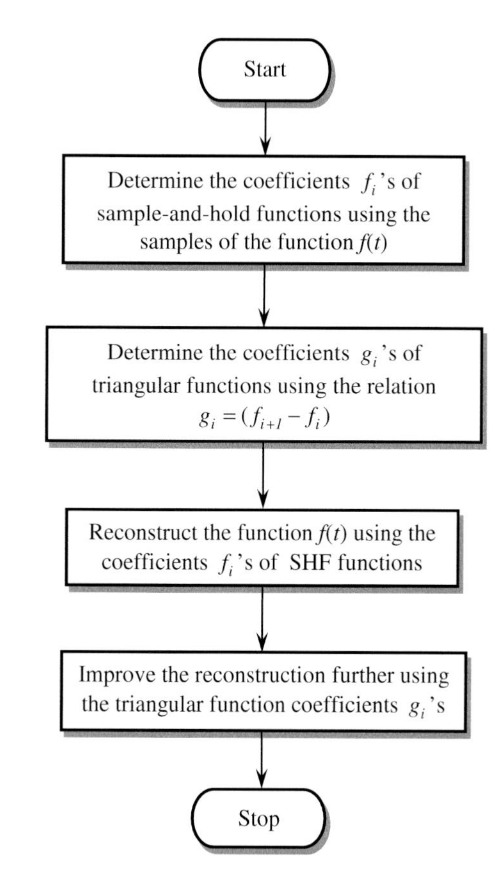

Fig. 3.3 The algorithm of function approximation via hybrid function domain

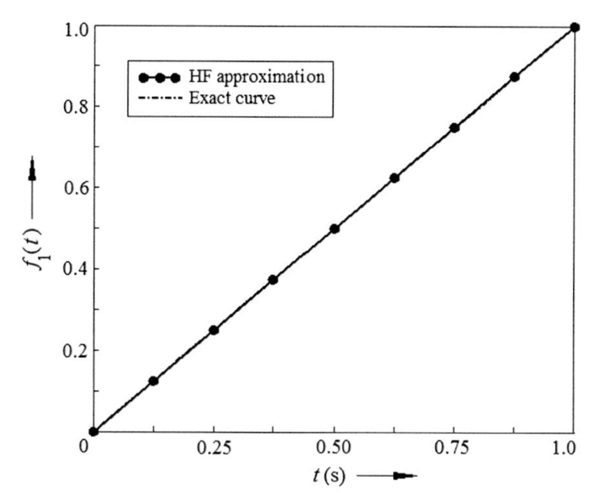

Fig. 3.4 Exact curve for $f_1(t) = t$ and its hybrid function approximation for $m = 8$ and $T = 1$ s. It is seen that the curves overlap

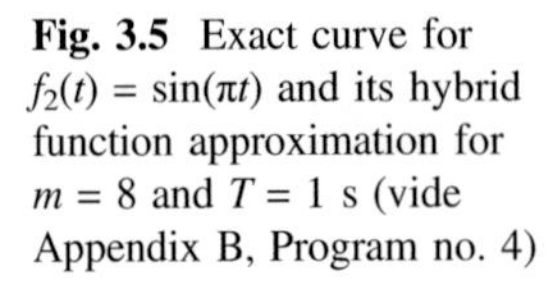

Fig. 3.5 Exact curve for $f_2(t) = \sin(\pi t)$ and its hybrid function approximation for $m = 8$ and $T = 1$ s (vide Appendix B, Program no. 4)

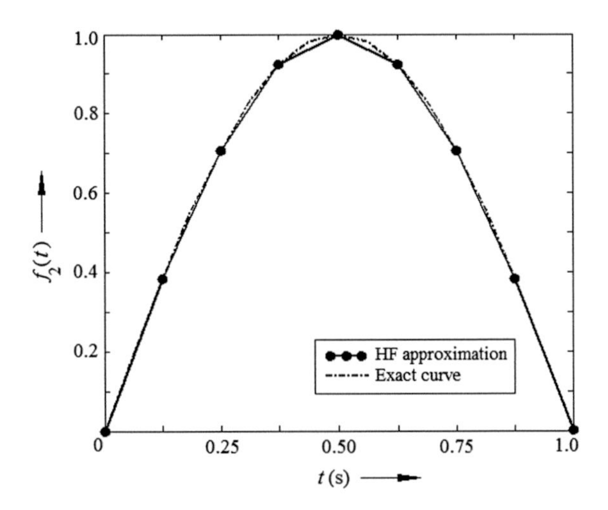

Figure 3.5 shows the plot of the function $f_2(t) = \sin(\pi t)$, and its hybrid function approximation. It is observed that the curve approximated by hybrid function is much closer to the exact curve compared to the BPF approximation of Fig. 3.2. This result may further be improved by increasing the number of samples. The closeness of the results with the exact curves and the pictorial presentation of the original functions along with their HF domain equivalent show the usefulness of the HF domain description.

3.4 Comparison Between BPF and HF Domain Approximations

Figures 3.6 and 3.7 show comparison of block pulse function and hybrid function based approximations for the functions $f_1(t)$ and $f_2(t)$. It is obvious from the figures that HF domain approximations are much better than BPF based approximations.

Quantitative estimates of MISE's of these two approximations are presented in Table 3.1. If we define an index Δ which is the ratio of the respective MISE's of BPF domain approximation and that of HF domain approximation, we see that, in case of ramp function, BPF domain approximation is no match for HF domain approximation. And for the time function $f_2(t)$, the MISE of BPF domain representation is about 25 times than that of HF domain. That is, for the sine wave, we have

$$\Delta = \frac{\text{MISE}_{\text{BPF}}}{\text{MISE}_{\text{HF}}} = 25.43420823 \tag{3.8}$$

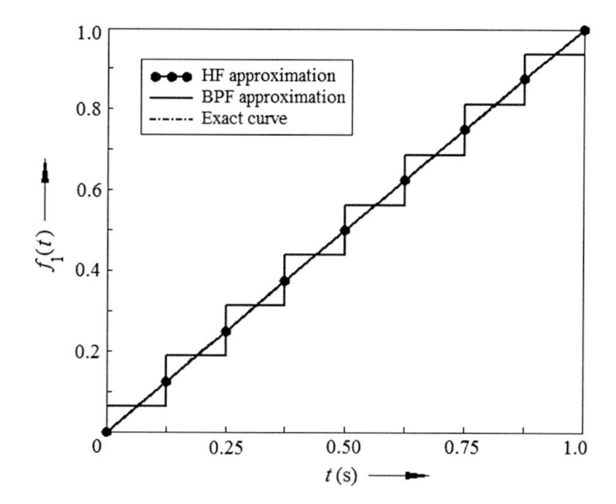

Fig. 3.6 Approximated curves for $f_1(t) = t$ in BPF domain and hybrid function domain for $m = 8$ and $T = 1$ s along with the exact curve. The HF domain approximation overlaps with the exact curve (vide Appendix B, Program no. 5)

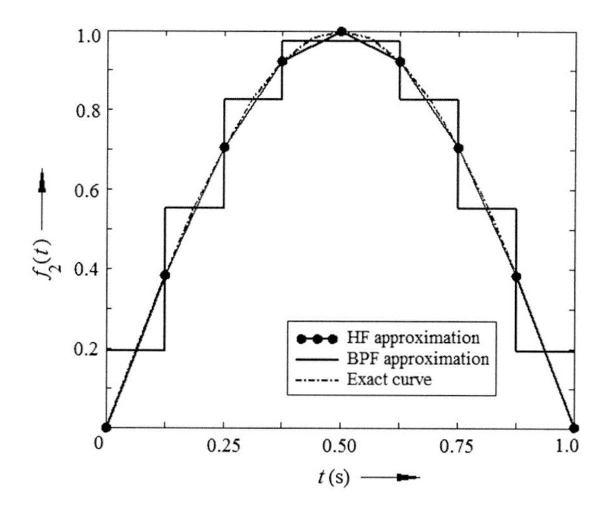

Fig. 3.7 Approximated curves for $f_2(t) = \sin(\pi t)$ in BPF domain and hybrid function domain for $m = 8$ and $T = 1$ s along with the exact curve

Table 3.1 Comparison of MISE's of function approximation via BPF and HF for two time functions, for $m = 10$ and $T = 2$ s

Function	MISE		$\Delta = \frac{\text{MISE}_{\text{BPF}}}{\text{MISE}_{\text{HF}}}$
	BPF	HF	
$f_1(t) = t$	0.00333333	0.00000000	–
$f_2(t) = \sin(\pi t)$	0.01623440	6.382897e−04	25.43420823

3.5 Approximation of Discontinuous Functions

In some cases, the value of a function changes rapidly, i.e., almost instantly, from one value to some other higher or lower value. This is termed as a 'jump discontinuity'. For such a discontinuity, both the left and right limits exist at the jump point, but obviously they are not equal. An attempt to analyse systems with such input discontinuities (say) in any orthogonal function domain framework, produces a large error at the function approximation stage. This error is propagated throughout the rest of the analysis.

For approximating a discontinuous function $g(t)$ having jumps at a finite number of points, i.e., $a_0, a_1, \ldots, a_i, \ldots, a_{m-1}$ over a semi-open interval [0, *T*) we can select the sampling interval h randomly and employ hybrid function approximation method to come up with a piecewise linear reconstruction of the function $g(t)$ (let us call it $\bar{g}(t)$). Obviously, the resulting function $\bar{g}(t)$ will be a piecewise linear function with its equidistant break points (i.e., sample points) evenly distributed over [0, *T*). Due to random or non-judicious selection of h, it may so happen that the sampling instants and jump points *may or may not coincide* on the time scale. Rather, coincidence of the sampling instants and the jump points, if any, will entirely depend on chance. However, judicious selection of the sampling period h may lead to the best approximation of a time function as far as jumps are concerned.

Figure 3.8 shows a piecewise continuous function $g(t)$ in [0, *T*) with three jump points (say) at t_1, t_2 and t_3 respectively. So, the four time intervals are:

(i) from the origin to the first jump point = $t_1 = a$(say)
(ii) from the first jump point to the second jump point = $t_2 - t_1 = b$
(iii) from the second jump point to the third jump point = $t_3 - t_2 = c$
(iv) from the third jump point to the end of the interval = $T - t_3 = d$

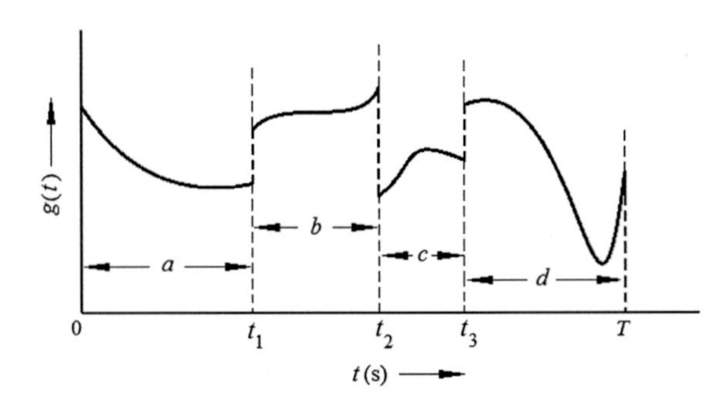

Fig. 3.8 A piecewise continuous function $g(t)$ having three jumps at points t_1, t_2 and t_3 over a period [0, *T*) seconds

The reconstruction of such a function in HF domain may be achieved in the following ways:

Case I For reconstructing this function via hybrid functions, we can take samples of $g(t)$ very closely with a small sampling period in the regions a, b, c and d, excluding the jump points. If these samples are now joined by straight lines to form a piecewise linear reconstruction, the function will be almost truthfully represented in HF domain in all the four regions. However, such reconstruction will involve a huge volume of data to represent the jump situations with reasonable accuracy.

Case II We can choose the sampling period h in such a fashion that the sample points always coincide with all the jump points. To make this happen, h should be the GCD of the four time intervals a, b, c and d, where $a + b + c + d = T$. Then, each of the intervals a, b, c and d will be divisible by h and the sampling points over $[0, T)$ will coincide with the three jump points. This is shown in Fig. 3.9. Under such circumstances, the HF reconstruction will not be able to indicate truthfully the jump points of the function $g(t)$. That is, hybrid function domain approximation will represent the piecewise continuous function $g(t)$ as shown in Fig. 3.9 in magenta.

Case III If we make h to be even smaller, say k, then if $k = \frac{h}{n}$, n being an integer, this will again make the sampling points almost coincide with the jump points and the HF domain reconstruction will show all the jump situations with reasonable accuracy, as shown in Fig. 3.10.

Case IV However, if h is not chosen judiciously, then the samples will not coincide with the jump points and HF domain reconstruction will be de-shaped compared to the original function and in the reconstructed function the jumps would be unrecognizable. This is shown in Fig. 3.11, where, the lines CD, EF and FG represent the function in the regions of discontinuities.

In the following, we discuss a *modified* HF domain approach for approximating such functions with jump discontinuities.

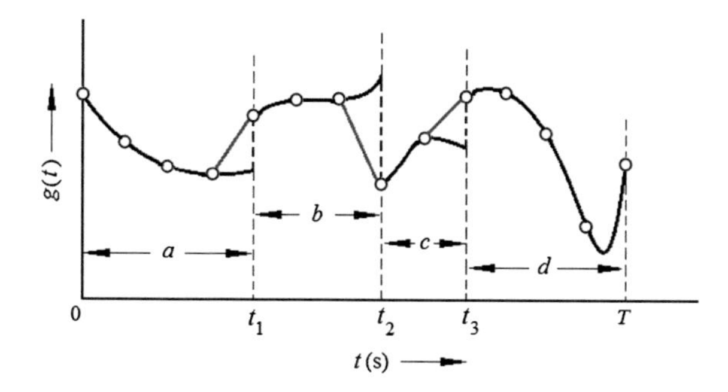

Fig. 3.9 The function $g(t)$ is sampled with a moderate sampling period h. it is noticed that the jumps are represented fairly accurately (shown in *magenta*) in the hybrid function reconstruction of the function

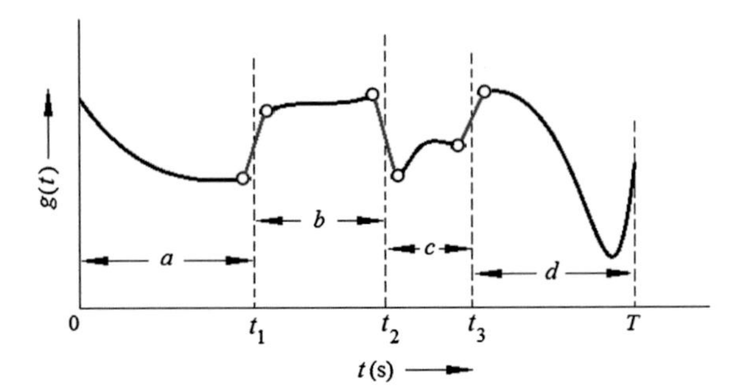

Fig. 3.10 The function $g(t)$ is sampled very closely with a small sampling period in the regions a, b, c and d, excluding the jump points. these samples are joined by straight lines to form a piecewise linear almost truthful reconstruction

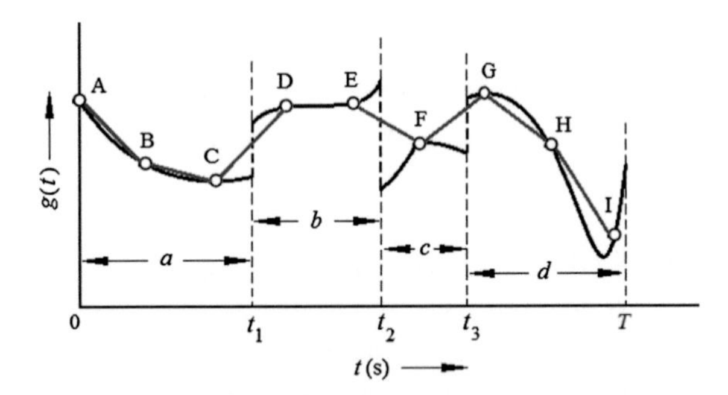

Fig. 3.11 The function $g(t)$ is sampled with a sampling period h chosen at random. It is noticed that the jumps are represented erroneously (shown in *magenta*) in the hybrid function reconstruction of the function. In the reconstructed function, no jumps are noticeable

3.5.1 Modified HF Domain Approach for Approximating Functions with Jump Discontinuities

We know that with conventional HF domain (HF_c, say) approximations, we end up with attractive piecewise linear reconstructions. But when the functions involve jump discontinuities, the approximations are not so attractive anymore, because a large amount of approximation error is introduced in the interval containing the jump, as is obvious from Fig. 3.9. This mars the quality of approximation, and this error is transmitted through the whole of the remaining analysis to infect the final results with unacceptable error.

To overcome such situations involving 'jump functions' (that is, functions with jump discontinuities), a modified approach in HF domain (HF_m domain, say) may be proposed. This approach is superior to the conventional HF_c approach.

For example, to approximate a function as shown in Fig. 3.12, having a jump discontinuity at $t = t_d$, the HF_m approach produces much better result than the HF_c

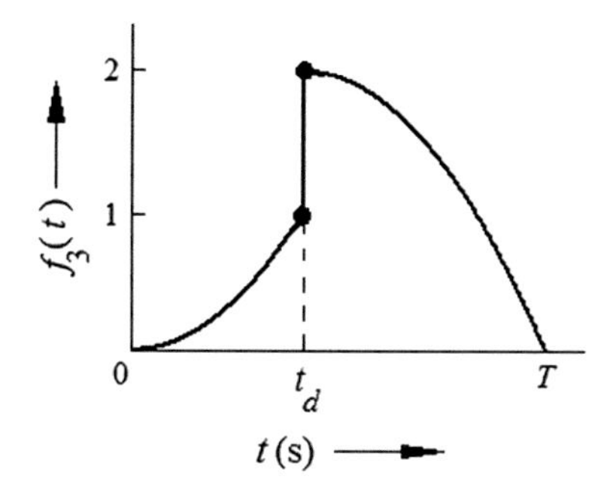

Fig. 3.12 A function $f_3(t)$ with jump discontinuity at $t = t_d$

approach. For most of the functions, this is so and will be illustrated later in Sect. 3.5.2. A 'better' approximation is always judged by the quantitative factor mean integral square error (MISE) [2], mentioned in Sect. 1.1.

To illustrate the superiority of the HF_m approach, we consider a simple delayed unit step function having a delay of kh seconds, shown in Fig. 3.13a. Using the HF_c approach, the HF domain approximation of $u(t - kh)$ is shown in Fig. 3.13b, whereas, Fig. 3.13c illustrates the approximation of $u(t - kh)$ using the modified HF_m approach. In this approach, we have dropped the triangular function component in the kth sub-interval (that is from $(k - 1)h$ to kh) by making the TF coefficient zero. This makes the approximation one hundred per cent accurate.

Comparing the figures, we see that while the conventional HF domain approximation produces a large error (the shaded triangular zone), the modified HF_m approach produces zero error.

The two approximations can be represented mathematically, as under.

Using HF_c approach, the approximation is given by

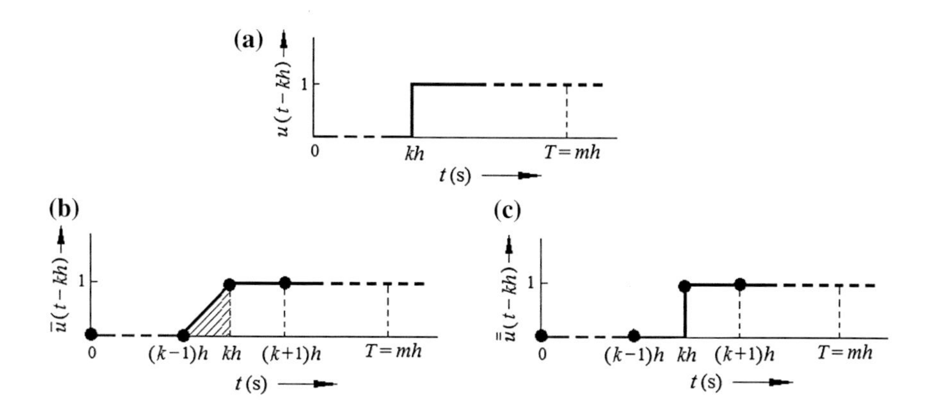

Fig. 3.13 **a** Delayed unit step function with a step change at $t = kh$ with **b** its approximation in HF domain using conventional approach and **c** a modified HF domain approach to tackle the step change leading to a better approximation

$$\bar{u}(t-kh) \approx \Big[\underbrace{0 \ \ldots \ 0}_{k \text{ zeros}} \quad \underbrace{1 \ \ldots \ 1}_{\text{all ones}}\Big]\mathbf{S}_{(m)} + \Big[\underbrace{0 \ \ldots \ 0}_{(k-1) \text{ zeros}} \quad \underset{k\text{th term}}{\overset{}{1}} \quad \underbrace{0 \ \ldots \ 0}_{\text{all zeros}}\Big]\mathbf{T}_{(m)} \tag{3.9}$$

This is shown in Fig. 3.13b.

If we approximate the same function using the $\mathrm{HF_m}$ approach, the result is

$$\bar{\bar{u}}(t-kh) = \Big[\underbrace{0 \ \ldots \ 0}_{k \text{ zeros}} \quad \underbrace{1 \ \ldots \ 1}_{\text{all ones}}\Big]\mathbf{S}_{(m)} + \Big[\underbrace{0 \ 0 \ \ldots \ 0}_{\text{all zeros}}\Big]\mathbf{T}_{(m)} \tag{3.10}$$

as illustrated in Fig. 3.13c.

Now consider the following function

$$f_4(t) = (c_0 + kt)u(t) + u(t-3h)$$

where, c_0 is the DC bias and k is the slope of the ramp function. The function is depicted in Fig. 3.14a. It is noted that the function has a jump of one unit at $t = 3\,h$ (say).

This function may be approximated via $\mathrm{HF_c}$ approach for $m = 6$ and $T = 1$ s as shown in Fig. 3.14b.

Mathematically, we can write

$$\begin{aligned} \bar{f}_4(t) &\approx [\,c_0 \quad c_1 \quad c_2 \quad c_3 \quad c_4 \quad c_5\,]\mathbf{S}_{(6)} \\ &\quad + [\,(c_1-c_0) \quad (c_2-c_1) \quad (c_3-c_2) \quad (c_4-c_3) \quad (c_5-c_4) \quad (c_6-c_5)\,]\mathbf{T}_{(6)} \\ &\triangleq \mathbf{C}_\mathrm{S}^\mathrm{T}\mathbf{S}_{(6)} + \mathbf{C}_\mathrm{T}^\mathrm{T}\mathbf{T}_{(6)} \end{aligned} \tag{3.11}$$

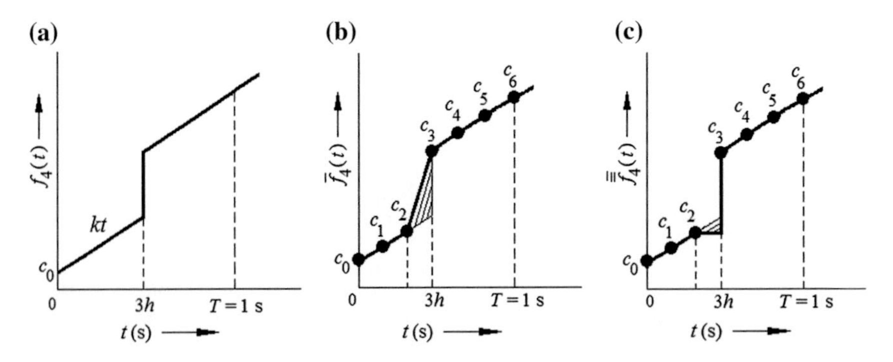

Fig. 3.14 **a** A typical time function $f_4(t)$ having a jump at $t = 3h$, **b** its approximation using the $\mathrm{HF_c}$ based approach and **c** the $\mathrm{HF_m}$ approach to approximate the jump discontinuity, for $m = 6$ and $T = 1$ s

Now, using the new approach $\mathrm{HF_m}$, we have

$$\begin{aligned}\bar{\bar{f}}_4(t) &\approx [c_0 \quad c_1 \quad c_2 \quad c_3 \quad c_4 \quad c_5]\mathbf{S}_{(6)} \\ &\quad + [(c_1-c_0) \quad (c_2-c_1) \quad 0 \quad (c_4-c_3) \quad (c_5-c_4) \quad (c_6-c_5)]\mathbf{T}_{(6)} \\ &\triangleq \mathbf{C}_\mathrm{S}^\mathrm{T}\mathbf{S}_{(6)} + \mathbf{C}_\mathrm{T}^\mathrm{T}\mathbf{J}_{3(6)}\mathbf{T}_{(6)}\end{aligned} \tag{3.12}$$

where, $\mathbf{J}_{3(6)}$ is a diagonal matrix, very much like the unit matrix $\mathbf{I}$, but having a zero as the third element of the diagonal. This special matrix is introduced to handle the jump discontinuity mathematically and it is defined as

$$\mathbf{J}_{3(6)} \triangleq \begin{bmatrix} 1 & & & & & \\ & 1 & & & & \\ & & 0 & & & \\ & & \uparrow & 1 & & \\ & & \text{3rd} & & 1 & \\ & & \text{term} & & & 1 \end{bmatrix}_{(6\times 6)}$$

According to Eq. (3.12), the approximated function $f_4(t)$ is shown in Fig. 3.14c. For any time function $f(t)$, if the jump occurs at $t = kh$, then the $\mathbf{J}$ matrix becomes

$$\mathbf{J}_{k(m)} = \begin{bmatrix} 1 & & & & & \\ & \ddots & & & & \\ & & 1 & & & \\ & & & 0 & & \\ & & & \uparrow & 1 & \\ & & & k\text{th} & & \ddots & \\ & & & \text{term} & & & 1 \end{bmatrix}_{(m\times m)} \tag{3.13}$$

If the function involved has more than one jump, obviously, the matrix $\mathbf{J}$ will have more than one zeroes in its diagonal at the locations, as discussed. If the function has no jump at all, the $\mathbf{J}$ matrix is simply replaced by the $\mathbf{I}$ matrix.

Consider a function with a downward jump at $t = kh$, shown in Fig. 3.15a. Figure 3.15b illustrates the approximation of the function using the $\mathrm{HF_c}$ approach. Here, the area of the shaded portion is some sort of indicator of the approximation error, and it is apparent that with respect to MISE, this approximation will incur more error compared to the approximation based upon the $\mathrm{HF_m}$ approach, shown in Fig. 3.15c.

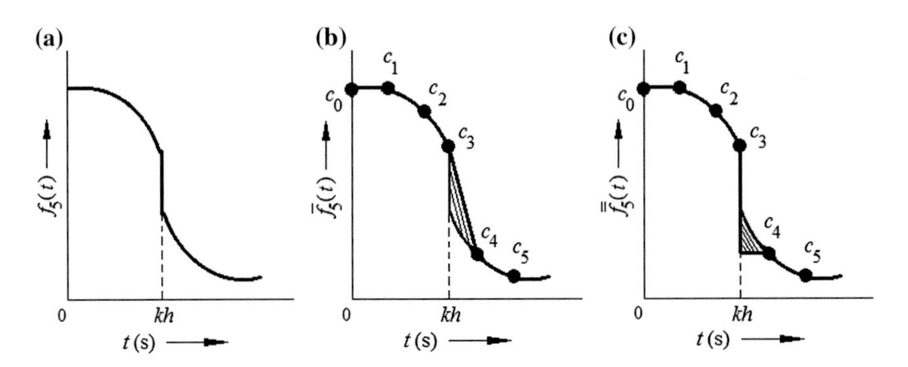

Fig. 3.15 **a** A typical time function having a downward jump at kh, with its approximation using **b** the HF_c approach and **c** the HF_m approach, for a typical value of $k = 3$

3.5.2 Numerical Examples

Example 3.5 Consider the function $f_4(t) = (0.2 + t)u(t) + u(t - 1)$ having a jump discontinuity at $t = 1$ s.

It may seem that the HF_m approach always provides a better approximation than the HF_c approach. But it is not *always* so.

Using Eqs. (3.11) and (3.12), approximations of this function via the HF_c and the HF_m approaches is compared graphically in Fig. 3.16a for $m = 10$ and in Fig. 3.16b for $m = 20$.

However, for linear or piecewise linear functions with jumps, we can improve the reconstruction further. To implement such improvement we need to use both the HF_c and HF_m approaches combined together. It should be kept in mind that a piecewise linear functions with jump is necessarily a combination of ramp and step functions. And a jump in such a function is always contributed by a delayed step function. The function $f_4(t)$ of Fig. 3.16 is one such example. For a flawless reconstruction, the jump points or break points (combination of ramp function having different slopes) have to coincide with the partition line of two adjacent sub-intervals, each of duration h.

Now let us again consider the function $f_4(t)$ of Fig. 3.16, which is

$$f_4(t) = (0.2 + t)u(t) + u(t - 1) \tag{3.14}$$

It is apparent that the function is a combination of step and ramp functions. We now express the component functions *without* delay by HF_c approach and the component functions *with* delay via HF_m approach. It is obvious that the reconstruction in such a fashion will be exact in nature. That is, for $m = 10$ and $T = 2$ s, we can express the component functions of $f_4(t)$ is HF domain as

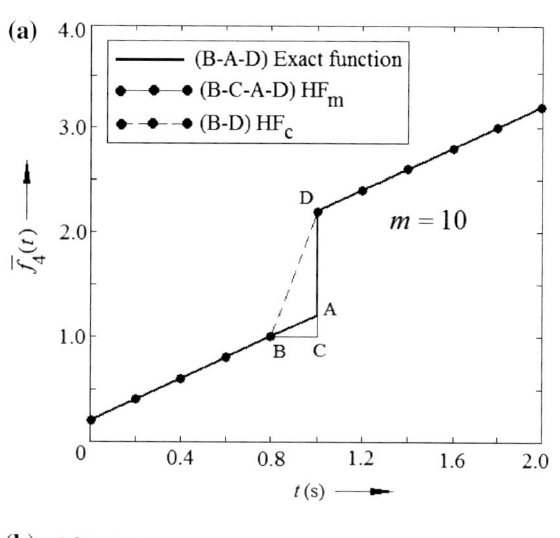

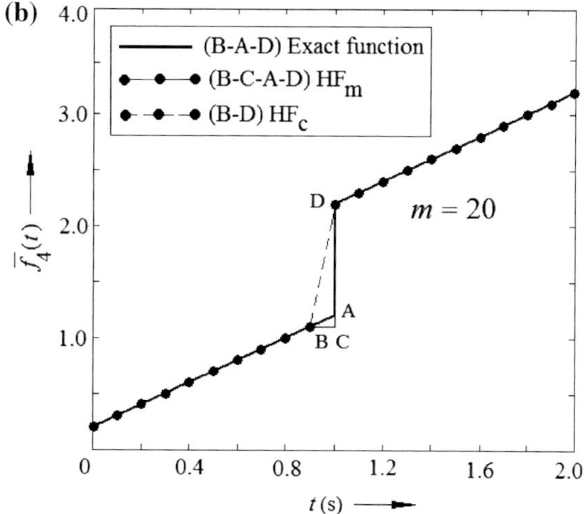

Fig. 3.16 Graphical comparison of HF domain approximation with the exact function, of function $f_4(t)$ of Example 3.5 using the HF$_c$ and the HF$_m$ approach, for **a** $m = 10$, $T = 2$ s and **b** $m = 20$, $T = 2$ s (vide Appendix B, Program no. 6)

$$\begin{aligned} 0.2u(t) = &[0.2 \quad 0.2 \quad 0.2 \quad 0.2 \quad 0.2 \quad 0.2 \quad 0.2 \quad 0.2 \quad 0.2 \quad 0.2]\mathbf{S}_{(10)} \\ &+[0 \quad 0 \quad 0 \quad 0 \quad 0 \quad 0 \quad 0 \quad 0 \quad 0 \quad 0]\mathbf{T}_{(10)} \end{aligned} \tag{3.15}$$

$$\begin{aligned} tu(t) = &[0 \quad 0.2 \quad 0.4 \quad 0.6 \quad 0.8 \quad 1.0 \quad 1.2 \quad 1.4 \quad 1.6 \quad 1.8]\mathbf{S}_{(10)} \\ &+[0.2 \quad 0.2 \quad 0.2 \quad 0.2 \quad 0.2 \quad 0.2 \quad 0.2 \quad 0.2 \quad 0.2 \quad 0.2]\mathbf{T}_{(10)} \end{aligned} \tag{3.16}$$

$$\begin{aligned} u(t-1) \approx &[0 \quad 0 \quad 0 \quad 0 \quad 0 \quad 1.0 \quad 1.0 \quad 1.0 \quad 1.0 \quad 1.0]\mathbf{S}_{(10)} \\ &+[0 \quad 0 \quad 0 \quad 0 \quad 1.0 \quad 0 \quad 0 \quad 0 \quad 0 \quad 0]\mathbf{J}_{5(10)}\mathbf{T}_{(10)} \end{aligned} \tag{3.17}$$

For Eqs. (3.15) and (3.16), we have used the HF_c approach, while for Eq. (3.17), we have employed the HF_m approach.

Combining the above three equations, we can write

$$\begin{aligned} f_4(t) = &[0.2 \quad 0.4 \quad 0.6 \quad 0.8 \quad 1.0 \quad 2.2 \quad 2.4 \quad 2.6 \quad 2.8 \quad 3.0]\mathbf{S}_{(10)} \\ &+[0.2 \quad 0.2 \quad 0.2 \quad 0.2 \quad 0.2 \quad 0.2 \quad 0.2 \quad 0.2 \quad 0.2 \quad 0.2]\mathbf{T}_{(10)} \\ &+[0 \quad 0 \quad 0 \quad 0 \quad 1.0 \quad 0 \quad 0 \quad 0 \quad 0 \quad 0]\mathbf{J}_{5(10)}\mathbf{T}_{(10)} \end{aligned} \tag{3.18}$$

Thus, the function $f_4(t)$ has been represented through HF domain in an exact manner. This representation has been shown in Fig. 3.17 along with the exact function.

Table 3.2 compares the MISE's of the function via the HF_c, the HF_m and the combined approaches for different values of m. It is noted that the MISE is much less for the HF_m approach for all values of m, but for the combined approach, the MISE is *zero*. Figure 3.18 depicts the comparison of MISE's for different values of m using all the three approaches.

Example 3.6 Consider the function of Fig. 3.12, having a jump discontinuity at t_d = 1 s (say). Calling this function $f_3(t)$, we write

$$f_3(t) = \begin{cases} t^2 & \text{for } t \le t_d \\ 2-(t-t_d)^2 & \text{for } t > t_d \end{cases} \tag{3.19}$$

Consider Fig. 3.19a. If we try to compare the amount of errors of approximation of the ramp function, it is noted that the areas of triangles ΔABD and ΔABC are rather head to head contenders. If CD = H and CA = xH, then the areas become equal only when x = ½. That is, CA = AD. At this point the errors are exactly equal

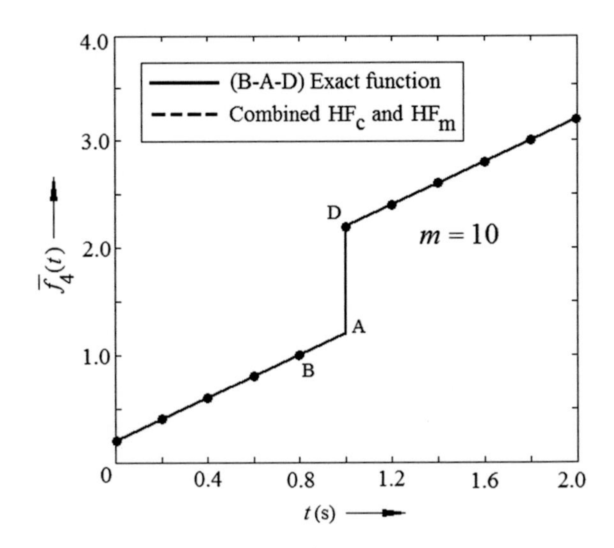

Fig. 3.17 Graphical comparison of HF domain approximation with the exact function, of function $f_4(t)$ of Example 3.5 using the combined HF_c and HF_m approach for m = 10 and T = 2 s

Table 3.2 Comparison of MISE's for HF domain approximation of the function $f_4(t)$ of Example 3.5 using the HF_c, the HF_m and the combined approaches for different values m for $T = 2$ s (vide Appendix B, Program no. 7)

Number of sub-intervals used (m)	MISE using HF_c approach	MISE using HF_m approach	MISE using combined HF_c and HF_m approaches
4	0.16666667	0.04166667	0.00000000
6	0.11111111	0.01234568	0.00000000
8	0.08333333	0.00520833	0.00000000
10	0.06666667	0.00266667	0.00000000
12	0.05555556	0.00154321	0.00000000
14	0.04761905	0.00097182	0.00000000
16	0.04166667	0.00065104	0.00000000
18	0.03703704	0.00045725	0.00000000
20	0.03333333	0.00033333	0.00000000
24	0.02777778	0.00019290	0.00000000
30	0.02222222	0.00009877	0.00000000

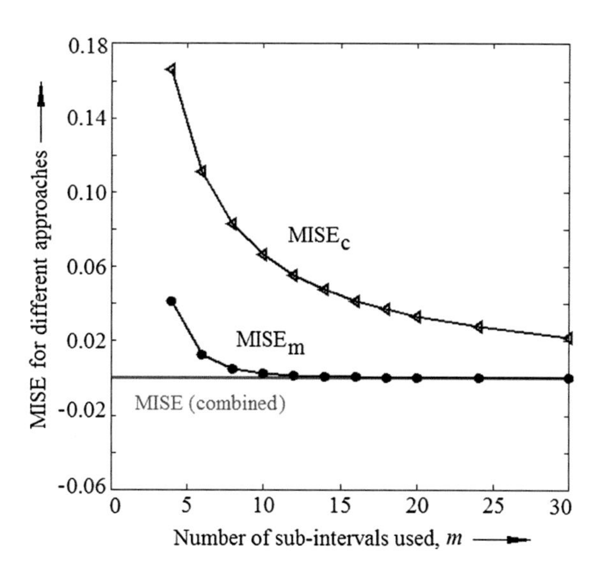

Fig. 3.18 Comparison of MISE's for different values of m for approximation of the function of Example 3.5, using the HF_c approach, the HF_m approach and the combined approach (vide Table 3.2 and Appendix B, Program no. 7)

and it is obvious that if CA > AD ($x > 0.5$) then the error of HF_m approach based approximation is larger than HF_c based approximation.

But since, for different practical applications we rarely have ramp type functions which are very steep, for most of the cases, HF_m wins the competition.

Table 3.3 computes the MISE's of the function $f_3(t)$ of Example 3.6 via the HF_c, HF_m and the combined approaches, for two different values of m ($m = 12$ and 24). It is noted that the MISE for the combined approach is significantly much less than that of the other two approaches.

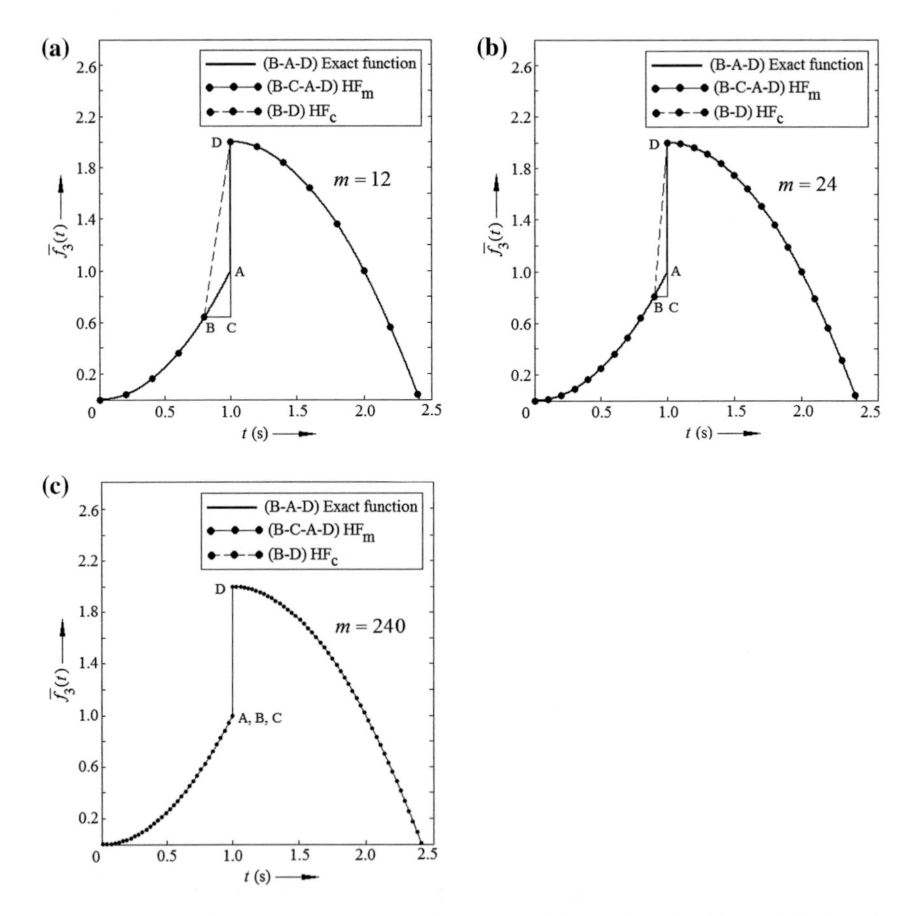

Fig. 3.19 Comparison of the exact function $f_3(t)$ of Example 3.6 with its HF domain approximations using the $\mathrm{HF_c}$ and the $\mathrm{HF_m}$ approaches, for **a** $m = 12$, **b** $m = 24$, and **c** $m = 240$, for $T = 2.4$ s. It is to be noted that for $m = 240$, both the approximations become indistinguishable from the exact function

Figure 3.19a, b show the HF domain approximation of $f_3(t)$ with different number of segments m and $T = 2.4$ s using the $\mathrm{HF_c}$ approach (along the dashed line BD) and $\mathrm{HF_m}$ approach (along the line BCAD).

In Fig. 3.19, with increase in m, both the areas of the triangles ΔABC and ΔBDA reduce and approach zero in the limit. When m is increased to 240, the ΔABC reduces almost to a point and the MISE is reduced almost to zero. This is shown in Fig. 3.19c for $m = 240$ and $T = 2.4$ s.

Table 3.3 Comparison of MISE's for HF domain approximation of the function $f_3(t)$ of Example 3.6 using the HF_c, HF_m and the combined approaches for different values m (m = 12 and 24) and T = 2.4 s

Number of sub-intervals used (m)	MISE using HF_c approach	MISE using HF_m approach	MISE using combined HF_c and HF_m approaches
12	2838.66666667e−05	345.33333333e−05	5.33333333e−05
24	13961.66666667e−06	491.52777778e−06	3.33333333e−06

3.6 Function Approximation: HF Versus Other Methods

The essence of function approximation by other functions, e.g., polynomial functions, orthogonal polynomials, orthogonal functions etc., is to satisfy the need of efficient computing keeping the mean integral square error (MISE) within tolerable limits. The general form to represent a square integrable function $f(t)$ via a set of orthogonal functions is given in Eq. (1.3). Such approximation via HF method may now be compared with a few other methods.

We consider block pulse functions [1] and Legendre polynomials [10, 11] to approximate a function of the form $f(t) = \exp(t-1)$ in $t \in [0, 2)$ and compare it with its HF equivalent.

To approximate $f(t)$ via Legendre polynomials [10, 11], we write

$$f(t) \approx \sum_{0}^{n-1} c_i P_i$$

where, P_i is the (i + 1)th Legendre polynomial and $c_0, c_1, \ldots, c_i$ are the respective coefficients given by

$$c_i = \frac{2i+1}{2} \int_{-1}^{1} f(t) P_i(t) \mathrm{d}t, \quad i = 0, 1, 2, 3, \ldots$$

For the example treated in the following, we consider up to sixth degree Legendre polynomial. The polynomials are

$$P_0(t) = 1, P_1(t) = t, P_2(t) = \frac{3t^2-1}{2}, P_3(t) = \frac{5t^3-3t}{2}, P_4(t) = \frac{35t^4-30t^2+3}{8},$$
$$P_5(t) = \frac{63t^5-70t^3+15t}{8}, P_6(t) = \frac{231t^6-315t^4+105t^2-5}{16}.$$

obtained from the well known recurrence formula for Legendre polynomials.

Figure 3.20 depicts the comparison of the function $f(t)$ with its approximation obtained via sixth degree Legendre polynomial. Figure 3.21 compares $f(t)$ with its

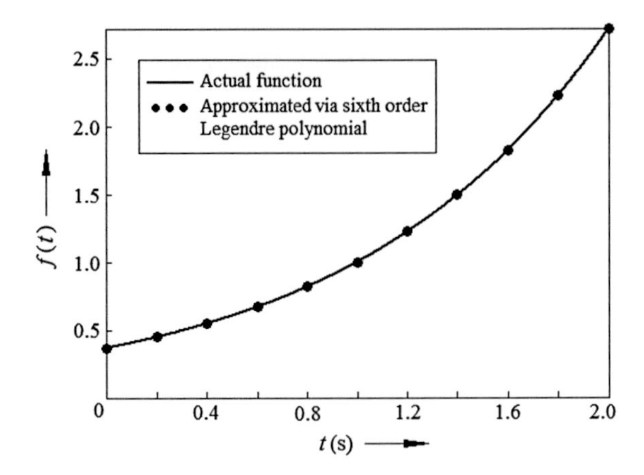

Fig. 3.20 Comparison of the function $f(t) = \exp(t - 1)$ with its Legendre polynomial approximated version using up to sixth degree polynomial with $T = 2$ s (vide Appendix B, Program no. 8)

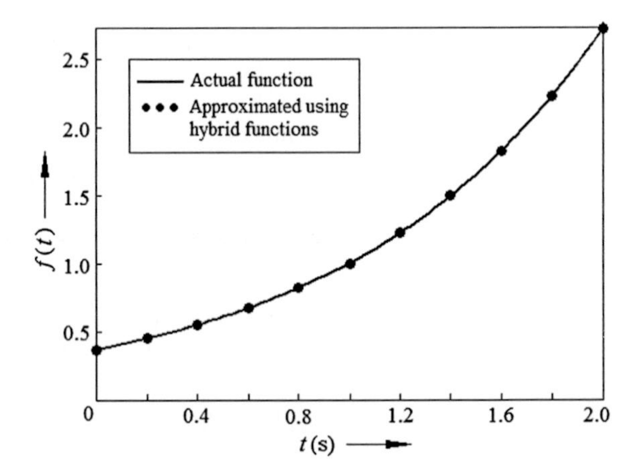

Fig. 3.21 Comparison of the function $f(t) = \exp(t - 1)$ with its HF approximated version using $m = 10$ and $T = 2$ s

HF domain approximation for $m = 10$. Finally, Fig. 3.22 shows the comparison of $f(t)$ with its BPF domain approximation for $m = 10$.

However, for function approximation via BPF or HF, the sampling theorem has to be satisfied while selecting the width of sub-interval, h. This takes care of the accuracy of approximation. And also, to approximate oscillatory functions—that is, high frequency signals—we need a smaller h to reach useful result, which in effect, is in compliance with the sampling theorem.

For function approximation via Legendre polynomials, the oscillatory property is inherently present in the polynomial set which helps in successful approximation of high frequency signals. Also, the Legendre approach and least squares approach usually come up with the same result.

In case of HF based approximation, we work with m number of samples and number of operations for function approximation is only m number of subtractions. In case of Legendre polynomial based approximation, we need to evaluate as many as n integrals (for n number of Legendre polynomials) numerically and also we

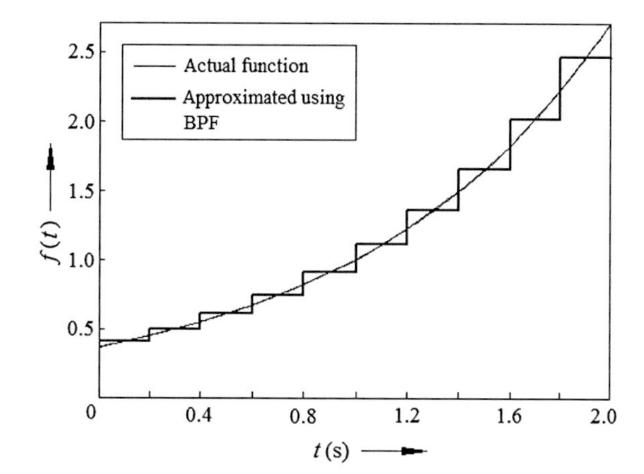

Fig. 3.22 Comparison of the function $f(t) = \exp(t - 1)$ with its BPF approximated version using $m = 10$ and $T = 2$ s

need many addition and multiplication operations. Thus, it is expected that its computational burden is much more compared to the HF method.

In case of block pulse based approximation, to evaluate each expansion coefficient we need to perform one numerical integration applying Simpson's 3/8th rule or the like. Such integration obviously add to the computational burden by way of memory as well as execute on time.

Also, hardware implementation of Legendre polynomials is hardly possible, and for generating BPF's, we need to take care of the 'orthogonality error' via hardware which is complex. Compared to these hassles, HF is much simpler to generate because we do not need to construct the sample-and-hold functions or the triangular functions. Instead, we take samples of the concerned function and produce its piecewise linear version.

Tables 3.4 and 3.5, along with Figs. 3.23 and 3.24, illustrate the essence of efficiency of approximation via three different methods.

From Table 3.4, Legendre polynomial based approximation with sixth degree polynomial fit incurs an error 3.88664563e−12, while for about the same MISE, the HF domain approximation needs $m = 500$. In Table 3.5, for BPF approximation method with $m = 128$, the MISE is 3.68934312e−005. This MISE is obtained for $m = 9$ in HF domain.

Data for Tables 3.4 and 3.5 are calculated via MATLAB 7.9 [12] and we have presented the result up to the eighth place of decimal. It is observed that with respect to MISE, fifth order polynomial fit of Legendre is somewhat equivalent to HF approximation with $m = 134$.

For BPF and HF approximation, such equivalence is obtained with $m = 158$ (BPF) and $m = 10$ (HF) for $T = 2$ s.

In Table 3.4, elapsed times are computed for evaluation of the expression $f(t) \approx \sum_0^{n-1} c_i P_i$ with Legendre polynomials for different values of n from 1 to 7, and also respective MISE's. It is observed that for computations with the index $n = 6$ and 7, elapsed times are much less compared to HF with $m = 134$ and 500 respectively.

Table 3.4 Comparison of MISE's for Legendre polynomial based approximation and HF based approximation of the function $f(t) = \exp(t - 1)$ with different degrees of Legendre polynomials and different number of HF component functions over a 2 s interval, along with respective elapsed times for computation (vide Appendix B, Program nos. 9 and 10)

S. no	Legendre polynomial based function approximation			Hybrid function based function approximation			$\Delta = \frac{MISE_{Legendre}}{MISE_{HF}}$
	Highest degree of polynomial used	$MISE_{Legendre}$	Elapsed time (s)	Number of SHF and TF components used (m)	$MISE_{HF}$	Elapsed time (s)	
1	P0(t)	0.43233236	2.141362	1	0.16336887	2.178037	2.64635704
2	P1(t)	0.02632651	2.810796	2	0.01353700	2.435132	1.94478171
3	P2(t)	7.20286766e−04	3.737012	5	3.79858533e−04	3.915994	1.89619741
4	P3(t)	1.11444352e−05	4.569169	13	8.44276211e−06	7.959310	1.31999872
5	P4(t)	1.10681987e−07	6.050759	39	1.04483986e−07	21.19464	1.05932010
6	P5(t)	7.64745535e−10	7.553432	134	7.49910190e−10	70.21305	1.01978283
7	P6(t)	3.88664563e−12	8.861730	500	3.86864398e−12	274.6432	1.00465322

Table 3.5 Comparison of MISE's for BPF based approximation and HF based approximation of the function $f(t) = \exp(t - 1)$ for different values of m with $T = 2$ s along with respective elapsed times for computation

S. no	BPF based function approximation			Hybrid function based function approximation			$\Delta_m = \frac{\text{MISE}_{\text{BPF}(m_1)}}{\text{MISE}_{\text{HF}(m_2)}}$
	Number of BPF component functions used ($m = m_1$, *say*)	$\text{MISE}_{\text{BPF}(m1)}$	Elapsed time (s)	Number of SHF and TF component functions used ($m = m_2$, *say*)	MISEHF(m2)	Elapsed time (s)	
1	39	3.97316113e−04	30.28443	5	3.79858533e−04	4.700064	1.04595811
2	57	1.86027184e−04	39.01264	6	1.84208601e−04	4.787076	1.00987241
3	77	1.01945683e−04	54.25152	7	9.97660910e−05	5.379379	1.02184702
4	101	5.92542920e−05	69.52123	8	5.86090963e−05	5.772918	1.01100845
5	128	3.68934312e−05	88.73480	9	3.66443586e−05	6.405623	1.00679702
6	158	2.42135494e−05	110.8396	10	2.40682301e−05	6.770899	1.00603780

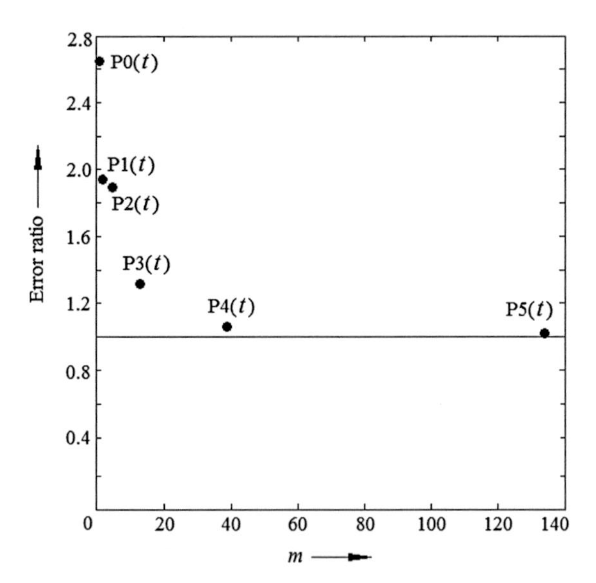

Fig. 3.23 Ratio of Legendre polynomial approximation MISE and HF approximation MISE for $T = 2$ s (refer to Table 3.4) for different values of m where $P_n(t)$ ($n = 0, \ldots, 5$) denotes the highest degree of Legendre polynomial used. Due to inconvenience in choosing the scale, the last entry of Table 3.4 is excluded from the figure

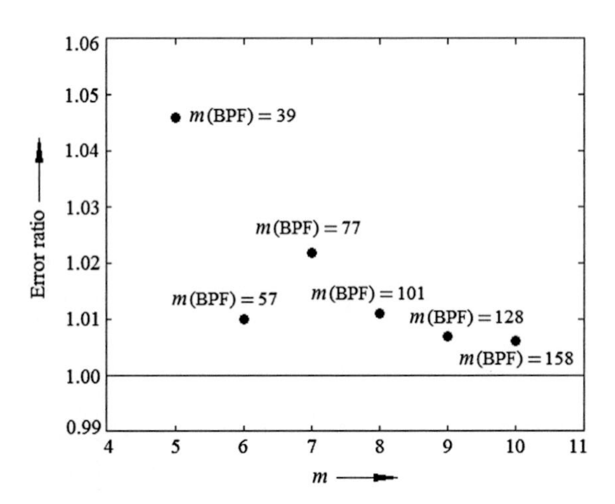

Fig. 3.24 Ratio of BPF approximation MISE and HF approximation MISE (refer to Table 3.5) for different values of m for $T = 2$ s

This is because, the evaluation of MISE for HF based approximation takes up much more time for m number of squaring operations and subsequent numerical integrations.

In fact, for any value of n, i.e., $n = 1$ to 5, this is true and the elapsed times are less compared to that of HF based procedure. Had we been able to represent the HF domain piecewise linear curve by a single analytical expression, suitable for MATLAB, it is possible that the HF based computations would have taken much less computational time.

Figure 3.25a, b show elapsed times for MISE computation for Legendre polynomial and HF based approximations. Also, from these two figures, for the same

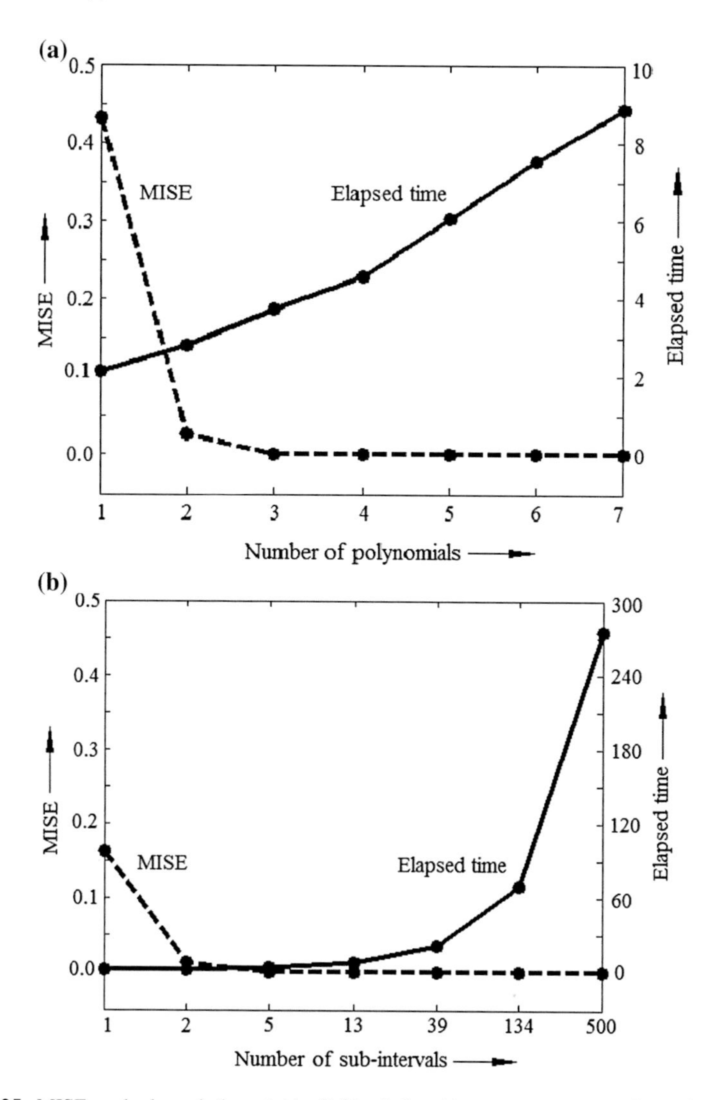

Fig. 3.25 MISE and elapsed time (vide Table 3.4) with respect to **a** number of Legendre polynomials and **b** number of sub-intervals m for HF approximation

range of MISE, number of Legendre polynomials required for the approximation may be compared with the number of sub-intervals (m) required for the HF based approach.

Also if we use only one sub-interval for HF domain approximation and one polynomial for Legendre, the MISE for HF based approach is noted as one-third of that of the Legendre approximation. It proves the effectiveness of HF based approximation even in smallest number of sub-interval.

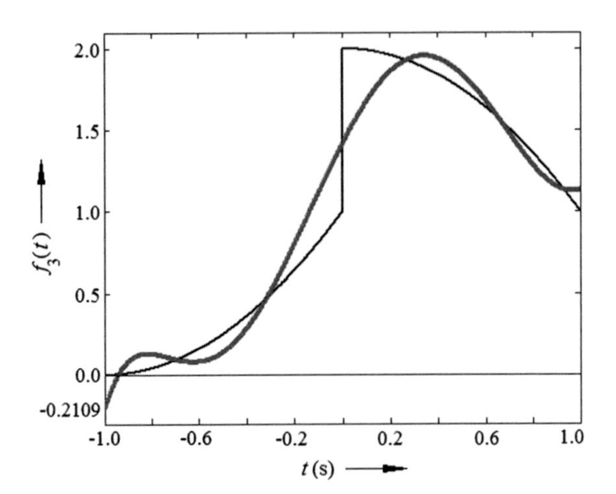

Fig. 3.26 Comparison of the function $f_3(t)$ of Example 3.6 with its Legendre polynomial approximated version using up to sixth degree polynomial with $T = 2$ s $(-1\ \text{s} \leq t \leq 1\ \text{s})$

However, for approximation of functions with jump discontinuities, the Legendre polynomial approach is not the best suited orthogonal polynomial set compared to hybrid function based approximation. This is because, the Legendre polynomial set contains only two members, one the constant $u(t)$ and the other unit ramp function, which are linear. All the other remaining polynomials being of curvy nature, it is a contention that approximation of 'jump functions' by Legendre polynomials will be met with difficulty.

That this is so, is proved by Fig. 3.26, where the approximation of the function of Example 3.6 is attempted with seven Legendre polynomials, that is P0(t) to P6(t). It is seen from Fig. 3.26 that such attempt turns out to be a fiasco, whereas with HF_c and HF_m, we have been able to approximate this function in a reasonably well manner, vide Fig. 3.17.

In Fig. 3.27a, b, comparison has been made between BPF approximation and HF approximation. It is noted that approximately same range of MISE can be obtained in HF based approximation using much lesser number of sub-intervals. Also the elapsed times for MISE computation for HF based computation are comparatively negligible as obtained in BPF based approximation.

3.7 Mean Integral Square Error (MISE) for HF Domain Approximations [8]

The representational error for equal width block pulse function expansion of any square integrable function of Lebesgue measure has been investigated by Rao and Srinivasan [13], while the error analysis for pulse-width modulated generalized block pulse function (PWM-GBPF) expansion has been carried out by Deb et al. [14].

Hybrid function approximation has two components: sample-and-hold function component and triangular function component. To present an error analysis in HF domain, it is obvious that the upper bound of representational error will be comprised of two parts: error in the component of sample-and-hold function based approximation and in the component of triangular function based approximation.

Any time function can be represented by Taylor series as an infinite sum of terms those are calculated from the values of the function's derivatives at a single point. In practice all the time functions are usually approximated using a finite number of terms of its Taylor series, which gives quantitative estimates on the error. Here the time function $f(t)$ first expanded into Taylor series polynomials. The maximum error that can incur in approximating a function within a time interval is termed as upper bound of error.

As sample-and-hold function gives a stair-case approximation, two terms of Taylor series i.e. two-degree Taylor polynomials are sufficient to approximate a function. Similarly the triangular function gives linear approximation, three terms of Taylor series or minimum three-degree Taylor polynomials are used to approximate a function.

3.7.1 *Error Estimate for Sample-and-Hold Function Domain Approximation [2]*

Let us consider m cells of equal width h spanning an interval $[0, T)$ such that $T = mh$. In the $(i + 1)$th interval, the representational error is

$$E_i(t) \triangleq |f(t) - f(ih)| \tag{3.20}$$

The mean integral square error (MISE) is given by

$$[E_i]^2 \triangleq \int_{ih}^{(i+1)h} [f(t) - f(ih)]^2 \mathrm{d}t \tag{3.21}$$

Now expanding $f(t)$ in the ith interval around any point μ_i, by Taylor series, we can write,

$$f(t) \approx f(\mu_i) + \dot{f}(\mu_i)(t - \mu_i) + \ddot{f}(\mu_i)(t - \mu_i)^2/2! + \cdots \tag{3.22}$$

where, $\mu_i \in [ih, (i+1)h]$

Neglecting second and higher order derivatives of $f(t)$ and substituting (3.22) in (3.21), we have

$$[E_i]^2 \triangleq \int_{ih}^{(i+1)h} \left\{f(\mu_i) + \dot{f}(\mu_i)(t-\mu_i) - f(ih)\right\}^2 \mathrm{d}t$$

Now considering

$$\begin{aligned}\dot{f}_{\max} &= \max\{\dot{f}(\mu_i)\}\\ d_{\max} &= f(\mu_i) - f(ih) = \max\{d_i\}\\ \mu_{\max} &= \max\{\mu_i\}\end{aligned}$$

Then the upper bound of MISE over the interval [0, T) is [2]

$$\sum_{i=0}^{m-1} E_i^2 = E^2 = mhd_{\max}\left[d_{\max} + mh\dot{f}_{\max} - 2\mu_{\max}\dot{f}_{\max}\right] + mh\dot{f}_{\max}^2\left[\frac{m^2h^2}{3} + \mu_{\max}(\mu_{\max} - mh)\right] \tag{3.23}$$

3.7.2 *Error Estimate for Triangular Function Domain Approximation [3–5]*

Let us consider (m + 1) sample points of the function $f(t)$, having a sampling period h, denoted by $f(ih)$, $i = 0, 1, 2, \ldots, m$. Then piecewise linear representation of the function $f(t)$ by triangular functions is obtained simply by joining these sample points. The equation of one such straight line $\hat{f}(t)$, approximating $f(t)$, in the (i + 1) th interval is

$$\hat{f}(t) = m_i t + f(ih) - i m_i h \tag{3.24}$$

where $m_i = \frac{f[(i+1)h] - f(ih)}{h}$.

Then integral square error (ISE) in the (i + 1)th interval is

$$[E_i]^2 \triangleq \int_{ih}^{(i+1)h} \left[f(t) - \hat{f}(t)\right]^2 \mathrm{d}t \tag{3.25}$$

Let the function $f(t)$ be expanded by Taylor series in the (i + 1)th interval around the point μ_i considering second order approximation. Then

$$f(t) \approx f(\mu_i) + \dot{f}(\mu_i)(t-\mu_i) + \ddot{f}(\mu_i)(t-\mu_i)^2/2! \tag{3.26}$$

Using Eqs. (3.26) and (3.25) may be written as

$$\begin{aligned}[E_i]^2 &= \int_{ih}^{(i+1)h} \Big[\{f(ih) - im_ih - f(\mu_i) + \mu_i\dot{f}(\mu_i)\} \\ &\quad + \{m_i - \dot{f}(\mu_i)\}t - \ddot{f}(\mu_i)(t-\mu_i)^2/2!\Big]^2 \mathrm{d}t \\ &= \int_{ih}^{(i+1)h} \left[A + Bt + C(t-\mu_i)^2\right]^2 \mathrm{d}t\end{aligned} \tag{3.27}$$

where

$$\left.\begin{aligned} A &\triangleq f(ih) - im_ih - f(\mu_i) + \mu_i\dot{f}(\mu_i) \\ B &\triangleq m_i - \dot{f}(\mu_i) \\ C &\triangleq -\ddot{f}(\mu_i)/2! \end{aligned}\right\} \tag{3.28}$$

Hence, Eq. (3.27) may be simplified to

$$\begin{aligned}[E_i]^2 &= A^2h + \frac{B^2h^3}{3}(3i^2+3i+1) + \frac{C^2h^5}{5}(5i^4+10i^3+10i^2+5i+1) - h^4\mu_iC^2(4i^3+6i^2+4i+1) \\ &\quad + 2h^3\mu_i^2C^2(3i^2+3i+1) - 2h^2\mu_i^3C^2(2i+1) + h\mu_i^4C^2 + ABh^2(2i+1) \\ &\quad + \frac{h^4BC}{2}(4i^3+6i^2+4i+1) - \frac{4h^3BC\mu_i}{3}(3i^2+3i+1) + h^2BC\mu_i^2(2i+1) \\ &\quad + \frac{2h^3CA}{3}(3i^2+3i+1) - 2h^2CA\mu_i(2i+1) + 2hCA\mu_i^2\end{aligned} \tag{3.29}$$

Then the upper bound of ISE over m subintervals is given by

$$\begin{aligned}\sum_{i=1}^{m-1}\left[E_{i\,\max}^2\right] &= \mathbf{E}^2 \\ &= \ddot{f}_{\max}^2\left[\frac{m^5h^5}{20} - \frac{m^4h^4}{4}\mu_{\max} + \frac{m^3h^3}{2}\mu_{\max}^2 - \frac{m^2h^2}{2}\mu_{\max}^3 + \frac{mh}{4}\mu_{\max}^4\right] \\ &\quad + \dot{f}_{\max}^2\left[\frac{(2m^3-3m^2+m)h^3}{6} - (m^2-m)h^2\mu_{\max} + mh\mu_{\max}^2\right] \\ &\quad + \ddot{f}_{\max}\dot{f}_{\max}\left[\frac{(3m^4-2m^3-m^2)h^4}{12} - \frac{(6m^3-3m^2-m)h^3}{6}\mu_{\max}\right. \\ &\quad \left. + \frac{(3m^2-m)h^2}{2}\mu_{\max}^2 - mh\mu_{\max}^3\right]\end{aligned} \tag{3.30}$$

where

$$f_{\max} \triangleq \max\{f(ih), f[(i+1)h], f(\mu_i)\},$$
$$\dot{f}_{\max} \triangleq \max\{\dot{f}(\mu_i), m_i\}$$
$$\ddot{f}_{\max} \triangleq \max\{\ddot{f}(\mu_i)\} \text{ and}$$
$$\mu_{\max} \triangleq \max\{\mu_i\}$$

These maximum values are considered to be the largest over the entire period. So, Eq. (3.30) gives the upper bound of ISE for the triangular function component.

For simplicity, assume that the function is approximated by first-order Taylor series expansion. In that case, $\ddot{f}(\mu_i) = 0$. Then from Eq. (3.28), $C = 0$.

Equation (3.29) now can be simplified to

$$[E_i]^2 = A^2 h + \frac{B^2 h^3}{3}\left(3i^2 + 3i + 1\right) + ABh^2(2i+1) \tag{3.31}$$

Case I Let us assume that the function $f(t)$ is a step function. Then,

$$f(ih) = f(\mu_i), m_i = 0 \quad \text{and} \quad \dot{f}(\mu_i) = 0.$$

This implies, from Eq. (3.29),

$A = 0$, and $B = 0$.

Hence, from Eq. (3.31), we have ISE in the *i*th interval, i.e.

$$[E_i]^2 = 0$$

Case II If $f(t)$ is a ramp function, then, $m_i = m_r$ (constant). Since, $m_r = \dot{f}(\mu_i)$, we have

$B = 0$.

Consider $\mu_i = ih + x, \quad 0 \le x \le h$.

Then,

$$f(\mu_i) = f[(ih) + x] = f(ih) + m_r x$$

Hence, from Eq. (3.22), we get

$$A = f(ih) - m_r ih - f(ih) - m_r x + (ih + x)m_r = 0$$

Then, from Eq. (3.31), ISE in the *i*th interval is zero.

Case III Let $f(t)$ be a piecewise ramp function having different slope in different sampling period. Then, considering the slope to be m_i in the *i*th interval, we have

$$m_i = \dot{f}(\mu_i) \quad \text{and} \quad \mu_i = ih + x, \;\; 0 \leq x \leq h$$

If follows from Eq. (3.28) that $A = B = 0$, implying zero ISE.

As the result of case II and case III are independent of x, we can conclude that for a ramp function, whether continuous or piecewise, ISE is zero irrespective of the magnitude of μ_i, the focal point of Taylor series expansion.

3.8 Comparison of Mean Integral Square Error (MISE) for Function Approximation via HF_c and HF_m Approaches

For the function of Example 3.6, the time interval under consideration is $T = 1 + \sqrt{2}$ s. We have considered an approximate interval of $T = 2.4$ s and have approximated the function for different number of m (for $m = 12$–240). But if we intend to improve the approximation even more, let us consider $T = 2.41$ s, and $m = 241$, so that each sub-interval matches with 0.01 s. It is found that the MISE remains the same as before, i.e., for $m = 240$. This indicates, the infinitesimal change in MISE is not reflected in the approximation.

For different values of m, as mentioned above, the MISE's for both the approaches within a specified time zone, are computed and the modified approach always comes up with less MISE. This has been studied deeply for no less than ten different values of m and two curves are drawn for approximation of $f_3(t)$ using the HF_c approach and HF_m approach, and as expected, the MISE for HF_m approach is always less than the HF_c approach. Figure 3.28 represents this fact visually.

Figure 3.27, $MISE_c$ and $MISE_m$, are computed for twelve different values of m. If we determine the ratio of the errors $R_{HFcm} \triangleq \frac{MISE_c}{MISE_m}$ for each value of m, we can plot a curve of R_{HFcm} with m. Figure 3.29a shows the variation where, with increasing m, the ratio R_{HFcm} increases parabolically. Study of Figs. 3.28 and 3.29a apparently presents a paradox: while both the MISEs converge to zero, with increasing m, their ratio R_{HFcm} increases. This is because the $MISE_m$ converges at a much faster rate compared to its counterpart $MISE_c$.

To explain this 'phenomenon' in more detail, we present Table 3.5 with different MISEs for approximations via different orthogonal function sets.

Remembering the fact that approximation of a function by conventional hybrid functions and orthogonal triangular functions yield identical results, in Table 3.6, MISEs for approximation via modified HF based method, conventional HF based method (equivalent to triangular function based approximation) and block pulse function domain approximation are tabulated for ten different values of m from 12 to 240.

Also, the ratios $\frac{MISE_{TF}}{MISE_m} \triangleq R_{TFm} = R_{HFcm}$ and $\frac{MISE_{BPF}}{MISE_m} \triangleq R_{BPFm}$ are defined.

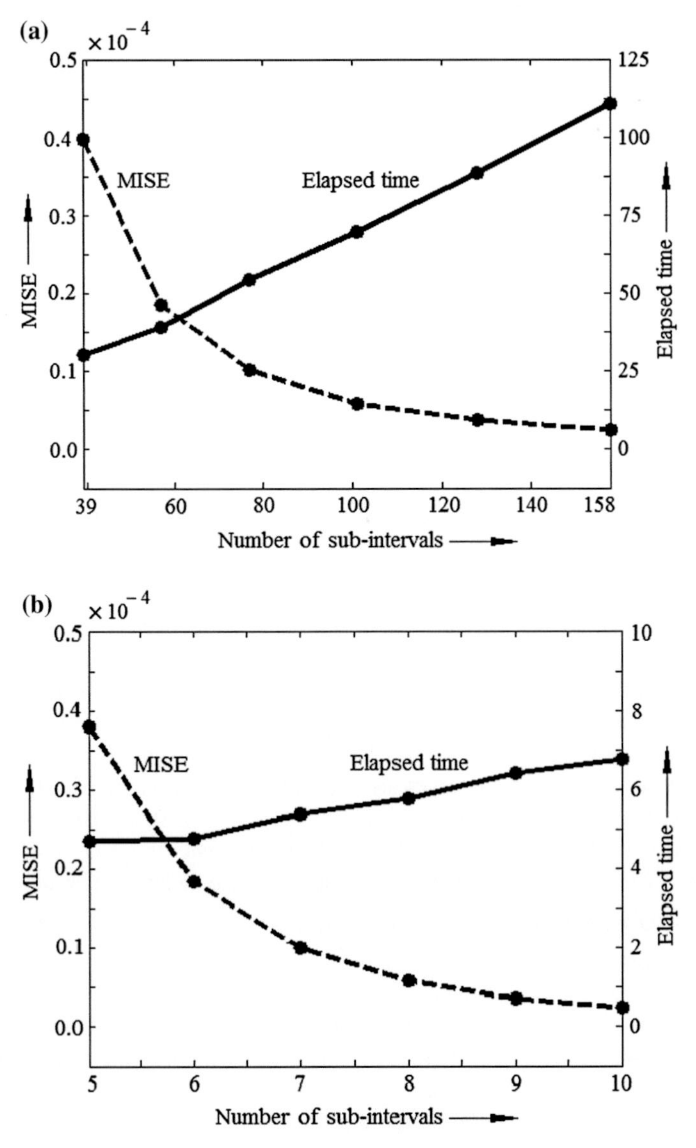

Fig. 3.27 Elapsed time and MISE (vide Table 3.5) with respect to number of sub-intervals used in **a** BPF based and **b** HF based approximation

Using the data of Table 3.6, we draw another curve, shown in Fig. 3.29b, to study the variation of the ratio of MISE_{BPF} and MISE_{m} with different values of m. It is observed that this ratio increases linearly with m.

Studying the two ratio curves, it is noted that for Fig. 3.29a, when m becomes greater than 100, the curve becomes acutely steep. At the end, when m reaches 240,

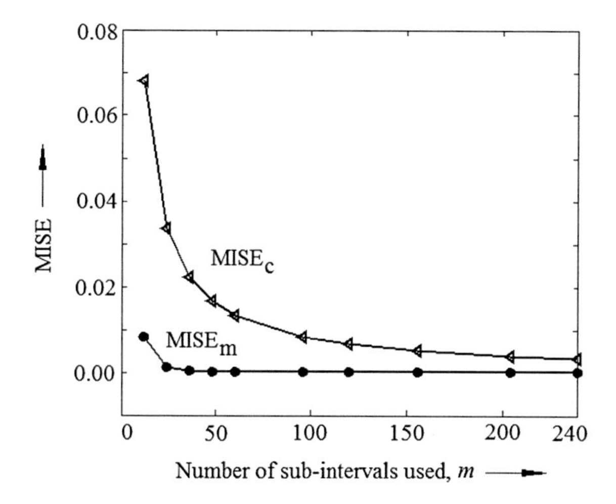

Fig. 3.28 Comparison of MISEs for different values of m, using the HF_c approach and the HF_m approach for approximating the function of Example 3.6, with $T = 2.4$ s

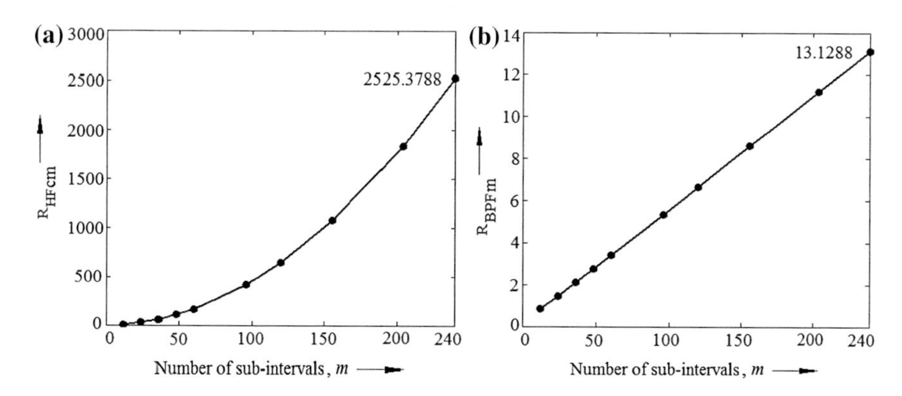

Fig. 3.29 Ratio of MISEs of **a** the HF_c approach and the HF_m approach, and **b** the BPF approach and the HF_m approach, for approximating the function of Example 3.6 for different values of m and $T = 2.4$ s

the magnitude of R_{HFcm} becomes 2525.3788, which is very high, tilting the case significantly in favour of HF based modified approach.

In contrast to the curve of Fig. 3.29a, the curve of Fig. 3.29b is linear, as indicated above. And at a value of $m = 240$, the magnitude of the ratio is only 13.1288. This apparently indicates, with increasing m, the rates of convergence to zero MISE for both the approaches are comparable. But this does not weaken the case for HF_m approach.

This small ratio 13.1288 keeps one wondering about the established superiority of HF based approximation over BPF based approximation in general. Though we

Table 3.6 Comparison of MISE's for HF domain approximation of function $f_3(t)$ of Example 3.6 using the HF_c and the HF_m approaches, for different number of segments m and $T = 2.4$ s

Number of sub-intervals used, m	MISE using HF_m approach ($MISE_m$)	MISE using HF_c or TF approach ($MISE_c$)	$MISE_{TF}/MISE_m = R_{TFm} = R_{HFcm}$	MISE using BPF approach ($MISE_{BPF}$)	$MISE_{BPF}/MISE_m = R_{BPFm}$
12	0.00828800	0.06812800	8.22007722	0.00689778	0.83226110
24	0.00117967	0.03350800	28.40455382	0.00173111	1.46745276
36	0.00036438	0.02227319	61.12626928	0.00076993	2.11298644
48	0.00015691	0.01668800	106.35396087	0.00043319	2.76075457
60	0.00008132	0.01334420	164.09493360	0.00027728	3.40973930
96	0.00002022	0.00833597	412.46759030	0.00010832	5.36021771
120	0.00001041	0.00666801	640.53890490	0.00006933	6.65994236
156	0.00000477	0.00512882	1075.22431866	0.00004102	8.59958071
204	0.00000214	0.00392184	1832.63551402	0.00002399	11.21028037
240	0.00000132	0.00333350	2525.37878788	0.00001733	13.12878788

Table 3.7 Cell wise comparison of MISEs for HF_m, HF_c and BPF based approximations of the function $f_3(t)$ of Example 3.6 for $m = 12$ and $T = 2.4$ s

Sub-interval no. [0 to $(m-1)$] $m = 12$, $h = 0.2$ s $T = 2.4$ s	Segment wise MISE using HF_m approach	MISE using HF_m approach ($MISE_m$) over the whole period	Segment wise MISE using HF_c or TF approach	$MISE_c$ or $MISE_{TF}$ over the whole period T	Segment wise MISE using BPF approach	MISE using BPF approach ($MISE_{BPF}$) over the whole period
0	0.00005333	0.00345333	0.00005333	**0.02838666**	0.00014222	0.00689778
1	0.00005333		0.00005333		0.00120889	
2	0.00005333		0.00005333		0.00334222	
3	0.00005333		0.00005333		0.00654222	
4	**0.04085333**		**0.34005333**		**0.01080889**	
5	0.00005333		0.00005333		0.00014222	
6	0.00005333		0.00005333		0.00120889	
7	0.00005333		0.00005333		0.00334222	
8	0.00005333		0.00005333		0.00654222	
9	0.00005333		0.00005333		0.01080889	
10	0.00005333		0.00005333		0.01614222	
11	0.00005333		0.00005333		0.02254222	

sub-interval containing the jump discontinuity.

have used HF based modified approach for this special kind of function $f_3(t)$ with one jump, the ratio is still more uncomfortable. For this reason, we investigated cell wise MISE for three approximations, namely, BPF based approximation and HF based approximation, both conventional and modified. The results are tabulated in Table 3.7.

It is noted from the table that for the cell immediately before the jump, MISE is maximum (=0.34005333) for $\mathrm{HF_c}$ and minimum (=0.01080889) for BPF. And for the same cell, for $\mathrm{HF_m}$ based approximation, MISE is 0.04085333 which is moderate. But for all other cells, $\mathrm{HF_m}$ and $\mathrm{HF_c}$ methods have the same MISE and its magnitude is much less than that of BPF method. In fact, the sum of MISEs of all cells for $\mathrm{HF_c}$ or $\mathrm{HF_m}$ methods, excluding the cell just before the jump, is 0.00058666 (for $\mathrm{HF_c}$ and $\mathrm{HF_m}$) and 0.06862221 for BPF. This proves the efficiency of HF based approximation and indicates its superiority over BPF technique.

Further, when m is increased from 12 to 24 or even higher values, it is noted from Table 3.6 that $\mathrm{HF_m}$ method is always more accurate than BPF based approximation, while $\mathrm{HF_c}$ method is not. This proves, $\mathrm{HF_m}$ method is the most competent for handling functions with jumps.

3.9 Conclusion

The orthogonal hybrid function (HF) set has been employed for piecewise linear approximation of time functions of Lebesgue measure. For HF domain approximation, the expansion coefficients are simply the samples of the function to be approximated, where as for BPF [1] domain approximation, each expansion coefficient is determined via integration of the function making the computation more complex. So is the case for approximation of a function using Legendre polynomial or any other orthogonal polynomial for that matter. Also, the block pulse function based approximation, being staircase in nature, incurs higher mean integral square error (MISE) compared to that via hybrid functions because HF domain approximation reconstructs a function in a piecewise linear manner.

For linear functions like $f_1(t) = t$, HF domain approximation comes up with zero mean integral square error (MISE) as expected. This fact is shown in Fig. 3.4. Also, HF domain approximation proved to be much more accurate compared to equivalent BPF domain approximation. For example, for the function $f_2(t) = \sin(\pi t)$, MISE for HF domain approximation is much less than BPF domain approximation. These facts are evident from Figs. 3.5, 3.6 and 3.7; Table 3.1. In Table 3.1 the ratio $\Delta = \frac{\mathrm{MISE_{BPF}}}{\mathrm{MISE_{HF}}}$ is found to be 25.43420823.

For approximation of discontinuous function, the outcome of HF domain approximation is illustrated via Fig. 3.8 through Fig. 3.11 qualitatively. It is noted that if the sampling period h is small enough, hybrid functions could turn out reasonably good approximation of discontinuous functions. However, the accuracy of approximation is much improved for most of the functions if we employ a modified HF domain approach. Calling this technique the HF_m approach, it has been established through numerical Examples 3.5 and 3.6 that HF_m approach is much more accurate than the conventional HF domain approach (named, HF_c approach). Table 3.2 is evidence enough to establish the superiority of the HF_m approach for eleven different values of m. Figure 3.18 provides the pictorial translation of Table 3.2 where the fineness of approximation using the HF_m approach is quite apparent.

For the numerical Example 3.6; Fig. 3.19 compares HF domain approximations with the exact function for three different values of m, namely, $m = 12$, 24 and 240 and thus compares effectiveness of approximation via HF_c and HF_m qualitatively.

Function approximation via Legendre polynomials are also compared with HF approximation. These are illustrated in Table 3.4 and Figs. 3.20 and 3.23. From Table 3.4, it is observed that with up to sixth degree Legendre polynomial approximation, the same order of MISE is achieved in HF domain approximation with $m = 500$. But with up to fifth degree polynomial approximation, the same order of MISE is achieved in HF domain approximation with $m = 134$. However, for BPF domain approximation with $m = 128$ and $T = 2$ s the MISE is 3.68934312e-05 and for the same order of MISE, HF domain approximation requires only $m = 9$, vide Table 3.5.

From Fig. 3.25a, b we find that the HF based approach proves to be more effective than Legendre polynomial based approximation, in the sense that HF based approximation provides a better deal even in smallest number of sub-interval.

For approximation of functions with jump discontinuities, again the HF domain approximation turns out to be the winner compared to Legendre polynomial approach.

In Fig. 3.27a, b, a comparison of BPF approximation and HF approximation shows that HF based approximation comes up with the same range of MISE using much lesser number of sub-intervals compared to its BPF equivalent with respect to MISE.

From the above discussion, it may be concluded that function approximation using hybrid functions is more simple as well as advantageous compared to block pulse function based equivalent approximation. Though Legendre polynomial based approximation produces good results, computation of its coefficients are much more tedious and complicated. Further, the HF based approximation works with function samples which is a great advantage in view of the present digital age. This advantage is offered neither by BPF approximation, nor Legendre polynomial based approximation.

For approximating functions with jump discontinuities, the modified HF domain approach seems to be more efficient than the conventional HF domain approach and examples have been provided to prove the point. This broadens application suitability of HF domain approach still more.

References

1. Jiang, J.H., Schaufelberger, W.: Block Pulse Functions and their Application in Control System, LNCIS, vol. 179. Springer, Berlin (1992)
2. Deb, A., Sarkar, G., Sen, S.K.: Block pulse functions, the most fun-damental of all piecewise constant basis functions. Int. J. Syst. Sci. **25**(2), 351–363 (1994)
3. Deb, A., Sarkar, G., Bhattacharjee, M., Sen, S.K.: A new set of piecewise constant orthogonal functions for the analysis of linear SISO systems with sample-and-hold. J. Franklin Inst. **335B** (2), 333–358 (1998)
4. Deb, A., Sarkar, G., Sengupta, A.: Triangular Orthogonal Functions for the Analysis of Continuous Time Systems. Anthem Press, London (2011)
5. Deb, A., Sarkar, G., Dasgupta, A.: A complementary pair of orthogonal triangular function sets and its application to the analysis of SISO control systems. J. Inst. Eng. (India) **84**, 120–129 (2003)
6. Deb, A., Dasgupta, A., Sarkar, G.: A complementary pair of orthogonal triangular function sets and its application to the analysis of dynamic systems. J. Franklin Inst. **343**(1), 1–26 (2006)
7. Rao, G.P.: Piecewise Constant Orthogonal Functions and their Application in Systems and Control, LNCIS, vol. 55. Springer, Berlin (1983)
8. Deb, A., Sarkar, G., Mandal, P., Biswas, A., Ganguly, A., Biswas, D.: Transfer function identification from impulse response via a new set of orthogonal hybrid function (HF). Appl. Math. Comput. **218**(9), 4760–4787 (2012)
9. Deb, A., Sarkar, G., Ganguly, A., Biswas, A.: Approximation, integration and differentiation of time functions using a set of orthogonal hybrid functions (HF) and their application to solution of first order differential equations. Appl. Math. Comput. **218**(9), 4731–4759 (2012)
10. Baranowski, J.: Legendre polynomial approximations of time delay systems. XII international Ph.D workshop OWD 2010, 23–26 Oct 2010
11. Tohidi, E., Samadi, O.R.N., Farahi, M.H.: Legendre approximation for solving a class of nonlinear optimal control problems. J. Math. Finance **1**, 8–13 (2011)
12. Mathews, J.H., Kurtis, D.F.: Numerical Methods using MATLAB, 4th edn. Prentice Hall of India Pvt. Ltd., New Delhi (2005)
13. Rao, G.P., Srinivasan, T.: Analysis and synthesis of dynamic systems containing time delays via block pulse functions. Proc. IEE **125**(9), 1064–1068 (1978)
14. Deb, A., Sarkar, G., Sen, SK.: Linearly pulse-width modulated block pulse functions and their application to linear SISO feedback control system identification. Proc. IEE, Part D, Control Theory Appl. **142**(1), 44–50 (1995)

Chapter 4
Integration and Differentiation Using HF Domain Operational Matrices

Abstract This chapter introduces the operational matrices for integration as well as differentiation. In such hybrid function domain integration or differentiation, the function to be integrated or differentiated is first expanded in hybrid function domain and then operated upon by some special matrices to achieve the result. These special matrices are the operational matrices for integration and differentiation and these are derived in this chapter. Also, the nature of accumulation of error at each stage of integration-differentiation dual operation is investigated. Four examples are treated to illustrate the operational methods. Three tables and fifteen figures are presented for user friendly clarity.

The proposed hybrid function (HF) set has been utilized for approximating square integrable functions in a piecewise linear manner. The spirit of such approximation was explained in the previous chapter. As was done with block pulse functions [1, 2], in this chapter, we use the complementary hybrid function sets in a similar fashion, to develop the operational matrices for integration [3, 4] and these new operational matrices are employed to integrate time functions in HF domain. These matrices are finally used for the analysis and synthesis of control systems, for solving the identification problem from state space description of systems, and parameter estimation of transfer functions from impulse response data.

4.1 Operational Matrices for Integration

A hybrid function set is a combination of a sample-and-hold function (SHF) [5] set and a triangular function set [6–8]. Thus, when we express a time function in HF domain, the result is comprised of two parts, namely, SHF part and TF part. So, to integrate such a function in HF domain, we need to integrate both the parts. Integration of each part would again produce SHF and TF parts, the combination of which is the result of integration in HF domain. This is achieved by means of operational matrices.

A. Deb et al., *Analysis and Identification of Time-Invariant Systems, Time-Varying Systems, and Multi-Delay Systems using Orthogonal Hybrid Functions*,
Studies in Systems, Decision and Control 46, DOI 10.1007/978-3-319-26684-8_4

In order to derive the operational matrices for integration in HF domain, we proceed in a manner adopted for Walsh and block pulse functions. Here, we consider for both the component function sets separately and develop the integration operational matrix for each of them to eventually derive the operational matrices for integration in HF domain [9].

4.1.1 Integration of Sample-and-Hold Functions [5]

First we take up the m-member sample-and-hold function set and integrate each of its components and consequently express the result in terms of hybrid function.

Figure 4.1a shows the first member S_0 of the SHF set and Fig. 4.1b shows its decomposition into two step functions.

Mathematically, S_0 can be expressed as

$$S_0 = u(t) - u(t - h)$$

Subsequent integration of S_0 produces two ramp functions as shown in Fig. 4.1c, while Fig. 4.1d depicts the resulting function $\int S_0 \, dt$. It is noted that the result is comprised of one triangular function and several sample-and-hold function blocks [9].

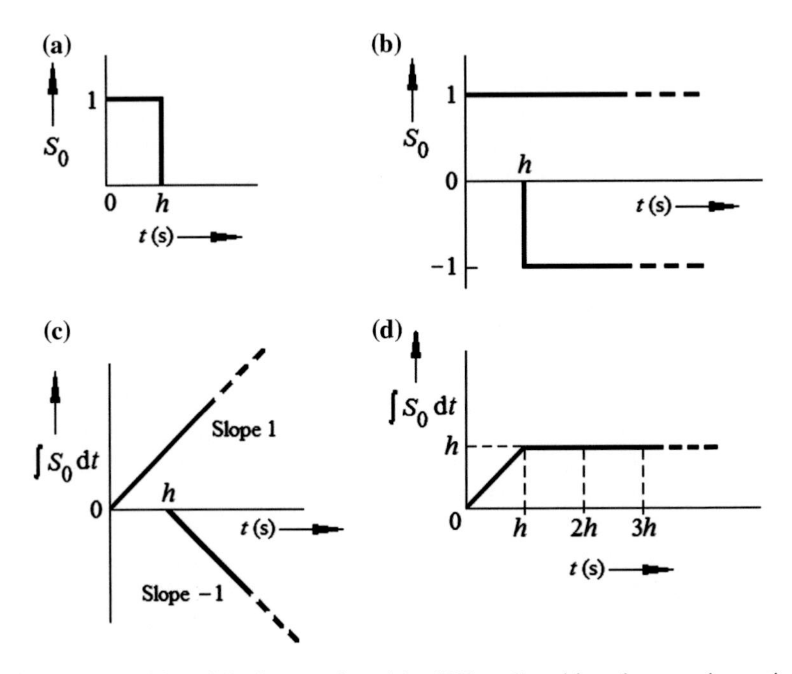

Fig. 4.1 Decomposition of the first member of the SHF set S_0 and its subsequent integration

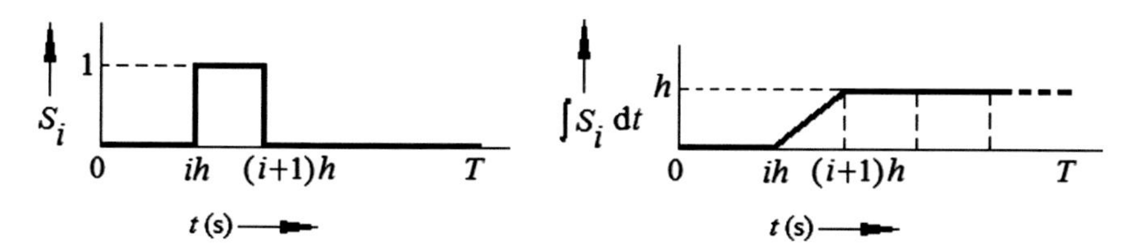

Fig. 4.2 Integration of the $(i + 1)$th member of the SHF set

Figure 4.2 shows the $(i + 1)$th member of the SHF set and its integration. Mathematically, the function $\int S_i \, \mathrm{d}t$ is given by

$$\int S_i(t)\mathrm{d}t = (t - ih)u(t - ih) - [t - (i+1)h]u[t - (i+1)h] \tag{4.1}$$

It is apparent that integration of the $(i + 1)$th member of the SHF set produces the same result of Fig. 4.1d, but with a shift of ih to the right as shown in Fig. 4.2. Putting different values of i in (4.1), e.g., 0, 1, 2, …, we can obtain expressions for integrations of different component SHF's.

Starting with the first member S_0 of the SHF set, we integrate each member and express the result [9] in hybrid function domain. Let us consider four members of the sample-and-hold function set, i.e. $m = 4$ and $T = 1$ s and $h = T/m$. Figure 4.3 shows the result of integration which is comprised of one triangular function and three sample-and-hold functions. The figure also contains the integration results of other three SHF's, namely, S_1, S_2 and S_3.

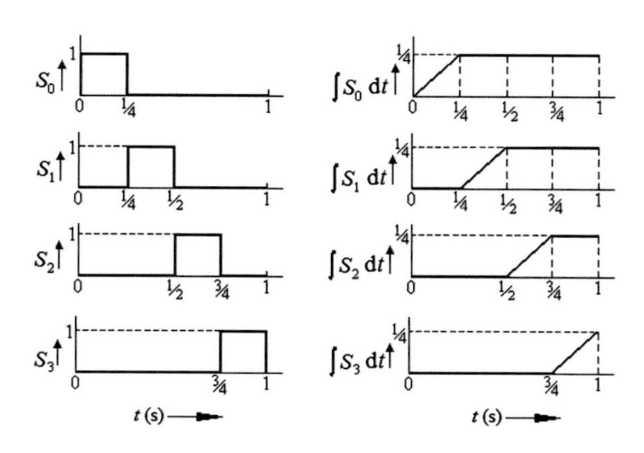

Fig. 4.3 Integration of first four members of the SHF set

Integration of the first member of the SHF set can be expressed mathematically as

$$\int S_0\,\mathrm{d}t = h\begin{bmatrix}0 & 1 & 1 & 1\end{bmatrix}\begin{bmatrix}S_0\\ S_1\\ S_2\\ S_3\end{bmatrix} + h\begin{bmatrix}1 & 0 & 0 & 0\end{bmatrix}\begin{bmatrix}T_0\\ T_1\\ T_2\\ T_3\end{bmatrix}$$

where, for convenience, we can write

$$\int S_0\,\mathrm{d}t = h\begin{bmatrix}0 & 1 & 1 & 1\end{bmatrix}\mathbf{S}_{(4)} + h\begin{bmatrix}1 & 0 & 0 & 0\end{bmatrix}\mathbf{T}_{(4)} \tag{4.2}$$

It is noted that, when the first member of the sample-and-hold function set is integrated and the result is expressed in HF domain, its approximation is comprised of two parts: sample-and-hold functions as well as triangular functions.

Following Eq. (4.2), the results of integration of the second, third and fourth members of the SHF set, as shown in Fig. 4.3, may be expressed as

$$\int S_1\,\mathrm{d}t = h\begin{bmatrix}0 & 0 & 1 & 1\end{bmatrix}\mathbf{S}_{(4)} + h\begin{bmatrix}0 & 1 & 0 & 0\end{bmatrix}\mathbf{T}_{(4)} \tag{4.3}$$

$$\int S_2\,\mathrm{d}t = h\begin{bmatrix}0 & 0 & 0 & 1\end{bmatrix}\mathbf{S}_{(4)} + h\begin{bmatrix}0 & 0 & 1 & 0\end{bmatrix}\mathbf{T}_{(4)} \tag{4.4}$$

$$\int S_3\,\mathrm{d}t = h\begin{bmatrix}0 & 0 & 0 & 0\end{bmatrix}\mathbf{S}_{(4)} + h\begin{bmatrix}0 & 0 & 0 & 1\end{bmatrix}\mathbf{T}_{(4)} \tag{4.5}$$

Expressing Eqs. (4.2)–(4.5) in matrix form, we have

$$\begin{bmatrix}\int S_0 \mathrm{d}t\\ \int S_1 \mathrm{d}t\\ \int S_2 \mathrm{d}t\\ \int S_3 \mathrm{d}t\end{bmatrix} = h\begin{bmatrix}0 & 1 & 1 & 1\\ 0 & 0 & 1 & 1\\ 0 & 0 & 0 & 1\\ 0 & 0 & 0 & 0\end{bmatrix}\mathbf{S}_{(4)} + h\begin{bmatrix}1 & 0 & 0 & 0\\ 0 & 1 & 0 & 0\\ 0 & 0 & 1 & 0\\ 0 & 0 & 0 & 1\end{bmatrix}\mathbf{T}_{(4)}$$

or,

$$\int \mathbf{S}_{(4)}\mathrm{d}t = h\sum_{i=1}^{3}\mathbf{Q}_{(4)}^{i}\mathbf{S}_{(4)} + h\mathbf{I}_{(4)}\mathbf{T}_{(4)} \tag{4.6}$$

where, $\mathbf{I}_{(4)}$ is the (4×4) identity matrix and $\mathbf{Q}_{(4)}$ is the delay matrix [3] of order 4, having a general structure

$$\mathbf{Q}^{i}_{(m)} = \begin{bmatrix} \mathbf{0}_{(m-i)\times i} & \vdots & \mathbf{I}_{(m-i)} \\ \cdots\cdots & \vdots & \cdots\cdots \\ \mathbf{0}_{(i)} & \vdots & \mathbf{0}_{i\times(m-i)} \end{bmatrix}_{(m\times m)} \tag{4.7}$$

where, $i = 1, 2, 3, \ldots, (m-1)$ and $\mathbf{Q}_{(\mathrm{m})}$ has the property such that

$$\mathbf{Q}^{m}_{(m)} = \mathbf{0}_{(\mathrm{m})}$$

Equation (4.6) may be expressed as

$$\int \mathbf{S}_{(4)}\,\mathrm{d}t = \mathbf{P1ss}_{(4)}\,\mathbf{S}_{(4)} + \mathbf{P1st}_{(4)}\,\mathbf{T}_{(4)} \tag{4.8}$$

where,

$$\mathbf{P1ss}_{(4)} \triangleq h\begin{bmatrix} 0 & 1 & 1 & 1 \\ 0 & 0 & 1 & 1 \\ 0 & 0 & 0 & 1 \\ 0 & 0 & 0 & 0 \end{bmatrix} \quad \text{and} \quad \mathbf{P1st}_{(4)} \triangleq h\begin{bmatrix} 1 & 0 & 0 & 0 \\ 0 & 1 & 0 & 0 \\ 0 & 0 & 1 & 0 \\ 0 & 0 & 0 & 1 \end{bmatrix}$$

The square matrices $\mathbf{P1ss}_{(4)}$ and $\mathbf{P1st}_{(4)}$ may be expressed in the following compact form

$$\mathbf{P1ss}_{(4)} = h[\![0 \quad 1 \quad 1 \quad 1]\!] \quad \text{and} \quad \mathbf{P1st}_{(4)} = h[\![1 \quad 0 \quad 0 \quad 0]\!]$$

in which $[\![\mathrm{a} \quad \mathrm{b} \quad \mathrm{c}]\!] \triangleq \begin{bmatrix} \mathrm{a} & \mathrm{b} & \mathrm{c} \\ 0 & \mathrm{a} & \mathrm{b} \\ 0 & 0 & \mathrm{a} \end{bmatrix}$

Following (4.7), in general, for m component functions in each of the SHF set and TF set, we can write

$$\int \mathbf{S}_{(\mathrm{m})}\mathrm{d}t = h\sum_{i=1}^{m-1}\mathbf{Q}^{i}_{(m)}\,\mathbf{S}_{(\mathrm{m})} + h\mathbf{I}_{(\mathrm{m})}\mathbf{T}_{(\mathrm{m})} = \mathbf{P1ss}_{(\mathrm{m})}\mathbf{S}_{(\mathrm{m})} + \mathbf{P1st}_{(\mathrm{m})}\mathbf{T}_{(\mathrm{m})} \tag{4.9}$$

where,

$$\mathbf{P1ss}_{(m)} \triangleq h \left[\left[\underbrace{0 \quad 1 \quad \ldots \quad \ldots \quad 1 \quad 1}_{m \text{ terms}}\right]\right] \quad \text{and}$$

$$\mathbf{P1st}_{(m)} \triangleq h \left[\left[\underbrace{1 \quad 0 \quad \ldots \quad \ldots \quad 0 \quad 0}_{m \text{ terms}}\right]\right]$$

Thus, $\mathbf{P1ss}_{(4)}$ and $\mathbf{P1st}_{(4)}$ are the first order operational matrices for integration for the sample-and-hold function part.

4.1.2 Integration of Triangular Functions [6]

Figure 4.4a shows the first member T_0 of the TF set and Fig. 4.4b shows its decomposition into two ramp functions and one delayed negative step function.

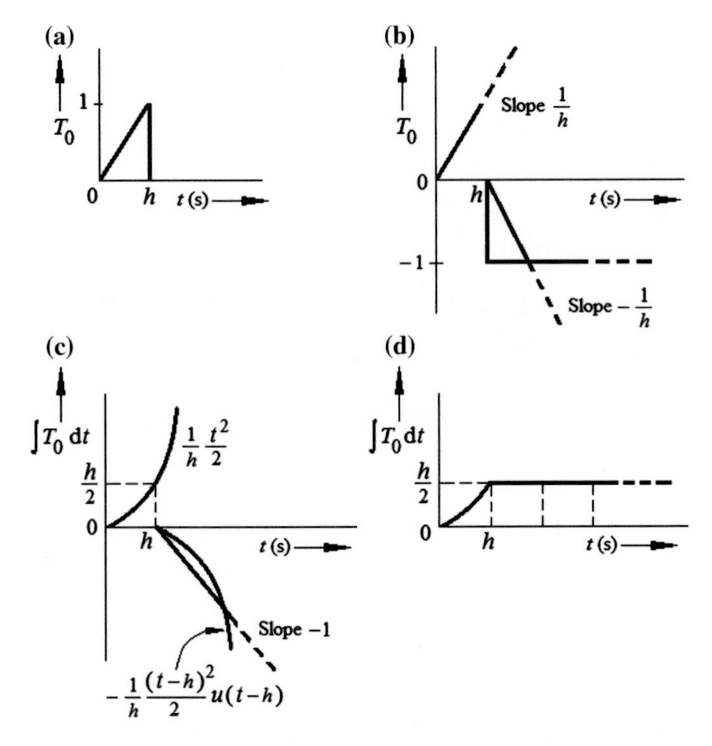

Fig. 4.4 Decomposition of the first member of the TF set and its subsequent integration. **a** the triangular function, **b** its decomposition into three functions, **c** integration of the component functions of (**b**), and (**d**) the result of integration after combining all the three component integrated functions of (**c**)

Mathematically, the function T_0 can be expressed as

$$T_0 = \frac{1}{h}t - \frac{1}{h}(t-h)u(t-h) - u(t-h)$$

Subsequent integration of the first member T_0 produces two parabolic functions and one ramp function as shown in Fig. 4.4c. The integrated function may be expressed as

$$\int T_0 \mathrm{d}t = \frac{1}{h}\frac{t^2}{2} - \frac{1}{h}\frac{(t-h)^2}{2}u(t-h) - (t-h)u(t-h)$$

Figure 4.4d depicts the resulting function $\int T_0 \, \mathrm{d}t$.

Similar to the case for sample-and-hold functions, the mathematical expression for the $(i+1)$th member of the TF set is

$$\int T_i \, \mathrm{d}t = \frac{1}{h}\frac{(t-ih)^2}{2}u(t-ih) - \frac{1}{h}\frac{[t-(i+1)h]^2}{2} u[t-(i+1)h] - [t-(i+1)h]u[t-(i+1)h] \tag{4.10}$$

Figure 4.5 shows pictorially the result integration of the $(i+1)$th member of the TF set.

Taking different values of i in (4.10), e.g., 0, 1, 2, …, we can obtain expressions for integrations of different component TF's.

Following a similar procedure [1, 6] for other components of the TF set and using Eq. (4.10), we integrate each member and express the result of integration in hybrid function domain.

The result of integration of T_0 is now to be expressed in HF domain. Since the HF technique works with function samples, it is apparent that the expansion will be comprised of one triangular function and three sample-and-hold functions, where, the triangular function represents the parabolic part of the exact integration.

This is shown in Fig. 4.6. The figure also depicts similar results for the next three members of the TF set.

This can be represented mathematically as

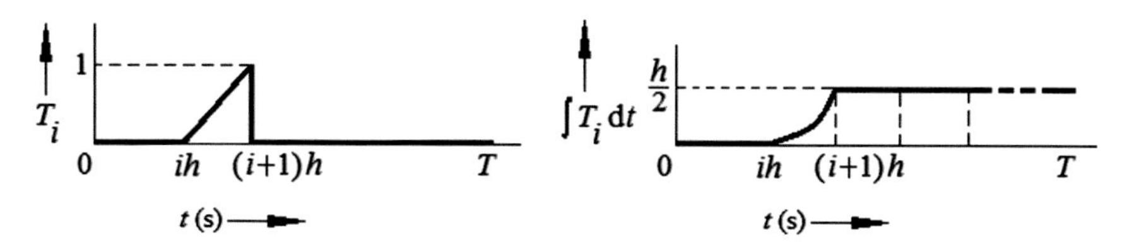

Fig. 4.5 Integration of the $(i+1)$th member of the TF set

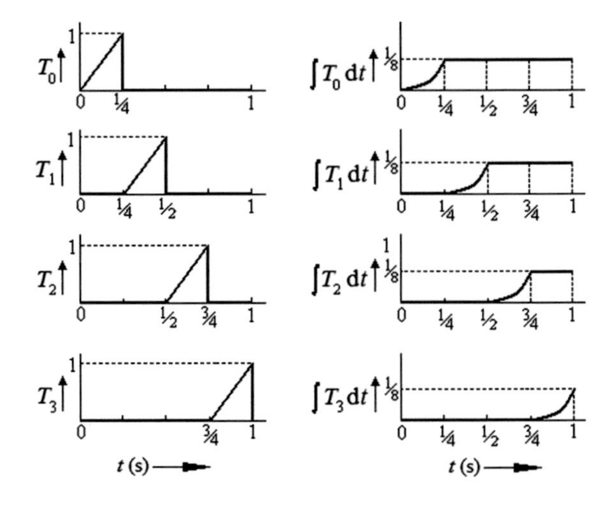

Fig. 4.6 Integration of first four members of the TF set

$$\int T_0 \mathrm{d}t \approx \frac{h}{2}[0 \quad 1 \quad 1 \quad 1]\begin{bmatrix} S_0 \\ S_1 \\ S_2 \\ S_3 \end{bmatrix} + \frac{h}{2}[1 \quad 0 \quad 0 \quad 0]\begin{bmatrix} T_0 \\ T_1 \\ T_2 \\ T_3 \end{bmatrix}$$
$$= \frac{h}{2}[0 \quad 1 \quad 1 \quad 1]\mathbf{S}_{(4)} + \frac{h}{2}[1 \quad 0 \quad 0 \quad 0]\mathbf{T}_{(4)} \tag{4.11}$$

Following a similar procedure as in Sect. 4.1.1, we integrate other three members of the triangular functions set and express the result in HF domain.

Following Eq. (4.11), integration of the second, third and fourth members of the TF set are given by

$$\int T_1 \mathrm{d}t = \frac{h}{2}[0 \quad 0 \quad 1 \quad 1]\mathbf{S}_{(4)} + \frac{h}{2}[0 \quad 1 \quad 0 \quad 0]\mathbf{T}_{(4)} \tag{4.12}$$

$$\int T_2 \mathrm{d}t = \frac{h}{2}[0 \quad 0 \quad 0 \quad 1]\mathbf{S}_{(4)} + \frac{h}{2}[0 \quad 0 \quad 1 \quad 0]\mathbf{T}_{(4)} \tag{4.13}$$

$$\int T_3 \, \mathrm{d}t = \frac{h}{2}[0 \quad 0 \quad 0 \quad 0]\,\mathbf{S}_{(4)} + \frac{h}{2}[0 \quad 0 \quad 0 \quad 1]\,\mathbf{T}_{(4)} \tag{4.14}$$

Expressing Eqs. (4.11)–(4.14) in matrix form, we have

$$\begin{bmatrix} \int T_0 \mathrm{d}t \\ \int T_1 \mathrm{d}t \\ \int T_2 \mathrm{d}t \\ \int T_3 \mathrm{d}t \end{bmatrix} = \frac{h}{2}\begin{bmatrix} 0 & 1 & 1 & 1 \\ 0 & 0 & 1 & 1 \\ 0 & 0 & 0 & 1 \\ 0 & 0 & 0 & 0 \end{bmatrix}\mathbf{S}_{(4)} + \frac{h}{2}\begin{bmatrix} 1 & 0 & 0 & 0 \\ 0 & 1 & 0 & 0 \\ 0 & 0 & 1 & 0 \\ 0 & 0 & 0 & 1 \end{bmatrix}\mathbf{T}_{(4)}$$

or

$$\int \mathbf{T}_{(4)} \mathrm{d}t = \frac{h}{2}\sum_{i=1}^{3} \mathbf{Q}^{i}_{(4)} \mathbf{S}_{(4)} + \frac{h}{2}\mathbf{I}_{(4)} \mathbf{T}_{(4)} \tag{4.15}$$

Equation (4.15) can be expressed as

$$\int \mathbf{T}_{(4)}\, \mathrm{d}t = \mathbf{P1ts}_{(4)}\, \mathbf{S}_{(4)} + \mathbf{P1tt}_{(4)}\, \mathbf{T}_{(4)} \tag{4.16}$$

where,

$$\mathbf{P1ts}_{(4)} \triangleq \frac{h}{2}\begin{bmatrix} 0 & 1 & 1 & 1 \\ 0 & 0 & 1 & 1 \\ 0 & 0 & 0 & 1 \\ 0 & 0 & 0 & 0 \end{bmatrix} \quad \text{and} \quad \mathbf{P1tt}_{(4)} \triangleq \frac{h}{2}\begin{bmatrix} 1 & 0 & 0 & 0 \\ 0 & 1 & 0 & 0 \\ 0 & 0 & 1 & 0 \\ 0 & 0 & 0 & 1 \end{bmatrix}$$

The square matrices $\mathbf{P1ts}_{(4)}$ and $\mathbf{P1tt}_{(4)}$ may be expressed in the following compact forms

$$\mathbf{P1ts}_{(4)} = \frac{h}{2}[\![0 \quad 1 \quad 1 \quad 1]\!] \quad \text{and} \quad \mathbf{P1tt}_{(4)} = \frac{h}{2}[\![1 \quad 0 \quad 0 \quad 0]\!]$$

Hence, as in Eq. (4.8), $\mathbf{P1ts}_{(4)}$ and $\mathbf{P1tt}_{(4)}$ are the first order integration operational matrices for integration of triangular function components of an HF domain expanded time function.

The following relations are noted amongst the operational matrices:

$$\left.\begin{aligned} \mathbf{P1ts} &= \frac{1}{2}\mathbf{P1ss} \\ \mathbf{P1tt} &= \frac{1}{2}\mathbf{P1st} \end{aligned}\right\} \tag{4.17}$$

In general, for an m-set function, similar to (4.9), we can write

$$\int \mathbf{T}_{(\mathrm{m})}\, \mathrm{d}t = \frac{h}{2}\sum_{i=1}^{m-1} \mathbf{Q}^{i}_{(m)}\, \mathbf{S}_{(m)} + \frac{h}{2}\mathbf{I}_{(m)} \mathbf{T}_{(m)} = \mathbf{P1ts}_{(m)}\, \mathbf{S}_{(m)} + \mathbf{P1tt}_{(m)}\, \mathbf{T}_{(m)} \tag{4.18}$$

where,

$$\mathbf{P1ts}_{(m)} \triangleq \frac{h}{2}\left[\left[\underbrace{0 \quad 1 \quad \ldots \quad \ldots \quad 1 \quad 1}_{m\ \text{terms}}\right]\right] \quad \text{and}$$

$$\mathbf{P1tt}_{(m)} \triangleq \frac{h}{2}\left[\left[\underbrace{1 \quad 0 \quad \ldots \quad \ldots \quad 0 \quad 0}_{m\ \text{terms}}\right]\right]$$

4.2 Integration of Functions Using Operational Matrices

The integration operational matrices developed for SHF and TF in Sects. 4.1.1 and 4.1.2 will now be used to integrate any time function in HF domain.

Let $f(t)$ be a square integrable function which can be expanded in hybrid function domain as

$$\begin{aligned} f(t) &\approx [\,c_0 \quad c_1 \quad c_2 \quad \ldots \quad c_i \quad \ldots \quad c_{m-1}\,]\mathbf{S}_{(m)} \\ &\quad + [\,(c_1 - c_0) \quad (c_2 - c_1) \quad (c_3 - c_2) \quad \ldots \quad (c_i - c_{i-1}) \quad \ldots \quad (c_m - c_{m-1})\,]\mathbf{T}_{(m)} \\ &\triangleq \mathbf{C}_{\mathrm{S}}^{\mathrm{T}}\mathbf{S}_{(m)} + \mathbf{C}_{\mathrm{T}}^{\mathrm{T}}\mathbf{T}_{(m)} \end{aligned} \tag{4.19}$$

where, $c_0, c_1, c_2, \ldots, c_m$ are $(m + 1)$ equidistant samples of $f(t)$ with a sampling period h and $[\ldots]^{\mathrm{T}}$ denotes transpose.

Integrating Eq. (4.19) with respect to t, we get

$$\begin{aligned} \int f(t)\mathrm{d}t &\approx \int \mathbf{C}_{\mathrm{S}}^{\mathrm{T}}\mathbf{S}_{(m)}\mathrm{d}t + \int \mathbf{C}_{\mathrm{T}}^{\mathrm{T}}\mathbf{T}_{(m)}\mathrm{d}t \\ &= \mathbf{C}_{\mathrm{S}}^{\mathrm{T}}\int \mathbf{S}_{(m)}\mathrm{d}t + \mathbf{C}_{\mathrm{T}}^{\mathrm{T}}\int \mathbf{T}_{(m)}\mathrm{d}t \\ &= \mathbf{C}_{\mathrm{S}}^{\mathrm{T}}\left[\mathbf{P1ss}_{(m)}\mathbf{S}_{(m)} + \mathbf{P1st}_{(m)}\mathbf{T}_{(m)}\right] + \mathbf{C}_{\mathrm{T}}^{\mathrm{T}}\left[\mathbf{P1ts}_{(m)}\mathbf{S}_{(m)} + \mathbf{P1tt}_{(m)}\mathbf{T}_{(m)}\right] \\ &= \left[\mathbf{C}_{\mathrm{S}}^{\mathrm{T}} + \frac{1}{2}\mathbf{C}_{\mathrm{T}}^{\mathrm{T}}\right]\left[\mathbf{P1ss}_{(m)}\mathbf{S}_{(m)} + \mathbf{P1st}_{(m)}\mathbf{T}_{(m)}\right] \end{aligned} \tag{4.20}$$

where we have been made of the relations (4.17).

Now we use (4.20) to perform integration of a few simple square integrable functions.

4.2.1 Numerical Examples

To show the validity of Eq. (4.20), we first compute the exact integration of time function $f(t)$ and then expand it directly in HF domain. Then we perform the same integration via operational technique using Eq. (4.20) and compare the results with the previous one.

Example 4.1 We integrate the function $f_1(t) = t$ via hybrid function method taking $T = 1$ s, $m = 8$ and $h = T/m$ s.

Exact integration of the given function is $\frac{t^2}{2}$ and direct expansion of $\frac{t^2}{2}$ in hybrid function domain, using Eq. (3.5), is

$$\int_0^t f_1(\tau)\mathrm{d}\tau = \frac{t^2}{2} \approx [0.00000000 \quad 0.00781250 \quad 0.03125000 \quad 0.07031250 \quad 0.12500000 \quad 0.19531250 \quad 0.28125000 \quad 0.38281250]\mathbf{S}_{(8)} + [0.00781250 \quad 0.02343750 \quad 0.3906250 \quad 0.05468750 \quad 0.07031250 \quad 0.08593750 \quad 0.10156250 \quad 0.11718750]\mathbf{T}_{(8)} \tag{4.21}$$

In hybrid function domain, the function $f_1(t) = t$ is expanded directly as

$$f_1(t) = t \approx [0.00000000 \quad 0.12500000 \quad 0.25000000 \quad 0.37500000 \quad 0.50000000 \quad 0.62500000 \quad 0.75000000 \quad 0.87500000]\mathbf{S}_{(8)} + [0.12500000 \quad 0.12500000 \quad 0.12500000 \quad 0.12500000 \quad 0.12500000 \quad 0.12500000 \quad 0.12500000 \quad 0.12500000]\mathbf{T}_{(8)} \tag{4.22}$$

Putting the values of $\mathbf{C}_S^T$ and $\mathbf{C}_T^T$ from Eq. (4.22) in Eq. (4.20), we perform operational integration of the function $f_1(t)$ in hybrid function domain to obtain

$$\int_0^t f_1(\tau)\mathrm{d}\tau = \frac{t^2}{2} \approx [0.00000000 \quad 0.00781250 \quad 0.03125000 \quad 0.07031250 \quad 0.12500000 \quad 0.19531250 \quad 0.28125000 \quad 0.38281250]\mathbf{S}_{(8)} + [0.00781250 \quad 0.02343750 \quad 0.03906250 \quad 0.05468750 \quad 0.07031250 \quad 0.08593750 \quad 0.10156250 \quad 0.11716750]\mathbf{T}_{(8)} \tag{4.23}$$

In Fig. 4.7, we compare the results obtained via direct expansion using Eq. (4.21) and using the HF domain operational technique via Eq. (4.23). It is observed that for the function $f_1(t) = t$, the result of direct integration and subsequent HF domain expansion, and integration by the operational method are identical. That is, in this case, percentage error is zero as shown in Table 4.1.

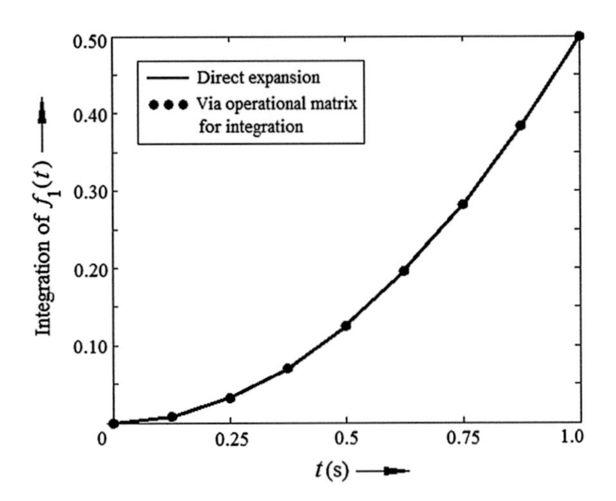

Fig. 4.7 Comparison of integration of the function $f_1(t) = t$ via (i) direct expansion of the integrated function in HF domain [Eq. (4.21)] and (ii) using HF domain integration operational matrices [Eq. (4.23)]. It is noted that the two curves overlap

Example 4.2 Now we integrate the function $f_2(t) = \sin(\pi t)$ in hybrid function domain taking $T = 1$ s, $m = 8$ and $h = T/m$ s. The exact integration of the given function is $[1 - \cos(\pi t)]/\pi$ and direct expansion of $[1 - \cos(\pi t)]/\pi$ in HF domain is

$$\begin{aligned}\int_0^t f_2(\tau)\mathrm{d}\tau &= [1 - \cos(\pi t)]/\pi \\ &\approx [0.00000000 \quad 0.02422989 \quad 0.09323080 \quad 0.19649796 \\ &\quad 0.31830988 \quad 0.44012180 \quad 0.54338896 \quad 0.61238987]\mathbf{S}_{(8)} \\ &\quad + [0.02422989 \quad 0.06900090 \quad 0.10326715 \quad 0.12181191 \\ &\quad 0.12181191 \quad 0.10326715 \quad 0.06900090 \quad 0.0242298]\mathbf{T}_{(8)}\end{aligned} \tag{4.24}$$

The function $f_2(t) = \sin(\pi t)$, when expanded directly in hybrid function domain, is expressed as

$$\begin{aligned}f_2(t) = \sin(\pi t) \approx [&0.00000000 \quad 0.38268343 \quad 0.70710678 \quad 0.92387953 \quad 1.00000000 \\ &0.92387953 \quad 0.70710678 \quad 0.38268343]\mathbf{S}_{(8)} \\ + [&0.38268343 \quad 0.32442334 \quad 0.21677275 \quad 0.07612046 - 0.07612046 \\ &- 0.21677275 - 0.32442334 - 0.38268343]\mathbf{T}_{(8)}\end{aligned} \tag{4.25}$$

Putting the values of $\mathbf{C}_S^T$ and $\mathbf{C}_T^T$ from Eq. (4.25) in Eq. (4.20), we perform operational integration of the function $f_2(t) = \sin(\pi t)$ in hybrid function domain to obtain

Table 4.1 Comparison of samples obtained via two methods and percentage error for (a) SHF coefficients and (b) TF coefficients for the function $f_1(t) = t$

t(s)	Direct expansion	Via operational method	% Error
a Sample-and-hold function domain coefficients			
0			
	0.00000000	0.00000000	–
$\frac{1}{8}$			
	0.00781250	0.00781250	0.00000000
$\frac{2}{8}$			
	0.03125000	0.03125000	0.00000000
$\frac{3}{8}$			
	0.07031250	0.07031250	0.00000000
$\frac{4}{8}$			
	0.12500000	0.12500000	0.00000000
$\frac{5}{8}$			
	0.19531250	0.19531250	0.00000000
$\frac{6}{8}$			
	0.28125000	0.28125000	0.00000000
$\frac{7}{8}$			
	0.38281250	0.38281250	0.00000000
$\frac{8}{8}$			
b Triangular function domain coefficients			
0			
	0.00781250	0.00781250	0.00000000
$\frac{1}{8}$			
	0.02343750	0.02343750	0.00000000
$\frac{2}{8}$			
	0.03906250	0.03906250	0.00000000
$\frac{3}{8}$			
	0.05468750	0.05468750	0.00000000
$\frac{4}{8}$			
	0.07031250	0.07031250	0.00000000
$\frac{5}{8}$			
	0.08593750	0.08593750	0.00000000
$\frac{6}{8}$			
	0.10156250	0.10156250	0.00000000
$\frac{7}{8}$			
	0.11718750	0.11718750	0.00000000
$\frac{8}{8}$			

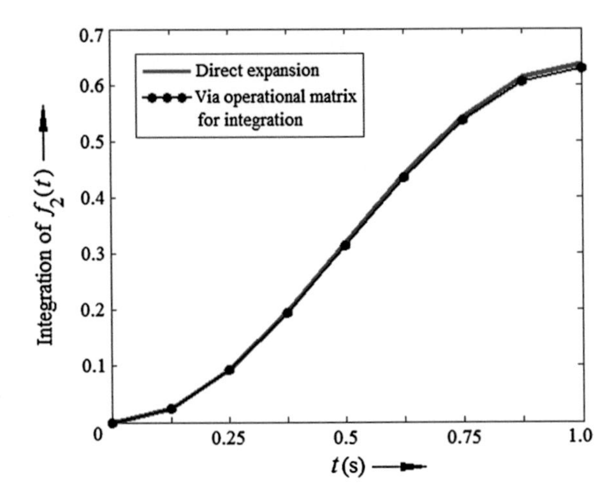

Fig. 4.8 Comparison of integration of the function $f_2(t) = \sin(\pi t)$ via (**a**) direct integration and subsequent expansion of the integrated function in HF domain [Eq. (4.24)] and (**b**) using HF domain integration operational matrices [Eq. (4.26)]. It is noted that the two curves almost overlap (vide Appendix B, Program no. 11)

$$\begin{aligned}\int_0^t f_2(\tau)\mathrm{d}\tau &= [1-\cos(\pi t)]/\pi \\ &\approx [0.00000000 \quad 0.02391771 \quad 0.09202960 \quad 0.19396624 \quad 0.31420871 \\ &\quad 0.43445118 \quad 0.53638783 \quad 0.60449972]\mathbf{S}_{(8)} \\ &\quad +[0.02391771 \quad 0.06811188 \quad 0.10193664 \quad 0.12024247 \quad 0.12024247 \\ &\quad 0.10193664 \quad 0.06811188 \quad 0.02391771]\mathbf{T}_{(8)}\end{aligned} \tag{4.26}$$

In Fig. 4.8, the result of above integration via direct expansion and by the operational method are plotted. It is noted that the two results are very close. This is also apparent from Table 4.2, which shows the results from Eqs. (4.24) and (4.26). It is seen from the table that percentage error is very small and reasonably constant over the time interval of interest.

4.3 Operational Matrices for Differentiation [10]

4.3.1 Differentiation of Time Functions Using Operational Matrices

Let a square integrable function $f(t)$ of Lebesgue measure be expressed in HF domain, for $m = 4$, as

Table 4.2 Comparison of samples via two methods and percentage error for (a) SHF coefficients and (b) TF coefficients for the function $f_2(t) = \sin(\pi t)$, (vide Appendix B, Program no. 11)

t(s)	Direct expansion	Via operational method	% Error
a Sample-and-hold function domain coefficients			
0			
	0.00000000	0.00000000	–
$\frac{1}{8}$			
	0.02422989	0.02391771	1.288408656
$\frac{2}{8}$			
	0.09323080	0.09202960	1.288415416
$\frac{3}{8}$			
	0.19649796	0.19396624	1.288420501
$\frac{4}{8}$			
	0.31830988	0.31420871	1.288420579
$\frac{5}{8}$			
	0.44012180	0.43445118	1.288420614
$\frac{6}{8}$			
	0.54338896	0.53638783	1.288419625
$\frac{7}{8}$			
	0.61238987	0.60449972	1.288419418
$\frac{8}{8}$			
b Triangular function domain coefficients			
0			
	0.02422989	0.02391771	1.288408656
$\frac{1}{8}$			
	0.06900090	0.06811188	1.288417977
$\frac{2}{8}$			
	0.10326715	0.10193664	1.288415532
$\frac{3}{8}$			
	0.12181191	0.12024247	1.288412603
$\frac{4}{8}$			
	0.12181191	0.12024247	1.288412603
$\frac{5}{8}$			
	0.10326715	0.10193664	1.288415532
$\frac{6}{8}$			
	0.06900090	0.06811188	1.288417977
$\frac{7}{8}$			
	0.02422989	0.02391771	1.288408656
$\frac{8}{8}$			

$$f(t) \approx [c_0 \quad c_1 \quad c_2 \quad c_3]\mathbf{S}_{(4)} + [(c_1 - c_0) \quad (c_2 - c_1) \quad (c_3 - c_2) \quad (c_4 - c_3)]\mathbf{T}_{(4)}$$
$$\triangleq \mathbf{C}_S^T \mathbf{S}_{(4)} + \mathbf{C}_T^T \mathbf{T}_{(4)} \tag{4.27}$$

When a function $f(t)$ is expressed in HF domain, it is converted to a piecewise linear function in $[0, T)$. If this converted function is differentiated, the result will be a staircase function. For such a function, any attempt to compute the higher derivatives will give rise to delta functions as well as double delta functions.

To avoid this difficulty, we compute the first derivative from the samples of the function $f(t)$ by taking appropriate first order differences. Thus, from Eq. (4.27), we can write

$$\begin{aligned}\dot{f}(t) \approx& \frac{1}{h}[(c_1 - c_0) \quad (c_2 - c_1) \quad (c_3 - c_2) \quad (c_4 - c_3)]\mathbf{S}_{(4)} \\ &+ \frac{1}{h}[\{(c_2 - c_1) - (c_1 - c_0)\} \quad \{(c_3 - c_2) - (c_2 - c_1)\} \\ &\{(c_4 - c_3) - (c_3 - c_2)\} \quad \{(c_5 - c_4) - (c_4 - c_3)\}]\mathbf{T}_{(4)}\end{aligned} \tag{4.28}$$

Let there be two square matrices $\mathbf{D}_{S(4)}$ and $\mathbf{D}_{T(4)}$ such that which, when operated upon the $\mathbf{S}_{(4)}$ vector and the $\mathbf{T}_{(4)}$ vector of Eq. (4.27) respectively, yield Eq. (4.28). That is, $\mathbf{D}_{S(4)}$ acts as the differentiation matrix in sample-and-hold function domain and $\mathbf{D}_{T(4)}$ acts as the differentiation matrix in triangular function domain. Thus, Eq. (4.28) may now be written as

$$\begin{aligned}\dot{f}(t) \approx& [c_0 \quad c_1 \quad c_2 \quad c_3]\mathbf{D}_{S(4)}\mathbf{S}_{(4)} + [(c_1 - c_0) \quad (c_2 - c_1) \quad (c_3 - c_2) \quad (c_4 - c_3)]\mathbf{D}_{T(4)}\mathbf{T}_{(4)} \\ =& \frac{1}{h}[(c_1 - c_0) \quad (c_2 - c_1) \quad (c_3 - c_2) \quad (c_4 - c_3)]\mathbf{S}_{(4)} \\ &+ \frac{1}{h}[\{(c_2 - c_1) - (c_1 - c_0)\} \quad \{(c_3 - c_2) - (c_2 - c_1)\} \\ &\{(c_4 - c_3) - (c_3 - c_2)\} \quad \{(c_5 - c_4) - (c_4 - c_3)\}]\mathbf{T}_{(4)}\end{aligned}$$

Thus, for $\mathbf{D}_{S(4)}$, we can write

$$[c_0 \quad c_1 \quad c_2 \quad c_3]\mathbf{D}_{S(4)} = \frac{1}{h}[(c_1 - c_0) \quad (c_2 - c_1) \quad (c_3 - c_2) \quad (c_4 - c_3)]$$

Solving for $\mathbf{D}_{S(4)}$ algebraically, we have

$$\mathbf{D}_{S(4)} = \frac{1}{h}\begin{bmatrix} -1 & 0 & 0 & 0 \\ 1 & -1 & 0 & 0 \\ 0 & 1 & -1 & 0 \\ 0 & 0 & 1 & \frac{(c_4 - c_3)}{c_3} \end{bmatrix} \tag{4.29}$$

Similarly, for the differentiation matrix $\mathbf{D}_{T(4)}$ in triangular function domain, we can write

$$\begin{aligned}&[(c_1-c_0)\quad(c_2-c_1)\quad(c_3-c_2)\quad(c_4-c_3)]\mathbf{D}_{\mathrm{T}(4)}\\&=\frac{1}{h}[\{(c_2-c_1)-(c_1-c_0)\}\quad\{(c_3-c_2)-(c_2-c_1)\}\\&\{(c_4-c_3)-(c_3-c_2)\}\quad\{(c_5-c_4)-(c_4-c_3)\}]\end{aligned}$$

Solving for $\mathbf{D}_{\mathrm{T}(4)}$, we have

$$\mathbf{D}_{\mathrm{T}(4)}=\frac{1}{h}\begin{bmatrix}-1&0&0&0\\1&-1&0&0\\0&1&-1&0\\0&0&1&\frac{(c_5-c_4)-(c_4-c_3)}{(c_4-c_3)}\end{bmatrix}\tag{4.30}$$

Following Eqs. (4.29) and (4.30), the generalized matrices of order m may be formed from the following equation, where m component functions have been used. That is

$$\begin{aligned}\dot{f}(t)\approx&\frac{1}{h}[(c_1-c_0)\quad(c_2-c_1)\quad(c_3-c_2)\quad\ldots\quad(c_m-c_{m-1})]\mathbf{S}_{(m)}\\&+\frac{1}{h}[\{(c_2-c_1)-(c_1-c_0)\}\quad\{(c_3-c_2)-(c_2-c_1)\}\quad\ldots\quad\{(c_{m+1}-c_m)-(c_m-c_{m-1})\}]\mathbf{T}_{(m)}\end{aligned}\tag{4.31}$$

Thus, the general forms of differentiation matrices, $\mathbf{D}_{\mathrm{S}(m)}$ and $\mathbf{D}_{\mathrm{T}(m)}$, are given by

$$\mathbf{D}_{\mathrm{S}(m)}=\frac{1}{h}\begin{bmatrix}-1&0&\cdots&0&0\\1&-1&\cdots&0&0\\0&1&\cdots&0&0\\\vdots&\vdots&\vdots&\vdots&\vdots\\0&0&\cdots&1&\frac{(c_m-c_{m-1})}{c_{m-1}}\end{bmatrix}_{(m\times m)}\tag{4.32}$$

$$\mathbf{D}_{\mathrm{T}(m)}=\frac{1}{h}\begin{bmatrix}-1&0&\ldots&0&0\\1&-1&\ldots&0&0\\0&1&\ldots&0&0\\\vdots&\vdots&\vdots&\vdots&\vdots\\0&0&\ldots&1&\frac{(c_{m+1}-c_m)-(c_m-c_{m-1})}{(c_m-c_{m-1})}\end{bmatrix}_{(m\times m)}\tag{4.33}$$

4.3.2 Numerical Examples

Example 4.3 Let us consider a function $f_3(t) = 1 - \exp(-t)$.

Expanding it in HF domain, for $m = 10$ and $T = 1$ s, we have

$$\begin{aligned} f_3(t) \approx & [0 \quad 0.09516258 \quad 0.18126924 \quad 0.25918177 \quad 0.32967995 \\ & 0.39346934 \quad 0.45118836 \quad 0.50341469 \quad 0.55067103 \quad 0.59343034]\mathbf{S}_{(10)} \\ & + [0.09516258 \quad 0.08610666 \quad 0.07791253 \quad 0.07049818 \quad 0.06378939 \\ & 0.05771902 \quad 0.05222633 \quad 0.04725634 \quad 0.04275931 \quad 0.03869021]\mathbf{T}_{(10)} \end{aligned} \tag{4.34}$$

Now we differentiate the function given in (4.34) using the matrices of Eqs. (4.32) and (4.33) for $m = 10$. The result of differentiation in HF domain is obtained as

$$\begin{aligned} \dot{f}_3(t) \approx & [0.95162581 \quad 0.86106664 \quad 0.77912532 \quad 0.70498174 \quad 0.63789386 \\ & 0.57719023 \quad 0.52226332 \quad 0.47256339 \quad 0.42759304 \quad 0.38690218]\mathbf{S}_{(10)} \\ & + [-0.09055917 \quad -0.08194132 \quad -0.07414358 \quad -0.06708788 \quad -0.06070363 \\ & -0.05492691 \quad -0.04969993 \quad -0.04497035 \quad -0.04069086 \quad -0.03681861]\mathbf{T}_{(10)} \end{aligned} \tag{4.35}$$

Direct expansion of the function $\exp(-t)$, that is $\dot{f}_3(t)$, in HF domain is

$$\begin{aligned} \dot{f}_3(t) \approx & [1.00000000 \quad 0.90483741 \quad 0.81873075 \quad 0.74081822 \quad 0.67032004 \\ & 0.60653065 \quad 0.54881163 \quad 0.49658530 \quad 0.44932896 \quad 0.40656965]\mathbf{S}_{(10)} \\ & + [-0.09516259 \quad -0.08610666 \quad -0.07791253 \quad -0.07049818 \quad -0.06378939 \\ & -0.05771902 \quad -0.05222633 \quad -0.04725633 \quad -0.04275930 \quad -0.03869021]\mathbf{T}_{(10)} \end{aligned} \tag{4.36}$$

Figure 4.9 shows the direct expansion of the original function $f_3(t)$ and its derivative $\dot{f}_3(t)$ in HF domain using Eqs. (4.34) and (4.36). The figure also includes HF domain representation of $\dot{f}_3(t)$ obtained via Eq. (4.35), using differentiation matrices.

From the curve, it is seen that at $t = 0$, the curve $\dot{f}_3(t)$ deviates from its exact value 1. This deviation may be reduced by increasing m. That is, an increased value of m will make the differentiated curve start from a value more close to 1 on the y axis at $t = 0$.

Example 4.4 Let us consider a function $f_4(t) = \sin(\pi t)/\pi$. Expanding it in HF domain, for $m = 10$ and $T = 1$ s, we have

$$\begin{aligned} f_4(t) \approx & [0.00000000 \quad 0.09836316 \quad 0.18709785 \quad 0.25751810 \quad 0.30273069 \\ & 0.31830988 \quad 0.30273069 \quad 0.25751810 \quad 0.18709785 \quad 0.09836316]\mathbf{S}_{(10)} \\ & + [0.09836316 \quad 0.08873469 \quad 0.07042025 \quad 0.04521258 \quad 0.01557919 \\ & -0.01557919 \quad -0.04521258 \quad -0.07042025 \quad -0.08873469 \quad -0.09836316]\mathbf{T}_{(10)} \end{aligned} \tag{4.37}$$

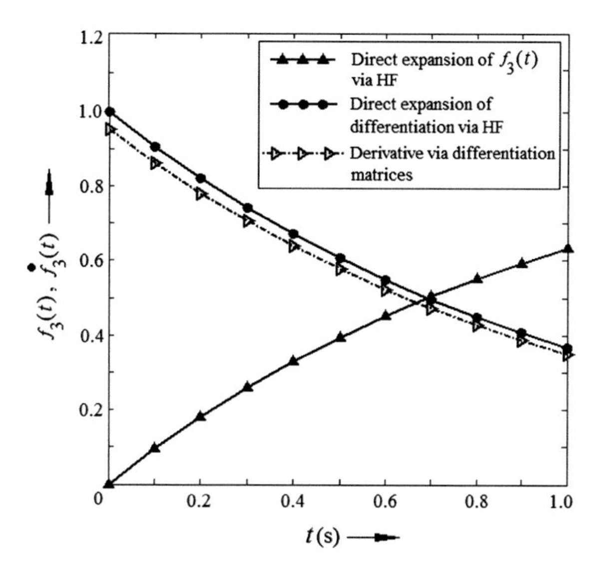

Fig. 4.9 HF domain direct expansion of the function $f_3(t) = 1 - \exp(-t)$, its exact derivative $\dot{f}_3(t)$ along with $\dot{f}_3(t)$ obtained using HF domain differentiation matrices for $m = 10$ and $T = 1$ s (vide Appendix B, Program no. 12)

Now we differentiate the function given in (4.37) using the matrices of Eqs. (4.32) and (4.33) for $m = 10$. The result of differentiation in HF domain is obtained as

$$\begin{aligned}\dot{f}_4(t) \approx\; & [0.98363164 \quad 0.88734692 \quad 0.70420250 \quad 0.45212584 \quad 0.15579194 \\ & -0.15579194 \quad -0.45212584 \quad -0.70420250 \quad -0.88734692 \quad -0.98363164]\mathbf{S}_{(10)} \\ & +[-0.09628471 \quad -0.18314441 \quad -0.25207666 \quad -0.29633389 \quad -0.31158389 \\ & -0.29633389 \quad -0.25207666 \quad -0.18314441 \quad -0.09628471 \quad -1.55431223\mathrm{e}-015]\mathbf{T}_{(10)}\end{aligned} \tag{4.38}$$

Direct expansion of the function $\cos(\pi t)$, that is $\dot{f}_4(t)$, in HF domain is

$$\begin{aligned}\dot{f}_4(t) \approx\; & [1.00000000 \quad 0.95105651 \quad 0.80901699 \quad 0.58778525 \quad 0.30901699 \\ & 0.00000000 \quad -0.30901699 \quad -0.58778525 \quad -0.80901699 \quad -0.95105651]\mathbf{S}_{(10)} \\ & +[-0.04894348 \quad -0.14203952 \quad -0.22123174 \quad -0.27876825 \quad -0.30901699 \\ & -0.30901699 \quad -0.27876825 \quad -0.22123174 \quad -0.14203952 \quad -0.04894348]\mathbf{T}_{(10)}\end{aligned} \tag{4.39}$$

Figure 4.10 shows both the original function $f_4(t)$ and the differentiated function $\dot{f}_4(t)$ expressed as HF domain direct expansions [using Eqs. (4.37) and (4.39)]. For comparison, $\dot{f}_4(t)$ obtained via HF domain differential matrices [Eq. (4.38)] is also plotted. On increasing sample density, that is m, the HF domain differentiated curve moves closer to the exact solution.

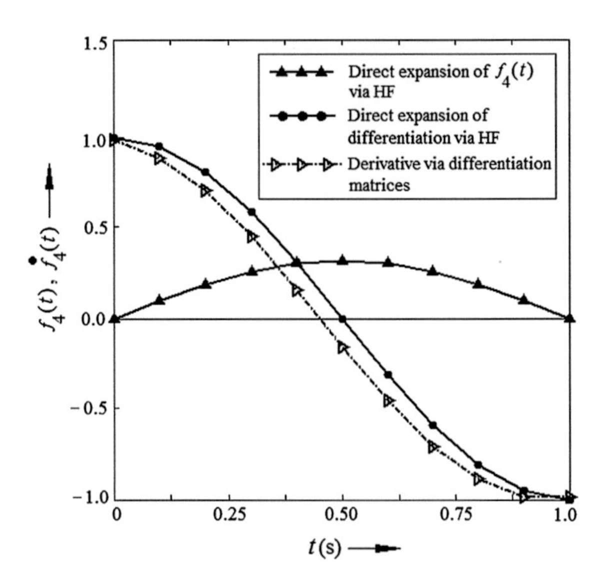

Fig. 4.10 HF domain direct expansion of the function $f_4(t) = \sin(\pi t)/\pi$, its derivative $\dot{f}_4(t)$ along with $\dot{f}_4(t)$ obtained using HF domain differential matrices for m = 10 and T = 1 s (vide Appendix B, Program no. 13)

4.4 Accumulation of Error for Subsequent Integration-Differentiation (I-D) Operation in HF Domain

It is obvious that integer integration using the integral operational matrices introduce error since the operation is approximate. So do the operational matrices for differentiation, when any time function is differentiated in HF domain. Hence, it is apparent that subsequent integration-differentiation (I-D) operation on a function in HF domain would fail to come up with the original function, unlike the exact I-D operation.

In the following, accumulation of error for subsequent I-D operation is studied in detail with table and characteristic curves.

For m = 4, we expand a function $f(t)$ in HF domain, as shown in Eq. (4.27), to write

$$\begin{aligned} f(t) &\approx [\,c_0 \quad c_1 \quad c_2 \quad c_3\,]\mathbf{S}_{(4)} + [\,(c_1 - c_0) \quad (c_2 - c_1) \quad (c_3 - c_2) \quad (c_4 - c_3)\,]\mathbf{T}_{(4)} \\ &\triangleq \mathbf{C}_{\mathrm{S}}^{\mathrm{T}}\mathbf{S}_{(4)} + \mathbf{C}_{\mathrm{T}}^{\mathrm{T}}\mathbf{T}_{(4)} \end{aligned}$$

Now integrating $f(t)$ using the operational matrices gives

$$\int f(t)\mathrm{d}t \triangleq F(t) \approx h\left[\mathbf{C}_{\mathrm{S}}^{\mathrm{T}} + \frac{1}{2}\mathbf{C}_{\mathrm{T}}^{\mathrm{T}}\right]\left[\mathbf{P}_{11}\,\mathbf{S}_{(4)} + \mathbf{I}\,\mathbf{T}_{(4)}\right] \triangleq \bar{F}(t) \quad \text{(say)}$$

where $\mathbf{P}_{11} = [\![0 \quad 1 \quad 1 \quad 1]\!]$ and $\mathbf{I}$ is an identity matrix of order 4.

It may be noted that $\mathbf{P1ss} = h\,\mathbf{P}_{11}$.

Thus

$$
\begin{aligned}
\bar{F}(t) &= \frac{h}{2}[(c_0+c_1) \quad (c_1+c_2) \quad (c_2+c_3) \quad (c_3+c_4)]\left[\mathbf{P}_{11}\,\mathbf{S}_{(4)} + \mathbf{I}\,\mathbf{T}_{(4)}\right] \\
&= \frac{h}{2}[(c_0+c_1) \quad (c_1+c_2) \quad (c_2+c_3) \quad (c_3+c_4)]\mathbf{P}_{11}\mathbf{S}_{(4)} \\
&\quad + \frac{h}{2}[(c_0+c_1) \quad (c_1+c_2) \quad (c_2+c_3) \quad (c_3+c_4)]\mathbf{T}_{(4)} \\
&= \frac{h}{2}[0 \quad (c_0+c_1) \quad \{(c_0+c_1)+(c_1+c_2)\} \quad \{(c_0+c_1)+(c_1+c_2)+(c_2+c_3)\}]\mathbf{S}_{(4)} \\
&\quad + \frac{h}{2}[(c_0+c_1) \quad (c_1+c_2) \quad (c_2+c_3) \quad (c_3+c_4)]\mathbf{T}_{(4)}
\end{aligned} \tag{4.40}
$$

Now, it is of interest to estimate the accumulation of error for subsequent integration-differentiation operation (I-D operation) on a function $f(t)$.

To achieve this end, we differentiate $\bar{F}(t)$ of Eq. (4.40) using $\mathbf{D}_{\mathrm{S}(4)}$ and $\mathbf{D}_{\mathrm{T}(4)}$. Usually, exact integration-differentiation always yields $f(t)$ itself. But since HF domain operational calculus is somewhat approximate, the resulting function is expected to deviate from the HF domain representation of $f(t)$.

Differentiating $\bar{F}(t)$ using $\mathbf{D}_{\mathrm{S}(4)}$ and $\mathbf{D}_{\mathrm{T}(4)}$, we have

$$
\begin{aligned}
f(t) &= \frac{\mathrm{d}}{\mathrm{d}t}[F(t)] \triangleq f(t)_{\mathrm{I-D},1} \\
&= \frac{h}{2}[0 \quad (c_0+c_1) \quad \{(c_0+c_1)+(c_1+c_2)\} \quad \{(c_0+c_1)+(c_1+c_2)+(c_2+c_3)\}]\mathbf{D}_{\mathrm{S}(4)}\mathbf{S}_{(4)} \\
&\quad + \frac{h}{2}[(c_0+c_1) \quad (c_1+c_2) \quad (c_2+c_3) \quad (c_3+c_4)]\mathbf{D}_{\mathrm{T}(4)}\mathbf{T}_{(4)}
\end{aligned}
$$

Substituting $\mathbf{D}_{\mathrm{S}(4)}$ and $\mathbf{D}_{\mathrm{T}(4)}$ from Eqs. (4.29) and (4.30), we have

$$
\begin{aligned}
f(t)_{\mathrm{I-D},1} &= \frac{1}{2}[(c_0+c_1) \quad (c_1+c_2) \quad (c_2+c_3) \quad (c_3+c_4)]\mathbf{S}_{(4)} \\
&\quad + \frac{1}{2}[\{(c_1+c_2)-(c_0+c_1)\} \quad \{(c_2+c_3)-(c_1+c_2)\} \quad \{(c_3+c_4)-(c_2+c_3)\} \\
&\qquad \{(c_4+c_5)-(c_3+c_4)\}]\mathbf{T}_{(4)} \\
&\triangleq [c'_0 \quad c'_1 \quad c'_2 \quad c'_3]\mathbf{S}_{(4)} + \left[(c'_1-c'_0) \quad (c'_2-c'_1) \quad (c'_3-c'_2) \quad (c'_4-c'_3)\right]\mathbf{T}_{(4)}
\end{aligned} \tag{4.41}
$$

where,

$$
\begin{aligned}
c'_0 &= \frac{1}{2}(c_0+c_1), c'_1 = \frac{1}{2}(c_1+c_2), c'_2 = \frac{1}{2}(c_2+c_3), c'_3 = \frac{1}{2}(c_3+c_4) \quad \text{and} \\
c'_4 &= \frac{1}{2}(c_4+c_5)
\end{aligned}
$$

The result obtained in Eq. (4.41) is somewhat deviated from $\bar{f}(t)$ the original HF domain expansion of $f(t)$.

Similarly, subsequent I-D operation upon the function $f(t)_{\mathrm{I-D},1}$ produces a function still more deviated from $\bar{f}(t)$.

For two subsequent I-D operations on the function $\bar{f}(t)$, namely $f(t)_{\mathrm{I-D},2}$, the result is

$$f(t)_{\mathrm{I-D},2} \triangleq [c_0'' \quad c_1'' \quad c_2'' \quad c_3'']\mathbf{S}_{(4)} + \left[\,(c_1''-c_0'') \quad (c_2''-c_1'') \quad (c_3''-c_2'') \quad (c_4''-c_3'')\,\right]\mathbf{T}_{(4)} \tag{4.42}$$

where,

$$\begin{aligned}
c_0'' &= \frac{1}{4}\{(c_0+c_1)+(c_1+c_2)\} = \frac{1}{4}(c_0+2c_1+c_2)\\
c_1'' &= \frac{1}{4}\{(c_1+c_2)+(c_2+c_3)\} = \frac{1}{4}(c_1+2c_2+c_3)\\
c_2'' &= \frac{1}{4}\{(c_2+c_3)+(c_3+c_4)\} = \frac{1}{4}(c_2+2c_3+c_4)\\
c_3'' &= \frac{1}{4}\{(c_3+c_4)+(c_4+c_5)\} = \frac{1}{4}(c_3+2c_4+c_5)\\
\text{and}\quad c_4'' &= \frac{1}{4}\{(c_4+c_5)+(c_5+c_6)\} = \frac{1}{4}(c_4+2c_5+c_6)
\end{aligned}$$

After three such operations, we have

$$\begin{aligned}
f(t)_{\mathrm{I-D},3} \triangleq\, &[c_0''' \quad c_1''' \quad c_2''' \quad c_3''']\mathbf{S}_{(4)}\\
&+\left[\,(c_1'''-c_0''') \quad (c_2'''-c_1''') \quad (c_3'''-c_2''') \quad (c_4'''-c_3''')\,\right]\mathbf{T}_{(4)}
\end{aligned} \tag{4.43}$$

where,

$$\begin{aligned}
c_0''' &= \frac{1}{2}\left[\frac{1}{4}\{(c_0+c_1)+(c_1+c_2)\}+\frac{1}{4}\{(c_1+c_2)+(c_2+c_3)\}\right] = \frac{1}{8}(c_0+3c_1+3c_2+c_3)\\
c_1''' &= \frac{1}{2}\left[\frac{1}{4}\{(c_1+c_2)+(c_2+c_3)\}+\frac{1}{4}\{(c_2+c_3)+(c_3+c_4)\}\right] = \frac{1}{8}(c_1+3c_2+3c_3+c_4)\\
c_2''' &= \frac{1}{2}\left[\frac{1}{4}\{(c_2+c_3)+(c_3+c_4)\}+\frac{1}{4}\{(c_3+c_4)+(c_4+c_5)\}\right] = \frac{1}{8}(c_2+3c_3+3c_4+c_5)\\
c_3''' &= \frac{1}{2}\left[\frac{1}{4}\{(c_3+c_4)+(c_4+c_5)\}+\frac{1}{4}\{(c_4+c_5)+(c_5+c_6)\}\right] = \frac{1}{8}(c_3+3c_4+3c_5+c_6)\\
c_4''' &= \frac{1}{2}\left[\frac{1}{4}\{(c_4+c_5)+(c_5+c_6)\}+\frac{1}{4}\{(c_5+c_6)+(c_6+c_7)\}\right] = \frac{1}{8}(c_4+3c_5+3c_6+c_7)
\end{aligned}$$

From inspection of Eqs. (4.41), (4.42) and (4.43), we can write down the expression for HF coefficients obtained after n-times repeated I-D operations in terms of the HF coefficients of the original function.

Thus, the kth coefficients of the SHF components are

$$c_j^n(k) = {}^nC_j = \frac{n!}{j!(n-j)!} \quad \text{for} \quad 0 \leq j \leq n \, \text{and} \, 1 \leq k \leq (n+1) \tag{4.44}$$

where, n is the number of I-D operations executed, and, c_j is the coefficient of jth element of the SHF coefficient matrix after n repeated I-D operations.

The coefficients for the TF components can easily be derived from Eq. (4.44).

Now, let us define an index called 'Average of Mod of Percentage' (AMP) error, which is given by

$$\text{AMP error}, \varepsilon_{\text{av(r)}} \triangleq \frac{\sum_{j=1}^{r} |\varepsilon_j|}{r} \tag{4.45}$$

where,

r is the number of sample points, or number of items, or elements considered,
ε_j is the percentage error at each sample point (or, percentage error for each item or element, as may be the case).

Figure 4.11 shows increasing trend of the *Average of Mod of Percentage* (AMP) error with subsequent I-D operations for a particular function $f(t) = t$, using only ten equidistant samples of the function. Though this pattern is somewhat function dependent, it is interesting to note that the pattern traces a ramp function. When the starting function is considered to be $f(t) = \sin(\pi t)$, once again the variation of the AMP error with number of I-D operations resembles the pattern of a sine wave. This is shown in Fig. 4.12.

Table 4.3 compares the SHF coefficients of the original function $f(t) = t$ with its SHF coefficients obtained after five subsequent I-D operations for $m = 10$ and $T = 1$ s. Also, the AMP error is computed.

From Table 4.3, the fourth sample was chosen arbitrarily and its shifting to more erroneous zone due to subsequent I-D operations in HF domain is depicted in Fig. 4.13.

Figure 4.14 shows the decaying nature of the AMP error with increasing number of sub-intervals, for four subsequent I-D operations of a typical ramp function over a time interval of $T = 1$ s.

Therefore, for a particular need of subsequent I-D operations, the error can be reduced by considering a larger number of samples within the particular time span.

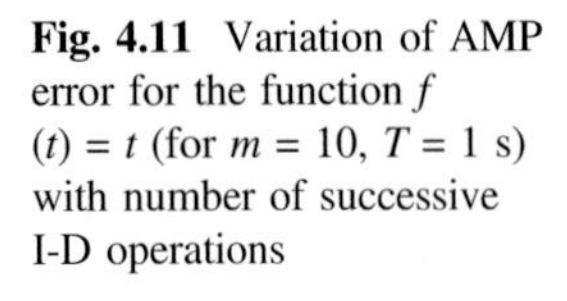

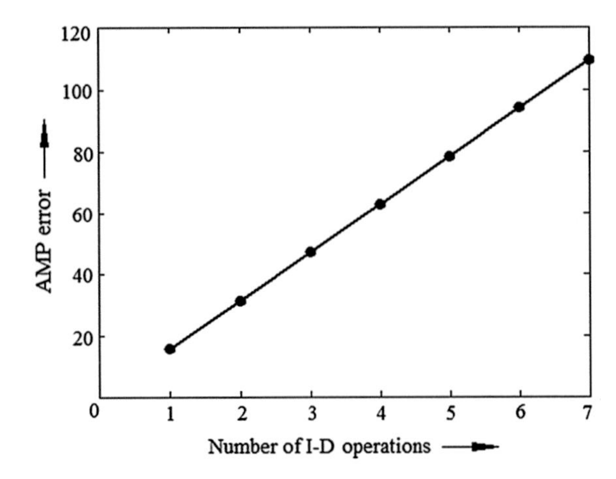

Fig. 4.11 Variation of AMP error for the function $f(t) = t$ (for $m = 10$, $T = 1$ s) with number of successive I-D operations

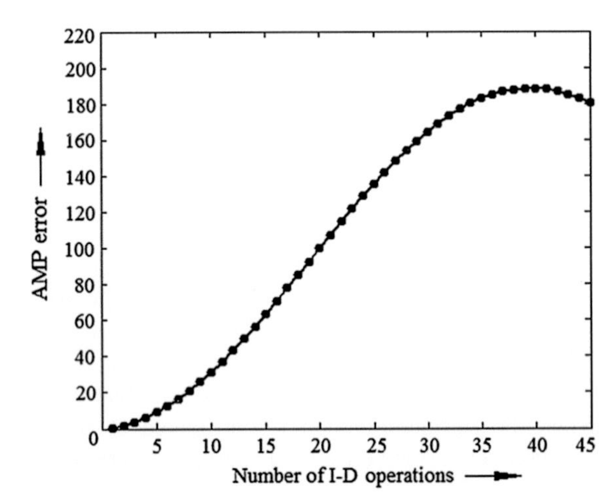

Fig. 4.12 Variation of AMP error for the function $f(t) = \sin(\pi t)$ ($m = 20$, $T = 1$ s) with number of successive I-D operations (vide Appendix B, Program no. 14)

Figure 4.15 shows the deviation of the function $f(t) = \sin(\pi t)$ from its original form with successive I-D operations.

4.5 Conclusion

In this chapter, the integration operational matrices for sample-and-hold functions and triangular functions are derived independently. Integration of the SHF part produces both SHF and TF components and the result is expressed in HF domain using two operational matrices, as shown in Eq. (4.8). Integration of TF components, like that of SHF part, also gives rise to both SHF and TF components, vide Eq. (4.16). In fact, a total of four operational matrices are used conjunctively to perform integration in the HF domain and the result of integration is again comprised of SHF and TF.

Table 4.3 Comparison of the SHF coefficients of the function $f(t) = t$ before and after five successive I-D operations for $m = 10$ and $T = 1$ s

t(s)	SHF coefficients of the original function $f(t) = t$	SHF coefficients of $f(t) = t$ after five subsequent I-D operations	% Error	AMP error
0				
	0	0.25	–	
0.1				
	0.1	0.35	−250.00	
0.2				
	0.2	0.45	−125.00	
0.3				
	0.3	0.55	−83.33	
0.4				
	0.4	0.65	−62.50	78.58
0.5				
	0.5	0.75	−50.00	
0.6				
	0.6	0.85	−41.67	
0.7				
	0.7	0.95	−35.71	
0.8				
	0.8	1.05	−31.25	
0.9				
	0.9	1.15	−27.78	
1.0				

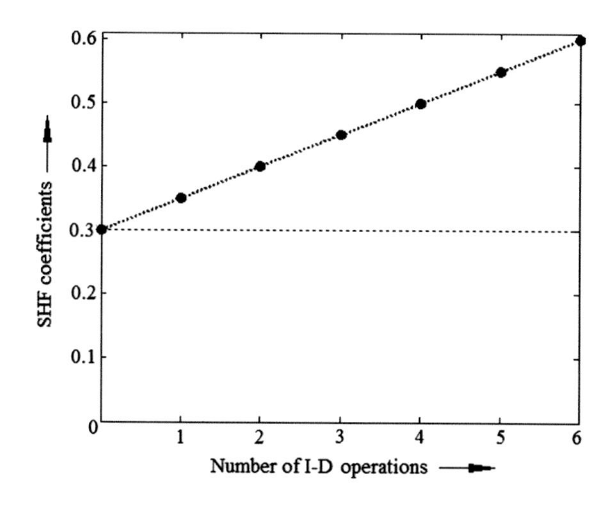

Fig. 4.13 Variation of a typical SHF coefficient 0.3 (vide Table 4.3) for the function $f(t) = t$ for $m = 10$, $T = 1$ s with number of successive I-D operations

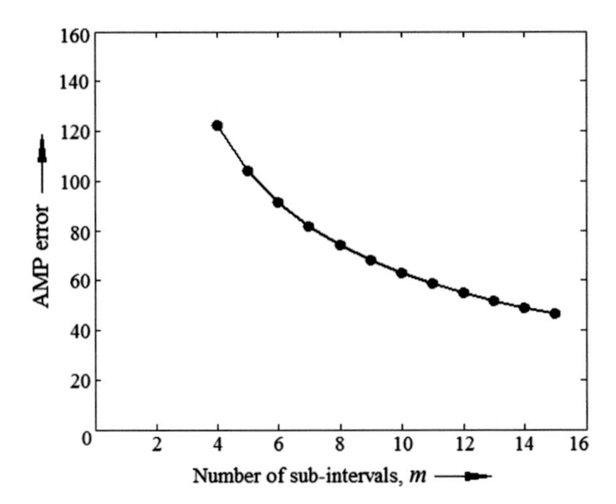

Fig. 4.14 Variation of AMP error for the function $f(t) = t$ for four subsequent I-D operations with increasing m and $T = 1$ s

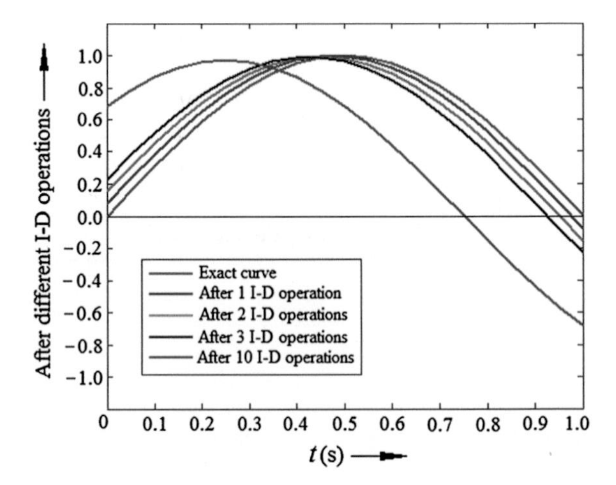

Fig. 4.15 Shifting of the original function $f(t) = \sin(\pi t)$ due to successive I-D operations for $m = 20$, $T = 1$ s (vide Appendix B, Program no. 15)

The integration operation is illustrated via a few examples. That is, the function $f_1(t) = t$ has been integrated in an exact manner and the result is expanded in HF domain.

Further, the function $f_1(t) = t$ is represented in HF domain and is then integrated using Eq. (4.20). These two results are compared in Table 4.1. Figure 4.7 also depicts this comparison for better clarity.

In Table 4.1, it is noted the percentage error is zero. This implies that for linear functions, HF domain integration results are identical with the exact solutions. However, for the second example, that is, the function $f_2(t) = \sin(\pi t)$, shown in Fig. 4.8, the results of integration via the two methods are not identical but very close. Both the results are tabulated and compared in Table 4.2.

The hybrid function (HF) set has been used to derive the operational matrices for differentiation as well. These matrices can be used for differentiation of functions in hybrid function domain. The operational matrix $\mathbf{D}_{S(m)}$ acts as the differentiation matrix in sample-and-hold function (SHF) domain while $\mathbf{D}_{T(m)}$ acts as the differentiation matrix in triangular function (TF) domain. These matrices are presented in Eqs. (4.32) and (4.33).

Figures 4.9 and 4.10 show graphically the application of differential operational matrices for two typical time functions and also compare the same with respective direct expansion in HF domain.

It is apparent that successive integration-differentiation (I-D) operation upon any time function in HF domain accumulates error in the result. That is, we do not get back the original time function as we do for exact I-D operation. The effect of HF domain I-D operation is thus of interest and has been studied. Figures 4.11 and 4.12 show typical curves for accumulation of errors for such repeated I-D operations of two different time functions, t and $\sin(\pi t)$.

The time function $f(t) = t$ has been subjected to five successive I-D operations considering ten sub-intervals over a period $T = 1$ s. For such successive operations Table 4.3 shows the error accumulated at different sample points, i.e., the SHF coefficients. As a typical case study, deviation of a sample 0.3 of the function after each I-D operation, has been tracked. Figure 4.13 shows the locus of the sample moving more and more into the erroneous zone. However, as is obvious, with increasing number of sub-intervals within a fixed time period, the error reduces. This is shown in Fig. 4.14 where for four successive I-D operations upon the function $f(t) = t$, the AMP error goes down exponentially with increasing m.

Figure 4.15 shows deviation of a function $f(t) = \sin(\pi t)$ with successive I-D operation for $m = 20$ and $T = 1$ s. It is noted that the original function shifts with each I-D operation, but reasonably maintains the shape of the original. However, this is merely function specific.

References

1. Jiang, J.H., Schaufelberger, W.: Block Pulse Functions and Their Application in Control System, LNCIS, vol. 179. Springer, Berlin (1992)
2. Deb, Anish, Sarkar, Gautam, Sen, Sunit K.: Linearly pulse-width modulated block pulse functions and their application to linear SISO feedback control system identification. IEE Proc. Control Theory Appl. **142**(1), 44–50 (1995)
3. Chen, C.F., Tsay, Y.T., Wu, T.T.: Walsh operational matrices for fractional calculus and their application to distributed systems. J. Franklin Inst. **303**(3), 267–284 (1977)
4. Rao, G.P., Srinivasan, T.: Analysis and synthesis of dynamic systems containing time delays via block pulse functions. Proc. IEE **125**(9), 1064–1068 (1978)
5. Deb, Anish, Sarkar, Gauatm, Bhattacharjee, Manabrata, Sen, Sunit K.: A new set of piecewise constant orthogonal functions for the analysis of linear SISO systems with sample-and-hold. J. Franklin Inst. **335B**(2), 333–358 (1998)
6. Deb, Anish, Sarkar, Gautam, Sengupta, Anindita: Triangular orthogonal functions for the analysis of continuous time systems. Anthem Press, London (2011)

7. Deb, Anish, Sarkar, Gautam, Dasgupta, Anindita: A complementary pair of orthogonal triangular function sets and its application to the analysis of SISO control systems. J. Inst. Eng. India **84**, 120–129 (2003)
8. Deb, Anish, Dasgupta, Anindita, Sarkar, Gautam: A complementary pair of orthogonal triangular function sets and its application to the analysis of dynamic systems. J Franklin Inst. **343**(1), 1–26 (2006)
9. Deb, Anish, Sarkar, Gautam, Mandal, Priyaranjan, Biswas, Amitava, Ganguly, Anindita, Biswas, Debasish: Transfer function identification from impulse response via a new set of orthogonal hybrid function (HF). Appl. Math. Comput. **218**(9), 4760–4787 (2012)
10. Deb, Anish, Sarkar, Gautam, Ganguly, Anindita, Biswas, Amitava: Approximation, integration and differentiation of time functions using a set of orthogonal hybrid functions (HF) and their application to solution of first order differential equations. Appl. Math. Comput. **218**(9), 4731–4759 (2012)

Chapter 5
One-Shot Operational Matrices for Integration

Abstract This chapter is devoted to develop the theory of one-shot operational matrices. These matrices are useful for multiple integration and, in general, are superior to repeated integration using the first order integration matrices. Theory of one-shot operational matrices is presented and the one-shot operational matrices of n-th order integration have been derived. Three examples with nine figures and four tables elucidate the technique.

In this chapter, the hybrid function set has been utilized to derive one-shot operational matrices [1, 2] for integration of different orders in HF domain. These matrices are employed for more accurate multiple integrations. That is, in case of repeated integrations, one may use the first order integration matrices, derived in Chap. 4, repeatedly. But this will lead to accumulation of errors at each stage, and finally such error may disqualify the result to be of any further use.

In case of Walsh [1] and block pulse functions [2, 3] such one-shot matrices were derived, and accumulation of errors was avoided. For hybrid functions, things are a bit different because the hybrid function domain theories always deals with function samples. But in this case also, the accumulation of errors is avoided and much better results are obtained.

After derivation of the one shot matrices of different orders in HF domain, they are used for many numerical examples to bring out the difference in computations of multiple integrals by using (a) the first order integration matrices repeatedly and (b) the one shot matrices only once.

First of all, second and third order one-shot operational matrices are derived and then the general form of $(m \times m)$ integration matrices for n times multiple integration are derived. Superiority of these matrices over the repeated use of first order integration matrices is strongly established from the examples treated here in.

A. Deb et al., *Analysis and Identification of Time-Invariant Systems, Time-Varying Systems, and Multi-Delay Systems using Orthogonal Hybrid Functions*,
Studies in Systems, Decision and Control 46, DOI 10.1007/978-3-319-26684-8_5

5.1 Integration Using First Order HF Domain Integration Matrices

For first-order integration of sample-and-hold [4] function component, referring to Eq. (4.9), we have

$$\int \mathbf{S}_{(m)}\mathrm{d}t = \mathbf{P1ss}_{(m)}\mathbf{S}_{(m)} + \mathbf{P1st}_{(m)}\mathbf{T}_{(m)}$$

where,

$$\left.\begin{aligned}\mathbf{P1ss}_{(m)} &\triangleq h[\![0 \quad 1 \quad \cdots \quad \cdots \quad 1 \quad 1]\!]_{(m\times m)}\\ \mathbf{P1st}_{(m)} &\triangleq h[\![1 \quad 0 \quad \cdots \quad \cdots \quad 0 \quad 0]\!]_{(m\times m)}\end{aligned}\right\} \tag{5.1}$$

Similarly, for first-order integration of triangular function component [5, 6], referring to Eq. (4.18), we have

$$\int \mathbf{T}_{(m)}\mathrm{d}t = \mathbf{P1ts}_{(m)}\mathbf{S}_{(m)} + \mathbf{P1tt}_{(m)}\mathbf{T}_{(m)}$$

where,

$$\left.\begin{aligned}\mathbf{P1ts}_{(m)} &\triangleq \tfrac{h}{2}[\![0 \quad 1 \quad \cdots \quad \cdots \quad 1 \quad 1]\!]_{(m\times m)}\\ \mathbf{P1tt}_{(m)} &\triangleq \tfrac{h}{2}[\![1 \quad 0 \quad \cdots \quad \cdots \quad 0 \quad 0]\!]_{(m\times m)}\end{aligned}\right\} \tag{5.2}$$

Using Eqs. (4.9), (4.17) and (4.18), we get

$$\int \mathbf{T}_{(m)}\mathrm{d}t = \frac{1}{2}\int \mathbf{S}_{(m)}\mathrm{d}t \tag{5.3}$$

We know that the operational matrix for integration in block pulse function domain is given by [2]

$$\mathbf{P}_{(m)} \triangleq h\left[\!\!\left[\underbrace{\frac{1}{2} \quad 1\cdots\cdots 1 \quad 1}_{m \text{ terms}}\right]\!\!\right] \tag{5.4}$$

Using Eqs. (5.1) and (5.2), we can write the following relations:

$$\mathbf{P1ss}_{(m)} + \frac{\mathbf{P1st}_{(m)}}{2} = 2\,\mathbf{P1ts}_{(m)} + \mathbf{P1tt}_{(m)} = \mathbf{P}_{(m)} \tag{5.5}$$

If a square integrable function $f(t)$ is expanded in hybrid function [7] domain as per Eq. (4.19), we can write

$$
\begin{aligned}
f(t) \approx & \begin{bmatrix} c_0 & c_1 & c_2 & \cdots & c_{(m-1)} \end{bmatrix} \mathbf{S}_{(m)}(t) \\
& + \begin{bmatrix} (c_1 - c_0) & (c_2 - c_1) & (c_3 - c_2) & \cdots & (c_m - c_{m-1}) \end{bmatrix} \mathbf{T}_{(m)}(t) \\
\triangleq & \, \mathbf{C}_{\mathrm{S}}^{\mathrm{T}} \mathbf{S}_{(m)}(t) + \mathbf{C}_{\mathrm{T}}^{\mathrm{T}} \mathbf{T}_{(m)}(t)
\end{aligned}
\tag{5.6}
$$

where, T denotes transpose.

Then integration of the time function $f(t)$ with respect to t, referring to Eq. (4.20), we have

$$
\int f(t)\, \mathrm{d}t \approx \left[\mathbf{C}_{\mathrm{S}}^{\mathrm{T}} + \frac{1}{2} \mathbf{C}_{\mathrm{T}}^{\mathrm{T}} \right] \left[\mathbf{P1ss}_{(m)} \mathbf{S}_{(m)} + \mathbf{P1st}_{(m)} \mathbf{T}_{(m)} \right] \tag{5.7}
$$

5.2 Repeated Integration Using First Order HF Domain Integration Matrices

Already we know that

$$
\int \mathbf{S}_{(m)} \mathrm{d}t = \mathbf{P1ss}_{(m)} \mathbf{S}_{(m)} + \mathbf{P1st}_{(m)} \mathbf{T}_{(m)}
$$

and

$$
\int \mathbf{T}_{(m)} \mathrm{d}t = \mathbf{P1ts}_{(m)} \mathbf{S}_{(m)} + \mathbf{P1tt}_{(m)} \mathbf{T}_{(m)}
$$

So we can write

$$
\begin{aligned}
\iint \mathbf{S}_{(m)} \mathrm{d}t &= \mathbf{P1ss}_{(m)} \int \mathbf{S}_{(m)} \mathrm{d}t + \mathbf{P1st}_{(m)} \int \mathbf{T}_{(m)} \mathrm{d}t \\
&= \mathbf{P1ss}_{(m)}^2 \mathbf{S}_{(m)} + \mathbf{P1ss}_{(m)} \mathbf{P1st}_{(m)} \mathbf{T}_{(m)} \\
&\quad + \mathbf{P1st}_{(m)} \mathbf{P1ts}_{(m)} \mathbf{S}_{(m)} + \mathbf{P1st}_{(m)} \mathbf{P1tt}_{(m)} \mathbf{T}_{(m)}
\end{aligned}
\tag{5.8}
$$

Using the relations (5.3) and (5.5) in (5.8), we get

$$
\iint \mathbf{S}_{(m)} \mathrm{d}t = \left[\mathbf{P1ss}_{(m)} + \frac{\mathbf{P1st}_{(m)}}{2} \right] \left[\mathbf{P1ss}_{(m)} \mathbf{S}_{(m)} + \mathbf{P1st}_{(m)} \mathbf{T}_{(m)} \right] = \mathbf{P}_{(m)} \int \mathbf{S}_{(m)} \mathrm{d}t \tag{5.9}
$$

With n times repeated integration of the $\mathbf{S}_{(m)}$ vector, we get

$$\underbrace{\iiint \cdots \int}_{n} \mathbf{S}_{(m)} \mathrm{d}t = \mathbf{P}_{(m)}^{(n-1)} \int \mathbf{S}_{(m)} \mathrm{d}t \qquad \text{where } n = 2, 3, 4, \ldots \tag{5.10}$$

Similarly, repeated integration of $\mathbf{T}_{(m)}$ vector gives

$$\underbrace{\iiint \cdots \int}_{n} \mathbf{T}_{(m)} \mathrm{d}t = \mathbf{P}_{(m)}^{(n-1)} \int \mathbf{T}_{(m)} \mathrm{d}t = \frac{\mathbf{P}_{(m)}^{(n-1)}}{2} \int \mathbf{S}_{(m)} \mathrm{d}t \quad \text{where } n = 2, 3, 4, \ldots \tag{5.11}$$

That is, for n times repeated integration, Eq. (5.11) takes the following form

$$\underbrace{\iiint \cdots \int}_{n} \mathbf{T}_{(m)} \mathrm{d}t = \frac{1}{2} \underbrace{\iiint \cdots \int}_{n} \mathbf{S}_{(m)} \mathrm{d}t \quad \text{for } n = 1, 2, 3, \ldots \tag{5.12}$$

5.3 One-Shot Integration Operational Matrices for Repeated Integration [8]

It is noted from Fig. 4.8 and Table 4.2, that the operation of first order integration using operational matrices **P1ss**, **P1st**, **P1ts**, **P1tt** the result of integration is somewhat approximate. If we carry on repeated integration using these matrices, error will be introduced at each integration stage and such accumulated error may corrupt the final result. Thus, higher order integrations in HF domain may become so corrupted that the effort may lead to a fiasco.

For this reason, we present in the following, one-shot operational matrices of different orders of integration suitable for computation of function integration with improved accuracy.

The basic principle of determination of one-shot operational matrices for integration is elaborated by the following steps:

(i) Integrate the sample-and-hold basis function set repeatedly n times. Find out the samples of the n times integrated curves.
(ii) From these samples, form corresponding sample-and-hold function coefficient row matrices as well as the triangular function coefficient row matrices. That is, the n times integrated sample-and-hold function is expressed in HF domain.
(iii) Integrate the triangular basis function set repeatedly n times. Find out the samples of the n times integrated curves.

(iv) From these samples, form corresponding sample-and-hold function coefficient row matrices and the triangular function coefficient row matrices. That is, the n times integrated triangular function is thus expressed in HF domain.
(v) From the above steps, form one-shot operational matrices of n-th order integration.

5.3.1 One-Shot Operational Matrices for Sample-and-Hold Functions

To improve accuracy of higher order integrations in hybrid function domain, we develop one-shot integration matrices both for the sample-and-hold function set and the triangular function set, since the hybrid function set is comprised of these sets.

As discussed in the earlier section, the integration matrices **P1ss** and **P1st** are essentially the 'one-shot operational matrices for single integration' from SHF set. For multiple integrations, instead of using these matrices repeatedly, one-shot matrices of different orders of integration are derived to obtain improved accuracy. These one-shot matrices from SHF set are presented in the following.

5.3.1.1 Second Order One-Shot Matrices

Referring to Sect. 4.1.1 and Fig. 4.1c, decomposition of first integration $\int S_0\,\mathrm{d}t$ into two ramp functions is shown in Fig. 5.1a. Their subsequent integration produces two parabolic functions as shown in Fig. 5.1b. Finally Fig. 5.2 depicts the resulting function $\iint S_0\,\mathrm{d}t$.

Mathematically it can be represented as,

$$\iint S_0\,\mathrm{d}t = \frac{t^2}{2} - \frac{(t-h)^2}{2}u(t-h).$$

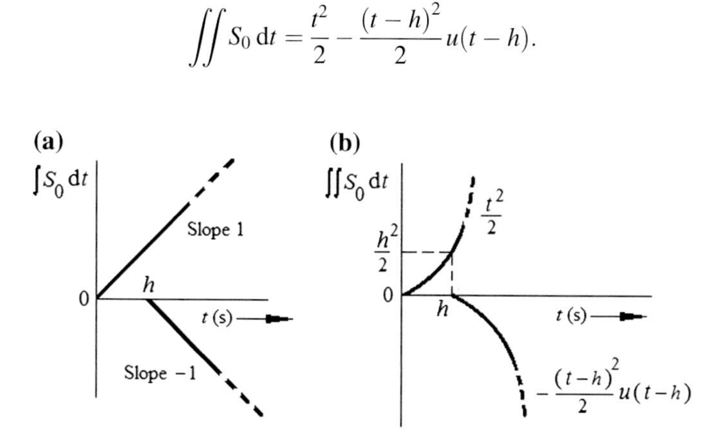

Fig. 5.1 Decomposition of **a** the first integration and **b** the double integration of the first member S_0

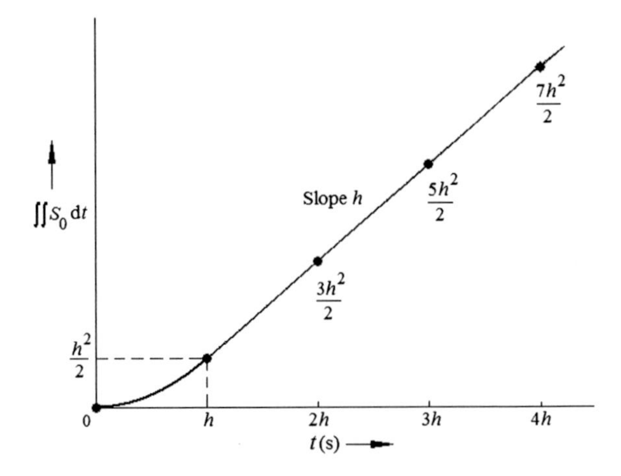

Fig. 5.2 Double integration of the first member S_0 of the SHF set

The samples of the above double integrated function at sampling instants 0, h, $2h$, $3h$ and $4h$ are 0, $\left\{\frac{h^2}{2}-\frac{(h-h)^2}{2}\right\}$, $\left\{\frac{(2h)^2}{2}-\frac{(2h-h)^2}{2}\right\}$, $\left\{\frac{(3h)^2}{2}-\frac{(3h-h)^2}{2}\right\}$, $\left\{\frac{(4h)^2}{2}-\frac{(4h-h)^2}{2}\right\}$ respectively.

As in Eq. (3.6), the first four samples of the function $f(t)=t$ are the coefficients of the SHF components while differences of the consecutive samples provide the coefficients of the TF components.

From these samples we develop the one-shot operational matrices **P2ss** and **P2st** for double integration considering $m = 4$ as

$$\iint \mathbf{S}_{(4)}\mathrm{d}t = \mathbf{P2ss}_{(4)}\mathbf{S}_{(4)} + \mathbf{P2st}_{(4)}\mathbf{T}_{(4)} \tag{5.13}$$

where

$$\mathbf{P2ss}_{(4)} \triangleq \frac{h^2}{2!}\left[\left[0 \quad (1^2-0^2) \quad (2^2-1^2) \quad (3^2-2^2)\right]\right]$$

and

$$\mathbf{P2st}_{(4)} \triangleq \frac{h^2}{2!}\left[\left[1 \quad \{(2^2-1^2)-(1^2-0^2)\} \quad \{(3^2-2^2)-(2^2-1^2)\} \quad \{(4^2-3^2)-(3^2-2^2)\}\right]\right]$$

Following the above pattern, the generalized one-shot operational matrices for m terms for double integration are

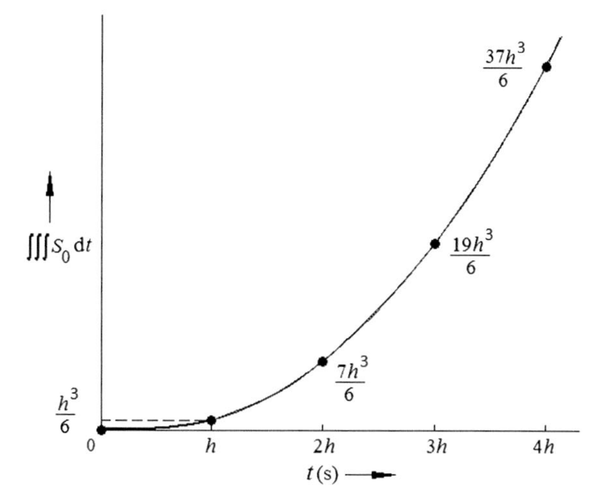

Fig. 5.3 Triple integration of the first member S_0 of the SHF set

$$\left.\begin{aligned}\mathbf{P2ss}_{(m)} &\triangleq \tfrac{h^2}{2!}\left[\left[0(1^2-0^2)(2^2-1^2)(3^2-2^2)\cdots\left\{(m-1)^2-(m-2)^2\right\}\right]\right]_{(m\times m)}\\ &\text{and}\\ \mathbf{P2st}_{(m)} &\triangleq \tfrac{h^2}{2!}\left[\left[1\left\{(2^2-1^2)-(1^2-0^2)\right\}\left\{(3^2-2^2)-(2^2-1^2)\right\}\cdots\right.\right.\\ &\qquad\left.\left.\left\{\left(m^2-(m-1)^2\right)-\left((m-1)^2-(m-2)^2\right)\right\}\right]\right]_{(m\times m)}\end{aligned}\right\} \tag{5.14}$$

5.3.1.2 Third Order One-Shot Matrices

The first member S_0 of the SHF set is integrated thrice and Fig. 5.3 shows the integrated function $\int s_0 \mathrm{d}t$. Mathematically it is represented as

$$\iiint S_0\,\mathrm{d}t = \frac{t^3}{6} - \frac{(t-h)^3}{6}u(t-h)$$

The samples of the resulting function at sampling instants 0, h, $2h$, $3h$ and $4h$ are given as 0, $\left\{\frac{h^3}{6}-\frac{(h-h)^3}{6}\right\}$, $\left\{\frac{(2h)^3}{6}-\frac{(2h-h)^3}{6}\right\}$, $\left\{\frac{(3h)^3}{6}-\frac{(3h-h)^3}{6}\right\}$ and $\left\{\frac{(4h)^3}{6}-\frac{(4h-h)^3}{6}\right\}$ respectively.

From these samples, the one-shot operational matrices **P3ss** and **P3st** for three consecutive integrations, considering $m = 4$ can be developed as follows:

$$\iiint \mathbf{S}_{(4)}\mathrm{d}t = \mathbf{P3ss}_{(4)}\mathbf{S}_{(4)} + \mathbf{P3st}_{(4)}\mathbf{T}_{(4)} \tag{5.15}$$

where,

$$\mathbf{P3ss}_{(4)} \triangleq \frac{h^3}{3!}\left[\!\left[0 \quad (1^3 - 0^3) \quad (2^3 - 1^3) \quad (3^3 - 2^3)\right]\!\right]$$

and

$$\mathbf{P3st}_{(4)} \triangleq \frac{h^3}{3!}\left[\!\left[1 \quad \{(2^3 - 1^3) - (1^3 - 0^3)\} \quad \{(3^3 - 2^3) - (2^3 - 1^3)\} \quad \{(4^3 - 3^3) - (3^3 - 2^3)\}\right]\!\right]$$

For m terms, the generalized one-shot operational matrices for triple integration are

$$\left.\begin{aligned} &\mathbf{P3ss}_{(m)} \triangleq \tfrac{h^3}{3!}\left[\!\left[0(1^3 - 0^3)(2^3 - 1^3)(3^3 - 2^3)\cdots\left\{(m-1)^3 - (m-2)^3\right\}\right]\!\right]_{(m\times m)} \\ &\text{and} \\ &\mathbf{P3st}_{(m)} \triangleq \tfrac{h^3}{3!}\left[\!\left[1\{(2^3 - 1^3) - (1^3 - 0^3)\}\{(3^3 - 2^3) - (2^3 - 1^3)\}\cdots \right.\right. \\ &\qquad\left.\left.\left\{\left(m^3 - (m-1)^3\right) - \left((m-1)^3 - (m-2)^3\right)\right\}\right]\!\right]_{(m\times m)} \end{aligned}\right\} \tag{5.16}$$

5.3.1.3 *n*-th Order One-Shot Matrices

Now considering n times repeated integration, and proceeding via a similar track, we can write the one-shot operational matrices for n times repeated integration for sample-and-hold functions as

$$\left.\begin{aligned} &\mathbf{Pnss}_{(m)} \triangleq \tfrac{h^n}{n!}[\![0(1^n - 0^n)(2^n - 1^n)(3^n - 2^n)\cdots\{(m-1)^n - (m-2)^n\}]\!]_{(m\times m)} \\ &\qquad\text{and} \\ &\mathbf{Pnst}_{(m)} \triangleq \tfrac{h^n}{n!}[\![1\{(2^n - 1^n) - (1^n - 0^n)\}\{(3^n - 2^n) - (2^n - 1^n)\}\cdots \\ &\qquad\{(m^n - (m-1)^n) - ((m-1)^n - (m-2)^n)\}]\!]_{(m\times m)} \end{aligned}\right\} \tag{5.17}$$

where, $n, m \geq 2$.

5.3.2 *One-Shot Operational Matrices for Triangular Functions*

To develop the one-shot integration matrices for the TF set, we proceed as in Sect. 5.3.1.

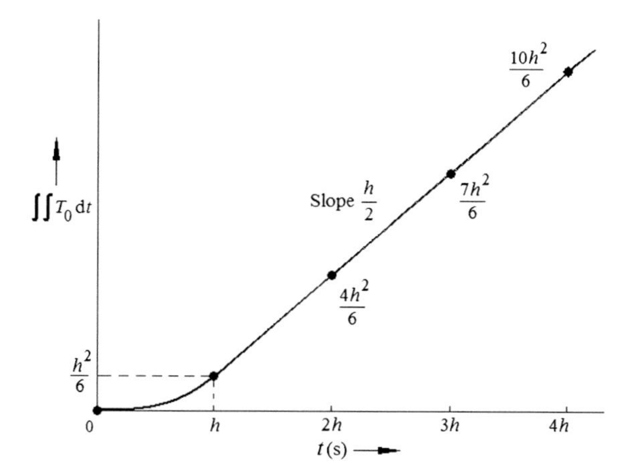

Fig. 5.4 Double integration of the first member T_0 of the triangular function set

The integration matrices **P1ts** and **P1tt** are essentially the 'one-shot operational matrices for single integration' for the TF set. For multiple integrations, instead of using these matrices repeatedly, one-shot matrices of different orders of integration are derived to obtain improved accuracy. These one-shot matrices from TF set are presented in the following.

5.3.2.1 Second Order One-Shot Matrices

The first member T_0 of the triangular function set is integrated twice and Fig. 5.4 shows the integrated function $\iint T_0\,dt$. Mathematically, it can be represented as,

$$\iint T_0\,dt = \frac{1}{h}\frac{t^3}{6} - \frac{1}{h}\frac{(t-h)^3}{6}u(t-h) - \frac{(t-h)^2}{2}u(t-h)$$

The samples of the resulting function at sampling instants 0, h, $2h$, $3h$ and $4h$ are 0, $\left\{\frac{1}{h}\frac{h^3}{6} - \frac{1}{h}\frac{(h-h)^3}{6} - \frac{(h-h)^2}{2}\right\}$, $\left\{\frac{1}{h}\frac{(2h)^3}{6} - \frac{1}{h}\frac{(2h-h)^3}{6} - \frac{(2h-h)^2}{2}\right\}$, $\left\{\frac{1}{h}\frac{(3h)^3}{6} - \frac{1}{h}\frac{(3h-h)^3}{6} - \frac{(3h-h)^2}{2}\right\}$, $\left\{\frac{1}{h}\frac{(4h)^3}{6} - \frac{1}{h}\frac{(4h-h)^3}{6} - \frac{(4h-h)^2}{2}\right\}$ respectively.

From these samples we develop the one-shot operational matrices **P2ts** and **P2tt** for double integration with $m = 4$, as follows.

$$\iint \mathbf{T}_{(4)}dt = \mathbf{P2ts}_{(4)}\mathbf{S}_{(4)} + \mathbf{P2tt}_{(4)}\mathbf{T}_{(4)} \tag{5.18}$$

where,

$$\mathbf{P2ts}_{(4)} \triangleq \frac{h^2}{(2+1)!}\left[\left[0 \quad 1 \quad (2^3-1^3-3.1^2) \quad (3^3-2^3-3.2^2)\right]\right]$$

and

$$\mathbf{P2tt}_{(4)} \triangleq \frac{h^2}{(2+1)!}\left[\left[1 \quad \{(2^3-1^3)-(1^3-0^3)-3.(1^2-0^2)\}\right.\right.$$
$$\left.\left.\{(3^3-2^3)-(2^3-1^3)-3.(2^2-1^2)\}\{(4^3-3^3)-(3^3-2^3)-3.(3^2-2^2)\}\right]\right]$$

For m terms, the generalized one-shot operational matrices for double integration are:

$$\left.\begin{aligned}
\mathbf{P2ts}_{(m)} &\triangleq \frac{h^2}{(2+1)!}\left[\left[0 \quad 1 \quad (2^3-1^3-3.1^2) \quad (3^3-2^3-3.2^2) \quad \cdots \right.\right.\\
&\left.\left.\left\{(m-1)^3-(m-2)^3-3.(m-2)^2\right\}\right]\right]_{(m\times m)}\\
&\text{and}\\
\mathbf{P2tt}_{(m)} &\triangleq \frac{h^2}{(2+1)!}\left[\left[1 \quad \{(2^3-1^3)-(1^3-0^3)-3.(1^2-0^2)\} \quad \{(3^3-2^3)-(2^3-1^3)-3.(2^2-1^2)\} \quad \cdots \right.\right.\\
&\left.\left.\left\{\left(m^3-(m-1)^3\right)-\left((m-1)^3-(m-2)^3\right)-3.\left((m-1)^2-(m-2)^2\right)\right\}\right]\right]_{(m\times m)}
\end{aligned}\right\} \tag{5.19}$$

5.3.2.2 Third Order One-Shot Matrices

The first member T_0 of the triangular function set is repeatedly integrated thrice and Fig. 5.5 shows the integrated function $\int T_0\,dt$, while its magnified view is shown in Fig. 5.6.

Mathematically, $\int T_0 dt$ can be represented as

$$\iiint T_0\,dt = \frac{1}{h}\frac{t^4}{24} - \frac{1}{h}\frac{(t-h)^4}{24}u(t-h) - \frac{(t-h)^3}{6}u(t-h)$$

The samples of the resulting function at sampling instants 0, h, $2h$, $3h$ and $4h$ are 0, $\left\{\frac{1}{h}\frac{h^4}{24} - \frac{1}{h}\frac{(h-h)^4}{24} - \frac{(h-h)^3}{6}\right\}$, $\left\{\frac{1}{h}\frac{(2h)^4}{24} - \frac{1}{h}\frac{(2h-h)^4}{24} - \frac{(2h-h)^3}{6}\right\}$, $\left\{\frac{1}{h}\frac{(3h)^4}{24} - \frac{1}{h}\frac{(3h-h)^4}{24} - \frac{(3h-h)^3}{6}\right\}$, $\left\{\frac{1}{h}\frac{(4h)^4}{24} - \frac{1}{h}\frac{(4h-h)^4}{24} - \frac{(4h-h)^3}{6}\right\}$ respectively.

The one-shot operational matrices **P3ts** and **P3tt** for three times repeated integrations with m = 4 can be developed from these samples as follows.

$$\iiint \mathbf{T}_{(4)}dt = \mathbf{P3ts}_{(4)}\mathbf{S}_{(4)} + \mathbf{P3tt}_{(4)}\mathbf{T}_{(4)} \tag{5.20}$$

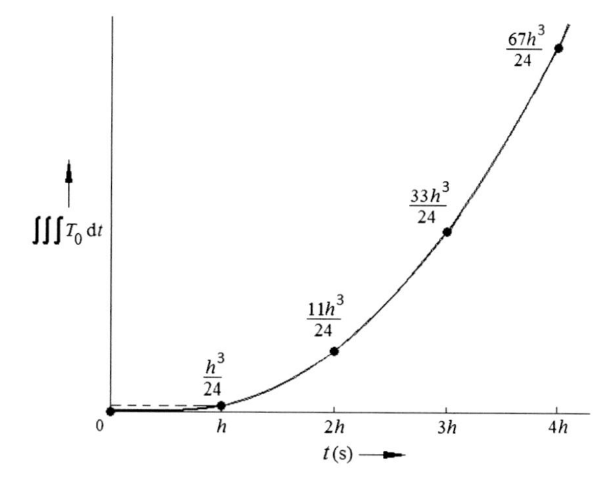

Fig. 5.5 Triple integration of the first member T_0 of the triangular function set

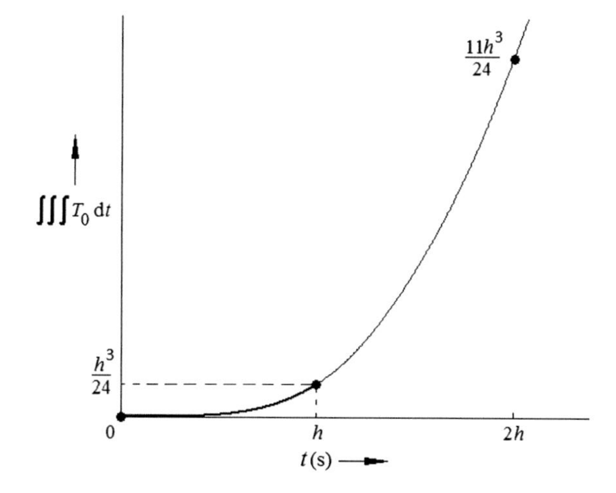

Fig. 5.6 Magnified view of the triple integration of the first member T_0 of the triangular function set

where

$$\mathbf{P3ts}_{(4)} \triangleq \frac{h^3}{(3+1)!}\left[\left[0 \quad 1 \quad (2^4 - 1^4 - 4.1^3) \quad (3^4 - 2^4 - 4.2^3)\right]\right]$$

and

$$\mathbf{P3tt}_4 \triangleq \frac{h^3}{(3+1)!}\left[\left[1 \quad \{(2^4 - 1^4) - (1^4 - 0^4) - 4.(1^3 - 0^3)\}\{(3^4 - 2^4) - (2^4 - 1^4) - 4.(2^3 - 1^3)\}\right.\right.$$
$$\left.\left.\{(4^4 - 3^4) - (3^4 - 2^4) - 4.(3^3 - 2^3)\}\right]\right]$$

For m terms, the generalized one-shot operational matrices for triple integration are

$$\left.\begin{aligned}
&\mathbf{P3ts}_{(m)} \triangleq \tfrac{h^3}{(3+1)!}\left[\left[0\ 1\ \ (2^4-1^4-4.1^3)\ \ (3^4-2^4-4.2^3)\ \cdots\ \left\{(m-1)^4-(m-2)^4-4.(m-2)^3\right\}\right]\right]_{(m\times m)}\\
&\text{and}\\
&\mathbf{P3tt}_{(m)} \triangleq \tfrac{h^3}{(3+1)!}\left[\left[1\ \left\{(2^4-1^4)-(1^4-0^4)-4.(1^3-0^3)\right\}\ \left\{(3^4-2^4)-(2^4-1^4)-4.(2^3-1^3)\right\}\cdots\right.\right.\\
&\qquad\left.\left.\left\{\left(m^4-(m-1)^4\right)-\left((m-1)^4-(m-2)^4\right)-4.\left((m-1)^3-(m-2)^3\right)\right\}\right]\right]_{(m\times m)}
\end{aligned}\right\} \tag{5.21}$$

5.3.2.3 n-th Order One-Shot Matrices

Now considering n times repeated integration, and following a similar track, we can write the one-shot operational matrices for n times repeated integration for triangular functions as

$$\left.\begin{aligned}
&\mathbf{Pnts}_{(m)} \triangleq \tfrac{h^n}{(n+1)!}\left[\left[01\left\{2^{(n+1)}-1^{(n+1)}-(n+1)0.1^n\right\}\ \left\{3^{(n+1)}-2^{(n+1)}-(n+1)0.2^n\right\}\ \cdots\right.\right.\\
&\qquad\left.\left.\left\{(m-1)^{(n+1)}-(m-2)^{(n+1)}-(n+1).(m-2)^n\right\}\right]\right]_{(m\times m)}\\
&\mathbf{Pntt}_{(m)} \triangleq \tfrac{h^n}{(n+1)!}\left[\left[1\left\{\left(2^{(n+1)}-1^{(n+1)}\right)-\left(1^{(n+1)}-0^{(n+1)}\right)-(n+1).(1^n-0^n)\right\}\ \cdots\ \cdots\right.\right.\\
&\left.\left.\left\{\left(m^{(n+1)}-(m-1)^{(n+1)}\right)-\left((m-1)^{(n+1)}-(m-2)^{(n+1)}\right)-(n+1).((m-1)^n-(m-2)^n)\right\}\right]\right]_{(m\times m)}
\end{aligned}\right\} \tag{5.22}$$

where, $n,\ m \geq 2$.

5.3.3 *One-Shot Integration Operational Matrices in HF Domain: A Combination of SHF Domain and TF Domain One-Shot Operational Matrices*

In the Sects. 5.3.1 and 5.3.2, we have constructed the one-shot integration operational matrices both for the sample-and-hold functions and the triangular functions. With the help of these one-shot operational matrices, we can perform repeated integration in HF domain with much better accuracy. First of all, the function to be integrated is described in HF domain and then the one-shot matrices are applied to obtain the desired degree of integration with higher accuracy.

Let us consider a function $f(t)$ to be integrated, is defined in HF domain as

$$f(t) \approx \mathbf{C}_{\mathrm{S}}^{\mathrm{T}}\mathbf{S}_{(m)} + \mathbf{C}_{\mathrm{T}}^{\mathrm{T}}\mathbf{T}_{(m)}$$

After referring to Sects. 5.3.1 and 5.3.2, for getting improved accuracy, using higher-order one-shot operational matrices, the repeated integrations of function $f(t)$ can be expressed as follows:

For double integration

$$\begin{aligned}\iint f(t)\,\mathrm{d}t &\approx \iint \left[\mathbf{C}_{\mathrm{S}}^{\mathrm{T}}\mathbf{S}_{(m)} + \mathbf{C}_{\mathrm{T}}^{\mathrm{T}}\mathbf{T}_{(m)}\right]\,\mathrm{d}t \\ &= \mathbf{C}_{\mathrm{S}}^{\mathrm{T}}\iint \mathbf{S}_{(m)}\,\mathrm{d}t + \mathbf{C}_{\mathrm{T}}^{\mathrm{T}}\iint \mathbf{T}_{(m)}\,\mathrm{d}t \\ &= \mathbf{C}_{\mathrm{S}}^{\mathrm{T}}\left[\mathbf{P2ss}_{(m)}\mathbf{S}_{(m)} + \mathbf{P2st}_{(m)}\mathbf{T}_{(m)}\right] + \mathbf{C}_{\mathrm{T}}^{\mathrm{T}}\left[\mathbf{P2ts}_{(m)}\mathbf{S}_{(m)} + \mathbf{P2tt}_{(m)}\mathbf{T}_{(m)}\right] \\ &= \left[\mathbf{C}_{\mathrm{S}}^{\mathrm{T}}\,\mathbf{P2ss}_{(m)} + \mathbf{C}_{\mathrm{T}}^{\mathrm{T}}\mathbf{P2ts}_{(m)}\right]\mathbf{S}_{(m)} + \left[\mathbf{C}_{\mathrm{S}}^{\mathrm{T}}\,\mathbf{P2st}_{(m)} + \mathbf{C}_{\mathrm{T}}^{\mathrm{T}}\,\mathbf{P2tt}_{(m)}\right]\mathbf{T}_{(m)}\end{aligned} \tag{5.23}$$

Similarly for triple integration of function $f(t)$ can be expressed as

$$\iiint f(t)\,\mathrm{d}t \approx \left[\mathbf{C}_{\mathrm{S}}^{\mathrm{T}}\,\mathbf{P3ss}_{(m)} + \mathbf{C}_{\mathrm{T}}^{\mathrm{T}}\mathbf{P3ts}_{(m)}\right]\mathbf{S}_{(m)} + \left[\mathbf{C}_{\mathrm{S}}^{\mathrm{T}}\,\mathbf{P3st}_{(m)} + \mathbf{C}_{\mathrm{T}}^{\mathrm{T}}\,\mathbf{P3tt}_{(m)}\right]\mathbf{T}_{(m)} \tag{5.24}$$

In a similar track, using higher-order one-shot matrices, the n-times repeated integration can be mathematically expressed as

$$\underbrace{\iiint \cdots \int}_{n} f(t)\,\mathrm{d}t \approx \left[\mathbf{C}_{\mathrm{S}}^{\mathrm{T}}\,\mathbf{Pnss}_{(m)} + \mathbf{C}_{\mathrm{T}}^{\mathrm{T}}\mathbf{Pnts}_{(m)}\right]\mathbf{S}_{(m)} + \left[\mathbf{C}_{\mathrm{S}}^{\mathrm{T}}\,\mathbf{Pnst}_{(m)} + \mathbf{C}_{\mathrm{T}}^{\mathrm{T}}\,\mathbf{Pntt}_{(m)}\right]\mathbf{T}_{(m)} \tag{5.25}$$

The process is illustrated in detail through the following numerical examples.

5.4 Two Theorems [8]

It should be noted that all the operational matrices **P**, **P1ss**, **P1st**, **P1ts**, **P1tt**, **P2ss**, **P2st**, **P2ts**, **P2tt**, **P3ss**, **P3st**, **P3ts**, **P3tt**, …, **Pnss**, **Pnst**, **Pnts**, **Pntt** are of regular upper triangular nature and may be represented by **S** having the following general form:

$$\mathbf{S} = \sum_{n=0}^{j} a_n\,\mathbf{Q}^n$$

where, the delay matrix **Q [12]** is given by

$$\mathbf{Q}_{(m)} \triangleq \underbrace{[\![0 \quad 1 \quad 0 \quad 0 \quad \cdots \quad 0]\!]}_{m\text{ terms}}$$

We present the following two theorems regarding commutative property of matrices of class **S** and its polynomials.

Theorem 1 *If a regular upper triangular matrix* **S** *of order m can be expressed as*

$$\mathbf{S}_{(m)} = \sum_{n=0}^{j} a_n \mathbf{Q}_{(m)}^n$$

*where, the coefficients a_n's are constants, $j \leq (m-1)$, then the product of two matrices **S1** and **S2**, similar to **S**, raised to different integral power p and q, is always commutative and of the form*

$$\mathbf{S1}_{(m)}^p \mathbf{S2}_{(m)}^q = \sum_{n=0}^{k} c_n \mathbf{Q}_{(m)}^n$$

where, the coefficients c_n's are constants and p, q, k are positive integers and $k \leq (m-1)$.

Proof Let,

$$\mathbf{S1}_{(m)} = \sum_{n=0}^{l} a_n \mathbf{Q}_{(m)}^n$$

and

$$\mathbf{S2}_{(m)} = \sum_{n=0}^{s} b_n \mathbf{Q}_{(m)}^n$$

where l, $s \leq (m-1)$ *and* a_n and b_n are constant coefficients.

Then the product $\left[\mathbf{S1}_{(m)}^p \mathbf{S2}_{(m)}^q\right]$ is given by

$$\mathbf{S1}_{(m)}^p \mathbf{S2}_{(m)}^q = \left[\sum_{n=0}^{l} a_n \mathbf{Q}_{(m)}^n\right]^p \left[\sum_{n=0}^{s} b_n \mathbf{Q}_{(m)}^n\right]^q \tag{5.26}$$

The resulting polynomial would contain different coefficients with different powers of $\mathbf{Q}_{(m)}$ from 0 to u (say) where $u \leq (m-1)$, as $\mathbf{Q}_{(m)}$ has the property **[12]**

$$\mathbf{Q}_{(m)}^n = \mathbf{0}_{(m)} \qquad \text{for} \quad n > (m-1)$$

Then Eq. (5.26) reduces to

$$\mathbf{S1}_{(m)}^p \mathbf{S2}_{(m)}^q = \sum_{n=0}^{k} c_n \mathbf{Q}_{(m)}^n \qquad \textit{for} \quad k \leq (m-1) \quad \square$$

Theorem 2 *If a regular upper triangular matrix* $\mathbf{S}_{(m)}$ *of order m can be expressed as*

$$\mathbf{S}_{(m)} = \sum_{n=0}^{v} a_n \mathbf{Q}_{(m)}^n$$

where, the coefficients a_n*'s are constants and* $v \le (m-1)$*, then any polynomial of* $\mathbf{S}_{(m)}$ *can be expressed as*

$$\sum_{n=0}^{j} c_n \mathbf{S}_{(m)}^n = \sum_{n=0}^{k} d_n \mathbf{Q}_{(m)}^n$$

where, c_n*'s,* d_n*'s are constants and* $j,\ k \le (m-1)$

Proof The $(r+1)$th term of the polynomial $\sum_{n=0}^{j} c_n \mathbf{S}_{(m)}^n$ is

$$c_r \mathbf{S}^r = c_r \left[\sum_{n=0}^{v} a_n \mathbf{Q}^n \right]^r = c_r \sum_{n=0}^{w} f_n \mathbf{Q}^n = \sum_{n=0}^{w} g_n \mathbf{Q}^n \tag{5.27}$$

Since $\mathbf{Q}$ has the property

$$\mathbf{Q}_{(m)}^n = \mathbf{0}_{(m)} \qquad \text{for } n > (m-1)$$

Hence, putting $r = n$, Eq. (5.27) can be written as

$$\sum_{0}^{j} c_n \boldsymbol{S}^n = \sum_{0}^{j} \sum_{0}^{w} g_n \boldsymbol{Q}^n = \sum_{0}^{k} d_n \boldsymbol{Q}^n \quad \square$$

Since all the HF domain integration operational matrices are of upper triangular nature having a form similar to $\mathbf{S1}_{(m)}$ or $\mathbf{S2}_{(m)}$ above, by virtue of Theorem 5.1, their products will always be commutative. Also, if higher power of any of the operational matrices is multiplied with any other operational matrix, or its higher power, the product is commutative as well.

These properties are frequently used in the derivations presented later in this chapter.

5.5 Numerical Examples

Let us consider few examples to compare the efficiencies of higher order one-shot integration matrices over the repeated use of first order integration matrices. Example 5.1 will illustrate the process of finding second order integration of function $f(t) = t$. Similarly, Example 5.2 will compare the effectiveness of higher order one-shot operational matrices in case of third order integration of function $f(t) = t$. Finally, Example 5.3 will show the cumulative effect of two higher order one-shot operational matrices for second and third order integrations and will compare the deviations of the samples obtained, with respect to exact values, using two different methods, as explained in previous sections.

5.5.1 Repeated Integration Using First Order Integration Matrices

Example 5.1 (vide ***Appendix B****, Program no.* ***16****)* Consider the function $f(t) = t$.

Integrating twice, we have $\iint f(t) = \frac{t^3}{6}$.

We expand this function directly in HF domain, for $m = 10$ and $T = 1$ s, to obtain

$$\begin{aligned}\iint f(t) \approx & [0.00000000 \quad 0.00016667 \quad 0.00133333 \quad 0.00450000 \quad 0.01066667 \\ & 0.02083333 \quad 0.03600000 \quad 0.05716667 \quad 0.08533333 \quad 0.12150000]\mathbf{S}_{(10)} \\ & + [0.00016667 \quad 0.00116667 \quad 0.00316667 \quad 0.00616667 \quad 0.01016667 \\ & 0.01516667 \quad 0.02116667 \quad 0.02816667 \quad 0.03616667 \quad 0.04516667]\mathbf{T}_{(10)}\end{aligned} \tag{5.28}$$

Now, the expansion of the function $f(t)$ in HF domain, for $m = 10$ and $T = 1$ s, results in

$$\begin{aligned}f(t) \approx & [0 \quad 0.1 \quad 0.2 \quad 0.3 \quad 0.4 \quad 0.5 \quad 0.6 \quad 0.7 \quad 0.8 \quad 0.9]\mathbf{S}_{(10)} \\ & + [0.1 \quad 0.1 \quad 0.1 \quad 0.1 \quad 0.1 \quad 0.1 \quad 0.1 \quad 0.1 \quad 0.1 \quad 0.1]\mathbf{T}_{(10)}\end{aligned}$$

Using second order one-shot integration operational matrices from Eq. (5.23), we obtain the results of double integration of $f(t)$ in HF domain as

$$\begin{aligned}\iint f(t) \approx & [0.00000000 \quad 0.00016667 \quad 0.00133333 \quad 0.00450000 \quad 0.01066667 \\ & 0.02083333 \quad 0.03600000 \quad 0.05716667 \quad 0.08533333 \quad 0.12150000]\mathbf{S}_{(10)} \\ & + [0.00016667 \quad 0.00116667 \quad 0.00316667 \quad 0.00616667 \quad 0.01016667 \\ & 0.01516667 \quad 0.02116667 \quad 0.02816667 \quad 0.03616667 \quad 0.04516667]\mathbf{T}_{(10)}\end{aligned} \tag{5.29}$$

It is noted that the results obtained in Eqs. (5.28) and (5.29) match exactly for this particular case. This is because, the function $f(t) = t$ is a linear function and hybrid functions represent any linear function in an exact manner. But had the function been non-linear, the results would have been very close, though not exact, indicating much less error for repeated integration by the use of one-shot matrices. This is also illustrated by examples to follow.

Let the derivation of the $(i + 1)$-th sample of an HF domain integrated function from its corresponding exact sample obtained via conventional integration, be Δ_i $(i = 0, 1, 2, \ldots, m)$. Then we can define the following two terms as indicators of the efficiency of multiple integration, calling each of them '*deviation index*', we can write

$$\delta_R \triangleq \frac{\sum_{i=1}^{m+1} |\Delta_{Ri}|}{(m+1)} \quad \text{and} \quad \delta_O \triangleq \frac{\sum_{i=1}^{m+1} |\Delta_{Oi}|}{(m+1)}$$

where, Δ_{Ri} is the deviation of the $(i + 1)$-th sample from its exact value for repeated integration, δ_R is the related '*deviation index*' and Δ_{Oi} is the deviation of the $(i + 1)$-th sample from its exact value for one-shot integration, and δ_O is the related '*deviation index*'.

In the following, computational efficiency of the second order one-shot integration operational matrices for different types of standard functions like t, $\exp(-t)$, $\sin(\pi t)$ and $\cos(\pi t)$ are studied rather closely. As expected, the higher order one-shot operational matrices provide better results compared to integration with repeated use of first order operational matrices. Table 5.1 tabulates the deviation indices for different types of standard functions, obtained using these two methods and has proved the effectiveness of using one-shot matrices.

Figure 5.7 translates Table 5.1 into visual form. It shows the deviation indices δ_R and δ_O for double integration of four functions t, $\exp(-t)$, $\sin(\pi t)$ and $\cos(\pi t)$ for $m = 10$ and $T = 1$ s. It is observed that there is difference in deviation indices (δ_R and δ_O) for repeated integration and one-shot integration for each of the four functions as expected. While the difference of the deviation indices is a maximum for the function $\cos(\pi t)$, δ_O being smaller, the same for the function $\sin(\pi t)$ is a minimum where δ_O is larger. This is an oddity which has been removed for triple integration illustrated later.

It is also seen from Fig. 5.7 that for the linear ramp function the deviation index for one-shot integration is zero. This is a specific case for linear functions.

Table 5.1 Deviation indices for double integration of four different functions for $m = 10$ and $T = 1$ s (vide **Appendix B**, Program no. **17**)

Method of integration	Deviation indices δ_R and δ_O for different functions			
	t	$exp(-t)$	$sin(\pi t)$	$cos(\pi t)$
Repeated integration (δ_R)	4.166667e−004	1.880368e−004	7.485343e−004	16.625382e−004
One-shot integration matrices (δ_O)	0.000000e−004	1.142047e−004	8.352104e−004	8.292048e−004

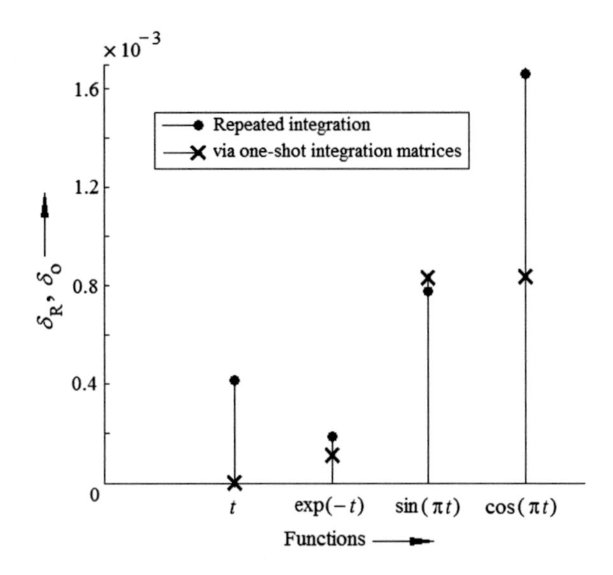

Fig. 5.7 Deviation indices (δ_R and δ_O) for double integration of four different functions, t, $\exp(-t)$, $\sin(\pi t)$ and $\cos(\pi t)$ for m = 10 and T = 1 s

5.5.2 *Higher Order Integration Using One-Shot Operational Matrices*

Example 5.2 Let us take up an example to compare the efficiencies of repeated use of first order integration matrices and third order one-shot integration matrices for the function $f(t)$.

Consider the function

$$f(t) = \iiint t\,\mathrm{d}t \tag{5.30}$$

Let

$$f(t) \approx \mathbf{D}_S^T \mathbf{S}_{(m)} + \mathbf{D}_T^T \mathbf{T}_{(m)} \tag{5.31}$$

where, $\mathbf{D}_S$ and $\mathbf{D}_T$ are HF domain coefficient vectors of $f(t)$ known from the actual samples of the function t.

Also, let

$$t \approx \mathbf{C}_S^T \mathbf{S}_{(m)} + \mathbf{C}_T^T \mathbf{T}_{(m)} \tag{5.32}$$

where, $\mathbf{C}_S$ and $\mathbf{C}_T$ are HF domain coefficient vectors known from actual samples of the function t.

Now we perform triple integration on the RHS of Eq. (5.32) via HF domain and obtain HF domain solution of $f(t)$ for Eq. (5.30).
Considering the discussion in earlier sections, we can determine the result by performing the integration in HF domain in the following two ways:

(i) Using the first order HF domain integration operational matrices $\mathbf{P1ss}_{(m)}$, $\mathbf{P1st}_{(m)}$, $\mathbf{P1ts}_{(m)}$ and $\mathbf{P1tt}_{(m)}$ of Eqs. (5.1) and (5.2).
(ii) Using HF domain one-shot integration operational matrices of third order from Eqs. (5.16) and (5.21).

Finally, the results obtained via above two integration methods are compared with the exact samples of the function $f(t)$ of Eq. (5.30).

5.5.2.1 By Repeated Use of HF Domain 1st Order Integration Matrices $\mathbf{P1ss}_{(m)}$, $\mathbf{P1st}_{(m)}$, $\mathbf{P1ts}_{(m)}$ and $\mathbf{P1tt}_{(m)}$

We know that $\int t\,\mathrm{d}t = \mathbf{C}_\mathrm{S}^\mathrm{T}\int \mathbf{S}_{(m)}\,\mathrm{d}t + \mathbf{C}_\mathrm{T}^\mathrm{T}\int \mathbf{T}_{(m)}\,\mathrm{d}t = \left[\mathbf{C}_\mathrm{S}^\mathrm{T} + \frac{1}{2}\mathbf{C}_\mathrm{T}^\mathrm{T}\right]\mathbf{P}^2 \int \mathbf{S}_{(m)}\,\mathrm{d}t$
Putting these results in Eq. (4.8), we obtain

$$\begin{aligned} f(t) &\approx \left[\mathbf{C}_\mathrm{S}^\mathrm{T} + \frac{1}{2}\mathbf{C}_\mathrm{T}^\mathrm{T}\right]\mathbf{P}^2\ \mathbf{P1ss}_{(m)}\mathbf{S}_{(m)} + \left[\mathbf{C}_\mathrm{S}^\mathrm{T} + \frac{1}{2}\mathbf{C}_\mathrm{T}^\mathrm{T}\right]\mathbf{P}^2\ \mathbf{P1st}_{(m)}\mathbf{T}_{(m)} \\ &\triangleq \mathbf{D}_\mathrm{1S}^\mathrm{T}\mathbf{S}_{(m)} + \mathbf{D}_\mathrm{1T}^\mathrm{T}\mathbf{T}_{(m)} \end{aligned} \tag{5.33}$$

From the two vectors $\mathbf{D}_\mathrm{1S}^\mathrm{T}$ and $\mathbf{D}_\mathrm{1T}^\mathrm{T}$, the samples of $f(t)$ can be computed easily.

5.5.2.2 By Use of HF Domain One-Shot Integration Operational Matrices

Knowing relations (4.8) and (4.15) and the one-shot operational matrices from Eqs. (5.16) and (5.21), we can express RHS of Eq. (5.30) as

$$\begin{aligned} f(t) &\approx \left[\mathbf{C}_\mathrm{S}^\mathrm{T}\mathbf{P3ss}_{(m)} + \mathbf{C}_\mathrm{T}^\mathrm{T}\mathbf{P3ts}_{(m)}\right]\mathbf{S}_{(m)} + \left[\mathbf{C}_\mathrm{S}^\mathrm{T}\mathbf{P3st}_{(m)} + \mathbf{C}_\mathrm{T}^\mathrm{T}\mathbf{P3tt}_{(m)}\right]\mathbf{T}_{(m)} \\ &\triangleq \mathbf{D}_\mathrm{2S}^\mathrm{T}\mathbf{S}_{(m)} + \mathbf{D}_\mathrm{2T}^\mathrm{T}\mathbf{T}_{(m)} \end{aligned} \tag{5.34}$$

From Eq. (5.34), we have

$$\mathbf{D}_\mathrm{2S}^\mathrm{T} = \left[\mathbf{C}_\mathrm{S}^\mathrm{T}\mathbf{P3ss}_{(m)} + \mathbf{C}_\mathrm{T}^\mathrm{T}\mathbf{P3ts}_{(m)}\right]$$

and

$$\mathbf{D}_{2T}^{T} = \left[\mathbf{C}_{S}^{T}\mathbf{P3st}_{(m)} + \mathbf{C}_{T}^{T}\mathbf{P3tt}_{(m)}\right]$$

From the two vectors $\mathbf{D}_{2S}^{T}$ and $\mathbf{D}_{2T}^{T}$ the samples of $f(t)$ can be computed.

After computation of $f(t)$ by the above three methods [using Eqs. (5.31), (5.33) and (5.34)], we get the solution for the coefficients $\mathbf{D}_{S}^{T}, \mathbf{D}_{T}^{T}, \mathbf{D}_{1S}^{T}, \mathbf{D}_{1T}^{T}, \mathbf{D}_{2S}^{T}$, and $\mathbf{D}_{2T}^{T}$ can easily find out the different sets of samples which are compared in Fig. 5.8, which shows that the application of one-shot operational matrices provides much more better approximation compared to the repeated use of first order integration matrices only as shown in Table 5.2.

Like in the case of second order one shot matrices, the computational efficiency of the third order one-shot integration operational matrices are studied for the same standard functions t, $\exp(-t)$, $\sin(\pi t)$ and $\cos(\pi t)$. As expected, the higher order one-shot operational matrices provide better results compared to integration with repeated use of first order operational matrices. Table 5.3 shows the deviation indices for different functions, obtained using these two methods.

Figure 5.9 translates Table 5.3 into visual form. It shows the deviation indices δ_R and δ_O for triple integration of four functions t, $\exp(-t)$, $\sin(\pi t)$ and $\cos(\pi t)$ for $m = 10$ and $T = 1$ s. It is observed that there is difference in deviation indices (δ_R and δ_O) for repeated integration and one-shot integration for each of the four functions as expected. While the difference of the deviation indices is a maximum for the function t, δ_R being larger, the same for the function $\sin(\pi t)$ is a minimum. It is observed that for all the cases, δ_R is larger than δ_O, proving the case for one-shot integration.

It is also seen from Fig. 5.9 that for the linear ramp function the deviation index for one-shot integration is zero. As mentioned earlier, this is a specific case for linear functions.

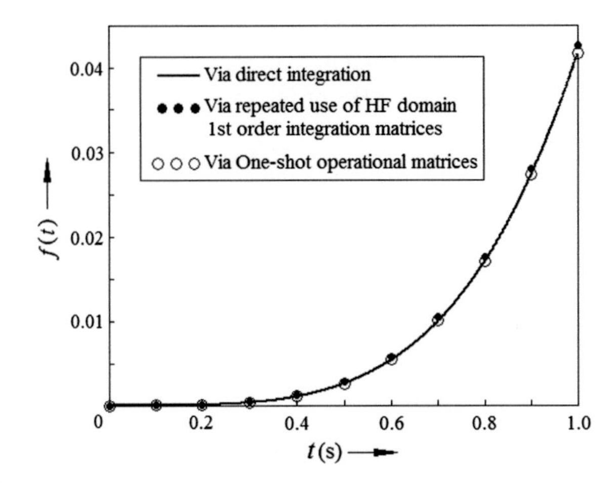

Fig. 5.8 Comparisons of three sets of solutions of the function $f(t)$ (**Example 5.2**) obtained (i) via direct expansion, (ii) via repeated application of integration operational matrices of first order only and (iii) via one-shot operational matrices of third order

Table 5.2 Comparison of the samples (**Example 5.2**) obtained via three repetitive integration and third order one-shot integration along with the exact samples for $m = 10$ and $T = 1$ s

t(s)	Exact samples (E)	Via repeated integration (R)	Via one-shot matrices (O)	Deviation $\Delta_{Ri} = E - R$	Deviation index δ_R	Deviation $\Delta_{Oi} = E - O$	Deviation index δ_O
0	0.000000	0.000000	0.000000	0.000000	2.917e−004	0.000000	0.000e−004
$\frac{1}{10}$	0.000004	0.000013	0.000004	−0.000008		0.000000	
$\frac{2}{10}$	0.000067	0.000100	0.000067	−0.000033		0.000000	
$\frac{3}{10}$	0.000338	0.000413	0.000338	−0.000075		0.000000	
$\frac{4}{10}$	0.001067	0.001200	0.001067	−0.000133		0.000000	
$\frac{5}{10}$	0.002604	0.002813	0.002604	−0.000208		0.000000	
$\frac{6}{10}$	0.005400	0.005700	0.005400	−0.000300		0.000000	
$\frac{7}{10}$	0.010004	0.010413	0.010004	−0.000408		0.000000	
$\frac{8}{10}$	0.017067	0.017600	0.017067	−0.000533		0.000000	
$\frac{9}{10}$	0.027338	0.028013	0.027338	−0.000675		0.000000	
$\frac{10}{10}$	0.041667	0.042500	0.041667	−0.000833		0.000000	

Table 5.3 Deviation indices for triple integration of four different functions for $m = 10$ and $T = 1$ s

Method of integration	Deviation indices δ_R and δ_O for different functions			
	t	$exp(-t)$	$sin(\pi t)$	$cos(\pi t)$
Repeated integration (δ_R)	2.916667e −004	2.195964e −004	3.316339e −004	4.571223e −004
One-shot integration matrices (δ_O)	0.000000e −004	0.312764e −004	1.945120e −004	2.628472e −004

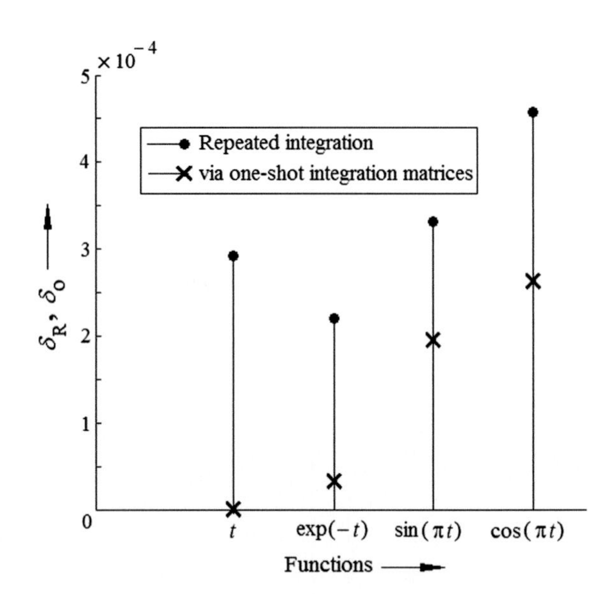

Fig. 5.9 Deviation indices (δ_R and δ_O) for tripple integration of four different functions, t, $\exp(-t)$, $\sin(\pi t)$ and $\cos(\pi t)$ for $m = 10$ and $T = 1$ s

5.5.3 Comparison of Two Integration Methods Involving First, Second and Third Order Integrations

Example 5.3 Now let us consider an example involving single integration, double integration and triple integration to study the overall effect and make comparisons of the results obtained via two integration methods explained earlier.

$$f(t) = \int t\,\mathrm{d}t + \iint t\,\mathrm{d}t + \iiint t\,\mathrm{d}t = \frac{t^2}{2} + \frac{t^3}{6} + \frac{t^4}{24} \tag{5.35}$$

Let

$$f(t) \approx \mathbf{D}_\mathrm{S}^\mathrm{T}\mathbf{S}_{(m)} + \mathbf{D}_\mathrm{T}^\mathrm{T}\mathbf{T}_{(m)} \tag{5.36}$$

where, $\mathbf{D}_S$ and $\mathbf{D}_T$ are the HF domain coefficient vectors of $f(t)$ known from its direct expansion.

Also, let

$$t \approx \mathbf{C}_S^T \mathbf{S}_{(m)} + \mathbf{C}_T^T \mathbf{T}_{(m)} \tag{5.37}$$

where, $\mathbf{C}_S$ and $\mathbf{C}_T$ are HF domain coefficient vectors known from actual samples of the function t.

Now we perform single, double and triple integration on the RHS of Eq. (5.37) via HF domain one-shot operational matrices and substitute the results in Eq. (5.35) to obtain HF domain representation of $f(t)$.

Finally, the results obtained via two integration methods (as discussed earlier) are compared with the exact samples of the function $f(t)$ of Eq. (5.36).

5.5.3.1 By Repeated Use of HF Domain 1st Order Integration Matrices $\mathbf{P1ss}_{(m)}$, $\mathbf{P1st}_{(m)}$, $\mathbf{P1ts}_{(m)}$ and $\mathbf{P1tt}_{(m)}$

We know that

$$\begin{aligned}
\int t\,\mathrm{d}t &= \mathbf{C}_S^T \int \mathbf{S}_{(m)}\,\mathrm{d}t + \mathbf{C}_T^T \int \mathbf{T}_{(m)}\,\mathrm{d}t = \left[\mathbf{C}_S^T + \frac{1}{2}\mathbf{C}_T^T\right] \int \mathbf{S}_{(m)}\,\mathrm{d}t \\
\iint t\,\mathrm{d}t &= \mathbf{C}_S^T \iint \mathbf{S}_{(m)}\,\mathrm{d}t + \mathbf{C}_T^T \iint \mathbf{T}_{(m)}\,\mathrm{d}t = \left[\mathbf{C}_S^T + \frac{1}{2}\mathbf{C}_T^T\right] \mathbf{P} \int \mathbf{S}_{(m)}\,\mathrm{d}t \\
\iiint t\,\mathrm{d}t &= \mathbf{C}_S^T \iiint \mathbf{S}_{(m)}\,\mathrm{d}t + \mathbf{C}_T^T \iiint \mathbf{T}_{(m)}\,\mathrm{d}t = \left[\mathbf{C}_S^T + \frac{1}{2}\mathbf{C}_T^T\right] \mathbf{P}^2 \int \mathbf{S}_{(m)}\,\mathrm{d}t
\end{aligned}$$

Putting these results in RHS of Eq. (5.35) and using Eqs. (5.10) and (5.11), we obtain

$$\begin{aligned}
f(t) &\approx \left[\mathbf{C}_S^T + \frac{1}{2}\mathbf{C}_T^T\right] \left[\mathbf{P}^2 + \mathbf{P} + \mathbf{I}\right] \mathbf{P1ss}_{(m)} \mathbf{S}_{(m)} + \left[\mathbf{C}_S^T + \frac{1}{2}\mathbf{C}_T^T\right] \left[\mathbf{P}^2 + \mathbf{P} + \mathbf{I}\right] \mathbf{P1st}_{(m)} \mathbf{T}_{(m)} \\
&\triangleq \mathbf{D}_{1S}^T \mathbf{S}_{(m)} + \mathbf{D}_{1T}^T \mathbf{T}_{(m)}
\end{aligned} \tag{5.38}$$

From the two vectors $\mathbf{D}_{1S}^T$ and $\mathbf{D}_{1T}^T$ the samples of $f(t)$ can be computed easily.

5.5.3.2 By Use of HF Domain One-Shot Integration Operational Matrices

Knowing the one-shot operational matrices from Eqs. (5.23) and (5.24), we can express RHS of Eq. (5.35) as

$$\begin{aligned} f(t) &\approx \left[\mathbf{C}_S^T\mathbf{P1ss}_{(m)} + \mathbf{C}_T^T\mathbf{P1ts}_{(m)}\right]\mathbf{S}_{(m)} + \left[\mathbf{C}_S^T\mathbf{P1st}_{(m)} + \mathbf{C}_T^T\mathbf{P1tt}_{(m)}\right]\mathbf{T}_{(m)} \\ &\quad + \left[\mathbf{C}_S^T\mathbf{P2ss}_{(m)} + \mathbf{C}_T^T\mathbf{P2ts}_{(m)}\right]\mathbf{S}_{(m)} + \left[\mathbf{C}_S^T\mathbf{P2st}_{(m)} + \mathbf{C}_T^T\mathbf{P2tt}_{(m)}\right]\mathbf{T}_{(m)} \\ &\quad + \left[\mathbf{C}_S^T\mathbf{P3ss}_{(m)} + \mathbf{C}_T^T\mathbf{P3ts}_{(m)}\right]\mathbf{S}_{(m)} + \left[\mathbf{C}_S^T\mathbf{P3st}_{(m)} + \mathbf{C}_T^T\mathbf{P3tt}_{(m)}\right]\mathbf{T}_{(m)} \\ &\triangleq \mathbf{D}_{2S}^T\mathbf{S}_{(m)} + \mathbf{D}_{2T}^T\mathbf{T}_{(m)} \end{aligned} \tag{5.39}$$

From Eq. (5.39), rearranging coefficients of $\mathbf{S}_{(m)}$, we have

$$\mathbf{D}_{2S}^T = \left[\mathbf{C}_S^T\mathbf{P1ss}_{(m)} + \mathbf{C}_T^T\mathbf{P1ts}_{(m)}\right] + \left[\mathbf{C}_S^T\mathbf{P2ss}_{(m)} + \mathbf{C}_T^T\mathbf{P2ts}_{(m)}\right] + \left[\mathbf{C}_S^T\mathbf{P3ss}_{(m)} + \mathbf{C}_T^T\mathbf{P3ts}_{(m)}\right]$$

Rearranging coefficients of $\mathbf{T}_{(m)}$, we get

$$\mathbf{D}_{2T}^T = \left[\mathbf{C}_S^T\mathbf{P1st}_{(m)} + \mathbf{C}_T^T\mathbf{P1tt}_{(m)}\right] + \left[\mathbf{C}_S^T\mathbf{P2st}_{(m)} + \mathbf{C}_T^T\mathbf{P2tt}_{(m)}\right] + \left[\mathbf{C}_S^T\mathbf{P3st}_{(m)} + \mathbf{C}_T^T\mathbf{P3tt}_{(m)}\right]$$

From the two vectors $\mathbf{D}_{2S}^T$ and $\mathbf{D}_{2T}^T$ the samples of $f(t)$ can be computed.

After computation of $f(t)$ by the above three methods [using Eqs. (5.36), (5.38) and (5.39)], we get the solution for the coefficients $\mathbf{D}_S^T$, $\mathbf{D}_T^T, \mathbf{D}_{1S}^T, \mathbf{D}_{1T}^T, \mathbf{D}_{2S}^T$, and $\mathbf{D}_{2T}^T$ and can easily find out the different sets of samples which are compared in Table 5.4.

5.6 Conclusion

In this chapter we have derived one-shot operational matrices of different orders in HF domain and the same have been employed for multiple integration. Finally, the generalized form of such matrices for n times repeated integration having the dimension $(m \times m)$ have been derived.

For evaluating multiple integrals, the one-shot operational matrices have been proved to be more efficient and they produced much more accurate results compared to the method of repeated use of the first order integration matrices.

Few examples are treated to compare the results obtained via repeated use of the first order operational matrices and using higher order one-shot operational matrices. The results are presented in Fig. 5.8 and Tables 5.2 and 5.4 to compare them closely. The maximum deviation with respect to exact solution for the samples obtained via one-shot integration matrices for second and third order integrations are found to be −0.138778e−016 and −0.111022e−015, vide Tables 5.2 and 5.4.

Table 5.4 Comparison of the samples (**Example 5.3**) obtained via repetitive integration and one-shot integration along with the exact samples for $m = 10$ and $T = 1$ s

t(s)	Exact samples (E)	Via repeated integration (R)	Via one-shot matrices (O)	Deviation $\Delta_{Ri} = E - R$	Deviation index δ_R	Deviation $\Delta_{Oi} = E - O$	Deviation index δ_O
0	0.000000	0.000000	0.000000	0.000000	7.083e−004	0.000000	0.000e−004
$\frac{1}{10}$	0.005171	0.005263	0.005171	−0.000092		0.000000	
$\frac{2}{10}$	0.021400	0.021600	0.021400	−0.000200		0.000000	
$\frac{3}{10}$	0.049838	0.050163	0.049838	−0.000325		0.000000	
$\frac{4}{10}$	0.091733	0.092200	0.091733	−0.000467		0.000000	
$\frac{5}{10}$	0.148438	0.149063	0.148438	−0.000625		0.000000	
$\frac{6}{10}$	0.221400	0.222200	0.221400	−0.000800		0.000000	
$\frac{7}{10}$	0.312171	0.313163	0.312171	−0.000992		0.000000	
$\frac{8}{10}$	0.422400	0.423600	0.422400	−0.001200		0.000000	
$\frac{9}{10}$	0.553838	0.555263	0.553838	−0.001425		0.000000	
$\frac{10}{10}$	0.708333	0.710000	0.708333	−0.001667		0.000000	

However, for the samples obtained via repeated use of first order integration operational matrices, maximum deviations, in terms of magnitudes, for second and third order integration turns out to be −0.833333e−003 and −1.666667e−003, vide Tables 5.2 and 5.4.

From Figs. 5.7 and 5.9, we observe that for most of the cases, while computing via first order integration operational matrices, the deviation indices for four different standard functions are much larger than that of one-shot matrices. Hence, it implies that for multiple integrations of any linear or non-linear function, the use of one-shot operational matrices provide highly accurate results.

References

1. Rao, G.P.: Piecewise constant orthogonal functions and their applications in systems and control, LNC1S, vol. 55. Springer, Berlin (1983)
2. Jiang, J.H., Schaufelberger, W.: Block pulse functions and their application in control system, LNCIS, vol. 179. Springer, Berlin (1992)
3. Deb, A., Sarkar, G., Bhattacharjee, M., Sen, S.: All integrator approach to linear SISO control system analysis using block pulse function (BPF). J. Franklin Instt. **334B**(2), 319–335 (1997)
4. Deb, A., Sarkar, G., Bhattacharjee, M., Sen, S.: A new set of piecewise constant orthogonal functions for the analysis of linear SISO systems with sample-and-hold. J. Franklin Instt. **335B** (2), 333–358 (1998)
5. Deb, A., Dasgupta, A., Sarkar, G.: A new set of orthogonal functions and its application to the analysis of dynamic systems. J. Franklin Instt. **343**(1), 1–26 (2006)
6. Deb, A., Sarkar, G., Sengupta, A.: Triangular orthogonal functions for the analysis of continuous time systems. Anthem Press, London (2011)
7. Deb, A., Sarkar, G., Ganguly, A., Biswas, A.: Approximation, integration and differentiation of time functions using a set of orthogonal hybrid functions (HF) and their application to solution of first order differential equations. Appl. Math. Comput. **218**(9), 4731–4759 (2012)
8. Deb, A., Ganguly, A., Sarkar, G., Biswas, A.: Numerical solution of third order linear differential equations using generalized one-shot operational matrices in orthogonal hybrid function (HF) domain. Appl. Math. Comput. **219**(4), 1485–1514 (2012)

Chapter 6
Linear Differential Equations

Abstract This chapter is devoted to linear differential equations. That is, it presents the solution of first order differential equations using both HF domain differentiation operational matrices and integration operational matrices. Higher order differential equations are also solved via the same first order operational matrices, and again employing one-shot integration matrices. The results are compared by way of treating five examples. Eleven figures are presented as illustration of the HF domain techniques.

The main tool for tackling differential equations in the modern age is the numerical analysis, and to be explicit, numerical integration. Differential equations, in general, have a wide range of varieties [1–3] along with different degrees of difficulties. For handling differential equations arising out of modern complex systems, numerical analysis is the forerunner of all solution techniques and modern day algorithms and number crunching capability of computers help in solving varieties of such equations to obtain practical solutions avoiding numerical instability. Work by Butcher [4] gives an exhaustive overview of numerical methods for solving ordinary differential equations. The 4th order Runge-Kutta method has undergone many improvements and modifications discussed by Butcher [2].

Differential equations having oscillatory solutions need special techniques for obtaining reasonable solution within tolerable error limits. Simos's [5] work on modified Runge-Kutta methods for the numerical solution of ODEs with oscillating solutions tackles simultaneous first order ODE's to obtain the required solution.

In control theory, essentially we handle differential equations of different forms and different orders. Any method based upon numerical techniques for solving such equations is of interest in modern control theory and applications.

For more than three decades, solution of differential equations as well as integral equations was also attempted by employing piecewise constant basis functions (PCBF) [6] like Walsh functions, block pulse functions [7] etc. In such attempts function approximation plays a pivotal role because the initial error in function approximation is propagated in a cumulative manner in different stages of computations. Apart from orthogonal functions, orthogonal polynomials have also played their important role [8] in this area.

A. Deb et al., *Analysis and Identification of Time-Invariant Systems, Time-Varying Systems, and Multi-Delay Systems using Orthogonal Hybrid Functions*,
Studies in Systems, Decision and Control 46, DOI 10.1007/978-3-319-26684-8_6

Now solution of differential equations is attempted using both differentiation and integration operational matrices.

6.1 Solution of Linear Differential Equations Using HF Domain Differentiation Operational Matrices

If we try to solve any differential equation with the operational matrices $\mathbf{D}_\mathrm{S}$ and $\mathbf{D}_\mathrm{T}$, (vide Eqs. 4.32 and 4.33), the attempt is met with a permanent difficulty: The samples of the unknown function, say $x(t)$, are required as elements of both the differentiation matrices. Obviously, any such attempt is certain to fail, because these samples of $x(t)$ are yet to be derived as the solution of the differential equation. However, the use of integral operational matrices to solve the problem do not have such difficulty.

Now, we employ the concept of numerical differentiation to solve a first order differential equation and derive the necessary theory.

Let us consider the following first order non-homogeneous differential equation.

$$\dot{g}(t) + ag(t) = b \tag{6.1}$$

where a and b are constants and $g(0) = 0$.

With m component functions in HF domain, we can express $g(t)$ in the following form as in Eq. (2.12). That is

$$\begin{aligned} g(t) &\approx [\,c_0 \quad c_1 \quad c_2 \quad \cdots \quad c_{m-1}\,]\mathbf{S}_{(m)} \\ &\quad + [\,(c_1 - c_0) \quad (c_2 - c_1) \quad (c_3 - c_2) \quad \cdots \quad (c_m - c_{m-1})\,]\mathbf{T}_{(m)} \\ &\triangleq \mathbf{C}_\mathrm{S}^\mathrm{T}\mathbf{S}_{(m)} + \mathbf{C}_\mathrm{T}^\mathrm{T}\mathbf{T}_{(m)} \end{aligned} \tag{6.2}$$

Also, following Eq. (4.31), $\dot{g}(t)$ may be expressed as

$$\begin{aligned} \dot{g}(t) &\approx \frac{1}{h}[\,(c_1 - c_0) \quad (c_2 - c_1) \quad (c_3 - c_2) \quad \cdots \quad (c_m - c_{m-1})\,]\mathbf{S}_{(m)} \\ &\quad + \frac{1}{h}[\,\{(c_2 - c_1) - (c_1 - c_0)\} \quad \{(c_3 - c_2) - (c_2 - c_1)\} \\ &\qquad \cdots \quad \{(c_{m+1} - c_m) - (c_m - c_{m-1})\}\,]\mathbf{T}_{(m)} \\ &\triangleq \frac{1}{h}\mathbf{C}_\mathrm{T}^\mathrm{T}\mathbf{S}_{(m)} + \frac{1}{h}\mathbf{C}_\mathrm{D}^\mathrm{T}\mathbf{T}_{(m)} \end{aligned} \tag{6.3}$$

where, $\mathbf{C}_\mathrm{D}^\mathrm{T} \triangleq [\,\{(c_2 - c_1) - (c_1 - c_0)\} \quad \{(c_3 - c_2) - (c_2 - c_1)\} \cdots \quad \{(c_{m+1} - c_m) - (c_m - c_{m-1})\}]$

Substituting (6.2) and (6.3) in (6.1), we get

$$\left[\frac{1}{h}\mathbf{C}_{\mathrm{T}}^{\mathrm{T}} + a\mathbf{C}_{\mathrm{S}}^{\mathrm{T}}\right]\mathbf{S}_{(m)} + \left[\frac{1}{h}\mathbf{C}_{\mathrm{D}}^{\mathrm{T}} + a\mathbf{C}_{\mathrm{T}}^{\mathrm{T}}\right]\mathbf{T}_{(m)} = [b \quad \cdots \quad b]\mathbf{S}_{(m)} + [0 \quad 0 \quad \cdots \quad 0]\mathbf{T}_{(m)} \tag{6.4}$$

Equating the like coefficients of the vectors in (6.4), we have

$$\left[\frac{1}{h}\mathbf{C}_{\mathrm{T}}^{\mathrm{T}} + a\,\mathbf{C}_{\mathrm{S}}^{\mathrm{T}}\right] = [b \quad b \quad \cdots \quad b] \tag{6.5}$$

and

$$\left[\frac{1}{h}\mathbf{C}_{\mathrm{D}}^{\mathrm{T}} + a\,\mathbf{C}_{\mathrm{T}}^{\mathrm{T}}\right] = [0 \quad 0 \quad \cdots \quad 0] \tag{6.6}$$

Proceeding further with (6.5), we get

$$\left[\mathbf{C}_{\mathrm{T}}^{\mathrm{T}} + a\,h\mathbf{C}_{\mathrm{S}}^{\mathrm{T}}\right] = [bh \quad bh \quad \cdots \quad bh]$$

or, $[(c_1 - c_0) \quad (c_2 - c_1) \quad \cdots \quad (c_m - c_{m-1})] + a\,h[c_0 \quad c_1 \quad \cdots \quad c_{m-1}] = [bh \quad bh \quad \cdots \quad bh]$ or, $(c_m - c_{m-1}) + ahc_{m-1} = bh$

Thus we obtain the following recursive equation as the solution for the HF domain coefficients of the unknown function $g(t)$ as

$$c_m = bh + (1 - ah)c_{m-1} \tag{6.7}$$

6.1.1 Numerical Examples

Example 6.1 (vide Appendix B, Program no. 18) Consider the non- homogeneous first order differential equation

$$\dot{g}_1(t) + 0.5g_1(t) = 1.25, \quad \text{where, } g_1(0) = 0 \tag{6.8}$$

The exact solution of (6.8) is

$$g_1(t) = 2.5[1 - \exp(-0.5t)] \tag{6.9}$$

The direct expansion of $g_1(t)$ in HF domain, for T = 1 s and m = 4, can be expressed as

$$g_1(t) \approx [0.00000000 \quad 0.29375774 \quad 0.55299804 \quad 0.78177680]\mathbf{S}_{(m)} + [0.29375774 \quad 0.25924030 \quad 0.22877876 \quad 0.20189655]\mathbf{T}_{(m)}$$

Whereas using the recursive relation of (6.7), for $T = 1$ s and $m = 4$, the HF domain expansion of $g_1(t)$ may be written as

$$g_1(t) \approx [0.00000000 \quad 0.31250000 \quad 0.58593750 \quad 0.82519531]\mathbf{S}_{(m)} + [0.31250000 \quad 0.27343750 \quad 0.23925781 \quad 0.20935059]\mathbf{T}_{(m)}$$

The exact solution of $g_1(t)$ expressed in HF domain is compared with the HF domain solution of (6.8) obtained using the recursive relation (6.7) in the above two expressions and in Fig. 6.1a. Figure 6.1b shows the same result with better accuracy due to increased m.

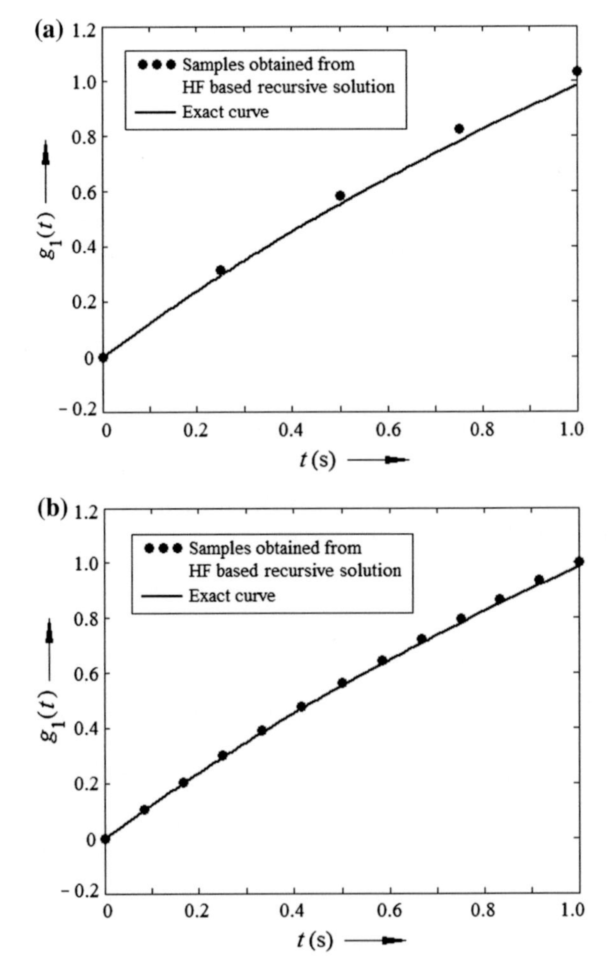

Fig. 6.1 Solution of Example 6.1 in hybrid function (HF) domain, using recursive relation (6.7) for **a** $m = 4$, $T = 1$ s and **b** $m = 12$, $T = 1$ s, along with the exact solution $g_1(t)$ of Eq. (6.8) (vide Appendix B, Program no. 18)

6.2 Solution of Linear Differential Equations Using HF Domain Integration Operational Matrices

Solving any differential equation in HF domain, provides the extra advantage that the differential equation is converted into a simple algebraic equation. This obviously reduces computational burden. Moreover, the HF domain analysis technique works with time samples of functions, meaning the whole analysis is carried out in time domain. So the final solution of the differential equation is also obtained directly in time domain.

We start with Eq. (6.1) and integrate it to get

$$g(t) - g(0)u(t) + a\int g(t)\,\mathrm{d}t = b\int u(t)\,\mathrm{d}t \tag{6.10}$$

Expanding each of the functions $g(t)$, $g(0)u(t)$ and $u(t)$ in hybrid function domain with m terms, we have

$$g(t) \approx \mathbf{C}_\mathrm{S}^\mathrm{T}\mathbf{S}_{(m)} + \mathbf{C}_\mathrm{T}^\mathrm{T}\mathbf{T}_{(m)} \tag{6.11}$$

$$\begin{aligned} g(0)u(t) &= g(0)\Big[\underbrace{1\quad 1\quad \cdots\quad 1\quad 1}_{m\ \text{terms}}\Big]\mathbf{S}_{(m)} + g(0)\Big[\underbrace{0\quad 0\quad \cdots\quad 0\quad 0}_{m\ \text{terms}}\Big]\mathbf{T}_{(m)} \\ &\triangleq g(0)\left[\mathbf{U}_\mathrm{S}^\mathrm{T}\mathbf{S}_{(m)} + \mathbf{Z}_\mathrm{T}^\mathrm{T}\mathbf{T}_{(m)}\right] \end{aligned} \tag{6.12}$$

where, $\mathbf{U}_\mathrm{S}^\mathrm{T} \triangleq [1\quad 1\quad \cdots\quad 1\quad 1]_{(1\times m)}$, $\mathbf{Z}_\mathrm{T}^\mathrm{T} \triangleq [0\quad 0\quad \cdots\quad 0\quad 0]_{(1\times m)}$ and

$$u(t) = \left[\mathbf{U}_\mathrm{S}^\mathrm{T}\mathbf{S}_{(m)} + \mathbf{Z}_\mathrm{T}^\mathrm{T}\mathbf{T}_{(m)}\right] \tag{6.13}$$

In the following, we drop the subscript (m) for simplicity.

Substituting Eqs. (6.11) to (6.13) in Eq. (6.10), we have

$$\mathbf{C}_\mathrm{S}^\mathrm{T}\mathbf{S} + \mathbf{C}_\mathrm{T}^\mathrm{T}\mathbf{T} - g(0)\left[\mathbf{U}_\mathrm{S}^\mathrm{T}\mathbf{S} + \mathbf{Z}_\mathrm{T}^\mathrm{T}\mathbf{T}\right] + a\int\left[\mathbf{C}_\mathrm{S}^\mathrm{T}\mathbf{S} + \mathbf{C}_\mathrm{T}^\mathrm{T}\mathbf{T}\right]\mathrm{d}t = b\int\left[\mathbf{U}_\mathrm{S}^\mathrm{T}\mathbf{S} + \mathbf{Z}_\mathrm{T}^\mathrm{T}\mathbf{T}\right]\mathrm{d}t \tag{6.14}$$

or

$$\mathbf{C}_\mathrm{S}^\mathrm{T}\mathbf{S} + \mathbf{C}_\mathrm{T}^\mathrm{T}\mathbf{T} - g(0)\mathbf{U}_\mathrm{S}^\mathrm{T}\mathbf{S} + a\,\mathbf{C}_\mathrm{S}^\mathrm{T}\int\mathbf{S}\,\mathrm{d}t + a\,\mathbf{C}_\mathrm{T}^\mathrm{T}\int\mathbf{T}\,\mathrm{d}t = b\mathbf{U}_\mathrm{S}^\mathrm{T}\int\mathbf{S}\,\mathrm{d}t \tag{6.15}$$

Using the operational matrices for integration in SHF and TF domain from Eqs. (4.9) and (4.18), we can write

$$\int \mathbf{S}_{(m)}\,\mathrm{d}t = \mathbf{P1ss}_{(m)}\mathbf{S}_{(m)} + \mathbf{P1st}_{(m)}\mathbf{T}_{(m)}$$

$$\int \mathbf{T}_{(m)}\,\mathrm{d}t = \mathbf{P1ts}_{(m)}\mathbf{S}_{(m)} + \mathbf{P1tt}_{(m)}\mathbf{T}_{(m)}$$

Employing (4.9) and (4.18) in (6.15) we get

$$\begin{aligned}&\mathbf{C}_{\mathrm{S}}^{\mathrm{T}}\mathbf{S} + \mathbf{C}_{\mathrm{T}}^{\mathrm{T}}\mathbf{T} - g(0)\mathbf{U}_{\mathrm{S}}^{\mathrm{T}}\mathbf{S} + a\,\mathbf{C}_{\mathrm{S}}^{\mathrm{T}}[\mathbf{P1ss}\,\mathbf{S} + \mathbf{P1st}\,\mathbf{T}] + a\,\mathbf{C}_{\mathrm{T}}^{\mathrm{T}}[\mathbf{P1ts}\,\mathbf{S} + \mathbf{P1tt}\,\mathbf{T}]\\&= b\mathbf{U}_{\mathrm{S}}^{\mathrm{T}}[\mathbf{P1ss}\,\mathbf{S} + \mathbf{P1st}\,\mathbf{T}]\end{aligned}$$

Equating the like coefficients of two vectors **S** and **T**, we have

$$\mathbf{C}_{\mathrm{S}}^{\mathrm{T}}[\mathbf{I} + a\,\mathbf{P1ss}] - g(0)\mathbf{U}_{\mathrm{S}}^{\mathrm{T}} + \frac{1}{2}a\,\mathbf{C}_{\mathrm{T}}^{\mathrm{T}}\,\mathbf{P1ss} = b\mathbf{U}_{\mathrm{S}}^{\mathrm{T}}\,\mathbf{P1ss} \tag{6.16}$$

where, **I** is the identity matrix of order m, and,

$$\left(1 + \frac{1}{2}ah\right)\mathbf{C}_{\mathrm{T}}^{\mathrm{T}} + ah\,\mathbf{C}_{\mathrm{S}}^{\mathrm{T}} = bh\,\mathbf{U}_{\mathrm{S}}^{\mathrm{T}} \tag{6.17}$$

Using these two Eqs. (6.16) and (6.17), we will solve for the two row matrices $\mathbf{C}_{\mathrm{S}}^{\mathrm{T}}$ and $\mathbf{C}_{\mathrm{T}}^{\mathrm{T}}$.

From Eq. (6.17), putting $\frac{2}{2+ah} = f$, we have

$$\frac{1}{f}\mathbf{C}_{\mathrm{T}}^{\mathrm{T}} + ah\,\mathbf{C}_{\mathrm{S}}^{\mathrm{T}} = bh\,\mathbf{U}_{\mathrm{S}}^{\mathrm{T}}$$

or

$$\mathbf{C}_{\mathrm{T}}^{\mathrm{T}} = bfh\,\mathbf{U}_{\mathrm{S}}^{\mathrm{T}} - afh\,\mathbf{C}_{\mathrm{S}}^{\mathrm{T}} \tag{6.18}$$

Substituting the expression for $\mathbf{C}_{\mathrm{T}}^{\mathrm{T}}$ from Eq. (6.18) into (6.16), we have,

$$\mathbf{C}_{\mathrm{S}}^{\mathrm{T}}[\![1 \quad ah \quad ah \quad \cdots \quad ah]\!] - g(0)\mathbf{U}_{\mathrm{S}}^{\mathrm{T}} + \frac{a}{2}\left[bfh\mathbf{U}_{\mathrm{S}}^{\mathrm{T}} - afh\mathbf{C}_{\mathrm{S}}^{\mathrm{T}}\right]\mathbf{P1ss} = b\mathbf{U}_{\mathrm{S}}^{\mathrm{T}}\mathbf{P1ss}$$

$$\text{or,}\quad \mathbf{C}_{\mathrm{S}}^{\mathrm{T}}[\![1 \quad ah \quad ah \quad \cdots \quad ah]\!] - \frac{a^2fh}{2}\mathbf{C}_{\mathrm{S}}^{\mathrm{T}}\mathbf{P1ss} = b\mathbf{U}_{\mathrm{S}}^{\mathrm{T}}\mathbf{P1ss} - \frac{abfh}{2}\mathbf{U}_{\mathrm{S}}^{\mathrm{T}}\mathbf{P1ss} + g(0)\mathbf{U}_{\mathrm{S}}^{\mathrm{T}}$$

$$\begin{aligned}\text{or,}\quad &\mathbf{C}_{\mathrm{S}}^{\mathrm{T}}[\![1 \quad ah \quad ah \quad \cdots \quad ah]\!] - \mathbf{C}_{\mathrm{S}}^{\mathrm{T}}\left[\!\left[0 \quad \tfrac{a^2fh^2}{2} \quad \tfrac{a^2fh^2}{2} \quad \cdots \quad \tfrac{a^2fh}{2}\right]\!\right]\\&= bh[0 \quad 1 \quad 2 \quad \cdots \quad (m-1)] - \frac{abfh^2}{2}[0 \quad 1 \quad 2 \quad \cdots \quad (m-1)] + g(0)\mathbf{U}_{\mathrm{S}}^{\mathrm{T}}\end{aligned}$$

$$\begin{aligned}\text{or,}\quad &\mathbf{C}_{\mathrm{S}}^{\mathrm{T}}\left[\!\left[1 \quad ah\left(1-\tfrac{afh}{2}\right) \quad ah\left(1-\tfrac{afh}{2}\right) \quad \cdots \quad ah\left(1-\tfrac{afh}{2}\right)\right]\!\right]\\&= bh\left(1 - \frac{afh}{2}\right)[0 \quad 1 \quad 2 \quad \cdots \quad (m-1)] + g(0)\mathbf{U}_{\mathrm{S}}^{\mathrm{T}}\end{aligned}$$

Now

$$\left(1 - \frac{afh}{2}\right) = f$$

Therefore, we can write,

$$\begin{aligned}
&\mathbf{C}_{\mathrm{S}}^{\mathrm{T}}[\![1 \quad afh \quad afh \quad \cdots \quad afh]\!] = bfh[0 \quad 1 \quad 2 \quad \cdots \quad (m-1)] + g(0)\mathbf{U}_{\mathrm{S}}^{\mathrm{T}} \\
&\mathbf{C}_{\mathrm{S}}^{\mathrm{T}} = bfh[0 \quad 1 \quad 2 \quad \cdots \quad (m-1)][\![1 \quad afh \quad afh \quad \cdots \quad afh]\!]^{-1} \\
&\quad + g(0)[1 \quad 1 \quad \cdots \quad 1][\![1 \quad afh \quad afh \quad \cdots \quad afh]\!]^{-1}
\end{aligned} \tag{6.19}$$

In (6.19), the inverse is given by

$$[\![1 \quad afh \quad afh \quad \cdots \quad afh]\!]^{-1} = \begin{bmatrix} 1 & -afh & -afh(1-afh) & -afh(1-afh)^2 & \cdots & -afh(1-afh)^{m-2} \\ & 1 & -afh & -afh(1-afh) & \cdots & -afh(1-afh)^{m-3} \\ & & 1 & -afh & \cdots & -afh(1-afh)^{m-4} \\ & & & \ddots & \cdots & \vdots \\ & & 0 & & 1 & -afh \\ & & & & & 1 \end{bmatrix}$$

Therefore, we can write,

$$\begin{aligned}
\mathbf{C}_{\mathrm{S}}^{\mathrm{T}} = {} & bfh\Big[0 \quad 1 \quad (1 + (1-afh)) \quad 1 + (1-afh) + (1-afh)^2 \quad \cdots \quad \left(1 + \cdots + (1-afh)^{m-3} + (1-afh)^{m-2}\right)\Big] \\
& + g(0)\left[1 \quad (1-afh) \quad (1-afh)^2 \quad \cdots \quad (1-afh)^{m-1}\right]
\end{aligned} \tag{6.20}$$

In (6.20), the r-th element of $\mathbf{C}_{\mathrm{S}}^{\mathrm{T}}$ can be expressed as

$$\mathbf{C}_{\mathrm{S}\,(1,r)}^{\mathrm{T}} = bfh \sum_{n=0}^{r-2} (1-afh)^n + g(0)(1-afh)^{r-1} \quad \text{where, } r = 1, 2, \ldots, m.$$

Now, we substitute the expression of $\mathbf{C}_{\mathrm{S}}^{\mathrm{T}}$ from (6.20) in (6.18) to obtain $\mathbf{C}_{\mathrm{T}}^{\mathrm{T}}$. Hence,

$$\begin{aligned}
\mathbf{C}_{\mathrm{T}}^{\mathrm{T}} &= fh\left[(b - ag(0)) \quad (b-ag(0))(1-afh) \quad (b-ag(0))(1-afh)^2 \quad \cdots \quad (b-ag(0))(1-afh)^{m-1}\right] \\
&= fh(b-ag(0))\left[1 \quad (1-afh) \quad (1-afh)^2 \quad \cdots \quad (1-afh)^{m-1}\right]
\end{aligned}$$

Now, the r-th element of $\mathbf{C}_{\mathrm{T}}^{\mathrm{T}}$ is $\mathbf{C}_{\mathrm{T}\,(1,r)}^{\mathrm{T}} = fh(b-ag(0))(1-afh)^r$ where, $r = 0$, 1, 2, …, m.

Therefore, finally we can write

$$\mathbf{C}_S^T = \Big[g(0) \quad \tfrac{b}{a} + \big(g(0) - \tfrac{b}{a}\big)(1 - afh) \quad \tfrac{b}{a} + \big(g(0) - \tfrac{b}{a}\big)(1 - afh)^2 \\ \cdots \quad \tfrac{b}{a} + \big(g(0) - \tfrac{b}{a}\big)(1 - afh)^{m-1} \Big] \tag{6.21}$$

and

$$\mathbf{C}_T^T = fh(b - ag(0))\left[1 \quad (1 - afh) \quad (1 - afh)^2 \quad \cdots \quad (1 - afh)^{m-1}\right] \tag{6.22}$$

Equations (6.21) and (6.22) provide the required solution for the samples of the unknown function $g(t)$ of Eq. (6.1). It is to be noted that for the solution to exist, h should be selected such that $ah \neq 2$.

From these two equations, we can derive a recursive formula for determining the samples of the solution. If we call the $(m+1)$-th sample of $g(t)$ to be $c_{S,m}$, and since the m-th sample of $g(t)$ is $c_{S,m-1}$, then according to Eqs. (6.20) we can write

$$c_{S,m} = bfh + (1 - afh)c_{S,m-1} \tag{6.23}$$

6.2.1 Numerical Examples

Example 6.2 (vide Appendix B, Program no. 18) Consider the non-homogeneous first order differential equation of Example 6.1 having the solution

$$g_1(t) = 2.5[1 - \exp(-0.5t)]$$

The samples of exact solution of $g_1(t)$ and the samples obtained using Eq. (6.23) in HF domain, for $T = 1$ s and $m = 12$, are tabulated and compared in Table 6.1.

Accuracy of the recursive relation (6.23) is apparent from the curves of the Fig. 6.2. While Fig. 6.3 compares sample values of the solution of Eq. (6.8) with those obtained with two HF domain solutions via relations (6.7) and (6.23), derived from application of the HF domain differential matrices and integration operational matrices respectively. In the latter case, since all the sample points almost overlap, it shows that the relation (6.23) is a shade better than (6.7). However, the choice of use of either of (6.7) or (6.23) depends entirely on the degree of accuracy desired for any first order differential equation.

In Fig. 6.3, since the solution points almost overlap, comparison of the samples of the function $g_1(t)$ is presented in Table 6.1 for better clarity.

Now, it seems fit that the solution obtained via the present method is compared with a standard proven method to assess its credibility. The method most proven and popular is the 4th order Runge-Kutta (RK4) method [5].

Table 6.1 Comparison of samples obtained from exact solution and the results obtained by using Eqs. (6.7) and (6.23) with respective percentage errors for SHF coefficients for Example 6.2 (vide Appendix B, Program no. 18)

t(s)	Direct expansion	SHF domain coefficients using Eq. (6.7)	% Error	SHF domain coefficients using Eq. (6.23)	% Error
0					
	0.00000000	0.00000000	–	0.00000000	–
$\frac{1}{12}$					
	0.10202636	0.10416667	−2.09780051	0.10204082	−0.01417193
$\frac{2}{12}$					
	0.19988896	0.20399306	−2.05318596	0.19991670	−0.01387668
$\frac{3}{12}$					
	0.29375774	0.29966001	−2.00922977	0.29379765	−0.01358562
$\frac{4}{12}$					
	0.38379569	0.39134084	−1.96593054	0.38384673	−0.01329874
$\frac{5}{12}$					
	0.47015913	0.47920164	−1.92328667	0.47022033	−0.01301604
$\frac{6}{12}$					
	0.55299804	0.56340157	−1.88129632	0.55306848	−0.01273750
$\frac{7}{12}$					
	0.63245625	0.64409318	−1.83995747	0.63253507	−0.01246314
$\frac{8}{12}$					
	0.70867172	0.72142263	−1.79926788	0.70875813	−0.01219292
$\frac{9}{12}$					
	0.78177680	0.79553002	−1.75922511	0.78187004	−0.01192685
$\frac{10}{12}$					
	0.85189842	0.86654960	−1.71982651	0.85199780	−0.01166491
$\frac{11}{12}$					
	0.91915834	0.93461003	−1.68106925	0.91926319	−0.01140708
$\frac{12}{12}$					

Its importance as well as span is apparent from the extensive insightful discussions presented in. Hence, RK4 is taken as the benchmark for comparison which is presented in Table 6.2.

It is observed that the order of accuracy for RK4 is slightly better than the proposed recursive Eq. (6.23). A point to be noted is that for RK4 method each iteration requires computation of four equations [2], where as Eq. (6.23) alone is competent to perform the iteration to produce updated values of the solution.

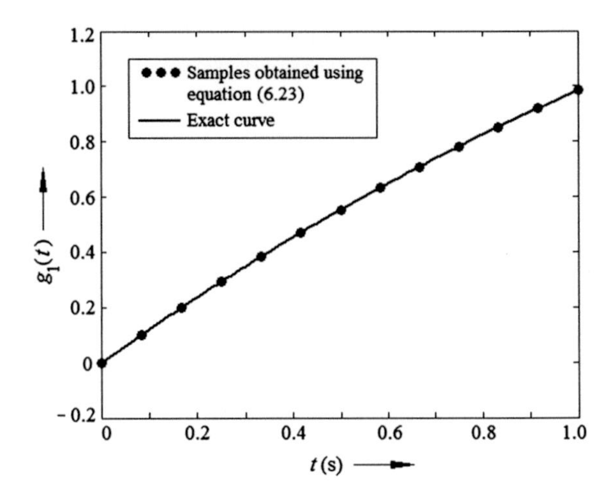

Fig. 6.2 Solution of Eq. (6.8) in hybrid function (HF) domain, using recursive relation (6.23) for $m = 12$, along with direct expansion of $g_1(t)$ via HF (vide Appendix B, Program no. 18)

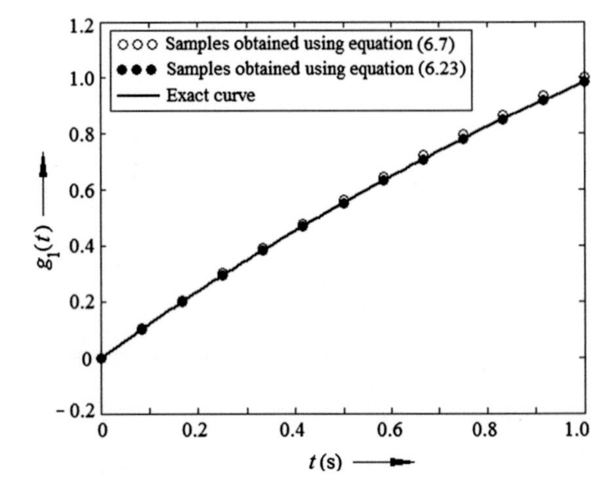

Fig. 6.3 Solution of Eq. (6.8) in hybrid function (HF) domain, using recursive relations (6.7) and (6.23) for $m = 12$, along with direct expansion of $g_1(t)$ via HF (vide Appendix B, Program no. 18)

For a homogeneous first order differential equation of the form

$$\dot{g}_2(t) + a\, g_2(t) = 0 \quad \text{with } g_2(0) = \frac{1}{a} \tag{6.24}$$

Its general solution is of the form $\frac{1}{a}\exp(-at)$.

Thus, for this case when $b = 0$, the recursive relations (6.7) and (6.23) are respectively modified as

$$c_m = (1 - ah)c_{m-1} \tag{6.25}$$

$$c_{\mathrm{S,m}} = (1 - afh)c_{\mathrm{S,m-1}} \tag{6.26}$$

Either of the relations (6.25) or (6.26) may be used for solution of Eq. (6.24).

Table 6.2 Exact samples of the solution $g_1(t)$ of Eq. (6.8) compared with its solutions obtained recursively from relations (6.7) and (6.23) in HF domain with $m = 12$ and $T = 1$ s. For further comparison, Eq. (6.8) has been solved via standard 4th order Runge-Kutta method with the results tabulated in the last column (vide Appendix B, Program no. 19)

t(s)	Exact samples of $g_1(t)$	$g_1(t)$ solved via Eq. (6.7)	$g_1(t)$ solved via Eq. (6.23)	$g_1(t)$ solved via 4th order Runge-Kutta method
0	0.00000000	0.00000000	0.00000000	0.00000000
$\frac{1}{12}$	0.10202636	0.10416667	0.10204082	0.10202635
$\frac{2}{12}$	0.19988896	0.20399306	0.19991670	0.19988896
$\frac{3}{12}$	0.29375774	0.29966001	0.29379765	0.29375774
$\frac{4}{12}$	0.38379569	0.39134084	0.38384673	0.38379568
$\frac{5}{12}$	0.47015913	0.47920164	0.47022033	0.47015912
$\frac{6}{12}$	0.55299804	0.56340157	0.55306848	0.55299803
$\frac{7}{12}$	0.63245625	0.64409318	0.63253507	0.63245624
$\frac{8}{12}$	0.70867172	0.72142263	0.70875813	0.70867171
$\frac{9}{12}$	0.78177680	0.79553002	0.78187004	0.78177678
$\frac{10}{12}$	0.85189842	0.86654960	0.85199780	0.85189841
$\frac{11}{12}$	0.91915834	0.93461003	0.91926319	0.91915833
$\frac{12}{12}$	0.98367335	0.99983461	0.98378306	0.98367333

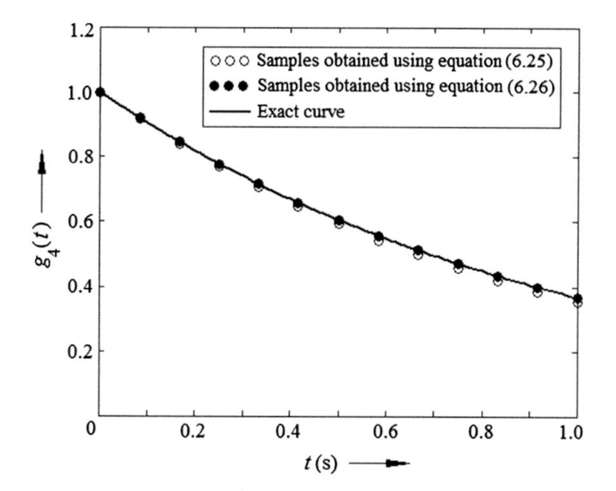

Fig. 6.4 Solution of Eq. (6.24) (homogeneous form of Example 6.2) in hybrid function (HF) domain for $a = 1$, using recursive relations (6.25) and (6.26) for $m = 12$ and $T = 1$ s, along with direct expansion of $g_2(t)$ via HF

Figure 6.4 shows the solution of Eq. (6.24) in hybrid function (HF) domain for $a = 1$, using both the recursive relations (6.25) or (6.26) for $m = 12$ and $T = 1$ s. The figure also plots the samples of the solution obtained via direct expansion in HF domain.

6.3 Solution of Second Order Linear Differential Equations

We present the two methods in the following based upon

(i) The repeated use of first order integration matrices.
(ii) The use of second order one-shot integration matrices.

6.3.1 *Using HF Domain First Order Integration Operational Matrices*

Consider the linear differential equation

$$\ddot{x}(t) + a\,\dot{x}(t) + b\,x(t) = d \tag{6.27}$$

where, a, b and d are positive constants.

Let the initial conditions be $x(0) = k_1$ and $\dot{x}(0) = k_2$.

Integrating Eq. (6.27) twice we get,

$$x(t) + a\int x(t)\,\mathrm{d}t + b\iint x(t)\,\mathrm{d}t = d\iint u(t)\,\mathrm{d}t + (ak_1 + k_2)\int u(t)\,\mathrm{d}t + k_1 u(t) \tag{6.28}$$

Let $(ak_1 + k_2) \triangleq r_2$ and $k_1 \triangleq r_3$.

So, Eq. (6.28) takes the form

$$x(t) + a\int x(t)\,\mathrm{d}t + b\iint x(t)\,\mathrm{d}t = d\iint u(t)\,\mathrm{d}t + r_2\int u(t)\,\mathrm{d}t + r_3\,u(t) \tag{6.29}$$

Expanding all the time functions in m-term HF domain, we have

$$\begin{aligned}&\mathbf{C}_\mathrm{S}^\mathrm{T}\mathbf{S} + \mathbf{C}_\mathrm{T}^\mathrm{T}\mathbf{T} + a\left[\mathbf{C}_\mathrm{S}^\mathrm{T}\int \mathbf{S}\,\mathrm{d}t + \mathbf{C}_\mathrm{T}^\mathrm{T}\int \mathbf{T}\,\mathrm{d}t\right] + b\left[\mathbf{C}_\mathrm{S}^\mathrm{T}\iint \mathbf{S}\,\mathrm{d}t + \mathbf{C}_\mathrm{T}^\mathrm{T}\iint \mathbf{T}\,\mathrm{d}t\right]\\ &= d\,\mathbf{U}_\mathrm{S}^\mathrm{T}\iint \mathbf{S}\,\mathrm{d}t + r_2\,\mathbf{U}_\mathrm{S}^\mathrm{T}\int \mathbf{S}\,\mathrm{d}t + r_3\,\mathbf{U}_\mathrm{S}^\mathrm{T}\mathbf{S}\end{aligned} \tag{6.30}$$

where, $\mathbf{U}_\mathrm{S}^\mathrm{T} = \Big[\underbrace{1 \quad 1 \quad \cdots \quad 1 \quad 1}_{m \text{ terms}}\Big]$

Using (5.10), (5.11) and (5.12), we can write

$$\begin{aligned}
&\mathbf{C}_S^T\mathbf{S} + \mathbf{C}_T^T\mathbf{T} + 2a\left[\mathbf{C}_S^T + \frac{1}{2}\mathbf{C}_T^T\right]\int \mathbf{T}\,dt + 2b\left[\mathbf{C}_S^T + \frac{1}{2}\mathbf{C}_T^T\right]\iint \mathbf{T}\,dt \\
&\quad = 2d\,\mathbf{U}_S^T \iint \mathbf{T}\,dt + 2r_2\,\mathbf{U}_S^T \int \mathbf{T}\,dt + r_3\,\mathbf{U}_S^T\mathbf{S} \\
\text{or,}\quad &\mathbf{C}_S^T\mathbf{S} + \mathbf{C}_T^T\mathbf{T} + 2a\left[\mathbf{C}_S^T + \frac{1}{2}\mathbf{C}_T^T\right]\int \mathbf{T}\,dt + 2b\left[\mathbf{C}_S^T + \frac{1}{2}\mathbf{C}_T^T\right]\mathbf{P}\int \mathbf{T}\,dt \\
&\quad = 2d\,\mathbf{U}_S^T\,\mathbf{P}\int \mathbf{T}\,dt + 2r_2\,\mathbf{U}_S^T \int \mathbf{T}\,dt + r_3\,\mathbf{U}_S^T\mathbf{S}
\end{aligned} \tag{6.31}$$

Rearranging the terms and using the first order integration matrices we can write

$$\begin{aligned}
&\left(\mathbf{C}_S^T - r_3\,\mathbf{U}_S^T\right)\mathbf{S} + \mathbf{C}_T^T\mathbf{T} = \left[2d\,\mathbf{U}_S^T\,\mathbf{P} + 2r_2\,\mathbf{U}_S^T\right]\int \mathbf{T}\,dt - \left[\mathbf{C}_S^T + \frac{1}{2}\mathbf{C}_T^T\right][2a\mathbf{I} + 2b\mathbf{P}]\int \mathbf{T}\,dt \\
\text{or,}\quad &\left(\mathbf{C}_S^T - r_3\,\mathbf{U}_S^T\right)\mathbf{S} + \mathbf{C}_T^T\mathbf{T} = \left[2d\,\mathbf{U}_S^T\,\mathbf{P} + 2r_2\,\mathbf{U}_S^T\right][\mathbf{P1ts}\,\mathbf{S} + \mathbf{P1tt}\,\mathbf{T}] \\
&\qquad - \left[\mathbf{C}_S^T + \frac{1}{2}\mathbf{C}_T^T\right][2a\mathbf{I} + 2b\mathbf{P}][\mathbf{P1ts}\,\mathbf{S} + \mathbf{P1tt}\,\mathbf{T}]
\end{aligned} \tag{6.32}$$

Equating the like coefficients of **S** from both the sides

$$\left(\mathbf{C}_S^T - r_3\,\mathbf{U}_S^T\right) = \left[2d\,\mathbf{U}_S^T\,\mathbf{P} + 2r_2\,\mathbf{U}_S^T\right]\mathbf{P1ts} - \left[\mathbf{C}_S^T + \frac{1}{2}\mathbf{C}_T^T\right][2a\mathbf{I} + 2b\mathbf{P}]\mathbf{P1ts}$$

Let $2d\,\mathbf{P} + 2r_2\,\mathbf{I} \triangleq \mathbf{L}$ and $-(a\mathbf{I} + b\mathbf{P}) \triangleq \mathbf{Q}$
Then

$$\left(\mathbf{C}_S^T - r_3\,\mathbf{U}_S^T\right) = \left[\mathbf{U}_S^T\,\mathbf{L} + 2\,\mathbf{C}_S^T\,\mathbf{Q} + \mathbf{C}_T^T\,\mathbf{Q}\right]\mathbf{P1ts} \tag{6.33}$$

Now, rearranging the coefficients of **T** of Eq. (6.32), we get

$$\begin{aligned}
&\mathbf{C}_T^T = \left[2d\,\mathbf{U}_S^T\,\mathbf{P} + 2r_2\,\mathbf{U}_S^T\right]\mathbf{P1tt} - \left[\mathbf{C}_S^T + \frac{1}{2}\mathbf{C}_T^T\right][2a\mathbf{I} + 2b\mathbf{P}]\mathbf{P1tt} \\
\text{or,}\quad &\mathbf{C}_T^T = \left[\mathbf{U}_S^T\,\mathbf{L} + 2\mathbf{C}_S^T\,\mathbf{Q} + \mathbf{C}_T^T\,\mathbf{Q}\right]\mathbf{P1tt}
\end{aligned} \tag{6.34}$$

From Eqs. (6.33) and (6.34), we can write

$$\left(\mathbf{C}_S^T - r_3\,\mathbf{U}_S^T\right) = \mathbf{C}_T^T\,\mathbf{P1tt}^{-1}\mathbf{P1ts} \tag{6.35}$$

Using the Eq. (5.2) in (6.35), we get

$$\left(\mathbf{C}_S^T - r_3\,\mathbf{U}_S^T\right) = \frac{2}{h}\mathbf{C}_T^T\,\mathbf{P1ts} \tag{6.36}$$

Solving the simultaneous Eqs. (6.33) and (6.36) for $\mathbf{C}_\mathrm{S}^\mathrm{T}$ and $\mathbf{C}_\mathrm{T}^\mathrm{T}$, we have

$$\mathbf{C}_\mathrm{T}^\mathrm{T} = \mathbf{U}_\mathrm{S}^\mathrm{T}[\mathbf{L} + 2r_3\,\mathbf{Q}]\left[\frac{2}{h}\mathbf{I} - \mathbf{Q} - \frac{4}{h}\mathbf{P1ts}\,\mathbf{Q}\right]^{-1} \tag{6.37}$$

$$\mathbf{C}_\mathrm{S}^\mathrm{T} = \frac{2}{h}\mathbf{U}_\mathrm{S}^\mathrm{T}[\mathbf{L} + 2r_3\,\mathbf{Q}]\left[\frac{2}{h}\mathbf{I} - \mathbf{Q} - \frac{4}{h}\mathbf{P1ts}\,\mathbf{Q}\right]^{-1}\mathbf{P1ts} + r_3\mathbf{U}_\mathrm{S}^\mathrm{T} \tag{6.38}$$

6.3.2 *Using HF Domain One-Shot Integration Operational Matrices*

We consider Eq. (6.27) and use one-shot operational matrices for integration of second order differential equation and to determine its solution.

After integrating the Eq. (6.27) twice, now we start from Eq. (6.29). We expand all the time functions in m-term HF domain and employ the one-shot integration matrices.

From Eq. (6.29), we can write

$$\begin{aligned}&\mathbf{C}_\mathrm{S}^\mathrm{T}\mathbf{S} + \mathbf{C}_\mathrm{T}^\mathrm{T}\mathbf{T} + a\left[\mathbf{C}_\mathrm{S}^\mathrm{T}\,\mathbf{P1ss} + \mathbf{C}_\mathrm{T}^\mathrm{T}\,\mathbf{P1ts}\right]\mathbf{S} + a\left[\mathbf{C}_\mathrm{S}^\mathrm{T}\,\mathbf{P1st} + \mathbf{C}_\mathrm{T}^\mathrm{T}\,\mathbf{P1tt}\right]\mathbf{T}\\ &\quad + b\left[\mathbf{C}_\mathrm{S}^\mathrm{T}\,\mathbf{P2ss} + \mathbf{C}_\mathrm{T}^\mathrm{T}\,\mathbf{P2ts}\right]\mathbf{S} + b\left[\mathbf{C}_\mathrm{S}^\mathrm{T}\,\mathbf{P2st} + \mathbf{C}_\mathrm{T}^\mathrm{T}\,\mathbf{P2tt}\right]\mathbf{T}\\ &\quad = d\,\mathbf{U}_\mathrm{S}^\mathrm{T}[\mathbf{P2ss}\,\mathbf{S} + \mathbf{P2st}\,\mathbf{T}] + r_2\,\mathbf{U}_\mathrm{S}^\mathrm{T}[\mathbf{P1ss}\,\mathbf{S} + \mathbf{P1st}\,\mathbf{T}] + r_3\,\mathbf{U}_\mathrm{S}^\mathrm{T}\,\mathbf{S}\end{aligned} \tag{6.39}$$

Rearranging the coefficients of $\mathbf{S}$, we have

$$\begin{aligned}&\mathbf{C}_\mathrm{S}^\mathrm{T} + a\,\mathbf{C}_\mathrm{S}^\mathrm{T}\,\mathbf{P1ss} + a\,\mathbf{C}_\mathrm{T}^\mathrm{T}\,\frac{\mathbf{P1ss}}{2} + b\,\mathbf{C}_\mathrm{S}^\mathrm{T}\,\mathbf{P2ss} + b\,\mathbf{C}_\mathrm{T}^\mathrm{T}\,\frac{\mathbf{P2ss}}{2} = \mathbf{U}_\mathrm{S}^\mathrm{T}[d\,\mathbf{P2ss} + r_2\,\mathbf{P1ss} + r_3\,\mathbf{I}]\\ \text{or,}\quad &\mathbf{C}_\mathrm{S}^\mathrm{T}[\mathbf{I} + a\,\mathbf{P1ss} + b\,\mathbf{P2ss}] + \mathbf{C}_\mathrm{T}^\mathrm{T}\left[a\,\frac{\mathbf{P1ss}}{2} + b\,\frac{\mathbf{P2ss}}{2}\right]\\ &\quad = \mathbf{U}_\mathrm{S}^\mathrm{T}[d\,\mathbf{P2ss} + r_2\,\mathbf{P1ss} + r_3\,\mathbf{I}]\end{aligned} \tag{6.40}$$

Rearranging the coefficients of $\mathbf{T}$, we get

$$\begin{aligned}&\mathbf{C}_\mathrm{T}^\mathrm{T} + a\,\mathbf{C}_\mathrm{S}^\mathrm{T}\,\mathbf{P1st} + a\,\mathbf{C}_\mathrm{T}^\mathrm{T}\,\frac{\mathbf{P1st}}{2} + b\,\mathbf{C}_\mathrm{S}^\mathrm{T}\,\mathbf{P2st} + b\,\mathbf{C}_\mathrm{T}^\mathrm{T}\,\frac{\mathbf{P2st}}{2} = \mathbf{U}_\mathrm{S}^\mathrm{T}[d\,\mathbf{P2st} + r_2\,\mathbf{P1st}]\\ \text{or,}\quad &\mathbf{C}_\mathrm{S}^\mathrm{T}[a\,\mathbf{P1st} + b\,\mathbf{P2st}] + \mathbf{C}_\mathrm{T}^\mathrm{T}\left[\mathbf{I} + a\,\frac{\mathbf{P1st}}{2} + b\,\frac{\mathbf{P2st}}{2}\right] = \mathbf{U}_\mathrm{S}^\mathrm{T}[d\,\mathbf{P2st} + r_2\,\mathbf{P1st}]\end{aligned} \tag{6.41}$$

In Eq. (6.40), let us define

$$\mathbf{I} + a\,\mathbf{P1ss} + b\,\mathbf{P2ss} \triangleq \mathbf{X} \text{ and } a\,\frac{\mathbf{P1ss}}{2} + b\,\frac{\mathbf{P2ss}}{2} \triangleq \mathbf{Y}.$$

Then Eq. (6.40) may be written as

$$\mathbf{C}_S^T\,\mathbf{X} + \mathbf{C}_T^T\,\mathbf{Y} = \mathbf{U}_S^T[d\,\mathbf{P2ss} + r_2\,\mathbf{P1ss} + r_3\,\mathbf{I}] \tag{6.42}$$

In Eq. (6.41), let us define

$$a\,\mathbf{P1st} + b\,\mathbf{P2st} \triangleq \mathbf{W} \text{ and } \mathbf{I} + a\,\frac{\mathbf{P1st}}{2} + b\,\frac{\mathbf{P2st}}{2} \triangleq \mathbf{Z}.$$

Then Eq. (6.41) may be expressed as

$$\text{or, } \mathbf{C}_S^T\,\mathbf{W} + \mathbf{C}_T^T\,\mathbf{Z} = \mathbf{U}_S^T[d\,\mathbf{P2st} + r_2\,\mathbf{P1st}] \tag{6.43}$$

Solving the matrix Eqs. (6.42) and (6.43) for $\mathbf{C}_S^T$ and $\mathbf{C}_T^T$, we get

$$\begin{aligned} &\mathbf{U}_S^T[d\,\mathbf{P2ss} + r_2\,\mathbf{P1ss} + r_3\,\mathbf{I}]\mathbf{X}^{-1} - \mathbf{C}_T^T\,\mathbf{Y}\,\mathbf{X}^{-1} = \mathbf{U}_S^T[d\,\mathbf{P2st} + r_2\,\mathbf{P1st}]\mathbf{W}^{-1} - \mathbf{C}_T^T\,\mathbf{Z}\,\mathbf{W}^{-1} \\ \text{or, } &\mathbf{C}_T^T\left[\mathbf{Y}\,\mathbf{X}^{-1} - \mathbf{Z}\,\mathbf{W}^{-1}\right] = \mathbf{U}_S^T[d\,\mathbf{P2ss} + r_2\,\mathbf{P1ss} + r_3\,\mathbf{I}]\mathbf{X}^{-1} - \mathbf{U}_S^T[d\,\mathbf{P2st} + r_2\,\mathbf{P1st}]\mathbf{W}^{-1} \end{aligned} \tag{6.44}$$

Let $\mathbf{Y}\,\mathbf{X}^{-1} - \mathbf{Z}\,\mathbf{W}^{-1} \triangleq \mathbf{M}_1$
and $\mathbf{U}_S^T[d\,\mathbf{P2ss} + r_2\,\mathbf{P1ss} + r_3\,\mathbf{I}]\mathbf{X}^{-1} - \mathbf{U}_S^T[d\,\mathbf{P2st} + r_2\,\mathbf{P1st}]\mathbf{W}^{-1} \triangleq \mathbf{M}_2$
So Eq. (6.44) becomes

$$\text{or, } \quad \mathbf{C}_T^T = \mathbf{M}_2\,\mathbf{M}_1^{-1} \tag{6.45}$$

Now substituting the expression of $\mathbf{C}_T^T$ in Eq. (6.43), we get

$$\text{or, } \quad \mathbf{C}_S^T = \mathbf{U}_S^T[d\,\mathbf{P2st} + r_2\,\mathbf{P1st}]\mathbf{W}^{-1} - \mathbf{M}_2\,\mathbf{M}_1^{-1}\mathbf{Z}\,\mathbf{W}^{-1} \tag{6.46}$$

Let $\mathbf{M}_2\,\mathbf{M}_1^{-1}\mathbf{Z}\,\mathbf{W}^{-1} \triangleq \mathbf{M}_3$ and $\mathbf{U}_S^T[d\,\mathbf{P2st} + r_2\,\mathbf{P1st}]\mathbf{W}^{-1} \triangleq \mathbf{M}_4$
Therefore Eq. (6.46) may be expressed as

$$\mathbf{C}_S^T = \mathbf{M}_4 - \mathbf{M}_3 \tag{6.47}$$

6.3.3 Numerical Examples

Example 6.3 (vide Appendix B, Program no. 20) Consider the non- homogeneous second order differential equation

$$\ddot{g}_3(t) + 3\dot{g}_3(t) + 2g_3(t) = 2, \text{ with, } \dot{g}_3(0) = -1 \text{ and } g_3(0) = 1 \tag{6.48}$$

The exact solution of (6.48) is

$$g_3(t) = \exp(-2t) - \exp(-t) + 1 \tag{6.49}$$

The samples of exact solution of $g_3(t)$ and the samples obtained using Eqs. (6.38) and (6.47) in HF domain, for $T = 1$ s and $m = 8$, are compared in Fig. 6.5.

Example 6.4 (vide Appendix B, Program no. 20) Consider the homogeneous second order differential equation

$$\ddot{g}_4(t) + 100g_4(t) = 0, \text{ with, } \dot{g}_4(0) = 0 \text{ and } g_4(0) = 2 \tag{6.50}$$

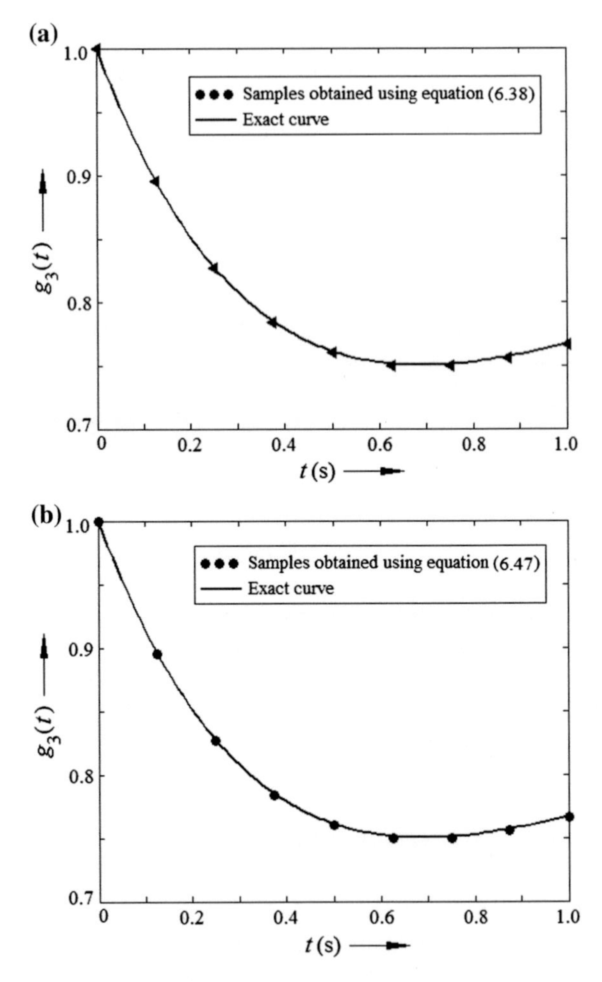

Fig. 6.5 Solution of Example 6.3 in hybrid function (HF) domain for, using **a** first order integration matrices (vide Eq. 6.38) and **b** using one-shot integration operational matrices (vide Eq. 6.47), for $m = 8$ and $T = 1$ s, along with direct expansion of $g_3(t)$ via HF (vide Appendix B, Program no. 20)

Table 6.3 Comparison of sample values of the function $g_3(t)$ of Example 6.3 and its solutions obtained via recursive relations (6.38) and (6.47) in HF domain (vide Appendix B, Program no. 20)

t(s)	Exact samples of $g_3(t)$	$g_3(t)$ solved via Eq. (6.38)	$g_3(t)$ solved via Eq. (6.47)
0	1.00000000	1.00000000	1.00000000
$\frac{1}{8}$	0.89630388	0.89542484	0.89542484
$\frac{2}{8}$	0.82772988	0.82639156	0.82639156
$\frac{3}{8}$	0.78507727	0.78355456	0.78355456
$\frac{4}{8}$	0.76134878	0.75981533	0.75981533
$\frac{5}{8}$	0.75124337	0.74980303	0.74980303
$\frac{6}{8}$	0.75076361	0.74947295	0.74947295
$\frac{7}{8}$	0.75691192	0.75579615	0.75579615
$\frac{8}{8}$	0.76745584	0.76652001	0.76652001

Table 6.4 Comparison of sample values of the function $g_4(t)$ of Example 6.4 and its solutions obtained via recursive relations (6.38) and (6.47) in HF domain (vide Appendix B, Program no. 20)

t(s)	Exact samples of $g_4(t)$	$g_4(t)$ solved via Eq. (6.38)	$g_4(t)$ solved via Eq. (6.47)
0	2.00000000	2.00000000	2.00000000
$\frac{1}{8}$	1.75516512	1.76470588	1.76470588
$\frac{2}{8}$	1.08060461	1.11418685	1.11418685
$\frac{3}{8}$	0.14147440	0.20150621	0.20150621
$\frac{4}{8}$	−0.83229367	−0.75858766	−0.75858766
$\frac{5}{8}$	−1.60228723	−1.54019031	−1.54019031
$\frac{6}{8}$	−1.97998499	−1.95939525	−1.95939525
$\frac{7}{8}$	−1.87291337	−1.91756601	−1.91756601
$\frac{8}{8}$	−1.30728724	−1.42454476	−1.42454476

The exact solution of (6.48) is

$$g_4(t) = 2\cos(10t) \tag{6.51}$$

Hybrid function domain solutions of Example 6.3 and Example 6.4, obtained via Eqs. (6.38) and (6.47) are presented in Tables 6.3 and 6.4 and parallely shown in Figs. 6.5 and 6.6. The results are contrary to the expectation that the use of second order one-shot integration operational matrices will yield better results. In fact, results obtained via repeated integration (vide Eq. 6.38) method and one-shot integration method (vide Eq. 6.47) are the same.

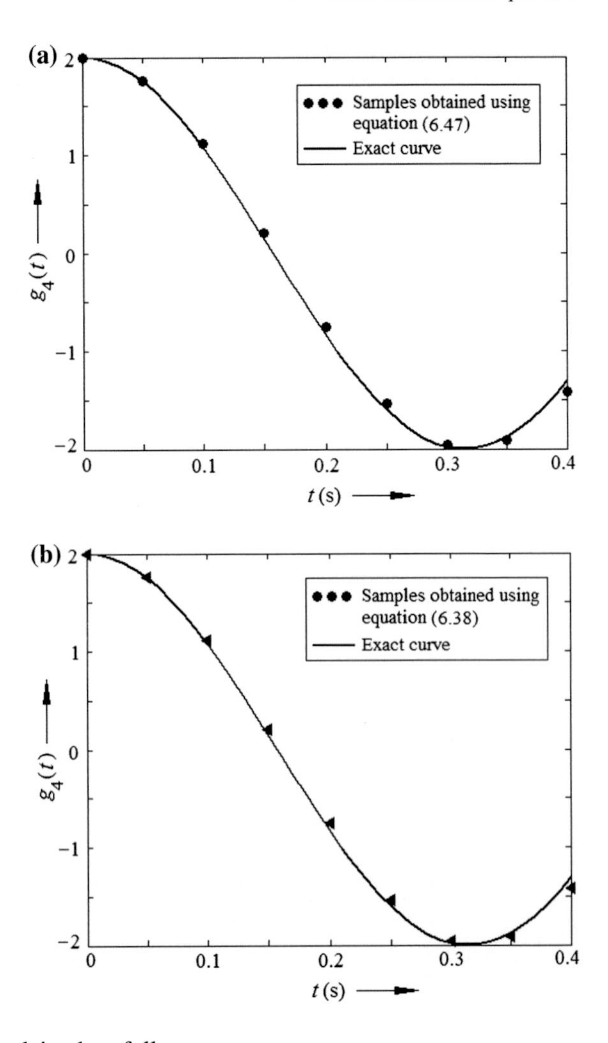

Fig. 6.6 Solution of Example 6.4 in hybrid function (HF) domain for, using **a** first order integration matrices (vide Eq. 6.38) and **b** using one-shot integration operational matrices (vide Eq. 6.47), for $m = 8$ and $T = 0.4$ s, along with direct expansion of $g_4(t)$ via HF (vide Appendix B, Program no. 20)

This paradox may be explained as follows:

(i) For second order repeated integration the result of exact integration at each stage is transformed to hybrid function domain. This should incur error at each stage. However, for the sample-and-hold component, the first stage integration *does not* incur any error, while the integration of the triangular function component does. That is, integration of SHF components incur error at one stage (i.e., the second stage) only, and integration of TF components incur error at both the stages. Thus, with respect to SHF component integration, the incurred error is the same for repeated integration method and one-shot integration method. This has been illustrated in Fig. 6.7.

(ii) For second order repeated integration, the expression for error for the SHF component is $\frac{h^3}{12}$ and that for the TF component is $\left(\frac{h^2}{12}+\frac{h^3}{24}\right)$, where, h is the width of the sub-interval. The error for one-shot integration is $\frac{h^3}{12}$ for the SHF

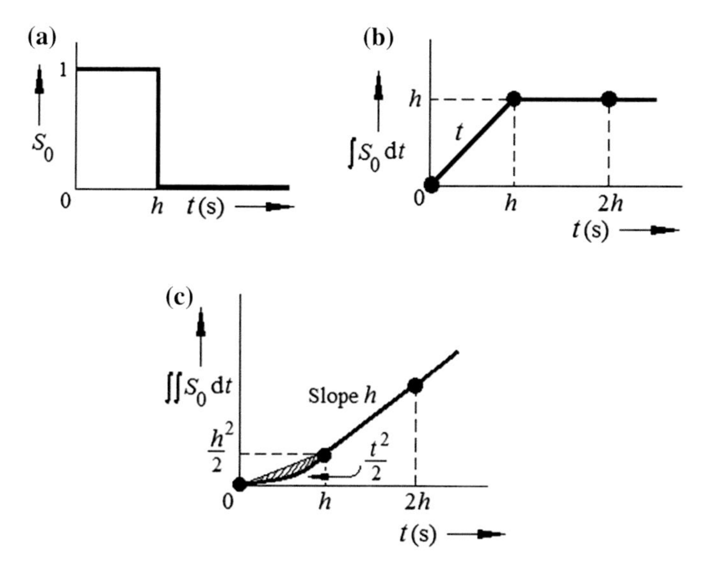

Fig. 6.7 Repeated integration of the first member of the hybrid function set: **a** first member S_0 of the HF set, **b** first integration of S_0 and **c** subsequent integration of the function of figure **b**

component and that for the TF component is $\frac{h^3}{24}$. It is noted that, h being small, the difference in error for second order integration using the two methods are really very small. Thus, the result obtained in Table 6.3 is found to be the same indicating non-superiority of the second order integration matrices in HF domain.

(iii) In view of the above, it is expected that for even higher order integrations, like third order integration, the results obtained via the above two methods will differ appreciably. This is because, for third order repeated integration (say), the integration of the SHF component will incur error at the second and third stages, while such integration of the TF components will incur error at all the three stages. However, for one-shot integration, the error is incurred at one go.

So, now we proceed to the task of comparing the repeated approach and the one-shot approach for third order integration in hybrid function domain.

6.4 Solution of Third Order Linear Differential Equations

We present the two methods in the following based upon

(i) The repeated use of first order integration matrices.
(ii) The use of second order one-shot integration matrices.

6.4.1 Using HF Domain First Order Integration Operational Matrices

Consider the third order linear differential equation

$$\dddot{x}(t) + a\ddot{x}(t) + b\dot{x}(t) + c\,x(t) = d \tag{6.52}$$

where, a, b, c and d are positive constants.

Let the initial conditions be $x(0) = k_1$, $\dot{x}(0) = k_2$ and $\ddot{x}(0) = k_3$.

Integrating Eq. (6.52) thrice we get,

$$\begin{aligned} & x(t) + a\int x(t)\,\mathrm{d}t + b\iint x(t)\,\mathrm{d}t + c\iiint x(t)\,\mathrm{d}t \\ & = d\iiint u(t)\,\mathrm{d}t + (bk_1 + ak_2 + k_3)\iint u(t)\,\mathrm{d}t + (ak_1 + k_2)\int u(t)\,\mathrm{d}t + k_1 u(t) \end{aligned} \tag{6.53}$$

Let $(bk_1 + ak_2 + k_3) \triangleq r_1$, $(ak_1 + k_2) \triangleq r_2$ and $k_1 \triangleq r_3$.

So, Eq. (6.53) takes the form

$$\begin{aligned} & x(t) + a\int x(t)\,\mathrm{d}t + b\iint x(t)\,\mathrm{d}t + c\iiint x(t)\,\mathrm{d}t \\ & = d\iiint u(t)\,\mathrm{d}t + r_1\iint u(t)\,\mathrm{d}t + r_2\int u(t)\,\mathrm{d}t + r_3 u(t) \end{aligned} \tag{6.54}$$

Expanding all the time functions in m-term HF domain, we have

$$\begin{aligned} & \mathbf{C}_\mathrm{S}^\mathrm{T}\mathbf{S}_{(m)} + \mathbf{C}_\mathrm{T}^\mathrm{T}\mathbf{T}_{(m)} + a\left[\mathbf{C}_\mathrm{S}^\mathrm{T}\int \mathbf{S}_{(m)}\,\mathrm{d}t + \mathbf{C}_\mathrm{T}^\mathrm{T}\int \mathbf{T}_{(m)}\,\mathrm{d}t\right] + b\left[\mathbf{C}_\mathrm{S}^\mathrm{T}\iint \mathbf{S}_{(m)}\,\mathrm{d}t + \mathbf{C}_\mathrm{T}^\mathrm{T}\iint \mathbf{T}_{(m)}\,\mathrm{d}t\right] \\ & \quad + c\left[\mathbf{C}_\mathrm{S}^\mathrm{T}\iiint \mathbf{S}_{(m)}\,\mathrm{d}t + \mathbf{C}_\mathrm{T}^\mathrm{T}\iiint \mathbf{T}_{(m)}\,\mathrm{d}t\right] \\ & = d\,\mathbf{U}_\mathrm{S}^\mathrm{T}\iiint \mathbf{S}_{(m)}\,\mathrm{d}t + r_1\,\mathbf{U}_\mathrm{S}^\mathrm{T}\iint \mathbf{S}_{(m)}\,\mathrm{d}t + r_2\,\mathbf{U}_\mathrm{S}^\mathrm{T}\int \mathbf{S}_{(m)}\,\mathrm{d}t + r_3\,\mathbf{U}_\mathrm{S}^\mathrm{T}\mathbf{S}_{(m)} \end{aligned} \tag{6.55}$$

where, $\mathbf{U}_\mathrm{S}^\mathrm{T} = \left[\underbrace{1 \quad 1 \quad \cdots \quad 1 \quad 1}_{m\text{ terms}}\right]$

Using (5.10), (5.11) and (5.12), we can write

$$\begin{aligned}
&\mathbf{C}_\mathrm{S}^\mathrm{T}\mathbf{S}_{(m)} + \mathbf{C}_\mathrm{T}^\mathrm{T}\mathbf{T}_{(m)} + 2a\left[\mathbf{C}_\mathrm{S}^\mathrm{T} + \frac{1}{2}\mathbf{C}_\mathrm{T}^\mathrm{T}\right]\int \mathbf{T}_{(m)}\,\mathrm{d}t + 2b\left[\mathbf{C}_\mathrm{S}^\mathrm{T} + \frac{1}{2}\mathbf{C}_\mathrm{T}^\mathrm{T}\right]\iint \mathbf{T}_{(m)}\,\mathrm{d}t \\
&\quad + 2c\left[\mathbf{C}_\mathrm{S}^\mathrm{T} + \frac{1}{2}\mathbf{C}_\mathrm{T}^\mathrm{T}\right]\iiint \mathbf{T}_{(m)}\,\mathrm{d}t \\
&= 2d\,\mathbf{U}_\mathrm{S}^\mathrm{T}\iiint \mathbf{T}_{(m)}\,\mathrm{d}t + 2r_1\,\mathbf{U}_\mathrm{S}^\mathrm{T}\iint \mathbf{T}_{(m)}\,\mathrm{d}t + 2r_2\,\mathbf{U}_\mathrm{S}^\mathrm{T}\int \mathbf{T}_{(m)}\,\mathrm{d}t + r_3\,\mathbf{U}_\mathrm{S}^\mathrm{T}\mathbf{S}_{(m)} \\
\text{or,}\quad &\mathbf{C}_\mathrm{S}^\mathrm{T}\mathbf{S}_{(m)} + \mathbf{C}_\mathrm{T}^\mathrm{T}\mathbf{T}_{(m)} + 2a\left[\mathbf{C}_\mathrm{S}^\mathrm{T} + \frac{1}{2}\mathbf{C}_\mathrm{T}^\mathrm{T}\right]\int \mathbf{T}_{(m)}\,\mathrm{d}t + 2b\left[\mathbf{C}_\mathrm{S}^\mathrm{T} + \frac{1}{2}\mathbf{C}_\mathrm{T}^\mathrm{T}\right]\mathbf{P}\int \mathbf{T}_{(m)}\,\mathrm{d}t \\
&\quad + 2c\left[\mathbf{C}_\mathrm{S}^\mathrm{T} + \frac{1}{2}\mathbf{C}_\mathrm{T}^\mathrm{T}\right]\mathbf{P}^2\int \mathbf{T}_{(m)}\,\mathrm{d}t \\
&= 2d\,\mathbf{U}_\mathrm{S}^\mathrm{T}\,\mathbf{P}^2\int \mathbf{T}_{(m)}\,\mathrm{d}t + 2r_1\,\mathbf{U}_\mathrm{S}^\mathrm{T}\,\mathbf{P}\int \mathbf{T}_{(m)}\,\mathrm{d}t + 2r_2\,\mathbf{U}_\mathrm{S}^\mathrm{T}\int \mathbf{T}_{(m)}\,\mathrm{d}t + r_3\,\mathbf{U}_\mathrm{S}^\mathrm{T}\mathbf{S}_{(m)}
\end{aligned} \tag{6.56}$$

Rearranging the terms and using the first order integration matrices we can write

$$\begin{aligned}
(\mathbf{C}_\mathrm{S}^\mathrm{T} - r_3\,\mathbf{U}_\mathrm{S}^\mathrm{T})\mathbf{S}_{(m)} + \mathbf{C}_\mathrm{T}^\mathrm{T}\mathbf{T}_{(m)} &= \left[2d\,\mathbf{U}_\mathrm{S}^\mathrm{T}\,\mathbf{P}^2 + 2r_1\,\mathbf{U}_\mathrm{S}^\mathrm{T}\,\mathbf{P} + 2r_2\,\mathbf{U}_\mathrm{S}^\mathrm{T}\right]\int \mathbf{T}_{(m)}\,\mathrm{d}t \\
&\quad - \left[\mathbf{C}_\mathrm{S}^\mathrm{T} + \frac{1}{2}\mathbf{C}_\mathrm{T}^\mathrm{T}\right]\left[2a\mathbf{I} + 2b\mathbf{P} + 2c\mathbf{P}^2\right]\int \mathbf{T}_{(m)}\,\mathrm{d}t \\
\text{or,}\quad (\mathbf{C}_\mathrm{S}^\mathrm{T} - r_3\,\mathbf{U}_\mathrm{S}^\mathrm{T})\mathbf{S}_{(m)} + \mathbf{C}_\mathrm{T}^\mathrm{T}\mathbf{T}_{(m)} &= \left[2d\,\mathbf{U}_\mathrm{S}^\mathrm{T}\,\mathbf{P}^2 + 2r_1\,\mathbf{U}_\mathrm{S}^\mathrm{T}\,\mathbf{P} + 2r_2\,\mathbf{U}_\mathrm{S}^\mathrm{T}\right]\left[\mathbf{P1ts}\,\mathbf{S}_{(m)} + \mathbf{P1tt}\,\mathbf{T}_{(m)}\right] \\
&\quad - \left[\mathbf{C}_\mathrm{S}^\mathrm{T} + \frac{1}{2}\mathbf{C}_\mathrm{T}^\mathrm{T}\right]\left[2a\mathbf{I} + 2b\mathbf{P} + 2c\mathbf{P}^2\right]\left[\mathbf{P1ts}\,\mathbf{S}_{(m)} + \mathbf{P1tt}\,\mathbf{T}_{(m)}\right]
\end{aligned} \tag{6.57}$$

Equating the like coefficients of $\mathbf{S}_{(m)}$ from both the sides

$$\begin{aligned}
\left(\mathbf{C}_\mathrm{S}^\mathrm{T} - r_3\,\mathbf{U}_\mathrm{S}^\mathrm{T}\right) &= \left[2d\,\mathbf{U}_\mathrm{S}^\mathrm{T}\,\mathbf{P}^2 + 2r_1\,\mathbf{U}_\mathrm{S}^\mathrm{T}\,\mathbf{P} + 2r_2\,\mathbf{U}_\mathrm{S}^\mathrm{T}\right]\mathbf{P1ts} \\
&\quad - \left[\mathbf{C}_\mathrm{S}^\mathrm{T} + \frac{1}{2}\mathbf{C}_\mathrm{T}^\mathrm{T}\right]\left[2a\mathbf{I} + 2b\mathbf{P} + 2c\mathbf{P}^2\right]\mathbf{P1ts}
\end{aligned}$$

Let $2d\,\mathbf{P}^2 + 2r_1\,\mathbf{P} + 2r_2\,\mathbf{I} \triangleq \mathbf{L}$ and $-\left(a\mathbf{I} + b\mathbf{P} + c\mathbf{P}^2\right) \triangleq \mathbf{Q}$
Then

$$\left(\mathbf{C}_\mathrm{S}^\mathrm{T} - r_3\,\mathbf{U}_\mathrm{S}^\mathrm{T}\right) = \left[\mathbf{U}_\mathrm{S}^\mathrm{T}\,\mathbf{L} + 2\,\mathbf{C}_\mathrm{S}^\mathrm{T}\,\mathbf{Q} + \mathbf{C}_\mathrm{T}^\mathrm{T}\,\mathbf{Q}\right]\mathbf{P1ts} \tag{6.58}$$

Now, rearranging the coefficients of $\mathbf{T}_{(m)}$ of Eq. (6.57), we get

$$\mathbf{C}_\mathrm{T}^\mathrm{T} = \left[2d\,\mathbf{U}_\mathrm{S}^\mathrm{T}\,\mathbf{P}^2 + 2r_1\,\mathbf{U}_\mathrm{S}^\mathrm{T}\,\mathbf{P} + 2r_2\,\mathbf{U}_\mathrm{S}^\mathrm{T}\right]\mathbf{P1tt} - \left[\mathbf{C}_\mathrm{S}^\mathrm{T} + \frac{1}{2}\mathbf{C}_\mathrm{T}^\mathrm{T}\right]\left[2a\mathbf{I} + 2b\mathbf{P} + 2c\mathbf{P}^2\right]\mathbf{P1tt}$$

or, $$\mathbf{C}_\mathrm{T}^\mathrm{T} = \left[\mathbf{U}_\mathrm{S}^\mathrm{T}\,\mathbf{L} + 2\mathbf{C}_\mathrm{S}^\mathrm{T}\,\mathbf{Q} + \mathbf{C}_\mathrm{T}^\mathrm{T}\,\mathbf{Q}\right]\mathbf{P1tt} \tag{6.59}$$

From Eqs. (6.58) and (6.59), we can write

$$\left(\mathbf{C}_\mathrm{S}^\mathrm{T} - r_3\,\mathbf{U}_\mathrm{S}^\mathrm{T}\right) = \mathbf{C}_\mathrm{T}^\mathrm{T}\,\mathbf{P1tt}^{-1}\mathbf{P1ts} \tag{6.60}$$

Using the Eq. (5.2) in (6.60), we get

$$\left(\mathbf{C}_\mathrm{S}^\mathrm{T} - r_3\,\mathbf{U}_\mathrm{S}^\mathrm{T}\right) = \frac{2}{h}\mathbf{C}_\mathrm{T}^\mathrm{T}\,\mathbf{P1ts} \tag{6.61}$$

Solving the simultaneous Eqs. (6.58) and (6.61) for $\mathbf{C}_\mathrm{S}^\mathrm{T}$ and $\mathbf{C}_\mathrm{T}^\mathrm{T}$, we have

$$\mathbf{C}_\mathrm{T}^\mathrm{T} = \mathbf{U}_\mathrm{S}^\mathrm{T}[\mathbf{L} + 2r_3\,\mathbf{Q}]\left[\frac{2}{h}\mathbf{I} - \mathbf{Q} - \frac{4}{h}\mathbf{P1ts}\,\mathbf{Q}\right]^{-1} \tag{6.62}$$

$$\mathbf{C}_\mathrm{S}^\mathrm{T} = \frac{2}{h}\mathbf{U}_\mathrm{S}^\mathrm{T}[\mathbf{L} + 2r_3\,\mathbf{Q}]\left[\frac{2}{h}\mathbf{I} - \mathbf{Q} - \frac{4}{h}\mathbf{P1ts}\,\mathbf{Q}\right]^{-1}\mathbf{P1ts} + r_3\mathbf{U}_\mathrm{S}^\mathrm{T} \tag{6.63}$$

6.4.2 Using HF Domain One-Shot Integration Operational Matrices

We consider Eq. (6.52) and use one-shot operational matrices for integration of second order differential equation and to determine its solution.

After integrating the Eq. (6.52) twice, now we start from Eq. (6.54). We expand all the time functions in m-term HF domain and employ the one-shot integration matrices.

From Eq. (6.54), we can write

$$\begin{aligned}
&\mathbf{C}_\mathrm{S}^\mathrm{T}\mathbf{S}_{(m)} + \mathbf{C}_\mathrm{T}^\mathrm{T}\mathbf{T}_{(m)} + a\left[\mathbf{C}_\mathrm{S}^\mathrm{T}\,\mathbf{P1ss} + \mathbf{C}_\mathrm{T}^\mathrm{T}\,\mathbf{P1ts}\right]\mathbf{S}_{(m)} + a\left[\mathbf{C}_\mathrm{S}^\mathrm{T}\,\mathbf{P1st} + \mathbf{C}_\mathrm{T}^\mathrm{T}\,\mathbf{P1tt}\right]\mathbf{T}_{(m)}\\
&\quad + b\left[\mathbf{C}_\mathrm{S}^\mathrm{T}\,\mathbf{P2ss} + \mathbf{C}_\mathrm{T}^\mathrm{T}\,\mathbf{P2ts}\right]\mathbf{S}_{(m)} + b\left[\mathbf{C}_\mathrm{S}^\mathrm{T}\,\mathbf{P2st} + \mathbf{C}_\mathrm{T}^\mathrm{T}\,\mathbf{P2tt}\right]\mathbf{T}_{(m)}\\
&\quad + c\left[\mathbf{C}_\mathrm{S}^\mathrm{T}\,\mathbf{P3ss} + \mathbf{C}_\mathrm{T}^\mathrm{T}\,\mathbf{P3ts}\right]\mathbf{S}_{(m)} + c\left[\mathbf{C}_\mathrm{S}^\mathrm{T}\,\mathbf{P3st} + \mathbf{C}_\mathrm{T}^\mathrm{T}\,\mathbf{P3tt}\right]\mathbf{T}_{(m)}\\
&= d\,\mathbf{U}_\mathrm{S}^\mathrm{T}\left[\mathbf{P3ss}\,\mathbf{S}_{(m)} + \mathbf{P3st}\,\mathbf{T}_{(m)}\right] + r_1\,\mathbf{U}_\mathrm{S}^\mathrm{T}\left[\mathbf{P2ss}\,\mathbf{S}_{(m)} + \mathbf{P2st}\,\mathbf{T}_{(m)}\right]\\
&\quad + r_2\,\mathbf{U}_\mathrm{S}^\mathrm{T}\left[\mathbf{P1ss}\,\mathbf{S}_{(m)} + \mathbf{P1st}\,\mathbf{T}_{(m)}\right] + r_3\,\mathbf{U}_\mathrm{S}^\mathrm{T}\,\mathbf{S}_{(m)}
\end{aligned} \tag{6.64}$$

Rearranging the coefficients of $\mathbf{S}_{(m)}$, we have

$$\begin{aligned}
&\mathbf{C}_\mathrm{S}^\mathrm{T} + a\,\mathbf{C}_\mathrm{S}^\mathrm{T}\,\mathbf{P1ss} + a\,\mathbf{C}_\mathrm{T}^\mathrm{T}\,\frac{\mathbf{P1ss}}{2} + b\,\mathbf{C}_\mathrm{S}^\mathrm{T}\,\mathbf{P2ss} + b\,\mathbf{C}_\mathrm{T}^\mathrm{T}\,\frac{\mathbf{P2ss}}{2} + c\,\mathbf{C}_\mathrm{S}^\mathrm{T}\,\mathbf{P3ss} + c\,\mathbf{C}_\mathrm{T}^\mathrm{T}\,\frac{\mathbf{P3ss}}{2}\\
&= \mathbf{U}_\mathrm{S}^\mathrm{T}[d\,\mathbf{P3ss} + r_1\,\mathbf{P2ss} + r_2\,\mathbf{P1ss} + r_3\,\mathbf{I}]\\
\text{or,}\quad &\mathbf{C}_\mathrm{S}^\mathrm{T}[\mathbf{I} + a\,\mathbf{P1ss} + b\,\mathbf{P2ss} + c\,\mathbf{P3ss}] + \mathbf{C}_\mathrm{T}^\mathrm{T}\left[a\,\frac{\mathbf{P1ss}}{2} + b\,\frac{\mathbf{P2ss}}{2} + c\,\frac{\mathbf{P3ss}}{2}\right]\\
&= \mathbf{U}_\mathrm{S}^\mathrm{T}[d\,\mathbf{P3ss} + r_1\,\mathbf{P2ss} + r_2\,\mathbf{P1ss} + r_3\,\mathbf{I}]
\end{aligned} \tag{6.65}$$

Rearranging the coefficients of $\mathbf{T}_{(m)}$, we get

$$\begin{aligned}
&\mathbf{C}_\mathrm{T}^\mathrm{T} + a\,\mathbf{C}_\mathrm{S}^\mathrm{T}\,\mathbf{P1st} + a\,\mathbf{C}_\mathrm{T}^\mathrm{T}\,\frac{\mathbf{P1st}}{2} + b\,\mathbf{C}_\mathrm{S}^\mathrm{T}\,\mathbf{P2st} + b\,\mathbf{C}_\mathrm{T}^\mathrm{T}\,\frac{\mathbf{P2st}}{2} + c\,\mathbf{C}_\mathrm{S}^\mathrm{T}\,\mathbf{P3st} + c\,\mathbf{C}_\mathrm{T}^\mathrm{T}\,\frac{\mathbf{P3st}}{2}\\
&= \mathbf{U}_\mathrm{S}^\mathrm{T}[d\,\mathbf{P3st} + r_1\,\mathbf{P2st} + r_2\,\mathbf{P1st}]\\
\text{or,}\quad &\mathbf{C}_\mathrm{S}^\mathrm{T}[a\,\mathbf{P1st} + b\,\mathbf{P2st} + c\,\mathbf{P3st}] + \mathbf{C}_\mathrm{T}^\mathrm{T}\left[\mathbf{I} + a\,\frac{\mathbf{P1st}}{2} + b\,\frac{\mathbf{P2st}}{2} + c\,\frac{\mathbf{P3st}}{2}\right]\\
&= \mathbf{U}_\mathrm{S}^\mathrm{T}[d\,\mathbf{P3st} + r_1\,\mathbf{P2st} + r_2\,\mathbf{P1st}]
\end{aligned} \tag{6.66}$$

In Eq. (6.65), let us define
$\mathbf{I} + a\,\mathbf{P1ss} + b\,\mathbf{P2ss} + c\,\mathbf{P3ss} \triangleq \mathbf{X}$ and $a\,\frac{\mathbf{P1ss}}{2} + b\,\frac{\mathbf{P2ss}}{2} + c\,\frac{\mathbf{P3ss}}{2} \triangleq \mathbf{Y}$.
Then Eq. (6.65) may be written as

$$\mathbf{C}_\mathrm{S}^\mathrm{T}\,\mathbf{X} + \mathbf{C}_\mathrm{T}^\mathrm{T}\,\mathbf{Y} = \mathbf{U}_\mathrm{S}^\mathrm{T}[d\,\mathbf{P3ss} + r_1\,\mathbf{P2ss} + r_2\,\mathbf{P1ss} + r_3\,\mathbf{I}] \tag{6.67}$$

In Eq. (6.66), let us define
$a\,\mathbf{P1st} + b\,\mathbf{P2st} + c\,\mathbf{P3st} \triangleq \mathbf{W}$ and $\mathbf{I} + a\,\frac{\mathbf{P1st}}{2} + b\,\frac{\mathbf{P2st}}{2} + c\,\frac{\mathbf{P3st}}{2} \triangleq \mathbf{Z}$.
Then Eq. (6.66) may be expressed as

$$\text{or,}\;\; \mathbf{C}_\mathrm{S}^\mathrm{T}\,\mathbf{W} + \mathbf{C}_\mathrm{T}^\mathrm{T}\,\mathbf{Z} = \mathbf{U}_\mathrm{S}^\mathrm{T}[d\,\mathbf{P3st} + r_1\,\mathbf{P2st} + r_2\,\mathbf{P1st}] \tag{6.68}$$

Solving the matrix Eqs. (6.67) and (6.68) for $\mathbf{C}_\mathrm{S}^\mathrm{T}$ and $\mathbf{C}_\mathrm{T}^\mathrm{T}$, we get

$$\begin{aligned}
&\mathbf{U}_\mathrm{S}^\mathrm{T}[d\,\mathbf{P3ss} + r_1\,\mathbf{P2ss} + r_2\,\mathbf{P1ss} + r_3\,\mathbf{I}]\mathbf{X}^{-1} - \mathbf{C}_\mathrm{T}^\mathrm{T}\,\mathbf{Y}\,\mathbf{X}^{-1}\\
&= \mathbf{U}_\mathrm{S}^\mathrm{T}[d\,\mathbf{P3st} + r_1\,\mathbf{P2st} + r_2\,\mathbf{P1st}]\mathbf{W}^{-1} - \mathbf{C}_\mathrm{T}^\mathrm{T}\,\mathbf{Z}\,\mathbf{W}^{-1}\\
\text{or,}\quad &\mathbf{C}_\mathrm{T}^\mathrm{T}\left[\mathbf{Y}\,\mathbf{X}^{-1} - \mathbf{Z}\,\mathbf{W}^{-1}\right] = \mathbf{U}_\mathrm{S}^\mathrm{T}[d\,\mathbf{P3ss} + r_1\,\mathbf{P2ss} + r_2\,\mathbf{P1ss} + r_3\,\mathbf{I}]\mathbf{X}^{-1}\\
&\qquad - \mathbf{U}_\mathrm{S}^\mathrm{T}[d\,\mathbf{P3st} + r_1\,\mathbf{P2st} + r_2\,\mathbf{P1st}]\mathbf{W}^{-1}
\end{aligned} \tag{6.69}$$

Let $\mathbf{Y}\,\mathbf{X}^{-1} - \mathbf{Z}\,\mathbf{W}^{-1} \triangleq \mathbf{M}_1$

and $\mathbf{U}_S^T[d\,\mathbf{P3ss} + r_1\,\mathbf{P2ss} + r_2\,\mathbf{P1ss} + r_3\,\mathbf{I}]\mathbf{X}^{-1} - \mathbf{U}_S^T[d\,\mathbf{P3st} + r_1\,\mathbf{P2st} + r_2\,\mathbf{P1st}]\mathbf{W}^{-1} \triangleq \mathbf{M}_2$

So Eq. (6.69) becomes

$$\text{or,}\ \mathbf{C}_T^T = \mathbf{M}_2\,\mathbf{M}_1^{-1} \tag{6.70}$$

Now substituting the expression of $\mathbf{C}_T^T$ in Eq. (6.68), we get

$$\text{or,}\quad \mathbf{C}_S^T = \mathbf{U}_S^T[d\,\mathbf{P3st} + r_1\,\mathbf{P2st} + r_2\,\mathbf{P1st}]\mathbf{W}^{-1} - \mathbf{M}_2\,\mathbf{M}_1^{-1}\mathbf{Z}\,\mathbf{W}^{-1} \tag{6.71}$$

Let $\mathbf{M}_2\,\mathbf{M}_1^{-1}\mathbf{Z}\,\mathbf{W}^{-1} \triangleq \mathbf{M}_3$ and $\mathbf{U}_S^T[d\,\mathbf{P3st} + r_1\,\mathbf{P2st} + r_2\,\mathbf{P1st}]\mathbf{W}^{-1} \triangleq \mathbf{M}_4$
Therefore Eq. (6.71) may be expressed as

$$\mathbf{C}_S^T = \mathbf{M}_4 - \mathbf{M}_3 \tag{6.72}$$

It is known that inversion of upper or lower triangular matrices can be computed by simple decomposition and multiplication.

6.4.3 Numerical Examples

Example 6.5 Consider the non- homogeneous third order differential equation

$$\begin{aligned} \dddot{g}_5(t) + 3\ddot{g}_5(t) - \dot{g}_5(t) - 3g_5(t) = 0,\\ \text{with,}\ \ddot{g}_5(0) = 22,\ \dot{g}_5(0) = -6 \text{ and } g_5(0) = 6 \end{aligned} \tag{6.73}$$

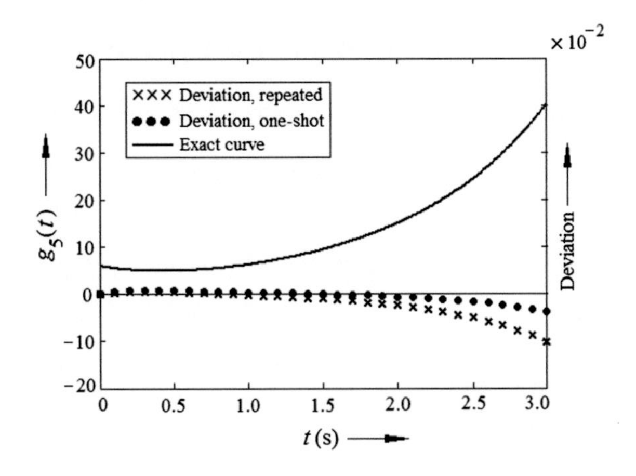

Fig. 6.8 Exact solution of Example 6.5 and comparison of deviation using first order integration matrices (vide Eq. 6.63) and one-shot integration operational matrices (vide Eq. 6.72), for m = 30 and T = 3 s

The exact solution of (6.73) is

$$g_5(t) = 2\exp(t) + 2\exp(-t) + 2\exp(-3t) \tag{6.74}$$

The samples of exact solution of $g_5(t)$ and deviations of the samples obtained using Eqs. (6.63) and (6.72) in HF domain, for $T = 3$ s and $m = 30$, are compared in Fig. 6.8.

6.5 Conclusion

One shot integration operational matrices like **P2ss**, **P2st**, **P2ts**, **P2tt**, **P3ss**, **P3st**, **P3ts, P3tt** for 2nd and 3rd order repeated integration and consequently the generalized one-shot matrices for n times repeated integration, have been used for solution of higher order differential equations. Some examples, separately for second order and third order differential equations, are treated to compare the results obtained via repeated use of 1st order operational matrices and using higher order one-shot operational matrices. The results are presented in Figs. 6.5, 6.6 and 6.8 to compare them graphically. It is observed that (vide Fig. 6.8) the method based upon one-shot operational matrices produces much accurate result compared to the method using only 1st order integration operational matrices.

One second order differential equation has been solved via the well established 4th order Runge-Kutta method and the results obtained via HF domain one-shot operational matrices and the results obtained via repeated use of 1st order integration operational matrices, are compared in Table 6.2. It is noted that 4th order Runge-Kutta method maintains its supremacy compared to HF domain analysis, as far as solution of 2nd order differential equation is concerned. But it should be kept in mind that while the 4th order Runge-Kutta method provides smart solution to differential equations only, the HF domain technique can (i) approximate square integrable time functions (ii) integrate time functions and (iii) can solve higher order differential equations with considerable accuracy. However, HF domain analysis with a higher value of m can produce more improved result to become a significant contender to 4th order Runge-Kutta method.

It is known that inversion of upper or lower triangular matrices can be computed by simple decomposition and multiplication. Hence the inversions in Eqs. (6.37), (6.38), (6.46), (6.47), (6.62), (6.63), (6.70) and (6.72) will not pose any computational burden while solving for the HF domain solution matrices $\mathbf{C}_\mathbf{s}^\mathrm{T}$ and $\mathbf{C}_\mathbf{T}^\mathrm{T}$.

Finally, an advantage of HF based analysis is, the sample-and-hold function based results may easily be obtained by simply dropping the triangular part of the hybrid function domain solution.

References

1. Tenenbaum, M., Pollard, H.: Ordinary differential equations. Dover publications, USA (1985)
2. Butcher, J.C.: Numerical methods for ordinary differential equations (2nd edn). Wiley, Hoboken (2008)
3. Coddington, E.A.: An introduction to differential equations. Dover publications, USA (1989)
4. Butcher, J.C.: Numerical methods for ordinary differential equations in the 20th century. J. Comput. Appl. Math. **125**, 1–29 (2000)
5. Simos, T.E.: Modified Runge-Kutta methods for the numerical solution of ODEs with oscillating solutions. Appl. Math. Comput. **84**, 131–143 (1997)
6. Rao, G.P.: Piecewise constant orthogonal functions and their applications in systems and control, LNC1S, vol. 55. Springer, Berlin (1983)
7. Jiang, J.H., Schaufelberger, W.: Block pulse functions and their application in control system, LNCIS, vol. 179. Springer, Berlin (1992)
8. Feng, Y.Y., Qi, D.X.: A sequence of piecewise orthogonal polynomials, SIAM J

Chapter 7
Convolution of Time Functions

Abstract In this chapter, theory of hybrid function domain convolution technique is presented. First, the rules for convolution for sample-and-hold functions and triangular functions are derived. Then, these two component results are combined to get the rules for convolution in HF domain. This idea is used to determine the result of convolution of two time functions in HF domain. One example and eleven figures are presented to illustrate the idea.

Having established the theoretical principles of the orthogonal hybrid function set, it is worthwhile to investigate the convolution operation of two real-valued functions, in hybrid function domain. This will later be useful for analysing control systems.

In control system analysis, the well-known relation [1] involving the input and the output of a linear time invariant system is given by

$$\mathrm{C(s)} = \mathrm{G(s)}\,\mathrm{R(s)} \tag{7.1}$$

where, C(s) is the Laplace transform of output, G(s) is the transfer function of the plant, R(s) is the Laplace transform of the input, and s is the Laplace operator.

In time domain, Eq. (7.1) takes the form

$$c(t) = g(t) * r(t) = \int_{0}^{\infty} g(\tau) r(t-\tau)\,\mathrm{d}\tau \tag{7.2}$$

That is, the output $c(t)$ in the time domain involves the convolution of the plant impulse response and the input function. The output $c(t)$ is determined by evaluating the convolution integral of the RHS of Eq. (7.2), where, it has been assumed that the integral exists.

Evaluation of this integral is frequently needed in the analysis of control systems. In what follows, such an integral is evaluated in its general form in hybrid function domain and the results are used to determine the output of a single input single output (SISO) linear control system.

A. Deb et al., *Analysis and Identification of Time-Invariant Systems, Time-Varying Systems, and Multi-Delay Systems using Orthogonal Hybrid Functions*,
Studies in Systems, Decision and Control 46, DOI 10.1007/978-3-319-26684-8_7

7.1 The Convolution Integral

Convolution [2] of two functions is a significant physical concept in many diverse scientific fields. However, as in the case of many important mathematical relationships, the convolution integral does not readily unveil itself as to its implications. The convolution integral of two convolving time function $x(t)$ and $h(t)$ over the entire time scale is given by

$$y(t) = x(t) * h(t) = \int_{-\infty}^{\infty} x(\tau)h(t - \tau)\,\mathrm{d}\tau \tag{7.3}$$

where $*$ indicates convolution.

Let $x(t)$ and $h(t)$ be two time functions represented by Figs. 7.1a, b, respectively.

To evaluate Eq. (7.3), functions $x(\tau)$ and $h(t - \tau)$ are required. The functions $x(\tau)$ and $h(\tau)$ are obtained by simply replacing the variable t to the variable τ. $h(-\tau)$ is the image of $h(\tau)$ about the orthogonal axis and $h(t - \tau)$ is the function $h(-\tau)$ shifted by the quantity t. Functions $x(\tau)$, $h(-\tau)$, and $h(t - \tau)$ are shown **in** Fig. 7.2. The resultant of convolution of $x(t)$ and $h(t)$, as per Eq. (7.3), is the triangular function shown in Fig. 7.3.

We can now summarize the steps for convolution as:

(i) *Folding*: Take the mirror image of $h(\tau)$ about the ordinate as shown in Fig. 7.2b.
(ii) *Shifting*: Shift $h(-\tau)$ by the amount t as shown in Fig. 7.2c.

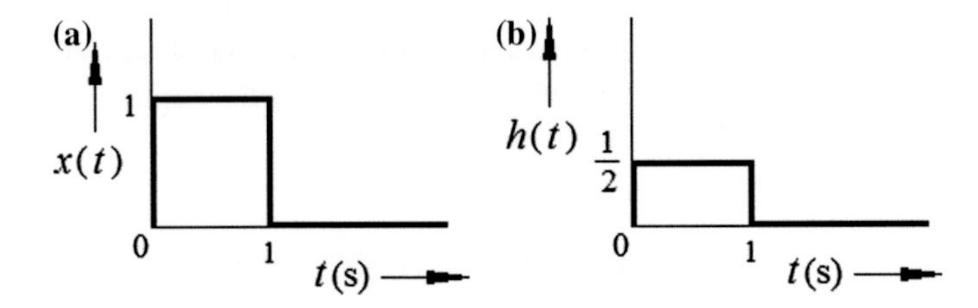

Fig. 7.1 Two typical waveforms for convolution

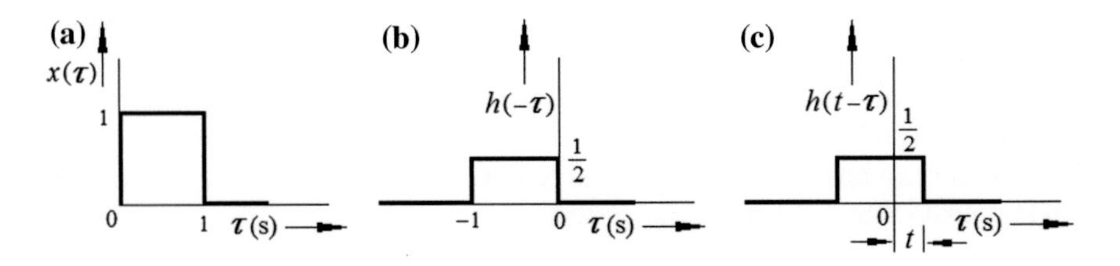

Fig. 7.2 Graphical illustration of folding and shifting operations

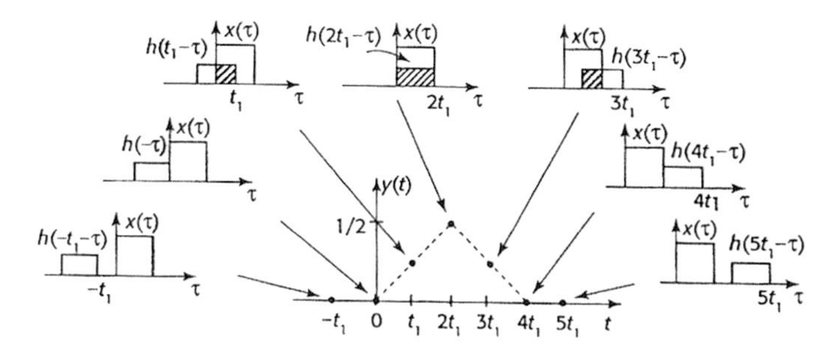

Fig. 7.3 Graphical example of convolution [2]

(iii) *Multiplication*: Multiply the shifted function $h(t - \tau)$ and $x(\tau)$.
(iv) *Integration*: The area under the product of $h(t - \tau)$ and $x(\tau)$ is the value of the convolution at time instant t.

7.2 Convolution of Basic Components of Hybrid Functions

The convolution process and 'deconvolution' in block pulse domain [3, 4] was introduced by Kwong and Chen [5] for identification of a system. We introduce the convolution as well as 'deconvolution' in hybrid function domain and subsequently use the results for control system analysis and synthesis.

Hybrid function expansion involves two kinds of basis functions: sample-and-hold function [6] and triangular function [7]. To derive the expression for convolution of two time functions in hybrid function domain, we consider convolution of different interactive components of hybrid functions [8]. That is, we need to compute the equidistant samples, having the same sampling period as the convolving functions, of the resulting function. These samples may be used for hybrid function expansion of the resulting function as per Eq. (2.13).

That is, the principle of HF domain convolution is:

(i) Expand the convolving functions in HF domain using their samples.
(ii) Convolve the component sample-and-hold functions [9] and triangular functions [7].
(iii) Express the result of convolution in hybrid function domain using the samples of the resulting function.

Let us consider two functions $r(t)$ and $g(t)$ and expand these functions into hybrid function domain as

$$r(t) \approx \left[\mathbf{R}_S^T \mathbf{S}_{(m)} + \mathbf{R}_T^T \mathbf{T}_{(m)}\right] \quad \text{and} \quad g(t) \approx \left[\mathbf{G}_S^T \mathbf{S}_{(m)} + \mathbf{G}_T^T \mathbf{T}_{(m)}\right]$$

Then the result of convolution $y(t)$ is given by

$$\begin{aligned} & y(t) = r(t) * g(t) \approx \left[\mathbf{R}_S^T \mathbf{S}_{(m)} + \mathbf{R}_T^T \mathbf{T}_{(m)}\right] * \left[\mathbf{G}_S^T \mathbf{S}_{(m)} + \mathbf{G}_T^T \mathbf{T}_{(m)}\right] \\ \text{or,} \quad & y(t) \approx \left[\mathbf{R}_S^T \mathbf{S}_{(m)}\right] * \left[\mathbf{G}_S^T \mathbf{S}_{(m)}\right] + \left[\mathbf{R}_S^T \mathbf{S}_{(m)}\right] * \left[\mathbf{G}_T^T \mathbf{T}_{(m)}\right] \\ & \quad + \left[\mathbf{R}_T^T \mathbf{T}_{(m)}\right] * \left[\mathbf{G}_S^T \mathbf{S}_{(m)}\right] + \left[\mathbf{R}_T^T \mathbf{T}_{(m)}\right] * \left[\mathbf{G}_T^T \mathbf{T}_{(m)}\right] \end{aligned} \tag{7.4}$$

where

$$\begin{aligned} \mathbf{R}_S^T &= \left[r_0 \quad r_1 \quad r_2 \quad \cdots \quad r_i \quad \cdots \quad r_{m-1}\right], \\ \mathbf{R}_T^T &= \left[(r_1 - r_0) \quad (r_2 - r_1) \quad (r_3 - r_2) \quad \cdots \quad (r_i - r_{i-1}) \quad \cdots \quad (r_m - r_{m-1})\right] \\ \mathbf{G}_S^T &= \left[g_0 \quad g_1 \quad g_2 \quad \cdots \quad g_i \quad \cdots \quad g_{m-1}\right] \quad \text{and} \\ \mathbf{G}_T^T &= \left[(g_1 - g_0) \quad (g_2 - g_1) \quad (g_3 - g_2) \quad \cdots \quad (g_i - g_{i-1}) \quad \cdots \quad (g_m - g_{m-1})\right] \end{aligned}$$

Inspection of Eq. (7.4) reveals that to determine $y(t)$, we need to compute the results of three types of convolution operations, namely

(i) Convolution between two sample-and-hold functions trains.
(ii) Convolution between a sample-and-hold function train and a triangular function train (or vice versa).
(iii) Convolution between two triangular function trains.

To achieve this end, we present below the convolution of all possible combinations of elementary sample-and-hold functions, triangular functions and subsequently their trains.

7.2.1 *Convolution of Two Elementary Sample-and-Hold Functions*

In Figs. 7.4a, b, $a_1(t)$ and $b_1(t)$ are two sample-and-hold functions of different amplitudes, both occurring at $t = 0$. The result of convolution of these two functions $a_1(t)$ and $b_1(t)$ is the triangular function $c_1(t)$ shown in Fig. 7.4c.

The function $c_1(t)$ may be expressed in hybrid function domain using its three samples, namely, 0, ha_1b_1 and 0.

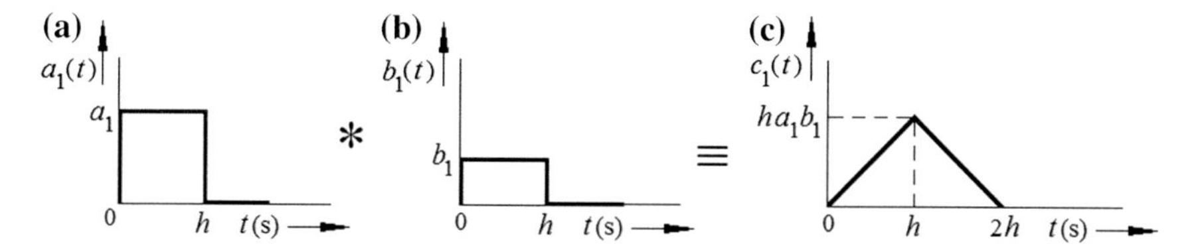

Fig. 7.4 Convolution of two elementary sample-and-hold functions (SHF)

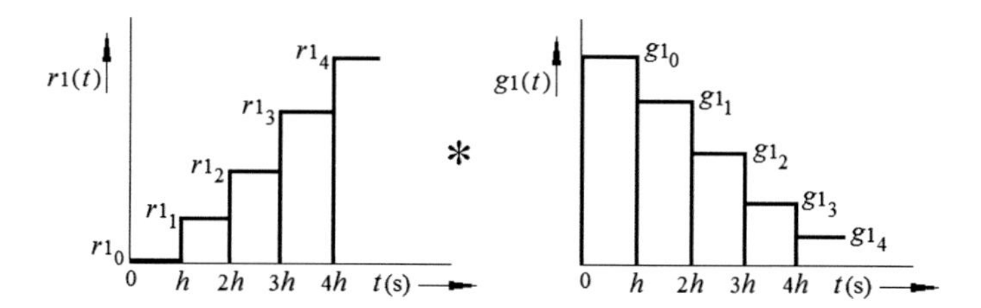

Fig. 7.5 Two trains of sample-and-hold functions

7.2.2 *Convolution of Two Sample-and-Hold Function Trains*

Now, we extend our idea to the convolution of two sample-and-hold function trains $r1(t)$ and $g1(t)$, comprised only of four component functions ($m = 4$), with different amplitudes. These function trains are shown in Fig. 7.5 along with their sample values. Finally, we represent the result in hybrid function domain.

Five samples of the resulting function $y1(t)$, with the sampling period h, are $0, hg1_0r1_0, h(g1_0r1_1 + g1_1r1_0)$, $h(g1_0r1_2 + g1_1r1_1 + g1_2r1_0)$, and $h(g1_0r1_3 + g1_1r1_2 + g1_2r1_1 + g1_3r1_0)$.

The resulting function $y1(t)$ may be described in the HF domain as

$$y1(t) = r1(t) * g1(t) \triangleq \mathbf{Y1}_S^T \mathbf{S}_{(4)} + \mathbf{Y1}_T^T \mathbf{T}_{(4)} \tag{7.5}$$

where

$$\begin{aligned}\mathbf{Y1}_S^T &= [0 \quad hg1_0r1_0 \quad h(g1_0r1_1 + g1_1r1_0) \quad h(g1_0r1_2 + g1_1r1_1 + g1_2r1_0)] \\ &\triangleq [y1_{S0} \quad y1_{S1} \quad y1_{S2} \quad y1_{S3}]\end{aligned}$$

and

$$\begin{aligned}\mathbf{Y1}_{\mathrm{T}}^{\mathrm{T}} = [&\{hg1_0r1_0 - 0\} \quad \{h(g1_0r1_1 + g1_1r1_0) - hg1_0r1_0\} \\ &\{h(g1_0r1_2 + g1_1r1_1 + g1_2r1_0) - h(g1_0r1_1 + g1_1r1_0)\} \\ &\{h(g1_0r1_3 + g1_1r1_2 + g1_2r1_1 + g1_3r1_0) - h(g1_0r1_2 + g1_1r1_1 + g1_2r1_0)\}] \\ \triangleq &[y1_{\mathrm{T0}} \quad y1_{\mathrm{T1}} \quad y1_{\mathrm{T2}} \quad y1_{\mathrm{T3}}]\end{aligned}$$

Writing Eq. (7.5) in matrix form, we have

$$\begin{aligned}y1(t) = h[g1_0 \quad g1_1 \quad g1_2 \quad g1_3]\begin{bmatrix}0 & r1_0 & r1_1 & r1_2 \\ 0 & 0 & r1_0 & r1_1 \\ 0 & 0 & 0 & r1_0 \\ 0 & 0 & 0 & 0\end{bmatrix}\mathbf{S}_{(4)} \\ + h[g1_0 \quad g1_1 \quad g1_2 \quad g1_3]\begin{bmatrix}r1_0 & (r1_1 - r1_0) & (r1_2 - r1_1) & (r1_3 - r1_2) \\ 0 & r1_0 & (r1_1 - r1_0) & (r1_2 - r1_1) \\ 0 & 0 & r1_0 & (r1_1 - r1_0) \\ 0 & 0 & 0 & r1_0\end{bmatrix}\mathbf{T}_{(4)}\end{aligned} \tag{7.6}$$

Writing (7.6) in a compact form, we get

$$\begin{aligned}y1(t) = h[g1_0 \quad g1_1 \quad g1_2 \quad g1_3][\![0 \quad r1_0 \quad r1_1 \quad r1_2]\!]\mathbf{S}_{(4)} \\ + h[g1_0 \quad g1_1 \quad g1_2 \quad g1_3][\![r1_0 \quad (r1_1 - r1_0) \quad (r1_2 - r1_1) \quad (r1_3 - r1_2)]\!]\mathbf{T}_{(4)}\end{aligned} \tag{7.7}$$

where $[\![a \quad b \quad c]\!] \triangleq \left[\begin{bmatrix}a & b & c \\ 0 & a & b \\ 0 & 0 & a\end{bmatrix}\right]$

7.2.3 Convolution of an Elementary Sample-and-Hold Function and an Elementary Triangular Function

The result of convolution of a sample-and-hold function and a triangular function is shown in Fig. 7.6.

The function $c_2(t)$ may now be expressed in hybrid function domain using its three equidistant samples 0, $\frac{h}{2}a_1b_2$ and 0.

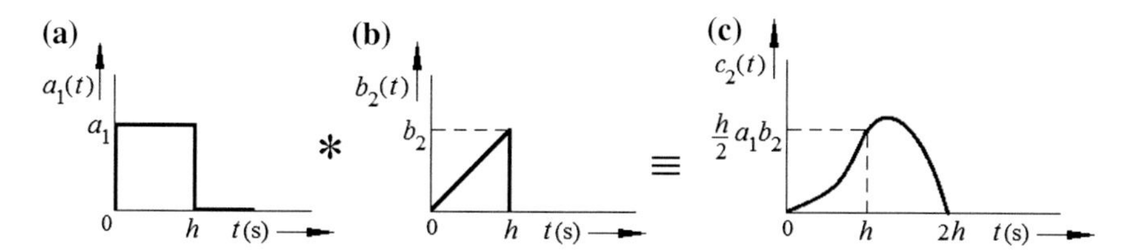

Fig. 7.6 Convolution of a sample-and-hold function and a right handed triangular function

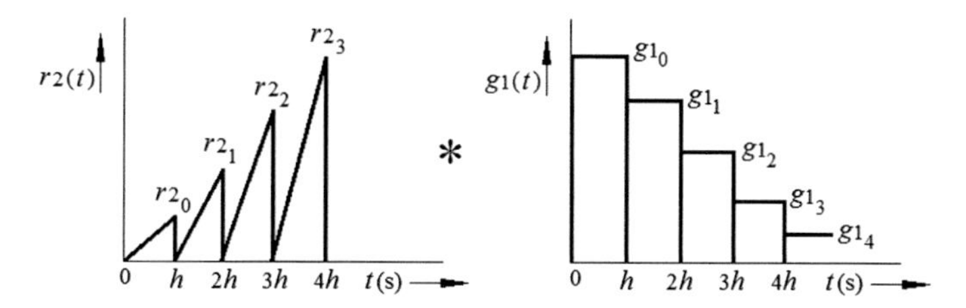

Fig. 7.7 Trains of triangular function and sample-and-hold functions

7.2.4 Convolution of a Triangular Function Train and a Sample-and-Hold Function Train

A triangular function train and a sample-and-hold function train of four component functions ($m = 4$) each, having different amplitudes, are shown in Fig. 7.7.

After convolution of these two trains, five samples of the resulting function $y2(t)$, with the sampling period h, are 0, $\frac{h}{2}g1_0r2_0$, $\frac{h}{2}(g1_0r2_1 + g1_1r2_0)$, $\frac{h}{2}(g1_0r2_2 + g1_1r2_1 + g1_2r2_0)$, and $\frac{h}{2}(g1_0r2_3 + g1_1r2_2 + g1_2r2_1 + g1_3r2_0)$.

The function $y2(t)$ may be described in HF domain as

$$y2(t) = r2(t) * g1(t) \triangleq \mathbf{Y2}_S^T\mathbf{S}_{(4)} + \mathbf{Y2}_T^T\mathbf{T}_{(4)} \tag{7.8}$$

where

$$\begin{aligned}\mathbf{Y2}_S^T &= \left[0 \quad \tfrac{h}{2}g1_0r2_0 \quad \tfrac{h}{2}(g1_0r2_1 + g1_1r2_0) \quad \tfrac{h}{2}(g1_0r2_2 + g1_1r2_1 + g1_2r2_0)\right] \\ &\triangleq [y2_{S0} \quad y2_{S1} \quad y2_{S2} \quad y2_{S3}]\end{aligned}$$

and

$$\begin{aligned}\mathbf{Y2}_T^T = \Big[&\frac{h}{2}\{g1_0r2_0 - 0\} \quad \frac{h}{2}\{(g1_0r2_1 + g1_1r2_0) - g1_0r2_0\} \\ &\frac{h}{2}\{(g1_0r2_2 + g1_1r2_1 + g1_2r2_0) - (g1_0r2_1 + g1_1r2_0)\} \\ &\frac{h}{2}\{(g1_0r2_3 + g1_1r2_2 + g1_2r2_1 + g1_3r2_0) - (g1_0r2_2 + g1_1r2_1 + g1_2r2_0)\}\Big] \\ \triangleq\ &[y2_{T0} \quad y2_{T1} \quad y2_{T2} \quad y2_{T3}]\end{aligned}$$

Writing Eq. (7.8) in matrix form, we get

$$y2(t) = \frac{h}{2}[g1_0 \quad g1_1 \quad g1_2 \quad g1_3] \begin{bmatrix} 0 & r2_0 & r2_1 & r2_2 \\ 0 & 0 & r2_0 & r2_1 \\ 0 & 0 & 0 & r2_0 \\ 0 & 0 & 0 & 0 \end{bmatrix} \mathbf{S}_{(4)}$$
$$+ \frac{h}{2}[g1_0 \quad g1_1 \quad g1_2 \quad g1_3] \begin{bmatrix} r2_0 & (r2_1 - r2_0) & (r2_2 - r2_1) & (r2_3 - r2_2) \\ 0 & r2_0 & (r2_1 - r2_0) & (r2_2 - r2_1) \\ 0 & 0 & r2_0 & (r2_1 - r2_0) \\ 0 & 0 & 0 & r2_0 \end{bmatrix} \mathbf{T}_{(4)} \tag{7.9}$$

Writing (7.9) in a compact form, we have

$$y2(t) = \frac{h}{2}[g1_0 \quad g1_1 \quad g1_2 \quad g1_3][\![0 \quad r2_0 \quad r2_1 \quad r2_2]\!]\mathbf{S}_{(4)}$$
$$+ \frac{h}{2}[g1_0 \quad g1_1 \quad g1_2 \quad g1_3][\![r2_0 \quad (r2_1 - r2_0) \quad (r2_2 - r2_1) \quad (r2_3 - r2_2)]\!]\mathbf{T}_{(4)} \tag{7.10}$$

7.2.5 Convolution of Two Elementary Triangular Functions

Let, $a_2(t)$ and $b_2(t)$ be two elementary triangular functions, as represented in Fig. 7.8a, b. Figure 7.8c shows the convolution result of these two functions. The function $c_3(t)$ may now be expressed in hybrid function domain using its three samples, namely, 0, $\frac{h}{6}a_2b_2$ and 0.

7.2.6 Convolution of Two Triangular Function Trains

Now we compute the result of convolution of two triangular function trains comprised of four component functions ($m = 4$) each, having different amplitudes. These trains are shown in Fig. 7.9.

Five samples of the resulting convolution function $y3(t)$, with the sampling period h, are 0, $\frac{h}{6}g2_0r2_0$, $\frac{h}{6}(g2_0r2_1 + g2_1r2_0)$, $\frac{h}{6}(g2_0r2_2 + g2_1r2_1 + g2_2r2_0)$, and $\frac{h}{6}(g2_0r2_3 + g2_1r2_2 + g2_2r2_1 + g2_3r2_0)$.

Hence, $y3(t)$, expressed in HF domain, is

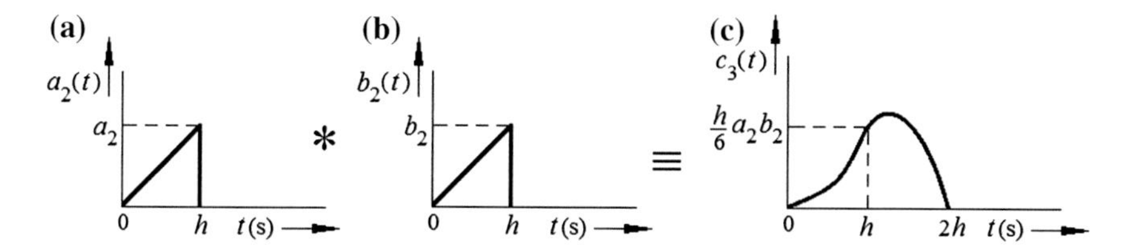

Fig. 7.8 Convolution of two elementary triangular functions

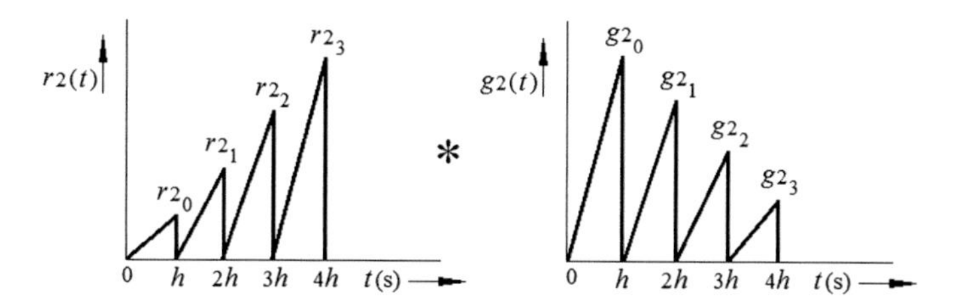

Fig. 7.9 Two trains of triangular functions

$$y3(t) = r2(t) * g2(t) \triangleq \mathbf{Y3}_{\mathrm{S}}^{\mathrm{T}}\mathbf{S}_{(4)} + \mathbf{Y3}_{\mathrm{T}}^{\mathrm{T}}\mathbf{T}_{(4)} \tag{7.11}$$

where

$$\begin{aligned}\mathbf{Y3}_{\mathrm{S}}^{\mathrm{T}} &= \left[0 \quad \tfrac{h}{6}g2_0r2_0 \quad \tfrac{h}{6}(g2_0r2_1 + g2_1r2_0) \quad \tfrac{h}{6}(g2_0r2_2 + g2_1r2_1 + g2_2r2_0)\right] \\ &\triangleq [y3_{\mathrm{S0}} \quad y3_{\mathrm{S1}} \quad y3_{\mathrm{S2}} \quad y3_{\mathrm{S3}}]\end{aligned}$$

and

$$\begin{aligned}\mathbf{Y3}_{\mathrm{T}}^{\mathrm{T}} = \Bigg[&\frac{h}{6}\{g2_0r2_0 - 0\} \quad \frac{h}{6}\{(g2_0r2_1 + g2_1r2_0) - g2_0r2_0\} \\ &\frac{h}{6}\{(g2_0r2_2 + g2_1r2_1 + g2_2r2_0) - (g2_0r2_1 + g2_1r2_0)\} \\ &\frac{h}{6}\{(g2_0r2_3 + g2_1r2_2 + g2_2r2_1 + g2_3r2_0) - (g2_0r2_2 + g2_1r2_1 + g2_2r2_0)\}\Bigg] \\ \triangleq &[y3_{\mathrm{T0}} \quad y3_{\mathrm{T1}} \quad y3_{\mathrm{T2}} \quad y3_{\mathrm{T3}}]\end{aligned}$$

Writing equation in matrix form, we get

$$y3(t) = \frac{h}{6}\begin{bmatrix} g2_0 & g2_1 & g2_2 & g2_3 \end{bmatrix}\begin{bmatrix} 0 & r2_0 & r2_1 & r2_2 \\ 0 & 0 & r2_0 & r2_1 \\ 0 & 0 & 0 & r2_0 \\ 0 & 0 & 0 & 0 \end{bmatrix}\mathbf{S}_{(4)}$$
$$+\frac{h}{6}\begin{bmatrix} g2_0 & g2_1 & g2_2 & g2_3 \end{bmatrix}\begin{bmatrix} r2_0 & (r2_1 - r2_0) & (r2_2 - r2_1) & (r2_3 - r2_2) \\ 0 & r2_0 & (r2_1 - r2_0) & (r2_2 - r2_1) \\ 0 & 0 & r2_0 & (r2_1 - r2_0) \\ 0 & 0 & 0 & r2_0 \end{bmatrix}\mathbf{T}_{(4)} \tag{7.12}$$

Writing in a compact form, we have

$$y3(t) = \frac{h}{6}\begin{bmatrix} g2_0 & g2_1 & g2_2 & g2_3 \end{bmatrix}[\![0 \quad r2_0 \quad r2_1 \quad r2_2]\!]\mathbf{S}_{(4)}$$
$$+\frac{h}{6}\begin{bmatrix} g2_0 & g2_1 & g2_2 & g2_3 \end{bmatrix}[\![r2_0 \quad (r2_1 - r2_0) \quad (r2_2 - r2_1) \quad (r2_3 - r2_2)]\!]\mathbf{T}_{(4)} \tag{7.13}$$

7.3 Convolution of Two Time Functions in HF Domain [8]

Consider two square integrable time functions $r(t)$ and $g(t)$. These two time functions are expressed in HF domain using their equidistant samples. Figure 7.10 shows these functions with their five time samples each. If we express these functions in hybrid function domain for $m = 4$, we have

$$r(t) \approx [r_0 \quad r_1 \quad r_2 \quad r_3]\mathbf{S}_{(4)}(t) + [(r_1 - r_0) \quad (r_2 - r_1) \quad (r_3 - r_2) \quad (r_4 - r_3)]\mathbf{T}_{(4)}(t)$$
$$g(t) \approx [g_0 \quad g_1 \quad g_2 \quad g_3]\mathbf{S}_{(4)}(t) + [(g_1 - g_0) \quad (g_2 - g_1) \quad (g_3 - g_2) \quad (g_4 - g_3)]\mathbf{T}_{(4)}(t)$$

Hence, the result of convolution in HF domain may be derived using the sub-results of convolutions between different sample-and-hold and triangular function trains of both r(t) and $g(t)$, deduced in Sect. 7.2.

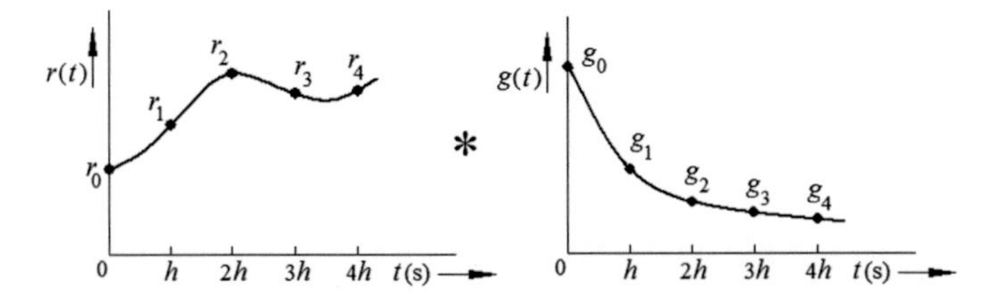

Fig. 7.10 Two time functions $r(t)$ and $g(t)$ and their equidistant samples

Using the results of Eqs. (7.7), (7.10) and (7.13), we can write

$$
\begin{aligned}
y(t) &= r(t) * g(t) \\
&\approx h[g_0 \quad g_1 \quad g_2 \quad g_3][\![0 \quad r_0 \quad r_1 \quad r_2]\!]\mathbf{S}_{(4)} \\
&\quad + h[g_0 \quad g_1 \quad g_2 \quad g_3][\![r_0 \quad (r_1-r_0) \quad (r_2-r_1) \quad (r_3-r_2)]\!]\mathbf{T}_{(4)} \\
&\quad + \frac{h}{2}[(g_1-g_0) \quad (g_2-g_1) \quad (g_3-g_2) \quad (g_4-g_3)][\![0 \quad r_0 \quad r_1 \quad r_2]\!]\mathbf{S}_{(4)} \\
&\quad + \frac{h}{2}[(g_1-g_0) \quad (g_2-g_1) \quad (g_3-g_2) \quad (g_4-g_3)][\![r_0 \quad (r_1-r_0) \quad (r_2-r_1) \quad (r_3-r_2)]\!]\mathbf{T}_{(4)} \\
&\quad + \frac{h}{2}[g_0 \quad g_1 \quad g_2 \quad g_3][\![0 \quad (r_1-r_0) \quad (r_2-r_1) \quad (r_3-r_2)]\!]\mathbf{S}_{(4)} \\
&\quad + \frac{h}{2}[g_0 \quad g_1 \quad g_2 \quad g_3][\![(r_1-r_0) \quad (r_2-2r_1+r_0) \quad (r_3-2r_2+r_1) \quad (r_4-2r_3+r_2)]\!]\mathbf{T}_{(4)} \\
&\quad + \frac{h}{6}[(g_1-g_0) \quad (g_2-g_1) \quad (g_3-g_2) \quad (g_4-g_3)][\![0 \quad (r_1-r_0) \quad (r_2-r_1) \quad (r_3-r_2)]\!]\mathbf{S}_{(4)} \\
&\quad + \frac{h}{6}[(g_1-g_0) \quad (g_2-g_1) \quad (g_3-g_2) \quad (g_4-g_3)] \\
&\quad \times [\![(r_1-r_0) \quad (r_2-2r_1+r_0) \quad (r_3-2r_2+r_1) \quad (r_4-2r_3+r_2)]\!]\mathbf{T}_{(4)} \\
&= \Big\{ h[g_0 \quad g_1 \quad g_2 \quad g_3][\![0 \quad r_0 \quad r_1 \quad r_2]\!] \\
&\quad + \frac{h}{2}[(g_1-g_0) \quad (g_2-g_1) \quad (g_3-g_2) \quad (g_4-g_3)][\![0 \quad r_0 \quad r_1 \quad r_2]\!] \\
&\quad + \frac{h}{2}[g_0 \quad g_1 \quad g_2 \quad g_3][\![0 \quad (r_1-r_0) \quad (r_2-r_1) \quad (r_3-r_2)]\!] \\
&\quad + \frac{h}{6}[(g_1-g_0) \quad (g_2-g_1) \quad (g_3-g_2) \quad (g_4-g_3)][\![0 \quad (r_1-r_0) \quad (r_2-r_1) \quad (r_3-r_2)]\!] \Big\}\mathbf{S}_{(4)} \\
&\quad + \Big\{ h[g_0 \quad g_1 \quad g_2 \quad g_3][\![r_0 \quad (r_1-r_0) \quad (r_2-r_1) \quad (r_3-r_2)]\!] \\
&\quad + \frac{h}{2}[(g_1-g_0) \quad (g_2-g_1) \quad (g_3-g_2) \quad (g_4-g_3)][\![r_0 \quad (r_1-r_0) \quad (r_2-r_1) \quad (r_3-r_2)]\!] \\
&\quad + \frac{h}{2}[g_0 \quad g_1 \quad g_2 \quad g_3][\![(r_1-r_0) \quad (r_2-2r_1+r_0) \quad (r_3-2r_2+r_1) \quad (r_4-2r_3+r_2)]\!] \\
&\quad + \frac{h}{6}[(g_1-g_0) \quad (g_2-g_1) \quad (g_3-g_2) \quad (g_4-g_3)] \\
&\quad \times [\![(r_1-r_0) \quad (r_2-2r_1+r_0) \quad (r_3-2r_2+r_1) \quad (r_4-2r_3+r_2)]\!] \Big\}\mathbf{T}_{(4)}
\end{aligned}
\tag{7.14}
$$

Equation (7.14) can be simplified to be arranged in the following form:

$$
\begin{aligned}
y(t) = \frac{h}{6}\Bigg\{ & [\,g_0 \;\; g_1 \;\; g_2 \;\; g_3\,] \begin{bmatrix} 0 & (2r_1+r_0) & (2r_2+r_1) & (2r_3+r_2) \\ 0 & 0 & (2r_1+r_0) & (2r_2+r_1) \\ 0 & 0 & 0 & (2r_1+r_0) \\ 0 & 0 & 0 & 0 \end{bmatrix} \\
& + [\,g_1 \;\; g_2 \;\; g_3 \;\; g_4\,] \begin{bmatrix} 0 & (r_1+2r_0) & (r_2+2r_1) & (r_3+2r_2) \\ 0 & 0 & (r_1+2r_0) & (r_2+2r_1) \\ 0 & 0 & 0 & (r_1+2r_0) \\ 0 & 0 & 0 & 0 \end{bmatrix} \Bigg\} \mathbf{S}_{(4)} \\
+ \frac{h}{6}\Bigg\{ & [\,g_0 \;\; g_1 \;\; g_2 \;\; g_3\,] \begin{bmatrix} (2r_1+r_0) & (2r_2-r_1-r_0) & (2r_3-r_2-r_1) & (2r_4-r_3-r_2) \\ 0 & (2r_1+r_0) & (2r_2-r_1-r_0) & (2r_3-r_2-r_1) \\ 0 & 0 & (2r_1+r_0) & (2r_2-r_1-r_0) \\ 0 & 0 & 0 & (2r_1+r_0) \end{bmatrix} \\
& + [\,g_1 \;\; g_2 \;\; g_3 \;\; g_4\,] \begin{bmatrix} (r_1+2r_0) & (r_2+r_1-2r_0) & (r_3+r_2-2r_1) & (r_4+r_3-2r_2) \\ 0 & (r_1+2r_0) & (r_2+r_1-2r_0) & (r_3+r_2-2r_1) \\ 0 & 0 & (r_1+2r_0) & (r_2+r_1-2r_0) \\ 0 & 0 & 0 & (r_1+2r_0) \end{bmatrix} \Bigg\} \mathbf{T}_{(4)} \\
= \frac{h}{6} & [\,g_0 \;\; g_1 \;\; g_2 \;\; g_3\,] \begin{bmatrix} 0 & (2r_1+r_0) & (2r_2+r_1) & (2r_3+r_2) \\ 0 & (r_1+2r_0) & (r_2+4r_1+r_0) & (r_3+4r_2+r_1) \\ 0 & 0 & (r_1+2r_0) & (r_2+4r_1+r_0) \\ 0 & 0 & 0 & (r_1+2r_0) \end{bmatrix} \mathbf{S}_{(4)} \\
+ \frac{h}{6}\Bigg\{ & [\,g_0 \;\; g_1 \;\; g_2 \;\; g_3\,] \begin{bmatrix} (2r_1+r_0) & (2r_2-r_1-r_0) & (2r_3-r_2-r_1) & (2r_4-r_3-r_2) \\ 0 & (2r_1+r_0) & (2r_2-r_1-r_0) & (2r_3-r_2-r_1) \\ 0 & 0 & (2r_1+r_0) & (2r_2-r_1-r_0) \\ 0 & 0 & 0 & (2r_1+r_0) \end{bmatrix} \\
& + [\,g_1 \;\; g_2 \;\; g_3 \;\; g_4\,] \begin{bmatrix} (r_1+2r_0) & (r_2+r_1-2r_0) & (r_3+r_2-2r_1) & (r_4+r_3-2r_2) \\ 0 & (r_1+2r_0) & (r_2+r_1-2r_0) & (r_3+r_2-2r_1) \\ 0 & 0 & (r_1+2r_0) & (r_2+r_1-2r_0) \\ 0 & 0 & 0 & (r_1+2r_0) \end{bmatrix} \Bigg\} \mathbf{T}_{(4)}
\end{aligned} \tag{7.15}
$$

Now let,

$$\left.\begin{aligned} R_0 &\triangleq 2r_1 + r_0 \\ R_1 &\triangleq 2r_2 + r_1 \\ R_2 &\triangleq 2r_3 + r_2 \\ R_3 &\triangleq 2r_4 + r_3 \\ R_4 &\triangleq r_1 + 2r_0 \\ R_5 &\triangleq r_2 + 4r_1 + r_0 \\ R_6 &\triangleq r_3 + 4r_2 + r_1 \\ R_7 &\triangleq r_4 + 4r_3 + r_2 \\ R_8 &\triangleq r_2 + r_1 - 2r_0 \\ R_9 &\triangleq r_3 + r_2 - 2r_1 \\ R_{10} &\triangleq r_4 + r_3 - 2r_2 \end{aligned}\right\} \tag{7.16}$$

Now, Eq. (7.15) can be written as follows

$$\begin{aligned} y(t) = {} & \frac{h}{6}[g_0 \quad g_1 \quad g_2 \quad g_3]\begin{bmatrix} 0 & R_0 & R_1 & R_2 \\ 0 & R_4 & R_5 & R_6 \\ 0 & 0 & R_4 & R_5 \\ 0 & 0 & 0 & R_4 \end{bmatrix}\mathbf{S}_{(4)} \\ & + \frac{h}{6}\left\{[g_0 \quad g_1 \quad g_2 \quad g_3]\begin{bmatrix} R_0 & (R_1 - R_0) & (R_2 - R_1) & (R_3 - R_2) \\ 0 & R_0 & (R_1 - R_0) & (R_2 - R_1) \\ 0 & 0 & R_0 & (R_1 - R_0) \\ 0 & 0 & 0 & R_0 \end{bmatrix}\right. \\ & \left. + \frac{h}{6}[g_1 \quad g_2 \quad g_3 \quad g_4]\begin{bmatrix} R_4 & R_8 & R_9 & R_{10} \\ 0 & R_4 & R_8 & R_9 \\ 0 & 0 & R_4 & R_8 \\ 0 & 0 & 0 & R_4 \end{bmatrix}\right\}\mathbf{T}_{(4)} \end{aligned} \tag{7.17}$$

$$\begin{aligned} y(t) = {} & \frac{h}{6}[0 \quad (g_0R_0 + g_1R_4) \quad (g_0R_1 + g_1R_5 + g_2R_4) \quad (g_0R_2 + g_1R_6 + g_2R_5 + g_3R_4)]\mathbf{S}_{(4)} \\ & + \frac{h}{6}[\{g_0R_0 + g_1R_4\} \quad \{g_0(R_1 - R_0) + g_1(R_0 + R_8) + g_2R_4\}\{g_0(R_2 - R_1) \\ & + g_1(R_1 - R_0 + R_9) + g_2(R_0 + R_8) + g_3R_4\}\{g_0(R_3 - R_2) \\ & + g_1(R_2 - R_1 + R_{10}) + g_2(R_1 - R_0 + R_9) + g_3(R_0 + R_8) + g_4R_4\}]\mathbf{T}_{(4)} \end{aligned} \tag{7.18}$$

Equation (7.18) can be modified to

$$
\begin{aligned}
y(t) = & \frac{h}{6}[0 \quad (g_0R_0 + g_1R_4) \quad (g_0R_1 + g_1R_5 + g_2R_4) \quad (g_0R_2 + g_1R_6 + g_2R_5 + g_3R_4)]\mathbf{S}_{(4)} \\
& + \frac{h}{6}[\{g_0R_0 + g_1R_4\} \quad \{g_0(R_1 - R_0) + g_1(R_5 - R_4) + g_2R_4\} \\
& \{g_0(R_2 - R_1) + g_1(R_6 - R_5) + g_2(R_5 - R_4) + g_3R_4\} \\
& \{g_0(R_3 - R_2) + g_1(R_7 - R_6) + g_2(R_6 - R_5) + g_3(R_5 - R_4) + g_4R_4\}]\mathbf{T}_{(4)}
\end{aligned}
\tag{7.19}
$$

Equation (7.19) represents the final output/result of the two convolving time functions, for $m = 4$, in hybrid function domain. In doing so, we have utilized the results of convolution of different possible combination of SHF and TF trains to yield the final result expressed in HF domain.

Direct expansion of the output $y(t)$ in HF domain is

$$
y(t) \triangleq [\,y_0 \quad y_1 \quad y_2 \quad y_3\,]\mathbf{S}_{(4)} + [\,(y_1 - y_0) \quad (y_2 - y_1) \quad (y_3 - y_2) \quad (y_4 - y_3)\,]\mathbf{T}_{(4)}
\tag{7.20}
$$

Comparing Eqs. (7.19) and (7.20), we get

$$
\begin{aligned}
y_0 &= 0 \\
y_1 &= \frac{h}{6}[g_0R_0 + g_1R_4] \\
y_2 &= \frac{h}{6}[g_0R_1 + g_1R_5 + g_2R_4] \\
y_3 &= \frac{h}{6}[g_0R_2 + g_1R_6 + g_2R_5 + g_3R_4]
\end{aligned}
$$

By following the pattern, we can write down the expression for y_4 as

$$
y_4 = \frac{h}{6}\,[g_0R_3 + g_1R_7 + g_2R_6 + g_3R_5 + g_4R_4]
$$

If we determine the term y_4 by adding the 4-th term of the row matrix for the $\mathbf{S}_{(4)}$ vector and the 4-th term of the row matrix for the $\mathbf{T}_{(4)}$ vector, the result turns out to be the same as above.

Hence, the generalized form of the i-th output coefficient is

$$
y_i = \frac{h}{6}\left[g_0R_{(i-1)} + \sum_{p=1}^{i} g_pR_{(m+i-p)}\right] \quad \text{for} \quad i = 1,2,3,\ldots,m
\tag{7.21}
$$

7.4 Numerical Example

To determine the convolution result using Eq. (7.21), we first compute the samples of both the functions and express the functions in HF domain. Then we use Eq. (7.21) to arrive at the results.

Example 7.1 (vide Appendix B, Program no. 21) Consider two time functions $r(t) = u(t)$ and $g(t) = \exp(-0.5t)\,(2\cos 2t - 0.5\sin 2t)$ and the exact convolution result of these two functions is $y(t) = \exp(-0.5t)\sin(2t)$ for $t \geq 0$.

To note the variation of the result for an appreciable time, we consider $T = 5$ s and for HF domain analysis, take $m = 25$, i.e., $h = T/m = 0.2$ s.

Then, in HF domain, $r(t)$ is

$$r(t) = \underbrace{[1 \quad 1 \quad \cdots \quad 1 \quad \cdots \quad 1 \quad 1]}_{25\ \text{terms}}\mathbf{S}_{(4)} + \underbrace{[0 \quad 0 \quad \cdots \quad 0 \quad \cdots \quad 0 \quad 0]}_{25\ \text{terms}}\mathbf{T}_{(4)}$$

and $g(t)$ is given by

$$\begin{aligned} g(t) \approx [&2.00000000 \quad 1.49064076 \quad 0.84716967 \quad 0.19164669 \quad -0.37416316 \\ &-0.78057001 \quad -0.99473153 \quad -1.01896263 \quad -0.88401092 \quad -0.63923184 \\ &-0.34171806 \quad -0.04622404 \quad 0.20272788 \quad 0.37575612 \quad 0.46033860 \\ &0.45965891 \quad 0.38927571 \quad 0.27251601 \quad 0.13552338 \quad 0.00277597 \\ &-0.10633011 \quad -0.17950603 \quad -0.21214867 \quad -0.20664673 \quad -0.17075045]\mathbf{S}_{(25)} \\ +[&-0.50935924 \quad -0.64347109 \quad -0.65552299 \quad -0.56580985 \quad -0.40640685 \\ &-0.21416151 \quad -0.02423110 \quad 0.13495171 \quad 0.24477908 \quad 0.29751377 \\ &0.29549402 \quad 0.24895193 \quad 0.17302824 \quad 0.08458248 \quad -0.00067969 \\ &-0.07038320 \quad -0.11675970 \quad -0.13699263 \quad -0.13274741 \quad -0.10910609 \\ &-0.07317592 \quad -0.03264264 \quad 0.00550194 \quad 0.03589628 \quad 0.05532806]\mathbf{T}_{(25)} \end{aligned}$$

Using Eqs. (7.19) or (7.21), convolution of $r(t)$ and $g(t)$ in HF domain yields

$$\begin{aligned} y_c(t) \approx [&0.00000000 \quad 0.34906408 \quad 0.58284512 \quad 0.68672675 \quad 0.66847511 \\ &0.55300179 \quad 0.37547163 \quad 0.17410222 \quad -0.01619514 \quad -0.16851941 \\ &-0.26661440 \quad -0.30540861 \quad -0.28975823 \quad -0.23190983 \quad -0.14830036 \\ &-0.05630060 \quad 0.02859286 \quad 0.09477203 \quad 0.13557597 \quad 0.14940591 \\ &0.13905049 \quad 0.11046688 \quad 0.07130141 \quad 0.02942187 \quad -0.00831785]\mathbf{S}_{(25)} \\ +[&0.34906408 \quad 0.23378104 \quad 0.10388164 \quad -0.01825165 \quad -0.11547332 \\ &-0.17753015 \quad -0.20136942 \quad -0.19029735 \quad -0.15232428 \quad -0.09809499 \\ &-0.03879421 \quad 0.01565038 \quad 0.05784839 \quad 0.08360947 \quad 0.09199975 \\ &0.08489346 \quad 0.06617917 \quad 0.04080394 \quad 0.01382994 \quad -0.01035541 \\ &-0.02858361 \quad -0.03916547 \quad -0.04187954 \quad -0.03773972 \quad -0.02884140]\mathbf{T}_{(25)} \end{aligned}$$

Direct expansion of $y(t)$, in HF domain, for $m = 25$ and $T = 5$ s, is given by

$$
\begin{aligned}
y_d(t) \approx [&0.00000000\quad 0.35236029\quad 0.58732149\quad 0.69047154\quad 0.67003422\\
&0.55151677\quad 0.37070205\quad 0.16635019\quad -0.02622919\quad -0.17991539\\
&-0.27841208\quad -0.31676081\quad -0.30003901\quad -0.24076948\quad -0.15566844\\
&-0.06234602\quad 0.02353088\quad 0.09026637\quad 0.13119242\quad 0.14477041\\
&0.13389508\quad 0.10465113\quad 0.06481067\quad 0.02234669\quad -0.01581457]\mathbf{S}_{(25)}\\
+[&0.35236029\quad 0.23496121\quad 0.10315004\quad -0.02043731\quad -0.11851746\\
&-0.18081471\quad -0.20435186\quad -0.19257939\quad -0.15368619\quad -0.09849669\\
&-0.03834873\quad 0.01672180\quad 0.05926953\quad 0.08510105\quad 0.09332241\\
&0.08587690\quad 0.06673549\quad 0.04092605\quad 0.01357799\quad -0.01087533\\
&-0.02924395\quad -0.03984046\quad -0.04246399\quad -0.03816125\quad -0.02884140]\mathbf{T}_{(25)}
\end{aligned}
$$

Figure 7.11 presents graphically the samples obtained through HF domain convolution of the functions $r(t)$ and $g(t)$ along with HF domain direct expansion of the result $y(t)$. These two results are compared for a typical of eleven samples in Table 7.1 and respective percentage errors are computed.

From Table 7.1, it is observed that error is quite large for the 17th sample ($t = \frac{80}{25}$ s) and the 24th sample ($t = \frac{115}{25}$ s). The reason for such sudden increase in error may be due to the fact that the sample values for these two cases are quite small, e.g., 0.02353088 and 0.02234669; in fact the lowest two of all the sample values, and computation of error needs the deviations ($y_d - y_c$) to be divided by these small sample values.

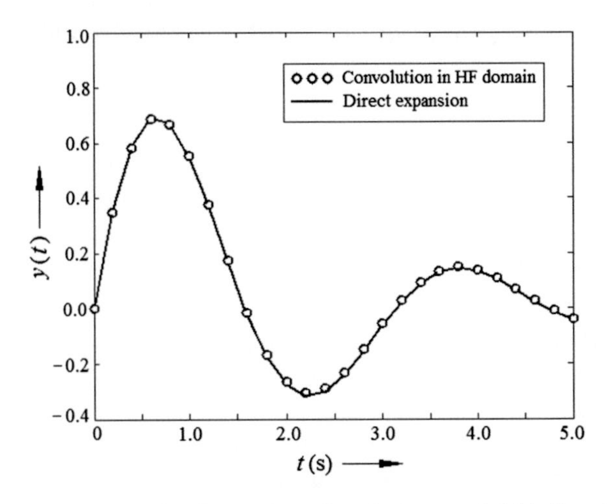

Fig. 7.11 Convolution of two functions $r(t) = u(t)$ and $g(t) = \exp(-0.5t)\ (2\cos 2t - 0.5\sin 2t)$, computed using Eq. (7.21) in HF domain and through direct expansion (y_d) for $T = 5$ s and $m = 25$. It is observed that that the two curves fairly overlap validating the HF domain convolution technique (vide Appendix B, Program no. 21)

Table 7.1 Convolution results via (a) HF domain direct expansion (y_d), and (b) HF domain convolution (y_c) along with percentage errors for eleven typical samples chosen randomly for Example 7.1 for $T = 5$ s and $m = 25$ (vide Appendix B, Program no. 21)

t (s)	Via direct expansion in HF domain (y_d)	Via convolution in HF Domain (y_c)	% Error $\varepsilon = \frac{(y_d - y_c)}{y_d} \times 100$
$\frac{5}{25}$	0.35236029	0.34906408	0.93546580
$\frac{20}{25}$	0.67003422	0.66847511	0.23269110
$\frac{25}{25}$	0.55151677	0.55300179	−0.26926108
$\frac{35}{25}$	0.16635019	0.17410222	−4.66006681
$\frac{50}{25}$	−0.27841208	−0.26661440	4.23748855
$\frac{60}{25}$	−0.30003901	−0.28975823	3.42648111
$\frac{70}{25}$	−0.15566844	−0.14830036	4.73318805
$\frac{80}{25}$	0.02353088	0.02859285	−21.51203015
$\frac{95}{25}$	0.14477041	0.14940591	−3.20196648
$\frac{105}{25}$	0.10465112	0.11046687	−5.55727449
$\frac{115}{25}$	0.02234669	0.02942187	−31.66097529

7.5 Conclusion

In this chapter we have introduced the idea of convolution in hybrid function domain. This idea has been built up in a step by step manner. That is, first of all convolution result for two elementary functions of the sample-and-hold function set and two elementary functions of the triangular function set are derived. Also, convolution of an elementary function of the SHF set with an elementary function of the TF set has also been treated. Then the idea of convolution of sample-and-hold function trains and triangular function trains are discussed with mathematical support for $m = 4$. All these results are transformed to hybrid function domain.

These sub-results of convolution, presented through Eqs. (7.7), (7.10) and (7.13), are the basic results involving different combinations of two function trains —SHF and TF. These three equations have been utilized to arrive at the general Eq. (7.21), giving the result of convolution of two time functions, expressed in HF domain.

Using the developed theory of HF domain convolution, an example has been treated to prove the viability of the method. Table 7.1 presents eleven typical sample values obtained via HF domain convolution technique and compares the same with the sample values obtained through direct HF domain expansion of the exact convolution result. It is noted that for two samples, the error is quite large, that is, −21.51203015 and −31.66097529 %. This may be due to low sample values as mentioned above. Since, the HF domain analysis uses only function samples, the numerical computation is rather simple, straight forward and computationally attractive.

References

1. Ogata, K.: Modern Control Engineering, 5th edn. Prentice-Hall of India Ltd., New Delhi (1997)
2. Brigham, E.O.: The Fast Fourier Transform and Its Applications. Prentice-Hall International Inc., New Jersey (1988)
3. Jiang, J.H., Schaufelberger, W.: Block Pulse Functions and their Application in Control System. LNCIS, vol. 179. Springer, Berlin (1992)
4. Deb, A., Sarkar, G., Sen, S.K.: Linearly pulse-width modulated block pulse functions and their application to linear SISO feedback control system identification. Proc. IEE, Part D Control Theory Appl. **142**(1), 44–50 (1995)
5. Kwong, C.P., Chen, C.F.: Linear feedback system identification via block pulse functions. Int. J. Syst. Sci. **12**(5), 635–642 (1981)
6. Deb, A., Sarkar, G., Bhattacharjee, M., Sen, S.K.: A new set of piecewise constant orthogonal functions for the analysis of linear SISO systems with sample-and-hold. J. Franklin Instt. **335B** (2), 333–358 (1998)
7. Deb, Anish, Sarkar, Gautam, Sengupta, Anindita: Triangular Orthogonal Functions for the Analysis of Continuous Time Systems. Anthem Press, London (2011)
8. Biswas, A.: Analysis and synthesis of continuous control systems using a set of orthogonal hybrid functions, Ph. D. dissertation, University of Calcutta (2015)
9. Deb, Anish, Sarkar, Gautam, Dasgupta, Anindita: A complementary pair of orthogonal triangular function sets and its application to the analysis of SISO control systems. J. Instt. Engrs. (India) **84**, 120–129 (2003)

Chapter 8
Time Invariant System Analysis: State Space Approach

Abstract This chapter is devoted to time invariant system analysis using state space approach in hybrid function platform. Both homogeneous and non-homogeneous systems are treated along with numerical examples. States and outputs of the systems are solved. Also, a non-homogeneous system with a jump discontinuity at input is analyzed. Exhaustive illustration has been provided with the support of nine examples, twenty two figures and twenty one tables.

In this chapter, we deal with linear time invariant (LTI) control systems [1]. A linear control system abides by the superposition law and being time invariant, its parameters do not vary with time. We intend to analyse two types of LTI control systems in hybrid function platform, namely non-homogeneous system and homogeneous system [1, 2].

Analysing a control system means, knowing the system parameters and the nature of input signal or forcing function, we determine the behavior of different system states over a common time frame. The output of the system may be any one of the states or a combination of two or many states. Therefore, after knowing all the states within the system, we can easily assess the performance of the system.

In practice, application of linear time-invariant systems may be found in circuits, control theory, NMR spectroscopy, signal processing, seismology and in many other areas.

Any LTI system can be classified into two broad categories: one is 'non-homogeneous' system and the other being 'homogeneous' system. In a homogeneous system, no external signal is applied and we look for behavior of the states due to the presence of initial condition only. So, in short, the analysis of a homogeneous system helps us to know about the internal behavior of the system.

In a non-homogeneous system, we deal with the presence of both the initial conditions and external input signals simultaneously.

In this chapter, the hybrid function set is employed for the analysis of non-homogeneous as well as homogeneous systems described as state space models [2].

A. Deb et al., *Analysis and Identification of Time-Invariant Systems, Time-Varying Systems, and Multi-Delay Systems using Orthogonal Hybrid Functions*,
Studies in Systems, Decision and Control 46, DOI 10.1007/978-3-319-26684-8_8

First we take up the problem of analysis of a non-homogeneous system in HF domain, because, after putting the specific condition of zero forcing function, we can arrive at the result of analysis of a homogeneous system.

8.1 Analysis of Non-homogeneous State Equations [3]

Consider the non-homogeneous state equation,

$$\dot{\mathbf{x}}(t) = \mathbf{A}\mathbf{x}(t) + \mathbf{B}u(t) \tag{8.1}$$

where, $\mathbf{A}$ is an $(n \times n)$ system matrix given by

$$\mathbf{A} \triangleq \begin{bmatrix} a_{11} & a_{12} & a_{13} & \cdots & a_{1n} \\ a_{21} & a_{22} & a_{23} & \cdots & a_{2n} \\ \vdots & \vdots & \vdots & & \vdots \\ \vdots & \vdots & \vdots & & \vdots \\ a_{n1} & a_{n2} & a_{n3} & \cdots & a_{nn} \end{bmatrix}$$

$\mathbf{B}$ is the $(n \times 1)$ input vector given by $\mathbf{B} = \begin{bmatrix} b_1 & b_2 & \cdots & b_n \end{bmatrix}^T$
$\mathbf{x}(t)$ is the state vector given by $\mathbf{x}(t) = \begin{bmatrix} x_1 & x_2 & \cdots & x_n \end{bmatrix}^{\mathrm{T}}$
with the initial conditions,

$$\mathbf{x}(0) = \begin{bmatrix} x_1(0) & x_2(0) & \cdots & x_n(0) \end{bmatrix}^{\mathrm{T}}$$

where, $[\cdots]^{\mathrm{T}}$ denotes transpose, and u is the forcing function.

Integrating Eq. (8.1) we have

$$\mathbf{x}(t) - \mathbf{x}(0) = \mathbf{A}\int \mathbf{x}(t)\mathrm{d}t + \mathbf{B}\int u(t)\mathrm{d}t \tag{8.2}$$

Expanding $\mathbf{x}(t)$, $\mathbf{x}(0)$ and $u(t)$ via an m-set hybrid function [4], we get

$$\begin{aligned} \mathbf{x}(t) &\triangleq \mathbf{C}_{\mathrm{Sx}}\mathbf{S}_{(m)} + \mathbf{C}_{\mathrm{Tx}}\mathbf{T}_{(m)} \\ \mathbf{x}(0) &\triangleq \mathbf{C}_{\mathrm{Sx0}}\mathbf{S}_{(m)} + \mathbf{C}_{\mathrm{Tx0}}\mathbf{T}_{(m)} \text{ and} \\ u(t) &\triangleq \mathbf{C}_{\mathrm{Su}}^{\mathrm{T}}\mathbf{S}_{(m)} + \mathbf{C}_{\mathrm{Tu}}^{\mathrm{T}}\mathbf{T}_{(m)} \end{aligned}$$

where,

$$\mathbf{C}_{\mathrm{Sx}} = \begin{bmatrix} \mathbf{C}_{\mathrm{Sx1}}^{\mathrm{T}} \\ \mathbf{C}_{\mathrm{Sx2}}^{\mathrm{T}} \\ \vdots \\ \vdots \\ \mathbf{C}_{\mathrm{Sx}i}^{\mathrm{T}} \\ \vdots \\ \vdots \\ \mathbf{C}_{\mathrm{Sx}n}^{\mathrm{T}} \end{bmatrix}, \quad \mathbf{C}_{\mathrm{Tx}} = \begin{bmatrix} \mathbf{C}_{\mathrm{Tx1}}^{\mathrm{T}} \\ \mathbf{C}_{\mathrm{Tx2}}^{\mathrm{T}} \\ \vdots \\ \vdots \\ \mathbf{C}_{\mathrm{Tx}i}^{\mathrm{T}} \\ \vdots \\ \vdots \\ \mathbf{C}_{\mathrm{Tx}n}^{\mathrm{T}} \end{bmatrix}, \quad \mathbf{C}_{\mathrm{Sx0}} = \begin{bmatrix} \mathbf{C}_{\mathrm{Sx01}}^{\mathrm{T}} \\ \mathbf{C}_{\mathrm{Sx02}}^{\mathrm{T}} \\ \vdots \\ \vdots \\ \mathbf{C}_{\mathrm{Sx0}i}^{\mathrm{T}} \\ \vdots \\ \vdots \\ \mathbf{C}_{\mathrm{Sx0}n}^{\mathrm{T}} \end{bmatrix} \quad \text{and} \quad \mathbf{C}_{\mathrm{Tx0}} = \begin{bmatrix} \mathbf{C}_{\mathrm{Tx01}}^{\mathrm{T}} \\ \mathbf{C}_{\mathrm{Tx02}}^{\mathrm{T}} \\ \vdots \\ \vdots \\ \mathbf{C}_{\mathrm{Tx0}i}^{\mathrm{T}} \\ \vdots \\ \vdots \\ \mathbf{C}_{\mathrm{Tx0}n}^{\mathrm{T}} \end{bmatrix}$$

$$\begin{aligned}
\mathbf{C}_{\mathrm{Sx}i} &= \begin{bmatrix} c_{\mathrm{Sx}i1} & c_{\mathrm{Sx}i2} & c_{\mathrm{Sx}i3} & \cdots & c_{\mathrm{Sx}im} \end{bmatrix}^{\mathrm{T}} \\
\mathbf{C}_{\mathrm{Tx}i} &= \begin{bmatrix} c_{\mathrm{Tx}i1} & c_{\mathrm{Tx}i2} & c_{\mathrm{Tx}i3} & \cdots & c_{\mathrm{Tx}im} \end{bmatrix}^{\mathrm{T}} \\
\mathbf{C}_{\mathrm{Sx0}i} &= \begin{bmatrix} c_{\mathrm{Sx0}i1} & c_{\mathrm{Sx0}i2} & c_{\mathrm{Sx0}i3} & \cdots & c_{\mathrm{Sx0}im} \end{bmatrix}^{\mathrm{T}}
\end{aligned}$$

$$\begin{aligned}
\mathbf{C}_{\mathrm{Tx0}i} &= \begin{bmatrix} c_{\mathrm{Tx0}i1} & c_{\mathrm{Tx0}i2} & c_{\mathrm{Tx0}i3} & \cdots & c_{\mathrm{Tx0}im} \end{bmatrix}^{\mathrm{T}} \\
\mathbf{C}_{\mathrm{su}} &= \begin{bmatrix} c_{\mathrm{su},1} & c_{\mathrm{su},2} & c_{\mathrm{su},3} & \cdots & c_{\mathrm{su},m} \end{bmatrix}^{\mathrm{T}} \\
\mathbf{C}_{\mathrm{Tu}} &= \begin{bmatrix} c_{\mathrm{Tu},1} & c_{\mathrm{Tu},2} & c_{\mathrm{Tu},3} & \cdots & c_{\mathrm{Tu},m} \end{bmatrix}^{\mathrm{T}}
\end{aligned}$$

Substituting in (8.1) and rearranging we have

$$(\mathbf{C}_{\mathrm{Sx}} - \mathbf{C}_{\mathrm{Sx0}})\mathbf{S}_{(m)} + (\mathbf{C}_{\mathrm{Tx}} - \mathbf{C}_{\mathrm{Tx0}})\mathbf{T}_{(m)} = \mathbf{A}\int \mathbf{x}\mathrm{d}t + \mathbf{B}\int \mathrm{u}\mathrm{d}t \tag{8.3}$$

We take up the first term on the RHS of (8.3) to write

$$\mathbf{A}\int \mathbf{x}\mathrm{d}t = \mathbf{A}\int \left[\mathbf{C}_{\mathrm{Sx}}\mathbf{S}_{(m)} + \mathbf{C}_{\mathrm{Tx}}\mathbf{T}_{(m)}\right]\mathrm{d}t = \mathbf{A}\mathbf{C}_{\mathrm{Sx}}\int \mathbf{S}_{(m)}\mathrm{d}t + \mathbf{A}\mathbf{C}_{\mathrm{Tx}}\int \mathbf{T}_{(m)}\mathrm{d}t$$

Using relations (4.9) and (4.18), we have

$$\begin{aligned}
\mathbf{A}\int \mathbf{x}\mathrm{d}t &= \mathbf{A}\mathbf{C}_{\mathrm{Sx}}\left[\mathbf{P1ss}_{(m)}\mathbf{S}_{(m)} + \mathbf{P1st}_{(m)}\mathbf{T}_{(m)}\right] + \mathbf{A}\mathbf{C}_{\mathrm{Tx}}\left[\mathbf{P1ts}_{(m)}\mathbf{S}_{(m)} + \mathbf{P1tt}_{(m)}\mathbf{T}_{(m)}\right] \\
&= \mathbf{A}\left[\mathbf{C}_{\mathrm{Sx}} + \frac{1}{2}\mathbf{C}_{\mathrm{Tx}}\right]\mathbf{P1ss}_{(m)}\mathbf{S}_{(m)} + h\mathbf{A}\left[\mathbf{C}_{\mathrm{Sx}} + \frac{1}{2}\mathbf{C}_{\mathrm{Tx}}\right]\mathbf{T}_{(m)}
\end{aligned} \tag{8.4}$$

and similarly for the second term on the RHS of (8.3) we have

$$\begin{aligned}\mathbf{B}\int u\mathrm{d}t &= \mathbf{B}\int \left[\mathbf{C}_{\mathrm{Su}}^{\mathrm{T}}\mathbf{S}_{(m)} + \mathbf{C}_{\mathrm{Tu}}^{\mathrm{T}}\mathbf{T}_{(m)}\right]\mathrm{d}t \\ &= \mathbf{B}\left[\mathbf{C}_{\mathrm{Su}}^{\mathrm{T}} + \frac{1}{2}\mathbf{C}_{\mathrm{Tu}}^{\mathrm{T}}\right]\mathbf{P1ss}_{(m)}\mathbf{S}_{(m)} + h\mathbf{B}\left[\mathbf{C}_{\mathrm{Su}}^{\mathrm{T}} + \frac{1}{2}\mathbf{C}_{\mathrm{Tu}}^{\mathrm{T}}\right]\mathbf{T}_{(m)}\end{aligned} \tag{8.5}$$

Hence we can rewrite Eq. (8.3) as (dropping the dimension argument m)

$$\begin{aligned}(\mathbf{C}_{\mathrm{Sx}} - \mathbf{C}_{\mathrm{Sx0}})\mathbf{S} + (\mathbf{C}_{\mathrm{Tx}} - \mathbf{C}_{\mathrm{Tx0}})\mathbf{T} &= \mathbf{A}\left[\mathbf{C}_{\mathrm{Sx}} + \frac{1}{2}\mathbf{C}_{\mathrm{Tx}}\right]\mathbf{P1ss}\ \mathbf{S} + h\mathbf{A}\left[\mathbf{C}_{\mathrm{Sx}} + \frac{1}{2}\mathbf{C}_{\mathrm{Tx}}\right]\mathbf{T} \\ &\quad + \mathbf{B}\left[\mathbf{C}_{\mathrm{Su}}^{\mathrm{T}} + \frac{1}{2}\mathbf{C}_{\mathrm{Tu}}^{\mathrm{T}}\right]\mathbf{P1ss}\ \mathbf{S} + h\mathbf{B}\left[\mathbf{C}_{\mathrm{Su}}^{\mathrm{T}} + \frac{1}{2}\mathbf{C}_{\mathrm{Tu}}^{\mathrm{T}}\right]\mathbf{T} \\ &= \left[\mathbf{A}\mathbf{C}_{\mathrm{Sx}} + \frac{1}{2}\mathbf{A}\mathbf{C}_{\mathrm{Tx}} + \mathbf{B}\mathbf{C}_{\mathrm{Su}}^{\mathrm{T}} + \frac{1}{2}\mathbf{B}\mathbf{C}_{\mathrm{Tu}}^{\mathrm{T}}\right]\mathbf{P1ss}\ \mathbf{S} \\ &\quad + h\left[\mathbf{A}\mathbf{C}_{\mathrm{Sx}} + \frac{1}{2}\mathbf{A}\mathbf{C}_{\mathrm{Tx}} + \mathbf{B}\mathbf{C}_{\mathrm{Su}}^{\mathrm{T}} + \frac{1}{2}\mathbf{B}\mathbf{C}_{\mathrm{Tu}}^{\mathrm{T}}\right]\mathbf{T}\end{aligned} \tag{8.6}$$

Now equating like coefficients of (8.6), we get

$$\mathbf{C}_{\mathrm{Sx}} - \mathbf{C}_{\mathrm{Sx0}} = \left[\mathbf{A}\mathbf{C}_{\mathrm{Sx}} + \frac{1}{2}\mathbf{A}\mathbf{C}_{\mathrm{Tx}} + \mathbf{B}\mathbf{C}_{\mathrm{Su}}^{\mathrm{T}} + \frac{1}{2}\mathbf{B}\mathbf{C}_{\mathrm{Tu}}^{\mathrm{T}}\right]\mathbf{P1ss} \tag{8.7}$$

$$\mathbf{C}_{\mathrm{Tx}} - \mathbf{C}_{\mathrm{Tx0}} = h\left[\mathbf{A}\mathbf{C}_{\mathrm{Sx}} + \frac{1}{2}\mathbf{A}\mathbf{C}_{\mathrm{Tx}} + \mathbf{B}\mathbf{C}_{\mathrm{Su}}^{\mathrm{T}} + \frac{1}{2}\mathbf{B}\mathbf{C}_{\mathrm{Tu}}^{\mathrm{T}}\right] \tag{8.8}$$

These two Eqs. (8.7) and (8.8) are to be solved for $\mathbf{C}_{\mathrm{Sx}}$ and $\mathbf{C}_{\mathrm{Tx}}$ respectively. We can solve for $\mathbf{C}_{\mathrm{Sx}}$ and $\mathbf{C}_{\mathrm{Tx}}$, either from sample-and-hold function vector, or from triangular function vector. Both these approaches are described below.

8.1.1 Solution from Sample-and-Hold Function Vectors

The initial values of all the states being constants, they always essentially represent step functions. Hence, HF domain expansions of the initial values will always yield null coefficient matrices for the **T** vectors.

So, from Eq. (8.8)

$$
\begin{aligned}
\mathbf{C}_{\mathrm{Tx}} &= h\left[\mathbf{AC}_{\mathrm{Sx}} + \frac{1}{2}\mathbf{AC}_{\mathrm{Tx}} + \mathbf{BC}_{\mathrm{Su}}^{\mathrm{T}} + \frac{1}{2}\mathbf{BC}_{\mathrm{Tu}}^{\mathrm{T}}\right] \\
\text{or,}\quad \mathbf{AC}_{\mathrm{Sx}} &= \frac{1}{h}\mathbf{C}_{\mathrm{Tx}} - \frac{1}{2}\mathbf{AC}_{\mathrm{Tx}} - \mathbf{B}\left(\mathbf{C}_{\mathrm{Su}}^{\mathrm{T}} + \frac{1}{2}\mathbf{C}_{\mathrm{Tu}}^{\mathrm{T}}\right)
\end{aligned} \tag{8.9}
$$

Substituting relation (8.8) into (8.7) and simplifying, we have

$$
\mathbf{C}_{\mathrm{Tx}}\mathbf{P1ss} = h(\mathbf{C}_{\mathrm{Sx}} - \mathbf{C}_{\mathrm{Sx0}}) \tag{8.10}
$$

From Eq. (8.7),

$$
\begin{aligned}
\mathbf{C}_{\mathrm{Sx}} - \mathbf{C}_{\mathrm{Sx0}} &= \left[\mathbf{AC}_{\mathrm{Sx}} + \frac{1}{2}\mathbf{AC}_{\mathrm{Tx}} + \mathbf{BC}_{\mathrm{Su}}^{\mathrm{T}} + \frac{1}{2}\mathbf{BC}_{\mathrm{Tu}}^{\mathrm{T}}\right]\mathbf{P1ss} \\
&= \mathbf{AC}_{\mathrm{Sx}}\mathbf{P1ss} + \frac{1}{2}\mathbf{AC}_{\mathrm{Tx}}\mathbf{P1ss} + \mathbf{B}\left(\mathbf{C}_{\mathrm{Su}}^{\mathrm{T}} + \frac{1}{2}\mathbf{C}_{\mathrm{Tu}}^{\mathrm{T}}\right)\mathbf{P1ss}
\end{aligned} \tag{8.11}
$$

Using (8.10) on the RHS of (8.11), we have

$$
\mathbf{C}_{\mathrm{Sx}} - \mathbf{C}_{\mathrm{Sx0}} = \mathbf{AC}_{\mathrm{Sx}}\mathbf{P1ss} + \frac{1}{2}\mathbf{A}h(\mathbf{C}_{\mathrm{Sx}} - \mathbf{C}_{\mathrm{Sx0}}) + \mathbf{B}\left(\mathbf{C}_{\mathrm{Su}}^{\mathrm{T}} + \frac{1}{2}\mathbf{C}_{\mathrm{Tu}}^{\mathrm{T}}\right)\mathbf{P1ss}
$$

Calling the operational matrix for integration in (BPF) domain **P** [5] we have the following relation:

$$
\mathbf{P1ss} = \mathbf{P} - \frac{h}{2}\mathbf{I} \tag{8.12}
$$

Replacing **P1ss** following (8.12), we get

$$
\begin{aligned}
\mathbf{C}_{\mathrm{Sx}} - \mathbf{C}_{\mathrm{Sx0}} &= \mathbf{AC}_{\mathrm{Sx}}\left(\mathbf{P} - \frac{h}{2}\mathbf{I}\right) + \frac{1}{2}\mathbf{A}h(\mathbf{C}_{\mathrm{Sx}} - \mathbf{C}_{\mathrm{Sx0}}) + \mathbf{B}\left(\mathbf{C}_{\mathrm{Su}}^{\mathrm{T}} + \frac{1}{2}\mathbf{C}_{\mathrm{Tu}}^{\mathrm{T}}\right)\mathbf{P1ss} \\
&= \mathbf{AC}_{\mathrm{Sx}}\mathbf{P} - \frac{h}{2}\mathbf{AC}_{\mathrm{Sx0}} + \mathbf{B}\left(\mathbf{C}_{\mathrm{Su}}^{\mathrm{T}} + \frac{1}{2}\mathbf{C}_{\mathrm{Tu}}^{\mathrm{T}}\right)\mathbf{P1ss}
\end{aligned}
$$

Therefore,

$$
\mathbf{C}_{\mathrm{Sx}} - \mathbf{AC}_{\mathrm{Sx}}\mathbf{P} = \left(\mathbf{I} - \frac{h}{2}\mathbf{A}\right)\mathbf{C}_{\mathrm{Sx0}} + \mathbf{B}\left(\mathbf{C}_{\mathrm{Su}}^{\mathrm{T}} + \frac{1}{2}\mathbf{C}_{\mathrm{Tu}}^{\mathrm{T}}\right)\mathbf{P1ss} \tag{8.13}
$$

Now subtracting the ith column from the $(i + 1)$th column, we get

$$
\begin{aligned}
&[\mathbf{C}_{\mathrm{Sx}}]_{i+1} - [\mathbf{C}_{\mathrm{Sx}}]_i - [\mathbf{A}\mathbf{C}_{\mathrm{Sx}}\mathbf{P}]_{i+1} + [\mathbf{A}\mathbf{C}_{\mathrm{Sx}}\mathbf{P}]_i \\
&\quad = \left[\left(\mathbf{I} - \frac{h}{2}\mathbf{A}\right)\mathbf{C}_{\mathrm{Sx0}}\right]_{i+1} - \left[\left(\mathbf{I} - \frac{h}{2}\mathbf{A}\right)\mathbf{C}_{\mathrm{Sx0}}\right]_i \\
&\qquad + \left[\mathbf{B}\left(\mathbf{C}_{\mathrm{Su}}^{\mathrm{T}} + \frac{1}{2}\mathbf{C}_{\mathrm{Tu}}^{\mathrm{T}}\right)\mathbf{P1ss}\right]_{i+1} - \left[\mathbf{B}\left(\mathbf{C}_{\mathrm{Su}}^{\mathrm{T}} + \frac{1}{2}\mathbf{C}_{\mathrm{Tu}}^{\mathrm{T}}\right)\mathbf{P1ss}\right]_i \\
&\quad = \left[\mathbf{B}\left(\mathbf{C}_{\mathrm{Su}}^{\mathrm{T}} + \frac{1}{2}\mathbf{C}_{\mathrm{Tu}}^{\mathrm{T}}\right)\mathbf{P1ss}\right]_{i+1} - \left[\mathbf{B}\left(\mathbf{C}_{\mathrm{Su}}^{\mathrm{T}} + \frac{1}{2}\mathbf{C}_{\mathrm{Tu}}^{\mathrm{T}}\right)\mathbf{P1ss}\right]_i
\end{aligned}
\tag{8.14}
$$

Now, the $(i+1)$th column of $\mathbf{A}\mathbf{C}_{\mathrm{Sx}}\mathbf{P}$ is

$$
[\mathbf{A}\mathbf{C}_{\mathrm{Sx}}\mathbf{P}]_{i+1} = \begin{bmatrix} a_{11} & a_{12} & \cdots & a_{1n} \\ a_{21} & a_{22} & \cdots & a_{2n} \\ \vdots & \vdots & & \vdots \\ a_{n1} & a_{n2} & \cdots & a_{nn} \end{bmatrix} \begin{bmatrix} c_{\mathrm{Sx}11} & c_{\mathrm{Sx}12} & \cdots & c_{\mathrm{Sx}1m} \\ c_{\mathrm{Sx}21} & c_{\mathrm{Sx}22} & \cdots & c_{\mathrm{Sx}2m} \\ \vdots & \vdots & & \vdots \\ c_{\mathrm{Sx}n1} & c_{\mathrm{Sx}n2} & \cdots & c_{\mathrm{Sx}nm} \end{bmatrix} h \overset{(i+1)\text{th column}}{\overset{\downarrow}{\begin{bmatrix} 1 \\ 1 \\ \vdots \\ \frac{1}{2} \\ 0 \\ \vdots \\ 0 \end{bmatrix}}} \begin{matrix} \\ \\ \\ \leftarrow (i+1)\text{th element} \\ \leftarrow (i+2)\text{th element} \\ \\ \\ \end{matrix}
$$

$$
= h \begin{bmatrix} a_{11} & a_{12} & \cdots & a_{1n} \\ a_{21} & a_{22} & \cdots & a_{2n} \\ \vdots & \vdots & & \vdots \\ a_{n1} & a_{n2} & \cdots & a_{nn} \end{bmatrix} \begin{bmatrix} c_{\mathrm{Sx}11} + c_{\mathrm{Sx}12} + \cdots + \frac{1}{2}c_{\mathrm{Sx}1(i+1)} \\ c_{\mathrm{Sx}21} + c_{\mathrm{Sx}22} + \cdots + \frac{1}{2}c_{\mathrm{Sx}2(i+1)} \\ \vdots \\ c_{\mathrm{Sx}n1} + c_{\mathrm{Sx}n2} + \cdots + \frac{1}{2}c_{\mathrm{Sx}n(i+1)} \end{bmatrix}
$$

$$
= h \begin{bmatrix} a_{11}\left(c_{\mathrm{Sx}11} + c_{\mathrm{Sx}12} + \cdots + \frac{1}{2}c_{\mathrm{Sx}1(i+1)}\right) + a_{12}\left(c_{\mathrm{Sx}21} + c_{\mathrm{Sx}22} + \cdots + \frac{1}{2}c_{\mathrm{Sx}2(i+1)}\right) + \cdots \\ \quad + a_{1n}\left(c_{\mathrm{Sx}n1} + c_{\mathrm{Sx}n2} + \cdots + \frac{1}{2}c_{\mathrm{Sx}n(i+1)}\right) \\ a_{21}\left(c_{\mathrm{Sx}11} + c_{\mathrm{Sx}12} + \cdots + \frac{1}{2}c_{\mathrm{Sx}1(i+1)}\right) + a_{22}\left(c_{\mathrm{Sx}21} + c_{\mathrm{Sx}22} + \cdots + \frac{1}{2}c_{\mathrm{Sx}2(i+1)}\right) + \cdots \\ \quad + a_{2n}\left(c_{\mathrm{Sx}n1} + c_{\mathrm{Sx}n2} + \cdots + \frac{1}{2}c_{\mathrm{Sx}n(i+1)}\right) \\ \vdots \\ a_{n1}\left(c_{\mathrm{Sx}11} + c_{\mathrm{Sx}12} + \cdots + \frac{1}{2}c_{\mathrm{Sx}1(i+1)}\right) + a_{n2}\left(c_{\mathrm{Sx}21} + c_{\mathrm{Sx}22} + \cdots + \frac{1}{2}c_{\mathrm{Sx}2(i+1)}\right) + \cdots \\ \quad + a_{nn}\left(c_{\mathrm{Sx}n1} + c_{\mathrm{Sx}n2} + \cdots + \frac{1}{2}c_{\mathrm{Sx}n(i+1)}\right) \end{bmatrix}
\tag{8.15}
$$

Similarly, the ith column of $\mathbf{AC}_{\mathrm{Sx}}\mathbf{P}$ is

$$[\mathbf{AC}_{\mathrm{Sx}}\mathbf{P}]_i = h\begin{bmatrix} a_{11}\left(c_{\mathrm{Sx}11}+c_{\mathrm{Sx}12}+\cdots+\frac{1}{2}c_{\mathrm{Sx}1i}\right)+a_{12}\left(c_{\mathrm{Sx}21}+c_{\mathrm{Sx}22}+\cdots+\frac{1}{2}c_{\mathrm{Sx}2i}\right)+\cdots \\ +a_{1n}\left(c_{\mathrm{Sx}n1}+c_{\mathrm{Sx}n2}+\cdots+\frac{1}{2}c_{\mathrm{Sx}ni}\right) \\ a_{21}\left(c_{\mathrm{Sx}11}+c_{\mathrm{Sx}12}+\cdots+\frac{1}{2}c_{\mathrm{Sx}1i}\right)+a_{22}\left(c_{\mathrm{Sx}21}+c_{\mathrm{Sx}22}+\cdots+\frac{1}{2}c_{\mathrm{Sx}2i}\right)+\cdots \\ +a_{2n}\left(c_{\mathrm{Sx}n1}+c_{\mathrm{Sx}n2}+\cdots+\frac{1}{2}c_{\mathrm{Sx}ni}\right) \\ \vdots \\ a_{n1}\left(c_{\mathrm{Sx}11}+c_{\mathrm{Sx}12}+\cdots+\frac{1}{2}c_{\mathrm{Sx}1i}\right)+a_{n2}\left(c_{\mathrm{Sx}21}+c_{\mathrm{Sx}22}+\cdots+\frac{1}{2}c_{\mathrm{Sx}2i}\right)+\cdots \\ +a_{nn}\left(c_{\mathrm{Sx}n1}+c_{\mathrm{Sx}n2}+\cdots+\frac{1}{2}c_{\mathrm{Sx}ni}\right) \end{bmatrix} \tag{8.16}$$

Subtracting (8.16) from (8.15), we can write

$$[\mathbf{AC}_{\mathrm{Sx}}\mathbf{P}]_{i+1}-[\mathbf{AC}_{\mathrm{Sx}}\mathbf{P}]_i = h\begin{bmatrix} \frac{a_{11}}{2}c_{\mathrm{Sx}1i}+\frac{a_{11}}{2}c_{\mathrm{Sx}1(i+1)}+\cdots+\frac{a_{1n}}{2}c_{\mathrm{Sx}ni}+\frac{a_{1n}}{2}c_{\mathrm{Sx}n(i+1)} \\ \frac{a_{21}}{2}c_{\mathrm{Sx}1i}+\frac{a_{21}}{2}c_{\mathrm{Sx}1(i+1)}+\cdots+\frac{a_{2n}}{2}c_{\mathrm{Sx}ni}+\frac{a_{2n}}{2}c_{\mathrm{Sx}n(i+1)} \\ \vdots \\ \frac{a_{n1}}{2}c_{\mathrm{Sx}1\mathrm{i}}+\frac{a_{n1}}{2}c_{\mathrm{Sx}1(i+1)}+\cdots+\frac{a_{nn}}{2}c_{\mathrm{Sx}ni}+\frac{a_{nn}}{2}c_{\mathrm{Sx}n(i+1)} \end{bmatrix}$$

$$= \frac{h}{2}\begin{bmatrix} a_{11} & a_{12} & \cdots & a_{1n} \\ a_{21} & a_{22} & \cdots & a_{2n} \\ \vdots & \vdots & & \vdots \\ a_{n1} & a_{n2} & \cdots & a_{nn} \end{bmatrix}\begin{bmatrix} c_{\mathrm{Sx}1i} \\ c_{\mathrm{Sx}2i} \\ \vdots \\ c_{\mathrm{Sx}ni} \end{bmatrix} + \frac{h}{2}\begin{bmatrix} a_{11} & a_{12} & \cdots & a_{1n} \\ a_{21} & a_{22} & \cdots & a_{2n} \\ \vdots & \vdots & & \vdots \\ a_{n1} & a_{n2} & \cdots & a_{nn} \end{bmatrix}\begin{bmatrix} c_{\mathrm{Sx}1(i+1)} \\ c_{\mathrm{Sx}2(i+1)} \\ \vdots \\ c_{\mathrm{Sx}n(i+1)} \end{bmatrix}$$

$$\text{or,}\quad [\mathbf{AC}_{\mathrm{Sx}}\mathbf{P}]_{i+1}-[\mathbf{AC}_{\mathrm{Sx}}\mathbf{P}]_i = \frac{h}{2}\mathbf{A}\begin{bmatrix} c_{\mathrm{Sx}1i} \\ c_{\mathrm{Sx}2\mathrm{i}} \\ \vdots \\ c_{\mathrm{Sx}ni} \end{bmatrix} + \frac{h}{2}\mathbf{A}\begin{bmatrix} c_{\mathrm{Sx}1(i+1)} \\ c_{\mathrm{Sx}2(\mathrm{i}+1)} \\ \vdots \\ c_{\mathrm{Sx}n(\mathrm{i}+1)} \end{bmatrix} \tag{8.17}$$

Substituting relation (8.17) in the LHS of Eq. (8.14), we get,

$$
\begin{aligned}
&[\mathbf{C}_{\mathrm{Sx}}]_{i+1} - [\mathbf{C}_{\mathrm{Sx}}]_i - [\mathbf{A}\mathbf{C}_{\mathrm{Sx}}\mathbf{P}]_{i+1} + [\mathbf{A}\mathbf{C}_{\mathrm{Sx}}\mathbf{P}]_i \\
&= \begin{bmatrix} c_{\mathrm{Sx}1(i+1)} \\ c_{\mathrm{Sx}2(i+1)} \\ \vdots \\ c_{\mathrm{Sx}n(i+1)} \end{bmatrix} - \begin{bmatrix} c_{\mathrm{Sx}1i} \\ c_{\mathrm{Sx}2i} \\ \vdots \\ c_{\mathrm{Sx}ni} \end{bmatrix} - \frac{h}{2}\mathbf{A} \begin{bmatrix} c_{\mathrm{Sx}1(i+1)} \\ c_{\mathrm{Sx}2(i+1)} \\ \vdots \\ c_{\mathrm{Sx}n(i+1)} \end{bmatrix} - \frac{h}{2}\mathbf{A} \begin{bmatrix} c_{\mathrm{Sx}1i} \\ c_{\mathrm{Sx}2i} \\ \vdots \\ c_{\mathrm{Sx}ni} \end{bmatrix} \\
&= \left[\mathbf{I} - \frac{h}{2}\mathbf{A}\right] \begin{bmatrix} c_{\mathrm{Sx}1(i+1)} \\ c_{\mathrm{Sx}2(i+1)} \\ \vdots \\ c_{\mathrm{Sx}n(i+1)} \end{bmatrix} - \left[\mathbf{I} + \frac{h}{2}\mathbf{A}\right] \begin{bmatrix} c_{\mathrm{Sx}1i} \\ c_{\mathrm{Sx}2i} \\ \vdots \\ c_{\mathrm{Sx}ni} \end{bmatrix}
\end{aligned}
\tag{8.18}
$$

Now from RHS of Eq. (8.14)

$$
\left[\mathbf{B}\left(\mathbf{C}_{\mathrm{Su}}^{\mathrm{T}} + \frac{1}{2}\mathbf{C}_{\mathrm{Tu}}^{\mathrm{T}}\right)\mathbf{P1ss}\right]_{i+1} = \mathbf{B}\left(\mathbf{C}_{\mathrm{Su}}^{\mathrm{T}} + \frac{1}{2}\mathbf{C}_{\mathrm{Tu}}^{\mathrm{T}}\right)[\mathbf{P1ss}]_{i+1}
$$

and,

$$
\left[\mathbf{B}\left(\mathbf{C}_{\mathrm{Su}}^{\mathrm{T}} + \frac{1}{2}\mathbf{C}_{\mathrm{Tu}}^{\mathrm{T}}\right)\mathbf{P1ss}\right]_{i} = \mathbf{B}\left(\mathbf{C}_{\mathrm{Su}}^{\mathrm{T}} + \frac{1}{2}\mathbf{C}_{\mathrm{Tu}}^{\mathrm{T}}\right)[\mathbf{P1ss}]_{i}
$$

We know from (8.12) that $\mathbf{P1ss} = \mathbf{P} - \frac{h}{2}\mathbf{I} = \begin{bmatrix} 0 & h & h & \cdots & h \\ 0 & 0 & h & \cdots & h \\ \vdots & \vdots & \vdots & & \vdots \\ 0 & 0 & 0 & \cdots & 0 \end{bmatrix}$

Therefore,

$$\mathbf{B}\left(\mathbf{C}_{\mathrm{Su}}^{\mathrm{T}}+\frac{1}{2}\mathbf{C}_{\mathrm{Tu}}^{\mathrm{T}}\right)[\mathbf{P1ss}]_{i+1}=\mathbf{B}\left(\mathbf{C}_{\mathrm{Su}}^{\mathrm{T}}+\frac{1}{2}\mathbf{C}_{\mathrm{Tu}}^{\mathrm{T}}\right)\overset{(i+1)\text{th column}}{\overset{\downarrow}{\begin{bmatrix} h \\ h \\ \vdots \\ h \\ 0 \\ \vdots \\ 0 \end{bmatrix}}}\begin{matrix} \\ \\ \\ \leftarrow \quad i\text{th element} \\ \leftarrow (i+1)\text{th element} \\ \\ \\ \end{matrix}$$

$$=[\mathbf{B}]_{n\times 1}\left[\mathbf{C}_{\mathrm{Su}}^{\mathrm{T}}+\frac{1}{2}\mathbf{C}_{\mathrm{Tu}}^{\mathrm{T}}\right]_{1\times m}\begin{bmatrix} h \\ h \\ \vdots \\ h \\ 0 \\ \vdots \\ 0 \end{bmatrix}_{m\times 1}=h[\mathbf{B}]_{n\times 1}\sum_{j=1}^{i}\left(\mathbf{C}_{\mathrm{Su}}^{\mathrm{T}}+\frac{1}{2}\mathbf{C}_{\mathrm{Tu}}^{\mathrm{T}}\right)_{j}$$

Similarly,

$$\mathbf{B}\left(\mathbf{C}_{\mathrm{Su}}^{\mathrm{T}}+\frac{1}{2}\mathbf{C}_{\mathrm{Tu}}^{\mathrm{T}}\right)[\mathbf{P1ss}]_{i}=\mathbf{B}\left(\mathbf{C}_{\mathrm{Su}}^{\mathrm{T}}+\frac{1}{2}\mathbf{C}_{\mathrm{Tu}}^{\mathrm{T}}\right)\overset{i\text{th element}}{\overset{\downarrow}{\begin{bmatrix} h \\ h \\ \vdots \\ h \\ 0 \\ \vdots \\ 0 \end{bmatrix}}}\begin{matrix} \\ \\ \\ \leftarrow (i-1)\text{th element} \\ \leftarrow \quad i\text{th element} \\ \\ \\ \end{matrix}$$

$$=h[\mathbf{B}]_{n\times 1}\sum_{j=1}^{i-1}\left(\mathbf{C}_{\mathrm{Su}}^{\mathrm{T}}+\frac{1}{2}\mathbf{C}_{\mathrm{Tu}}^{\mathrm{T}}\right)_{j}$$

Therefore,

$$\begin{aligned}
&\left[\mathbf{B}\left(\mathbf{C}_{\mathrm{Su}}^{\mathrm{T}}+\frac{1}{2}\mathbf{C}_{\mathrm{Tu}}^{\mathrm{T}}\right)\mathbf{P1ss}\right]_{i+1}-\left[\mathbf{B}\left(\mathbf{C}_{\mathrm{Su}}^{\mathrm{T}}+\frac{1}{2}\mathbf{C}_{\mathrm{Tu}}^{\mathrm{T}}\right)\mathbf{P1ss}\right]_{i}\\
&=h[\mathbf{B}]_{n\times1}\sum_{j=1}^{i}\left(\mathbf{C}_{\mathrm{Su}}^{\mathrm{T}}+\frac{1}{2}\mathbf{C}_{\mathrm{Tu}}^{\mathrm{T}}\right)_{j}-h[\mathbf{B}]_{n\times1}\sum_{j=1}^{i-1}\left(\mathbf{C}_{\mathrm{Su}}^{\mathrm{T}}+\frac{1}{2}\mathbf{C}_{\mathrm{Tu}}^{\mathrm{T}}\right)_{j} \qquad (8.19)\\
&=h[\mathbf{B}]_{n\times1}\left[\mathbf{C}_{\mathrm{Su}}^{\mathrm{T}}+\frac{1}{2}\mathbf{C}_{\mathrm{Tu}}^{\mathrm{T}}\right]_{i}
\end{aligned}$$

After substituting the expressions from Eqs. (8.14), (8.18) and (8.19), we can write the following recursive structure of system states

$$\left[\mathbf{I}-\frac{h}{2}\mathbf{A}\right]\begin{bmatrix} c_{\mathrm{Sx}1(i+1)}\\ c_{\mathrm{Sx}2(i+1)}\\ \vdots\\ c_{\mathrm{Sx}n(i+1)}\end{bmatrix}-\left[\mathbf{I}+\frac{h}{2}\mathbf{A}\right]\begin{bmatrix} c_{\mathrm{Sx}1i}\\ c_{\mathrm{Sx}2i}\\ \vdots\\ c_{\mathrm{Sx}ni}\end{bmatrix}=h\mathbf{B}\left[\mathbf{C}_{\mathrm{Su}}^{\mathrm{T}}+\frac{1}{2}\mathbf{C}_{\mathrm{Tu}}^{\mathrm{T}}\right]_{i}$$

$$\text{or,}\quad \left[\frac{2}{h}\mathbf{I}-\mathbf{A}\right]\begin{bmatrix} c_{\mathrm{Sx}1(i+1)}\\ c_{\mathrm{Sx}2(i+1)}\\ \vdots\\ c_{\mathrm{Sx}n(i+1)}\end{bmatrix}-\left[\frac{2}{h}\mathbf{I}+\mathbf{A}\right]\begin{bmatrix} c_{\mathrm{Sx}1i}\\ c_{\mathrm{Sx}2i}\\ \vdots\\ c_{\mathrm{Sx}ni}\end{bmatrix}=2\mathbf{B}\left[\mathbf{C}_{\mathrm{Su}}^{\mathrm{T}}+\frac{1}{2}\mathbf{C}_{\mathrm{Tu}}^{\mathrm{T}}\right]_{i} \qquad (8.20)$$

From Eq. (8.20), using matrix inversion, we have

$$\begin{bmatrix} c_{\mathrm{Sx}1(i+1)}\\ c_{\mathrm{Sx}2(i+1)}\\ \vdots\\ c_{\mathrm{Sxn}(i+1)}\end{bmatrix}=\left[\frac{2}{h}\mathbf{I}-\mathbf{A}\right]^{-1}\left[\frac{2}{h}\mathbf{I}+\mathbf{A}\right]\begin{bmatrix} c_{\mathrm{Sx}1i}\\ c_{\mathrm{Sx}2i}\\ \vdots\\ c_{\mathrm{Sx}ni}\end{bmatrix}+2\left[\frac{2}{h}\mathbf{I}-\mathbf{A}\right]^{-1}\mathbf{B}\left[\mathbf{C}_{\mathrm{Su}}^{\mathrm{T}}+\frac{1}{2}\mathbf{C}_{\mathrm{Tu}}^{\mathrm{T}}\right]_{i} \qquad (8.21)$$

The inverse in (8.21) can always be made to exist by judicious choice of h.

Equation (8.21) provides a simple recursive solution of the states of a non-homogeneous system, or, in other words, time samples of the states, with a sampling period of h knowing the system matrix **A**, the input matrix **B**, the input signal u, and the initial values of the states.

8.1.2 Solution from Triangular Function Vectors

Now from Eq. (8.8), we have

$$\mathbf{C}_{\mathrm{Tx}} = h\left[\mathbf{AC}_{\mathrm{Sx}} + \frac{1}{2}\mathbf{AC}_{\mathrm{Tx}} + \mathbf{BC}_{\mathrm{Su}}^{\mathrm{T}} + \frac{1}{2}\mathbf{BC}_{\mathrm{Tu}}^{\mathrm{T}}\right]$$

And from Eq. (8.7), using (8.12), we have

$$\begin{aligned}\mathbf{C}_{\mathrm{Sx}} - \mathbf{C}_{\mathrm{Sx0}} &= \left[\mathbf{AC}_{\mathrm{Sx}} + \frac{1}{2}\mathbf{AC}_{\mathrm{Tx}}\right]\left[\mathbf{P} - \frac{h}{2}\mathbf{I}\right] + \mathbf{B}\left[\mathbf{C}_{\mathrm{Su}}^{\mathrm{T}} + \frac{1}{2}\mathbf{C}_{\mathrm{Tu}}^{\mathrm{T}}\right]\mathbf{P1ss} \\ &= \mathbf{AC}_{\mathrm{Sx}}\mathbf{P} - \frac{h}{2}\mathbf{AC}_{\mathrm{Sx}} + \frac{1}{2}\mathbf{AC}_{\mathrm{Tx}}\mathbf{P} - \frac{h}{4}\mathbf{AC}_{\mathrm{Tx}} + \mathbf{B}\left[\mathbf{C}_{\mathrm{Su}}^{\mathrm{T}} + \frac{1}{2}\mathbf{C}_{\mathrm{Tu}}^{\mathrm{T}}\right]\mathbf{P1ss} \\ \text{or,}\quad & (\mathbf{C}_{\mathrm{Sx}} - \mathbf{AC}_{\mathrm{Sx}}\mathbf{P}) - \mathbf{C}_{\mathrm{Sx0}} + \frac{h}{2}\mathbf{AC}_{\mathrm{Sx}} \\ &= \frac{1}{2}\mathbf{AC}_{\mathrm{Tx}}\mathbf{P} - \frac{h}{4}\mathbf{AC}_{\mathrm{Tx}} + \mathbf{B}\left[\mathbf{C}_{\mathrm{Su}}^{\mathrm{T}} + \frac{1}{2}\mathbf{C}_{\mathrm{Tu}}^{\mathrm{T}}\right]\mathbf{P1ss}\end{aligned} \tag{8.22}$$

From Eqs. (8.13) and (8.22), we have

$$\begin{aligned}&\left[\mathbf{I} - \frac{h}{2}\mathbf{A}\right]\mathbf{C}_{\mathrm{Sx0}} + \mathbf{B}\left[\mathbf{C}_{\mathrm{Su}}^{\mathrm{T}} + \frac{1}{2}\mathbf{C}_{\mathrm{Tu}}^{\mathrm{T}}\right]\mathbf{P1ss} - \mathbf{C}_{\mathrm{Sx0}} + \frac{h}{2}\mathbf{AC}_{\mathrm{Sx}} \\ &= \frac{1}{2}\mathbf{AC}_{\mathrm{Tx}}\mathbf{P} - \frac{h}{4}\mathbf{AC}_{\mathrm{Tx}} + \mathbf{B}\left[\mathbf{C}_{\mathrm{Su}}^{\mathrm{T}} + \frac{1}{2}\mathbf{C}_{\mathrm{Tu}}^{\mathrm{T}}\right]\mathbf{P1ss} \\ &\text{or,}\quad \mathbf{AC}_{\mathrm{Sx}} = \mathbf{AC}_{\mathrm{Sx0}} + \frac{1}{h}\mathbf{AC}_{\mathrm{Tx}}\mathbf{P} - \frac{1}{2}\mathbf{AC}_{\mathrm{Tx}}\end{aligned} \tag{8.23}$$

The initial values of all the states being constants, they always essentially represent step functions. Hence, HF domain expansions of the initial values will always yield null coefficient matrices for the **T** vectors. That means $\mathbf{C}_{\mathrm{Tx0}} = 0$.

Using (8.23) in (8.8) we have

$$\begin{aligned}&\mathbf{C}_{\mathrm{Tx}} = h\left[\mathbf{AC}_{\mathrm{Sx0}} + \frac{1}{h}\mathbf{AC}_{\mathrm{Tx}}\mathbf{P} - \frac{1}{2}\mathbf{AC}_{\mathrm{Tx}} + \frac{1}{2}\mathbf{AC}_{\mathrm{Tx}} + \mathbf{B}\left(\mathbf{C}_{\mathrm{Su}}^{\mathrm{T}} + \frac{1}{2}\mathbf{C}_{\mathrm{Tu}}^{\mathrm{T}}\right)\right] \\ &\text{or,}\quad \mathbf{C}_{\mathrm{Tx}} - \mathbf{AC}_{\mathrm{Tx}}\mathbf{P} = h\mathbf{AC}_{\mathrm{Sx0}} + h\mathbf{B}\left(\mathbf{C}_{\mathrm{Su}}^{\mathrm{T}} + \frac{1}{2}\mathbf{C}_{\mathrm{Tu}}^{\mathrm{T}}\right)\end{aligned} \tag{8.24}$$

Subtracting the ith column from the $(i + 1)$th column, we have

$$\begin{aligned}&[\mathbf{C}_{\mathrm{Tx}}]_{i+1}-[\mathbf{C}_{\mathrm{Tx}}]_i-[\mathbf{A}\mathbf{C}_{\mathrm{Tx}}\mathbf{P}]_{i+1}+[\mathbf{A}\mathbf{C}_{\mathrm{Tx}}\mathbf{P}]_i\\&=\left[h\mathbf{B}\left(\mathbf{C}_{\mathrm{Su}}^{\mathrm{T}}+\frac{1}{2}\mathbf{C}_{\mathrm{Tu}}^{\mathrm{T}}\right)\right]_{i+1}-\left[h\mathbf{B}\left(\mathbf{C}_{\mathrm{Su}}^{\mathrm{T}}+\frac{1}{2}\mathbf{C}_{\mathrm{Tu}}^{\mathrm{T}}\right)\right]_i\end{aligned}\tag{8.25}$$

Similar to Eqs. (8.18) and (8.25) can be written as

$$\begin{aligned}&[\mathbf{C}_{\mathrm{Tx}}]_{i+1}-[\mathbf{C}_{\mathrm{Tx}}]_i-[\mathbf{A}\mathbf{C}_{\mathrm{Tx}}\mathbf{P}]_{i+1}+[\mathbf{A}\mathbf{C}_{\mathrm{Tx}}\mathbf{P}]_i\\&=\left[\mathbf{I}-\frac{h}{2}\mathbf{A}\right]\begin{bmatrix}c_{\mathrm{Tx}1(i+1)}\\c_{\mathrm{Tx}2(i+1)}\\\vdots\\c_{\mathrm{Tx}n(i+1)}\end{bmatrix}-\left[\mathbf{I}+\frac{h}{2}\mathbf{A}\right]\begin{bmatrix}c_{\mathrm{Tx}1i}\\c_{\mathrm{Tx}2i}\\\vdots\\c_{\mathrm{Tx}ni}\end{bmatrix}\end{aligned}\tag{8.26}$$

From Eqs. (8.25) and (8.26), we have

$$\begin{aligned}&\left[\mathbf{I}-\frac{h}{2}\mathbf{A}\right]\begin{bmatrix}\mathbf{C}_{\mathrm{Tx}1(i+1)}\\\mathbf{C}_{\mathrm{Tx}2(i+1)}\\\vdots\\\mathbf{C}_{\mathrm{Tx}n(i+1)}\end{bmatrix}-\left[\mathbf{I}+\frac{h}{2}\mathbf{A}\right]\begin{bmatrix}c_{\mathrm{Tx}1i}\\c_{\mathrm{Tx}2i}\\\vdots\\c_{\mathrm{Tx}ni}\end{bmatrix}\\&=h\left[\mathbf{B}\left(\mathbf{C}_{\mathrm{Su}}^{\mathrm{T}}+\frac{1}{2}\mathbf{C}_{\mathrm{Tu}}^{\mathrm{T}}\right)\right]_{i+1}-h\left[\mathbf{B}\left(\mathbf{C}_{\mathrm{Su}}^{\mathrm{T}}+\frac{1}{2}\mathbf{C}_{\mathrm{Tu}}^{\mathrm{T}}\right)\right]_i\end{aligned}\tag{8.27}$$

or,

$$\begin{aligned}&\begin{bmatrix}c_{\mathrm{Tx}1(i+1)}\\c_{\mathrm{Tx}2(i+1)}\\\vdots\\c_{\mathrm{Tx}n(i+1)}\end{bmatrix}=\left[\frac{2}{h}\mathbf{I}-\mathbf{A}\right]^{-1}\left[\frac{2}{h}\mathbf{I}+\mathbf{A}\right]\begin{bmatrix}c_{\mathrm{Tx}1i}\\c_{\mathrm{Tx}2i}\\\vdots\\c_{\mathrm{Tx}ni}\end{bmatrix}\\&+2\left[\frac{2}{h}\mathbf{I}-\mathbf{A}\right]^{-1}\left\{\left[\mathbf{B}\left(\mathbf{C}_{\mathrm{Su}}^{\mathrm{T}}+\frac{1}{2}\mathbf{C}_{\mathrm{Tu}}^{\mathrm{T}}\right)\right]_{i+1}-\left[\mathbf{B}\left(\mathbf{C}_{\mathrm{Su}}^{\mathrm{T}}+\frac{1}{2}\mathbf{C}_{\mathrm{Tu}}^{\mathrm{T}}\right)\right]_i\right\}\end{aligned}\tag{8.28}$$

Equation (8.28) provides an alternative recursive solution of the states of a non-homogeneous system, knowing the system matrix **A**, the input matrix **B**, the input signal u, and the initial values of the states. The solution as obtained via Eq. (8.21) can be verified by Eq. (8.28) as well. But the only thing we have to remember that, in case of Eq. (8.28), to know the initial value of **T** matrix, we should have first two samples of the states. Whereas the second sample of the state can be determined with the help of Eq. (8.21) only. So when only first sample of the states are given, the system states can be solved only by Eq. (8.21). And if fortunately first two samples of states are available, then only with the help of **T** matrix i.e. equation (8.28), the system states can be solved.

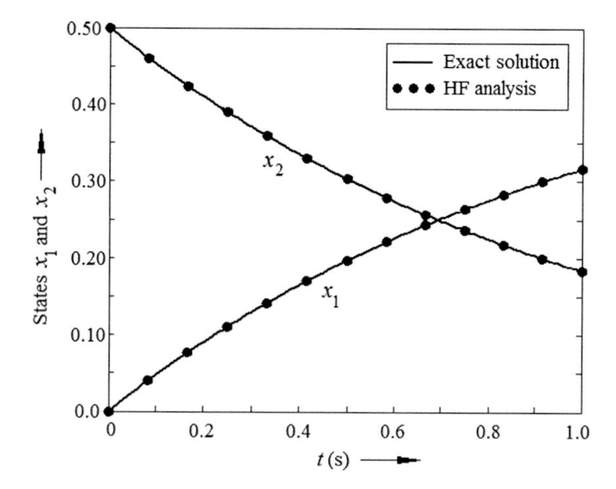

Fig. 8.1 Comparison of HF based recursive solution of Example 8.1, for $m = 12$ with the exact solutions of state x_1 and state x_2 (vide Appendix B, Program no. 22)

8.1.3 Numerical Examples

Example 8.1 [1] (vide Appendix B, Program no. 22) Consider a non-homogeneous system given by $\dot{\mathbf{x}}(t) = \mathbf{A}\mathbf{x}(t) + \mathbf{B}u(t)$ where

$$\mathbf{A} = \begin{bmatrix} 0 & 1 \\ -2 & -3 \end{bmatrix}, \quad \mathbf{B} = \begin{bmatrix} 0 \\ 1 \end{bmatrix}, \quad \mathbf{x}_0 = \begin{bmatrix} 0 \\ 0.5 \end{bmatrix} \quad \text{and} \quad u(t) = 1$$

having the solution $x_1(t) = 0.5(1 - exp(-t))$ and $x_2(t) = 0.5\,exp(-t)$.
The graphical comparison of the system states of Example 8.1, obtained via HF domain analysis (for $m = 12$) with their direct expansion are presented in Fig. 8.1, whereas in Table 8.1 we compare the results obtained in HF domain using direct expansion for $m = 8$.

Figure 8.2 is proof enough that the percentage error of HF based recursive solution decreases drastically as the number of segments m increases. With the increase in m, it is observed that the number of zero error points has increased.

8.2 Determination of Output of a Non-homogeneous System [3]

Consider the output of a non-homogeneous system described by

$$\mathbf{y}(t) = \mathbf{C}\mathbf{x}(t) + \mathbf{D}\mathbf{u}(t) \tag{8.29}$$

where,

$\mathbf{x}$ is the *state vector* given by $\mathbf{x} = \begin{bmatrix} x_1 & x_2 & \cdots & x_n \end{bmatrix}^T$

Table 8.1 Solution of states x_1 and x_2 of the non-homogeneous system of Example 8.1 with comparison of exact samples and corresponding samples obtained via HF domain with percentage error at different sample points for m = 8 and T = 1 s (vide Appendix B, Program no. 22)

(a)

t(s)	System state x_1		
	Exact samples of the state $x_{1,d}$	Samples from HF analysis, using Eq. (8.21), $x_{1,h}$	% error $\epsilon_1 = \frac{x_{1,d}-x_{1,h}}{x_{1,d}} \times 100$
0	0.00000000	0.00000000	–
$\frac{1}{8}$	0.05875155	0.05882353	−0.12251592
$\frac{2}{8}$	0.11059961	0.11072664	−0.11485574
$\frac{3}{8}$	0.15635536	0.15652351	−0.10754348
$\frac{4}{8}$	0.19673467	0.19693251	−0.10056184
$\frac{5}{8}$	0.23236929	0.23258751	−0.09391086
$\frac{6}{8}$	0.26381672	0.26404780	−0.08759111
$\frac{7}{8}$	0.29156899	0.29180688	−0.08158961
$\frac{8}{8}$	0.31606028	0.31630019	−0.07590641

(b)

t(s)	System state x_2		
	Exact samples of the state $x_{2,d}$	Samples from HF analysis, using Eq. (8.21), $x_{2,h}$	% error $\epsilon_2 = \frac{x_{2,d}-x_{2,h}}{x_{2,d}} \times 100$
0	0.50000000	0.50000000	0.00000000
$\frac{1}{8}$	0.44124845	0.44117647	0.01631281
$\frac{2}{8}$	0.38940039	0.38927336	0.03262195
$\frac{3}{8}$	0.34364464	0.34347649	0.04893136
$\frac{4}{8}$	0.30326533	0.30306749	0.06523660
$\frac{5}{8}$	0.26763071	0.26741249	0.08153773
$\frac{6}{8}$	0.23618328	0.23595220	0.09783927
$\frac{7}{8}$	0.20843101	0.20819312	0.11413369
$\frac{8}{8}$	0.18393972	0.18369981	0.13042860

$\mathbf{y}(t)$ is the *output vector*, is expressed by $\mathbf{y}(t) \triangleq [y_1 \quad y_2 \quad \cdots \quad y_v]^T$

$\mathbf{u}(t)$ is the *input vector*, is expressed by $\mathbf{u}(t) \triangleq [u_1 \quad u_2 \quad \cdots \quad u_r]^T$

$\mathbf{C}$ is the *output matrix* given by $\mathbf{C} \triangleq \begin{bmatrix} c_{11} & c_{12} & c_{13} & \cdots & c_{1n} \\ c_{21} & c_{22} & c_{23} & \cdots & c_{2n} \\ \vdots & \vdots & \vdots & & \vdots \\ \vdots & \vdots & \vdots & & \vdots \\ c_{v1} & c_{v2} & c_{v3} & \cdots & c_{vn} \end{bmatrix}$

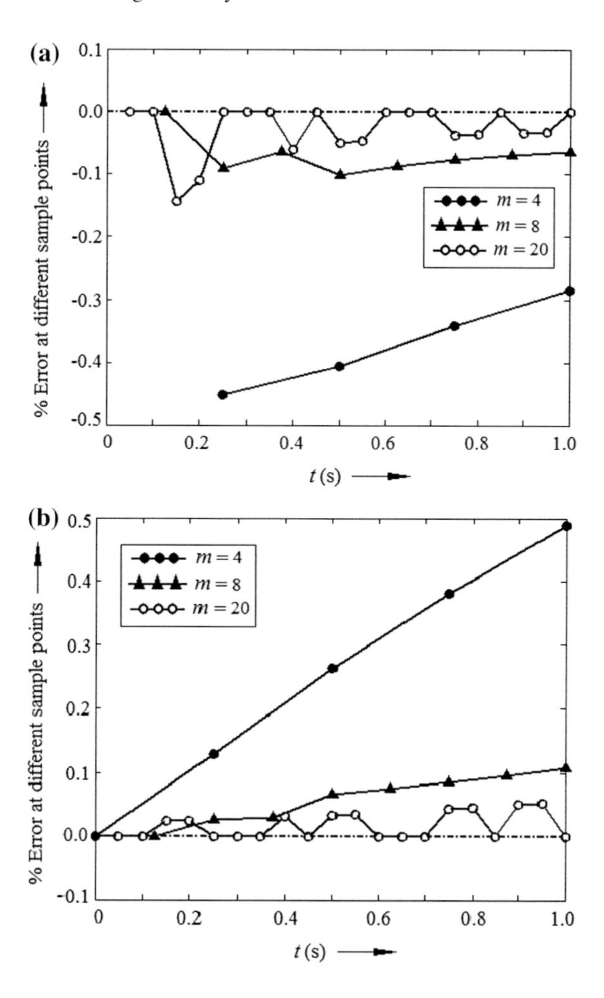

Fig. 8.2 Percentage error for three different values of m (m = 4, 8 and 20) and T = 1 s for **a** state x_1 and **b** state x_2 of Example 8.1

D is the *direct transmission matrix* given by $\mathbf{D} \triangleq \begin{bmatrix} d_{11} & d_{12} & d_{13} & \cdots & d_{1r} \\ d_{21} & d_{22} & d_{23} & \cdots & d_{2r} \\ \vdots & \vdots & \vdots & & \vdots \\ \vdots & \vdots & \vdots & & \vdots \\ d_{v1} & d_{v2} & d_{v3} & \cdots & d_{vr} \end{bmatrix}$

As before, expanding state vector **x**, output vector $\boldsymbol{y}(t)$ and the forcing function $u(t)$ via an m-set hybrid function set, we get

$$\mathbf{x}(t) \triangleq \mathbf{C}_{\mathrm{Sx}}\mathbf{S}_{(m)} + \mathbf{C}_{\mathrm{Tx}}\mathbf{T}_{(m)} \quad \mathbf{y}(t) \triangleq \mathbf{y}_{\mathrm{S}}\mathbf{S}_{(m)} + \mathbf{y}_{\mathrm{T}}\mathbf{T}_{(m)} \quad \mathbf{u}(t) \triangleq \mathbf{C}_{\mathrm{Su}}\mathbf{S}_{(m)} + \mathbf{C}_{\mathrm{Tu}}\mathbf{T}_{(m)}$$

where,

$$\mathbf{C}_{\mathrm{Sx}} = \begin{bmatrix} \mathbf{C}_{\mathrm{Sx}1}^{\mathrm{T}} \\ \mathbf{C}_{\mathrm{Sx}2}^{\mathrm{T}} \\ \vdots \\ \vdots \\ \mathbf{C}_{\mathrm{Sx}i}^{\mathrm{T}} \\ \vdots \\ \vdots \\ \mathbf{C}_{\mathrm{Sx}n}^{\mathrm{T}} \end{bmatrix}_{n\times m}, \quad \mathbf{C}_{\mathrm{Tx}} = \begin{bmatrix} \mathbf{C}_{\mathrm{Tx}1}^{\mathrm{T}} \\ \mathbf{C}_{\mathrm{Tx}2}^{\mathrm{T}} \\ \vdots \\ \vdots \\ \mathbf{C}_{\mathrm{Tx}i}^{\mathrm{T}} \\ \vdots \\ \vdots \\ \mathbf{C}_{\mathrm{Tx}n}^{\mathrm{T}} \end{bmatrix}_{n\times m}, \quad \mathbf{C}_{\mathrm{Su}} = \begin{bmatrix} \mathbf{C}_{\mathrm{Su}1}^{\mathrm{T}} \\ \mathbf{C}_{\mathrm{Su}2}^{\mathrm{T}} \\ \vdots \\ \vdots \\ \mathbf{C}_{\mathrm{Su}i}^{\mathrm{T}} \\ \vdots \\ \vdots \\ \mathbf{C}_{\mathrm{Su}r}^{\mathrm{T}} \end{bmatrix}_{r\times m},$$

$$\mathbf{C}_{\mathrm{Tu}} = \begin{bmatrix} \mathbf{C}_{\mathrm{Tu}1}^{\mathrm{T}} \\ \mathbf{C}_{\mathrm{Tu}2}^{\mathrm{T}} \\ \vdots \\ \vdots \\ \mathbf{C}_{\mathrm{Tu}i}^{\mathrm{T}} \\ \vdots \\ \vdots \\ \mathbf{C}_{\mathrm{Tu}r}^{\mathrm{T}} \end{bmatrix}_{r\times m} \quad \mathbf{y}_{\mathrm{S}} = \begin{bmatrix} \mathbf{y}_{\mathrm{S}1}^{\mathrm{T}} \\ \mathbf{y}_{\mathrm{S}2}^{\mathrm{T}} \\ \vdots \\ \vdots \\ \mathbf{y}_{\mathrm{S}i}^{\mathrm{T}} \\ \vdots \\ \vdots \\ \mathbf{y}_{\mathrm{S}v}^{\mathrm{T}} \end{bmatrix}_{v\times m}, \mathbf{y}_{\mathrm{T}} = \begin{bmatrix} \mathbf{y}_{\mathrm{T}1}^{\mathrm{T}} \\ \mathbf{y}_{\mathrm{T}2}^{\mathrm{T}} \\ \vdots \\ \vdots \\ \mathbf{y}_{\mathrm{T}i}^{\mathrm{T}} \\ \vdots \\ \vdots \\ \mathbf{y}_{\mathrm{T}v}^{\mathrm{T}} \end{bmatrix}_{v\times m}$$

and,

$$\begin{aligned}
&\mathbf{C}_{\mathrm{Sx}i}^{\mathrm{T}} = \begin{bmatrix} c_{\mathrm{Sx}i1} & c_{\mathrm{Sx}i2} & c_{\mathrm{Sx}i3} & \cdots & c_{\mathrm{Sx}im} \end{bmatrix}, \quad \mathbf{C}_{\mathrm{Tx}i}^{\mathrm{T}} = \begin{bmatrix} c_{\mathrm{Tx}i1} & c_{\mathrm{Tx}i2} & c_{\mathrm{Tx}i3} & \cdots & c_{\mathrm{Tx}im} \end{bmatrix} \\
&\mathbf{C}_{\mathrm{Su}i}^{\mathrm{T}} = \begin{bmatrix} c_{\mathrm{Su}i1} & c_{\mathrm{Su}i2} & c_{\mathrm{Su}i3} & \cdots & c_{\mathrm{Su}im} \end{bmatrix}, \quad \mathbf{C}_{\mathrm{Tu}i}^{\mathrm{T}} = \begin{bmatrix} c_{\mathrm{Tu}i1} & c_{\mathrm{Tu}i2} & c_{\mathrm{Tu}i3} & \cdots & c_{\mathrm{Tu}im} \end{bmatrix} \\
&\mathbf{y}_{\mathrm{S}i}^{\mathrm{T}} = \begin{bmatrix} c_{\mathrm{S}i1} & c_{\mathrm{S}i2} & c_{\mathrm{S}i3} & \cdots & c_{\mathrm{S}im} \end{bmatrix}, \quad \mathbf{y}_{\mathrm{T}i}^{\mathrm{T}} = \begin{bmatrix} c_{\mathrm{T}i1} & c_{\mathrm{T}i2} & c_{\mathrm{T}i3} & \cdots & c_{\mathrm{T}im} \end{bmatrix}
\end{aligned}$$

Substituting in (8.29) we have

$$\begin{aligned}
\mathbf{y}_{\mathrm{S}}\mathbf{S}_{(m)} + \mathbf{y}_{\mathrm{T}}\mathbf{T}_{(m)} &= \mathbf{C}\left(\mathbf{C}_{\mathrm{Sx}}\mathbf{S}_{(m)} + \mathbf{C}_{\mathrm{Tx}}\mathbf{T}_{(m)}\right) + \mathbf{D}\left(\mathbf{C}_{\mathrm{Su}}\mathbf{S}_{(m)} + \mathbf{C}_{\mathrm{Tu}}\mathbf{T}_{(m)}\right) \\
&= (\mathbf{C}\mathbf{C}_{\mathrm{Sx}} + \mathbf{D}\mathbf{C}_{\mathrm{Su}})\mathbf{S}_{(m)} + (\mathbf{C}\mathbf{C}_{\mathrm{Tx}} + \mathbf{D}\mathbf{C}_{\mathrm{Tu}})\mathbf{T}_{(m)}
\end{aligned} \tag{8.30}$$

From Eq. (8.30) we can write

$$\mathbf{y}_{\mathrm{S}} = \mathbf{C}\mathbf{C}_{\mathrm{Sx}} + \mathbf{D}\mathbf{C}_{\mathrm{Su}} \tag{8.31}$$

$$\mathbf{y}_{\mathrm{T}} = \mathbf{C}\mathbf{C}_{\mathrm{Tx}} + \mathbf{D}\mathbf{C}_{\mathrm{Tu}} \tag{8.32}$$

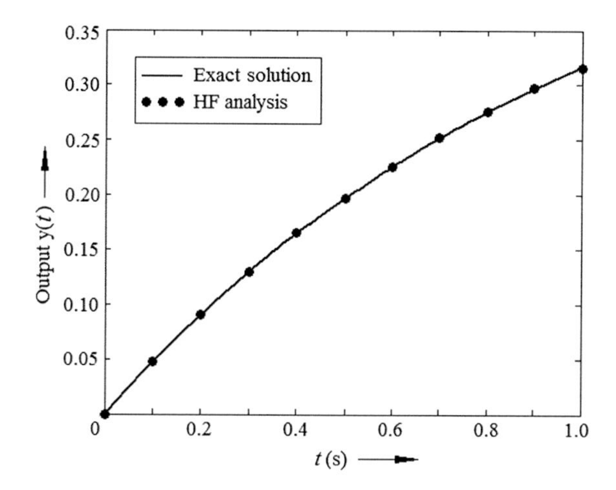

Fig. 8.3 Hybrid function based analysis of system output with $m = 10$ and its comparison with the exact output of Example 8.2 (vide Appendix B, Program no. 23)

Table 8.2 Solution of output of the non-homogeneous system of Example 8.2 with comparison of exact samples and corresponding samples obtained via HF domain with percentage error at different sample points for $m = 8$ and $T = 1$ s (vide Appendix B, Program no. 23)

t(s)	System output $y(t)$		
	Direct expansion	HF coefficients using Eq. (8.31)	% error
0	0.00000000	0.00000000	–
$\frac{1}{8}$	0.05875155	0.05882353	−0.12251592
$\frac{2}{8}$	0.11059961	0.11072664	−0.11485574
$\frac{3}{8}$	0.15635536	0.15652351	−0.10754348
$\frac{4}{8}$	0.19673467	0.19693251	−0.10056184
$\frac{5}{8}$	0.23236929	0.23258751	−0.09391086
$\frac{6}{8}$	0.26381672	0.26404780	−0.08759111
$\frac{7}{8}$	0.29156899	0.29180688	−0.08158961
$\frac{8}{8}$	0.31606028	0.31630019	−0.07590641

8.2.1 Numerical Examples

Example 8.2 (vide Appendix B, Program no. 23) Consider a non-homogeneous system $\dot{\mathbf{x}}(t) = \mathbf{A}\mathbf{x}(t) + \mathbf{B}u(t), \mathrm{y}(t) = \mathbf{C}\mathbf{x}(t)$ with unit step forcing function, where $\mathbf{A} = \begin{bmatrix} 0 & 1 \\ -2 & -3 \end{bmatrix}, \mathbf{B} = \begin{bmatrix} 0 \\ 1 \end{bmatrix}, \mathbf{C} = \begin{bmatrix} 1 & 0 \end{bmatrix}, \mathrm{D} = 0$ and $\mathbf{x}_0 = \begin{bmatrix} 0 \\ 0.5 \end{bmatrix}$

The time variation of the output $y(t)$ is shown in Fig. 8.3 and the respective sample values are compared in Table 8.2.

8.3 Analysis of Homogeneous State Equation [4]

For a homogeneous system, **B** is zero and Eq. (8.21) will be reduced to

$$\begin{bmatrix} c_{Sx1(i+1)} \\ c_{Sx2(i+1)} \\ \vdots \\ c_{Sxn(i+1)} \end{bmatrix} = \left[\frac{2}{h}\mathbf{I} - \mathbf{A}\right]^{-1} \left[\frac{2}{h}\mathbf{I} + \mathbf{A}\right] \begin{bmatrix} c_{Sx1i} \\ c_{Sx2i} \\ \vdots \\ c_{Sxni} \end{bmatrix} \tag{8.33}$$

Whereas Eq. (8.33) provides a simple recursive solution of the states of homogeneous system, or, in other words, time samples of the states, with a sampling period of h knowing the system matrix **A** and the initial values of the states.

8.3.1 Numerical Examples

Example 8.3 (vide Appendix B, Program no. 24) Consider a homogeneous system $\dot{\mathbf{x}}(t) = \mathbf{A}\mathbf{x}(t)$, where $\mathbf{A} = \begin{bmatrix} 0 & 1 \\ -1 & -2 \end{bmatrix}$ and $\mathbf{x}(0) = \begin{bmatrix} 0 \\ 1 \end{bmatrix}$ having the solution $x_1(t) = t\,exp(-t)$ and $x_2(t) = (1-t)\,exp(-t)$.

The results of analysis of the given system are presented in Table 8.3 and Fig. 8.4 compares the result obtained in HF domain for $m = 4$ with its direct expansion.

Example 8.4 [6] Consider a homogeneous system $\dot{\mathbf{x}}(t) = \begin{bmatrix} 0 & \omega \\ \omega & 0 \end{bmatrix}, \mathbf{x}(t) \triangleq \mathbf{A}\mathbf{x}(t)$, with initial condition $\mathbf{x}(0) = [1 \quad 0]^{\mathrm{T}}$ and the exact solution is given by $\mathbf{x}(t) = \begin{bmatrix} cos(\omega t) \\ -sin(\omega t) \end{bmatrix}$.

It is observed that the use of only Eq. (8.33) provides the complete solution of the states $x_1(t)$ and $x_2(t)$ in hybrid function domain as the method is recursive and we can solve for any sample point using the previous sample.

From Eq. (8.33), we solve for the vector $\mathbf{x}(t)$ and study four cases with $h = 0.1$, 0.01, 0.001 and 0.0001 s, $\omega = 10$ and $m = 10$. The Tables 8.4, 8.5, 8.6, 8.7, 8.8, 8.9, 8.10 and 8.11 compares the HF domain results with the exact solution. The last column in each table contain the percentage errors for different samples of $x_1(t)$ and $x_2(t)$. Also, respective figures are shown from Figs. 8.5, 8.6, 8.7, 8.8, 8.9, 8.10, 8.11 and 8.12.

Since we have considered values of h from 0.1 to 0.0001, it may be expected that for smaller h, HF domain solutions will match almost *exactly* with the exact sample values of the states $x_1(t)$ and $x_2(t)$. For this reason, to bring out the difference between these two solutions, whatever less it may be, we have used MATLAB *long*

Table 8.3 Solution of states x_1 and x_2 of the homogeneous system of Example 8.3 with comparison of exact samples and corresponding samples obtained via HF domain with percentage error at different sample points for $m = 4$ and $T = 1$ s (vide Appendix B, Program no. 24)

(a)

t(s)	System state x_1		
	Exact samples of the state $x_{1,d}$	Samples from HF domain, using Eq. (8.33), $x_{1,h}$	% error $\epsilon_1 = \frac{x_{1,d} - x_{1,h}}{x_{1,d}} \times 100$
0	0.00000000	0.00000000	–
$\frac{1}{4}$	0.19470020	0.19753086	−1.45385572
$\frac{2}{4}$	0.30326533	0.30727023	−1.32059276
$\frac{3}{4}$	0.35427491	0.35848194	−1.18750436
$\frac{4}{4}$	0.36787944	0.37175905	−1.05458734

(b)

t(s)	System state x_2		
	Exact samples of the state $x_{2,d}$	Samples from HF domain, using Eq. (8.33), $x_{2,h}$	% error $\epsilon_2 = \frac{x_{2,d} - x_{2,h}}{x_{2,d}} \times 100$
0	1.00000000	1.00000000	0.00000000
$\frac{1}{4}$	0.58410059	0.58024691	0.65976307
$\frac{2}{4}$	0.30326533	0.29766804	1.84567421
$\frac{3}{4}$	0.11809164	0.11202561	5.13671417
$\frac{4}{4}$	0.00000000	−0.00580874	–

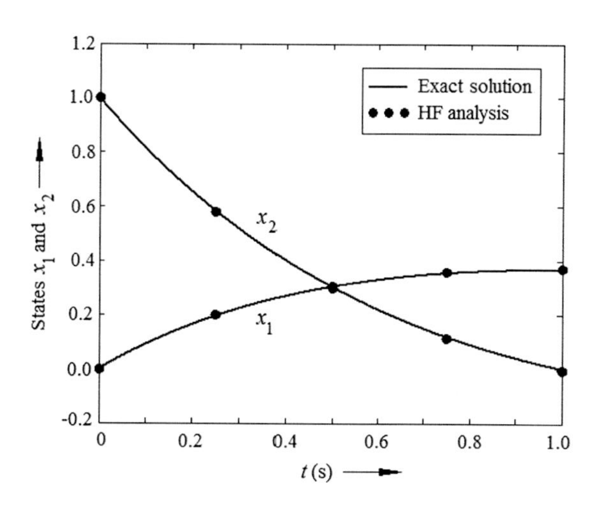

Fig. 8.4 Comparison of HF domain recursive solution of Example 8.3, for $m = 4$ and $T = 1$ s with the exact solutions of states x_1 and x_2 (vide Appendix B, Program no. 24)

format computations for computing the HF domain as well as exact values of the samples of the states tabulated Tables 8.8, 8.9, 8.10 and 8.11.

Table 8.12 [6] compare the maximum absolute errors of five different methods including the HF domain approach.

Table 8.4 Solution of states x_1 of the homogeneous system of Example 8.4 with comparison of exact samples and corresponding samples obtained via HF domain with percentage error at different sample points for $m = 10$ and $T = 1$ s

t(s)	System state x_1		
	Exact samples of the state $x_{1,\mathrm{d}}$	Samples from HF domain, using Eq. (8.33), $x_{1,\mathrm{h}}$	% error $\epsilon_1 = \frac{x_{1,\mathrm{d}} - x_{1,\mathrm{h}}}{x_{1,\mathrm{d}}} \times 100$
0	1.00000000	1.00000000	0.00000000
$\frac{1}{10}$	0.54030231	0.60000000	−11.04894221
$\frac{2}{10}$	−0.41614684	−0.28000000	32.71605763
$\frac{3}{10}$	−0.98999250	−0.93600000	5.45382920
$\frac{4}{10}$	−0.65364362	−0.84320000	−28.99995872
$\frac{5}{10}$	0.28366219	−0.07584000	126.73602710
$\frac{6}{10}$	0.96017029	0.75219200	21.66056294
$\frac{7}{10}$	0.75390225	0.97847040	−29.78743597
$\frac{8}{10}$	−0.14550003	0.42197248	390.01539037
$\frac{9}{10}$	−0.91113026	−0.47210342	48.18485998
$\frac{10}{10}$	−0.83907153	−0.98849659	−17.80838160

Table 8.5 Solution of state x_2 of the homogeneous system of Example 8.4 with comparison of exact samples and corresponding samples obtained via HF domain with percentage error at different sample points for $m = 10$ and $T = 1$ s

t(s)	System state x_2		
	Exact samples of the state $x_{2,\mathrm{d}}$	Samples from HF domain, using Eq. (8.33), $x_{2,\mathrm{h}}$	% error $\epsilon_2 = \frac{x_{2,\mathrm{d}} - x_{2,\mathrm{h}}}{x_{2,\mathrm{d}}} \times 100$
0	0.00000000	0.00000000	–
$\frac{1}{10}$	−0.84147098	−0.80000000	4.92839099
$\frac{2}{10}$	−0.90929743	−0.96000000	−5.57601598
$\frac{3}{10}$	−0.14112001	−0.35200000	−149.43308890
$\frac{4}{10}$	0.75680250	0.53760000	28.96429385
$\frac{5}{10}$	0.95892427	0.99712000	−3.98318524
$\frac{6}{10}$	0.27941550	0.65894400	−135.82943681
$\frac{7}{10}$	−0.65698660	−0.20638720	68.58578242
$\frac{8}{10}$	−0.98935825	−0.90660864	8.36396826
$\frac{9}{10}$	−0.41211849	−0.88154317	−113.90527030
$\frac{10}{10}$	0.54402111	−0.15124316	127.80097265

Example 8.5 Consider another homogeneous system $\dot{\mathbf{x}}(t) = \mathbf{A}\mathbf{x}(t)$, where $\mathbf{A} = \begin{bmatrix} 0 & 1 & 0 \\ 0 & 0 & 1 \\ -6 & -11 & -6 \end{bmatrix}$ and $\mathbf{x}(0) = \begin{bmatrix} 1 \\ 0 \\ 0 \end{bmatrix}$ having the

Table 8.6 Solution of state x_1 of the homogeneous system of Example 8.4 with comparison of exact samples and corresponding samples obtained via HF domain with percentage error at different sample points for $m = 10$ and $T = 0.1$ s

t(s)	System state x_1		
	Exact samples of the state $x_{1,d}$	Samples from HF domain, using Eq. (8.33), $x_{1,h}$	% error $\epsilon_1 = \frac{x_{1,d}-x_{1,h}}{x_{1,d}} \times 100$
0	1.00000000	1.00000000	0.00000000
$\frac{1}{100}$	0.99500417	0.99501247	−0.00083417
$\frac{2}{100}$	0.98006658	0.98009963	−0.00337222
$\frac{3}{100}$	0.95533649	0.95541023	−0.00771875
$\frac{4}{100}$	0.92106099	0.92119055	−0.01406639
$\frac{5}{100}$	0.87758256	0.87778195	−0.02272037
$\frac{6}{100}$	0.82533561	0.82561741	−0.03414369
$\frac{7}{100}$	0.76484219	0.76521729	−0.04904280
$\frac{8}{100}$	0.69670671	0.69718408	−0.06851807
$\frac{9}{100}$	0.62160997	0.62219641	−0.09434212
$\frac{10}{100}$	0.54030231	0.54100229	−0.12955340

Table 8.7 Solution of state x_2 of the homogeneous system of Example 8.4 with comparison of exact samples and corresponding samples obtained via HF domain with percentage error at different sample points for $m = 10$ and $T = 0.1$ s

t(s)	System state x_2		
	Exact samples of the state $x_{2,d}$	Samples from HF domain, using Eq. (8.33), $x_{2,h}$	% error $\epsilon_2 = \frac{x_{2,d}-x_{2,h}}{x_{2,d}} \times 100$
0	0.00000000	0.00000000	–
$\frac{1}{100}$	−0.09983342	−0.09975062	0.08293816
$\frac{2}{100}$	−0.19866933	−0.19850623	0.08209621
$\frac{3}{100}$	−0.29552021	−0.29528172	0.08070176
$\frac{4}{100}$	−0.38941834	−0.38911176	0.07872767
$\frac{5}{100}$	−0.47942554	−0.47906039	0.07616407
$\frac{6}{100}$	−0.56464247	−0.56423035	0.07298778
$\frac{7}{100}$	−0.64421769	−0.64377209	0.06916917
$\frac{8}{100}$	−0.71735609	−0.71689216	0.06467220
$\frac{9}{100}$	−0.78332691	−0.78286118	0.05945538
$\frac{10}{100}$	−0.84147098	−0.84102112	0.05346114

$$x_1(t) = 3\exp(-t) - 3\exp(-2t) + \exp(-3t),$$

solution $x_2(t) = -3\exp(-t) + 6\exp(-2t) - 3\exp(-3t)$ and

$$x_3(t) = 3\exp(-t) - 12\exp(-2t) + 9\exp(-3t)$$

Figure 8.13 shows the solution of the system states $x_1(t)$, $x_2(t)$ and $x_3(t)$ and the results obtained for $m = 8$ in HF domain are compared with the direct expansion.

Table 8.8 Solution of state x_1 of the homogeneous system of Example 8.4 with comparison of exact samples and corresponding samples obtained via HF domain with percentage error at different sample points for $m = 10$ and $T = 0.01$ s

t(s)	System state x_1		
	Exact samples of the state $x_{1,d}$	Samples from HF domain, using Eq. (8.33), $x_{1,h}$	% error $\epsilon_1 = \frac{x_{1,d}-x_{1,h}}{x_{1,d}} \times 100$
0	1.000000000000000	1.000000000000000	0.000000000000000
$\frac{1}{1000}$	0.999950000416665	0.999950001249969	−8.333450931694e−8
$\frac{2}{1000}$	0.999800006666578	0.999800009999625	−3.333713781104e−7
$\frac{3}{1000}$	0.999550033748988	0.999550041247719	−7.502106664312e−7
$\frac{4}{1000}$	0.999200106660978	0.999200119990500	−1.334019299526e−6
$\frac{5}{1000}$	0.998750260394966	0.998750281219221	−2.085031178134e−6
$\frac{6}{1000}$	0.998200539935204	0.998200569916632	−3.003547586832e−6
$\frac{7}{1000}$	0.997551000253280	0.997551041052492	−4.089937456679e−6
$\frac{8}{1000}$	0.996801706302619	0.996801759578061	−5.344637849625e−6
$\frac{9}{1000}$	0.995952733011994	0.995952800419614	−6.768154521393e−6
$\frac{10}{1000}$	0.995004165278026	0.995004248470945	−8.361062463701e−6

Table 8.9 Solution of state x_2 of the homogeneous system of Example 8.4 with comparison of exact samples and corresponding samples obtained via HF domain with percentage error at different sample points for $m = 10$ and $T = 0.01$ s

t(s)	System state x_2		
	Exact samples of the state $x_{2,d}$	Samples from HF domain, using Eq. (8.33), $x_{2,h}$	% error $\epsilon_2 = \frac{x_{2,d}-x_{2,h}}{x_{2,d}} \times 100$
0	0.000000000000000	0.000000000000000	–
$\frac{1}{1000}$	−0.009999833334167	−0.009999750006250	8.332930563493e−4
$\frac{2}{1000}$	−0.019998666693333	−0.019998500062498	8.332097225479e−4
$\frac{3}{1000}$	−0.029995500202496	−0.029995250318735	8.330708254490e−4
$\frac{4}{1000}$	−0.039989334186634	−0.039989001124926	8.328763539592e−4
$\frac{5}{1000}$	−0.049979169270678	−0.049978753130974	8.326262924834e−4
$\frac{6}{1000}$	−0.059964006479445	−0.059963507386653	8.323206210049e−4
$\frac{7}{1000}$	−0.069942847337533	−0.069942265441499	8.319593150895e−4
$\frac{8}{1000}$	−0.079914693969173	−0.079914029444652	8.315423457460e−4
$\frac{9}{1000}$	−0.089878549198011	−0.089877802244640	8.310696795552e−4
$\frac{10}{1000}$	−0.099833416646828	−0.099832587489093	8.305412786219e−4

The results of analysis of the system are presented in Table 8.13 and we have compared the samples obtained via HF domain analysis with its direct expansion for $m = 8$.

Table 8.10 Solution of state x_1 of the homogeneous system of Example 8.4 with comparison of exact samples and corresponding samples obtained via HF domain with percentage error at different sample points for m = 10 and T = 0.001 s

t(s)	System state x_1		
	Exact samples of the state $x_{1,\mathrm{d}}$	Samples from HF domain, using Eq. (8.33), $x_{1,\mathrm{h}}$	% error $\epsilon_1 = \frac{x_{1,\mathrm{d}} - x_{1,\mathrm{h}}}{x_{1,\mathrm{d}}} \times 100$
0	1.000000000000000	1.000000000000000	0.000000000000000
$\frac{1}{10000}$	0.999999500000042	0.999999500000125	−8.337779083824e−12
$\frac{2}{10000}$	0.999998000000667	0.999998000001000	−3.335116636205e−11
$\frac{3}{10000}$	0.999995500003375	0.999995500004125	−7.501810735515e−11
$\frac{4}{10000}$	0.999992000010667	0.999992000012000	−1.333499542859e−10
$\frac{5}{10000}$	0.999987500026042	0.999987500028125	−2.083581595030e−10
$\frac{6}{10000}$	0.999982000054000	0.999982000057000	−3.000320707358e−10
$\frac{7}{10000}$	0.999975500100042	0.999975500104125	−4.083722379966e−10
$\frac{8}{10000}$	0.999968000170666	0.999968000176000	−5.333904138926e−10
$\frac{9}{10000}$	0.999959500273374	0.999959500280125	−6.750762460663e−10
$\frac{10}{10000}$	0.999950000416665	0.999950000424999	−8.334194818315e−10

Table 8.11 Solution of state x_2 of the homogeneous system of Example 8.4 with comparison of exact samples and corresponding samples obtained via HF domain with percentage error at different sample points for m = 10 and T = 0.001 s

t(s)	System state x_2		
	Exact samples of the state $x_{2,\mathrm{d}}$	Samples from HF domain, using Eq. (8.33), $x_{2,\mathrm{h}}$	% error $\epsilon_2 = \frac{x_{2,\mathrm{d}} - x_{2,\mathrm{h}}}{x_{2,\mathrm{d}}} \times 100$
0	0.000000000000000	0.000000000000000	–
$\frac{1}{10000}$	−0.000999999833333	−0.000999999750000	8.333329318508e−6
$\frac{2}{10000}$	−0.001999998666667	−0.001999998500001	8.333320995157e−6
$\frac{3}{10000}$	−0.002999995500002	−0.002999995250003	8.333307115657e−6
$\frac{4}{10000}$	−0.003999989333342	−0.003999989000011	8.333287679974e−6
$\frac{5}{10000}$	−0.004999979166693	−0.004999978750031	8.333262666380e−6
$\frac{6}{10000}$	−0.005999964000065	−0.005999963500074	8.333232118187e−6
$\frac{7}{10000}$	−0.006999942833473	−0.006999942250154	8.333196004346e−6
$\frac{8}{10000}$	−0.007999914666940	−0.007999914000294	8.333154330969e−6
$\frac{9}{10000}$	−0.008999878500492	−0.008999877750523	8.333107091766e−6
$\frac{10}{10000}$	−0.009999833334167	−0.009999832500875	8.333054300258e−6

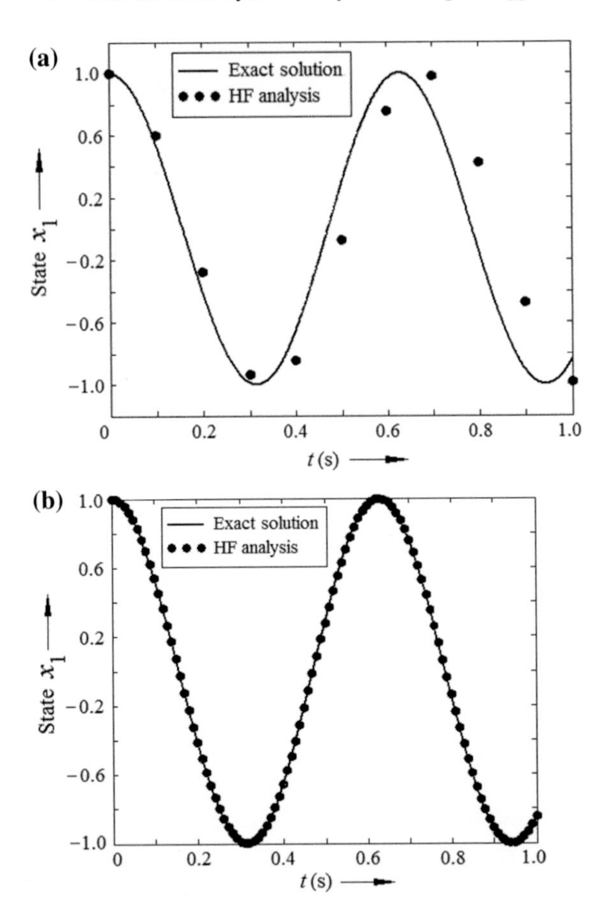

Fig. 8.5 Solution of state x_1 in HF domain **a** using step size $h = 0.1$ s, $m = 10$, $T = 1$ s and **b** using step size $h = 0.01$ s, $m = 100$, $T = 1$ s, along with the exact solution of Example 8.4

8.4 Determination of Output of a Homogeneous System [3]

For an $n \times n$ homogeneous system, **D** will be zero and Eqs. (8.31) and (8.32) will be reduced to,

$$\mathbf{y}_S = \mathbf{C}\mathbf{C}_{Sx} \tag{8.34}$$

$$\mathbf{y}_T = \mathbf{C}\mathbf{C}_{Tx} \tag{8.35}$$

These Eqs. (8.34) and (8.35) provide simple solution of the output of homogeneous system.

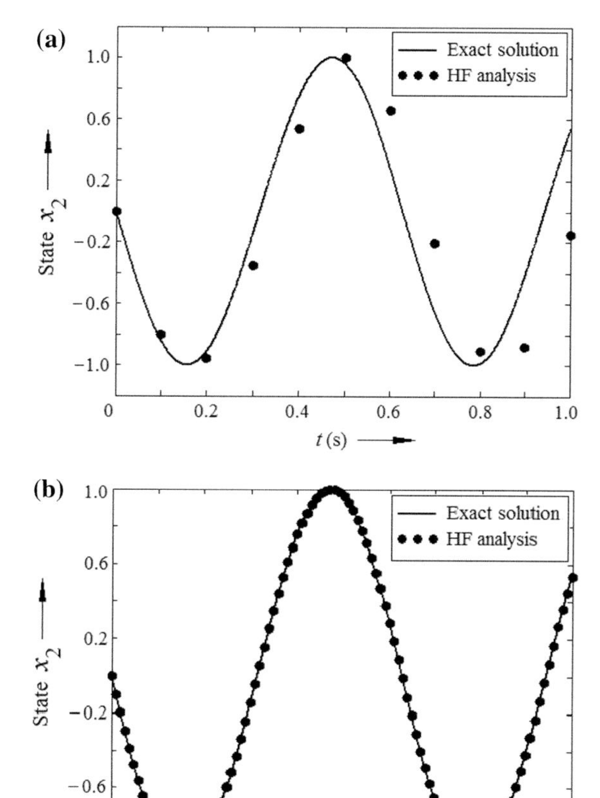

Fig. 8.6 Solution of state x_2 in HF domain **a** using step size $h = 0.1$ s, $m = 10$, $T = 1$ s and **b** using step size $h = 0.01$ s, $m = 100$, $T = 1$ s, along with the exact solution of Example 8.4

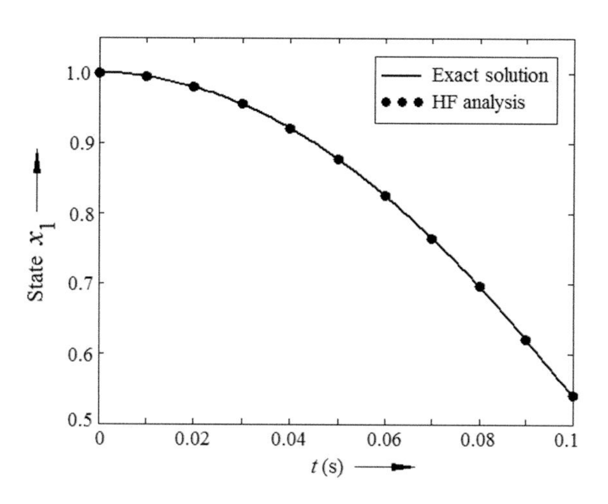

Fig. 8.7 Solution of state x_1 in HF domain, using step size $h = 0.01$ s, for $m = 10$ and $T = 0.1$ s along with the exact solution of Example 8.4

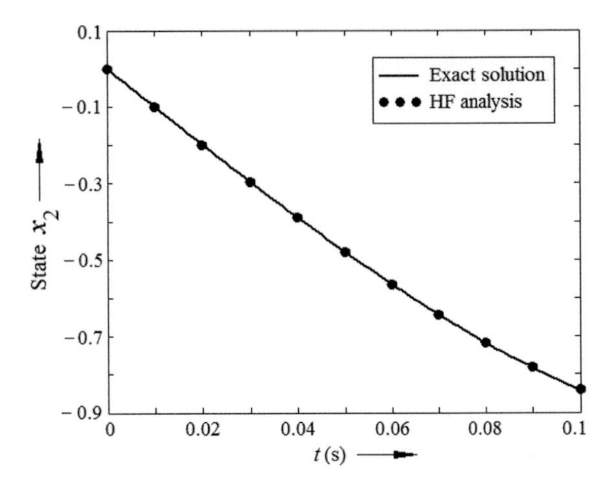

Fig. 8.8 Solution of state x_2 in HF domain, using step size $h = 0.01$ s, for $m = 10$ and $T = 0.1$ s along with the exact solution of Example 8.4

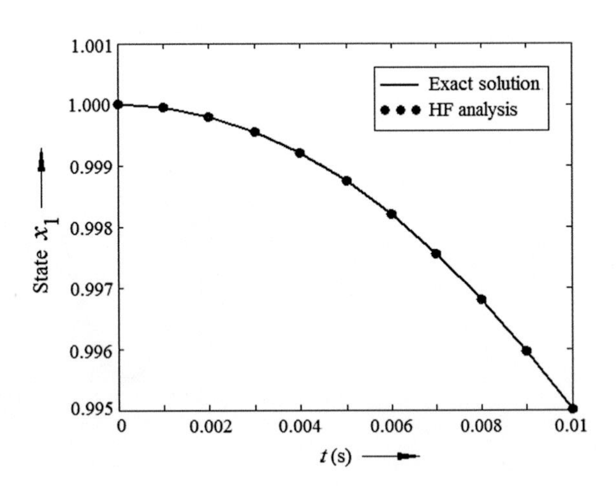

Fig. 8.9 Solution of state x_1 in HF domain, using step size $h = 0.001$ s, for $m = 10$ and $T = 0.01$ s along with the exact solution of Example 8.4

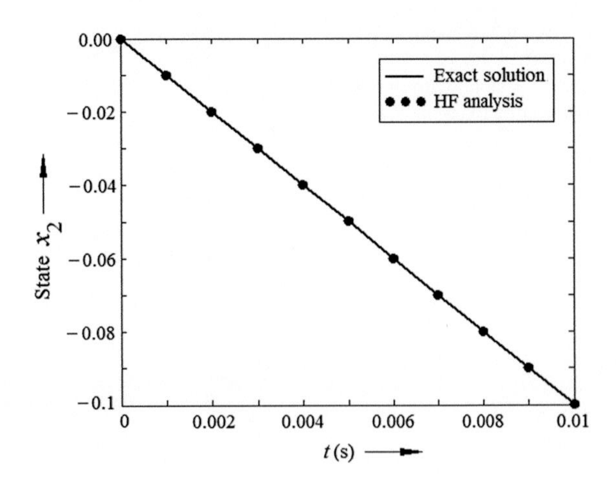

Fig. 8.10 Solution of state x_2 in HF domain, using step size $h = 0.001$ s, for $m = 10$ and $T = 0.01$ s along with the exact solution of Example 8.4

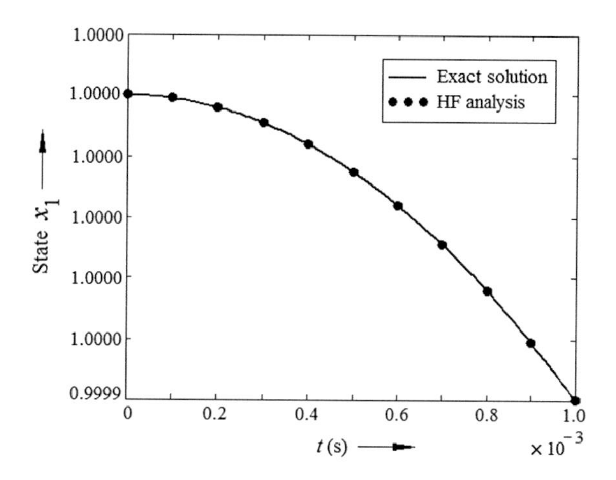

Fig. 8.11 Solution of state x_1 in HF domain, using step size $h = 0.0001$ s, for $m = 10$ and $T = 0.001$ s along with the exact solution of Example 8.4

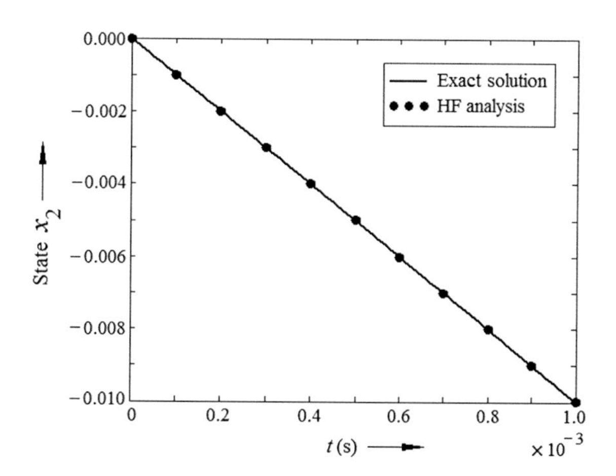

Fig. 8.12 Solution of state x_2 in HF domain, using step size $h = 0.0001$ s, for $m = 10$ and $T = 0.001$ s along with the exact solution of Example 8.4

Table 8.12 Comparison of maximum absolute error obtained via MERKDP510, MERKDP512, MERKDP514 [6, 7] and HF based approaches for step sizes h of 0.1, 0.01, 0.001 and 0.0001 s respectively, for Example 8.4

h(s)	Maximum absolute error			
	MERKDP510	MERKDP512	MERKDP514	HF domain
0.1	1.6×10^{-1}	4.4×10^{-2}	1.8×10^{-3}	390.01539037
0.01	3.1×10^{-3}	1.1×10^{-3}	7.4×10^{-4}	0.12955340
0.001	2.9×10^{-4}	2.8×10^{-5}	1.9×10^{-6}	8.332930563493e−4
0.0001	2.9×10^{-5}	1.0×10^{-6}	2.9×10^{-7}	8.333329318508e−6

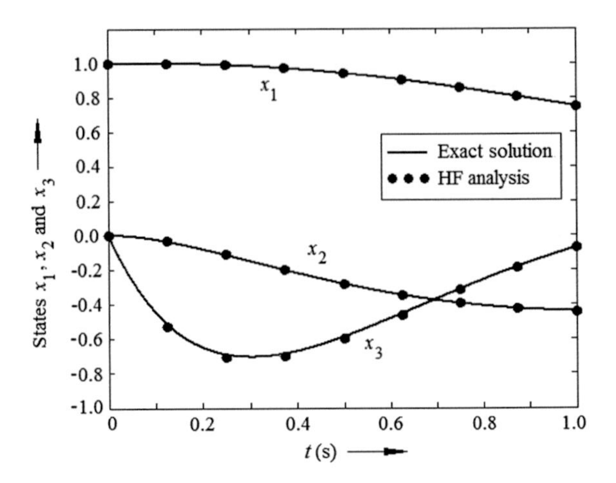

Fig. 8.13 Comparison of HF domain recursive solution for $m = 8$ and $T = 1$ s with the exact solutions of state x_1, state x_2 and state x_3 of Example 8.5

8.4.1 Numerical Examples

Example 8.6 Consider a homogeneous system $\dot{\mathbf{x}}(t) = \mathbf{A}\mathbf{x}(t), \mathrm{y}(t) = \mathbf{C}\mathbf{x}(t)$where, $\mathbf{A} = \begin{bmatrix} 0 & 1 \\ -1 & -2 \end{bmatrix}$ $\mathbf{C} = [1 \quad 0]$ $\mathbf{x}_0 = \begin{bmatrix} 0 \\ 1 \end{bmatrix}$

The samples of the system output $y(t)$ is presented in Table 8.14 and its time variation is shown in Fig. 8.14 for $m = 10$ and $T = 1$ s.

Example 8.7 Consider a homogeneous system $\dot{\mathbf{x}}(t) = \mathbf{A}\mathbf{x}(t), \mathrm{y}(t) = \mathbf{C}\mathbf{x}(t)$where $\mathbf{A} = \begin{bmatrix} 0 & 1 & 0 \\ 0 & 0 & 1 \\ -6 & -11 & -6 \end{bmatrix}, \mathbf{C} = [4 \quad 5 \quad 1], \mathbf{x}_0 = \begin{bmatrix} 1 \\ 0 \\ 0 \end{bmatrix}$

The system output $y(t)$ is presented in Table 8.15 and its time variation is shown in Fig. 8.15 for $m = 10$ and $T = 1$ s.

8.5 Analysis of a Non-homogeneous System with Jump Discontinuity at Input

We can modify Eq. (8.21) to make it suitable for the $\mathrm{HF_m}$ based approach (as described in Chap. 3), so that it can come up with good results in spite of the jump discontinuities. The modification is quite simple in the sense that in the RHS of Eq. (8.21), all the triangular function coefficient matrices associated with the matrix **B** have to be modified as discussed in Chap. 3, Sect. 3.5.1. That is, all the $\mathbf{C}_{\mathrm{Tu}}^{\mathrm{T}}$'s in (8.21) are to be replaced by $\mathbf{C}_{\mathrm{Tu}}^{\prime\mathrm{T}}$, where $\mathbf{C}_{\mathrm{Tu}}^{\prime\mathrm{T}} \triangleq \mathbf{C}_{\mathrm{Tu}}^{\mathrm{T}}\mathbf{J}_{k(m)}$. This modification will come up with good results for system analysis, shown in the numerical section. The modified form for analyzing the system with jump discontinuities, is given by

Table 8.13 Solution of states x_1, x_2 and x_3 of the homogeneous system of Example 8.5 with comparison of exact samples and corresponding samples obtained via HF domain with percentage error at different sample points for $m = 8$ and $T = 1$ s

(a)

t(s)	System state x_1		
	Exact samples of the state $x_{1,d}$	Samples from HF domain, using Eq. (8.33), $x_{1,h}$	% error $\epsilon_1 = \frac{x_{1,d} - x_{1,h}}{x_{1,d}} \times 100$
0	1.00000000	1.00000000	0.00000000
$\frac{1}{8}$	0.99837764	0.99793602	0.04423376
$\frac{2}{8}$	0.98917692	0.98896937	0.02098209
$\frac{3}{8}$	0.96942065	0.96964539	−0.02318292
$\frac{4}{8}$	0.93908382	0.93971286	−0.06698444
$\frac{5}{8}$	0.89962486	0.90054168	−0.10191137
$\frac{6}{8}$	0.85310840	0.85417906	−0.12550105
$\frac{7}{8}$	0.80170398	0.80281012	−0.13797362
$\frac{8}{8}$	0.74741954	0.74847056	−0.14061982

(b)

t(s)	System state x_2		
	Exact samples of the state $x_{2,d}$	Samples from HF domain, using Eq. (8.33), $x_{2,h}$	% error $\epsilon_2 = \frac{x_{2,d} - x_{2,h}}{x_{2,d}} \times 100$
0	0.00000000	0.00000000	–
$\frac{1}{8}$	−0.03655385	−0.03302374	9.65728644
$\frac{2}{8}$	−0.11431805	−0.11044264	3.39002458
$\frac{3}{8}$	−0.20162592	−0.19874093	1.43086265
$\frac{4}{8}$	−0.28170581	−0.28017962	0.54176731
$\frac{5}{8}$	−0.34682040	−0.34655920	0.07531276
$\frac{6}{8}$	−0.39451637	−0.39524283	−0.18413938
$\frac{7}{8}$	−0.42526167	−0.42666011	−0.32884224
$\frac{8}{8}$	−0.44098783	−0.44277287	−0.40478215

(c)

t(s)	System state x_3		
	Exact samples of the state $x_{3,d}$	Samples from HF domain, using Eq. (8.33), $x_{3,h}$	% Error $\epsilon_3 = \frac{x_{3,d} - x_{3,h}}{x_{3,d}} \times 100$
0	0.00000000	0.00000000	–
$\frac{1}{8}$	−0.51251518	−0.52837977	−3.09543807
$\frac{2}{8}$	−0.69066659	−0.71032272	−2.84596508
$\frac{3}{8}$	−0.68465859	−0.70244984	−2.59855792
$\frac{4}{8}$	−0.58678987	−0.60056918	−2.34825288
$\frac{5}{8}$	−0.45207858	−0.46150419	−2.08494948
$\frac{6}{8}$	−0.31186924	−0.31743382	−1.78426702
$\frac{7}{8}$	−0.18274345	−0.18524277	−1.36766598
$\frac{8}{8}$	−0.07230146	−0.07256132	−0.35941183

Table 8.14 Solution of output of the homogeneous system of Example 8.6 with comparison of exact samples and corresponding samples obtained via HF domain with percentage error at different sample points for m = 10 and T = 1 s

t(s)	System output $y(t)$		
	Direct expansion	HF coefficients using Eq. (8.34)	% error
0	0.00000000	0.00000000	–
$\frac{1}{10}$	0.09048374	0.09070295	−0.24226452
$\frac{2}{10}$	0.16374615	0.16412914	−0.23389252
$\frac{3}{10}$	0.22224547	0.22274670	−0.22552991
$\frac{4}{10}$	0.26812802	0.26871030	−0.21716492
$\frac{5}{10}$	0.30326533	0.30389855	−0.20880066
$\frac{6}{10}$	0.32928698	0.32994700	−0.20043914
$\frac{7}{10}$	0.34760971	0.34827739	−0.19207749
$\frac{8}{10}$	0.35946317	0.36012356	−0.18371562
$\frac{9}{10}$	0.36591269	0.36655434	−0.17535604
$\frac{10}{10}$	0.36787944	0.36849378	−0.16699493

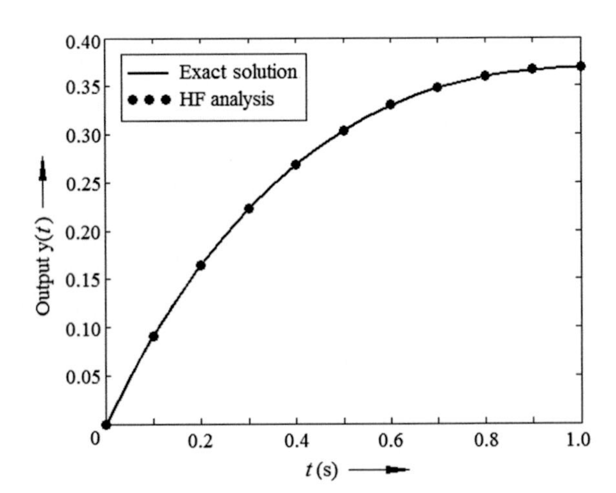

Fig. 8.14 Hybrid function based solution of system output with m = 10, T = 1 s, and its comparison with exact output of Example 8.6

$$\begin{bmatrix} c_{\mathrm{Sx1}(i+1)} \\ c_{\mathrm{Sx2}(i+1)} \\ \vdots \\ c_{\mathrm{Sxn}(i+1)} \end{bmatrix} = \left[\frac{2}{h}\mathbf{I} - \mathbf{A}\right]^{-1}\left[\frac{2}{h}\mathbf{I} + \mathbf{A}\right]\begin{bmatrix} c_{\mathrm{Sx1}i} \\ c_{\mathrm{Sx2}i} \\ \vdots \\ c_{\mathrm{Sx}ni} \end{bmatrix} + 2\left[\frac{2}{h}\mathbf{I} - \mathbf{A}\right]^{-1}\mathbf{B}\left[\mathbf{C}_{\mathrm{Su}}^{\mathrm{T}} + \frac{1}{2}\mathbf{C}_{\mathrm{Tu}}^{\prime T}\right]_i \tag{8.36}$$

The inverse in (8.36) can always be made to exist by judicious choice of h.

Table 8.15 Solution of output of the non-homogeneous system of Example 8.7 with comparison of exact samples and corresponding samples obtained via HF domain with percentage error at different sample points for m = 10 and T = 1 s

t(s)	System output $y(t)$		
	Direct expansion	HF coefficients using Eq. (8.34)	% error
0	4.00000000	4.00000000	0.00000000
$\frac{1}{10}$	3.43074808	3.43083004	−0.00238898
$\frac{2}{10}$	2.92429700	2.92390133	0.01353043
$\frac{3}{10}$	2.47973050	2.47865663	0.04330592
$\frac{4}{10}$	2.09358536	2.09183323	0.08369040
$\frac{5}{10}$	1.76101633	1.75868707	0.13226794
$\frac{6}{10}$	1.47656750	1.47380325	0.18720783
$\frac{7}{10}$	1.23466893	1.23161802	0.24710348
$\frac{8}{10}$	1.02994320	1.02674151	0.31086083
$\frac{9}{10}$	0.85738230	0.85414466	0.37761917
$\frac{10}{10}$	0.71243756	0.70925511	0.44669880

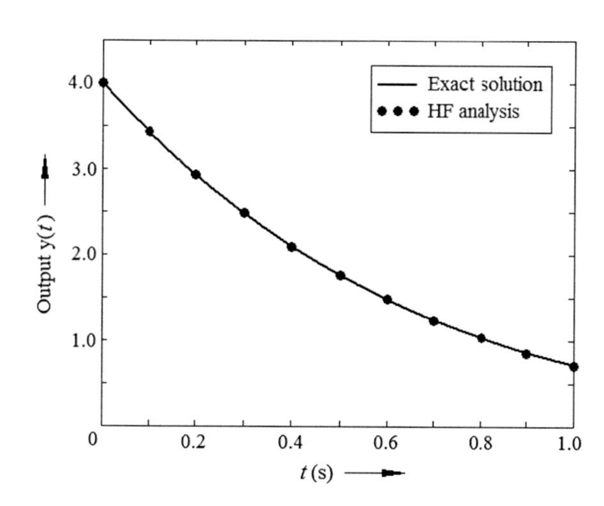

Fig. 8.15 Hybrid function based solution of system output with m = 10, T = 1 s, and its comparison with exact output of Example 8.7

8.5.1 Numerical Example

Example 8.8 (vide Appendix B, Program no. 25)

Consider the non-homogeneous system $\dot{\mathbf{x}}(t) = \mathbf{A}\mathbf{x}(t) + \mathbf{B}u(t) + \mathbf{B}u(t-a)$ where $\mathbf{A} = \begin{bmatrix} 0 & 1 \\ -2 & -3 \end{bmatrix}$, $\mathbf{B} = \begin{bmatrix} 0 \\ 1 \end{bmatrix}$, $\mathbf{x}(0) = \begin{bmatrix} 0 \\ 0.5 \end{bmatrix}$, $u(t)$ is the unit step function and $u(t-a)$ is the delayed unit step function, having the solution

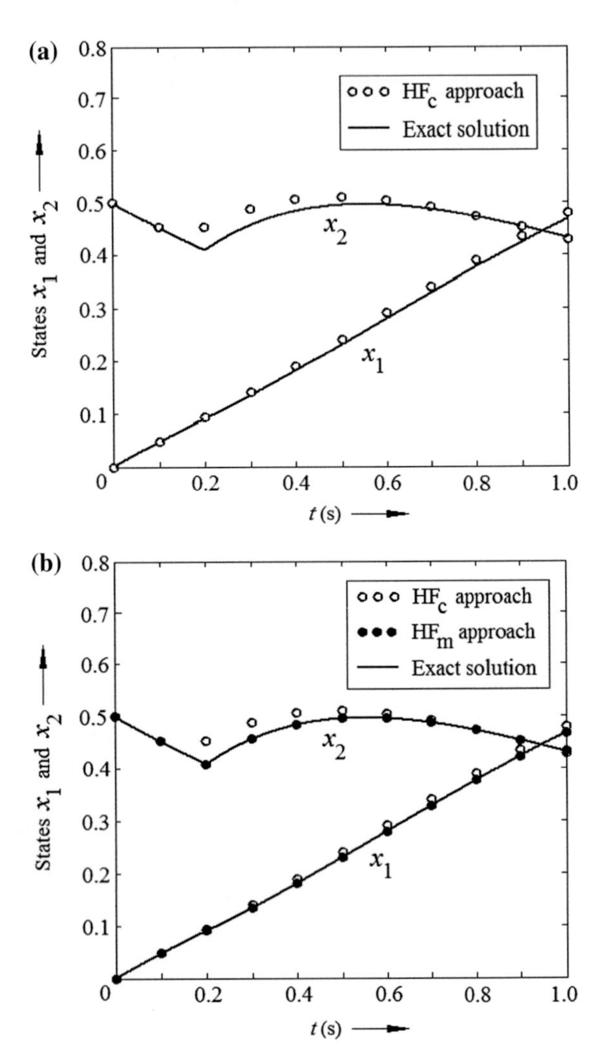

Fig. 8.16 Comparison of exact samples of the states x_1 and x_2 of the non-homogeneous system of Example 8.8 with the samples obtained using **a** the HF$_c$ approach and **b** the HF$_m$ approach, for $m = 10$ and $T = 1$ s (vide Appendix B, Program no. 25)

$$x_1(t) = 1 - \frac{1}{2}exp(-t) - \left[exp(-(t-a)) - \frac{1}{2}exp(-2(t-a))\right]u(t-a)$$

$$x_2(t) = \frac{1}{2}exp(-t) + [exp(-(t-a)) - exp(-2(t-a))]u(t-a)$$

Considering $a = 0.2$ s, the exact solution of the states $x_1(t)$ and $x_2(t)$ are shown in Fig. 8.16a along with results computed using the HF$_c$ approach, with the effect of combined presence of non-delayed and delayed inputs to the system for $m = 10$ and $T = 1$ s. Figure 8.16b shows similar results using the HF$_m$ approach and presents a visual comparison of the results with the exact solution and the results obtained via the HF$_c$ approach. It is evident that the HF$_m$ approach produces much better result compared to the HF$_c$ based analysis.

To have an idea about the variation of error with increasing m for both of $\mathrm{HF_c}$ and $\mathrm{HF_m}$ approaches, we compute percentage errors of $(m + 1)$ number of samples of each an individual state and calculate AMP error, vide Eq. (4.45). Only in this case, number of elements in the denominator will be $r = m + 1$.

When we compute the AMP error using the conventional HF based approach, let us call it ε_{avic}, and when we compute the same using the modified HF based approach, we call it ε_{avim}.

Table 8.16 presents the percentage error at different sample points in a tabular form. It is observed that the error is much less for the $\mathrm{HF_m}$ approach. Also, the ratio of the AMP errors of the states x_1 and x_2, via $\mathrm{HF_c}$ approach and the $\mathrm{HF_m}$ approach, are given by

$$\frac{\varepsilon_{\mathrm{av1c}}}{\varepsilon_{\mathrm{av1m}}} = 46.23544541 \quad (\text{for } x_1) \text{ and}$$
$$\frac{\varepsilon_{\mathrm{av2c}}}{\varepsilon_{\mathrm{av2m}}} = 24.26089295 \quad (\text{for } x_2)$$

This indicates superiority of the $\mathrm{HF_m}$ approach beyond any doubt.

Example 8.9 Consider the non-homogeneous system $\dot{\mathbf{x}}(t) = \mathbf{A}\mathbf{x}(t) + \mathbf{B}u_1(t) + \mathbf{B}u_2(t-a)$where $\mathbf{A} = \begin{bmatrix} 0 & 1 \\ -2 & -3 \end{bmatrix}$, $\mathbf{B} = \begin{bmatrix} 0 \\ 1 \end{bmatrix}$, $\mathbf{x}(0) = \begin{bmatrix} 0 \\ 0.5 \end{bmatrix}$, $u_1(t)$ is a unit ramp function and $u_2(t-a)$ is a delayed unit step function.

The system has the following solution

$$x_1(t) = -\frac{1}{4} + \frac{1}{2}t + \frac{3}{2}\exp(-t) - \frac{3}{4}\exp(-2t) - \left[\exp(-(t-a)) - \frac{1}{2}\exp(-2(t-a))\right]u(t-a)$$
$$x_2(t) = \frac{1}{2} - \frac{3}{2}\exp(-t) + \frac{3}{2}\exp(-2t) + [\exp(-(t-a)) - \exp(-2(t-a))]u(t-a)$$

Considering $a = 0.2$ s, Fig. 8.17a shows the solution of the system states $x_1(t)$ and $x_2(t)$, along with the effect of the combined presence of non-delayed and delayed inputs to the system, using the $\mathrm{HF_c}$ approach.

Figure 8.17b represents the effectiveness of the $\mathrm{HF_m}$ approach compared to the $\mathrm{HF_c}$ approach. It is noted that for the state x_2, which is affected by the jump input, the samples derived using the $\mathrm{HF_m}$ approach, are very close to the exact samples of the states, while that obtained via the $\mathrm{HF_c}$ approach are not that close to the exact samples. This indicates reasonably less MISE with the $\mathrm{HF_m}$ based analysis compared to the $\mathrm{HF_c}$ based analysis.

Table 8.16 Percentage errors at different sample points of the (a) state x_1 and (b) state x_2 of Example 8.8 computed using the HF_c based approach and the HF_m based approach, for $m = 10$, $T = 1$ s. The results are compared with the exact samples (vide Appendix B, Program no. 25)

(a)

t(s)	Exact samples of the state $x_{1,d}$	Samples from HF_c approach, $x_{1,hc}$	Samples from HF_m approach, $x_{1,hm}$	% error $\varepsilon_{x1,c} = \frac{x_{1,d}-x_{1,hc}}{x_{1,d}} \times 100$	ε_{av1c}	% error $\varepsilon_{x1,m} = \frac{x_{1,d}-x_{1,hm}}{x_{1,d}} \times 100$	ε_{av1m}
0	0.00000000	0.00000000	0.00000000	–	3.23226287	–	0.06990876
$\frac{1}{10}$	0.04758129	0.04761905	0.04761905	−0.07935892		−0.07935892	
$\frac{2}{10}$	0.09063462	0.09286745	0.09070295	−2.46355090		−0.07539062	
$\frac{3}{10}$	0.13411885	0.13990644	0.13401262	−4.31526963		0.07920587	
$\frac{4}{10}$	0.18126925	0.18962091	0.18106849	−4.60732308		0.11075238	
$\frac{5}{10}$	0.23032227	0.24045506	0.23008268	−4.39939655		0.10402381	
$\frac{6}{10}$	0.27993862	0.29123784	0.27969781	−4.03632053		0.08602243	
$\frac{7}{10}$	0.32911641	0.34110322	0.32889867	−3.64211860		0.06615896	
$\frac{8}{10}$	0.37712099	0.38942602	0.37694088	−3.26288653		0.04775921	
$\frac{9}{10}$	0.42342835	0.43577015	0.42329350	−2.91473162		0.03184718	
$\frac{10}{10}$	0.46767957	0.47984706	0.46759273	−2.60167234		0.01856827	

(b)

t(s)	Exact samples of the state $x_{2,d}$	Samples from HF_c approach, $x_{2,hc}$	Samples from HF_m approach, $x_{2,hm}$	% Error $\varepsilon_{x2,c} = \frac{x_{2,d}-x_{2,hc}}{x_{2,d}} \times 100$	ε_{av2c}	% error $\varepsilon_{x2,m} = \frac{x_{2,d}-x_{2,hm}}{x_{2,d}} \times 100$	ε_{av2m}
0	0.50000000	0.50000000	0.50000000	0.00000000	2.94188446	0.00000000	0.12126035
$\frac{1}{10}$	0.45241871	0.45238095	0.45238095	0.00834625		0.00834625	
$\frac{2}{10}$	0.40936538	0.45258710	0.40929705	−10.55822551		0.01669169	
$\frac{3}{10}$	0.45651578	0.48819273	0.45689647	−6.93885105		−0.08339033	
$\frac{4}{10}$	0.48357073	0.50609660	0.48422077	−4.65823686		−0.13442501	
$\frac{5}{10}$	0.49527191	0.51058653	0.49606308	−3.09216406		−0.15974457	
$\frac{6}{10}$	0.49539690	0.50506892	0.49623962	−1.95237798		−0.17011007	
$\frac{7}{10}$	0.48694387	0.49223868	0.48777742	−1.08735530		−0.17117989	
$\frac{8}{10}$	0.47228191	0.47421735	0.47306683	−0.40980608		−0.16619735	
$\frac{9}{10}$	0.45327317	0.45266533	0.45398554	0.13410014		−0.15716130	
$\frac{10}{10}$	0.43137217	0.42887288	0.43199920	0.57938137		−0.14535708	

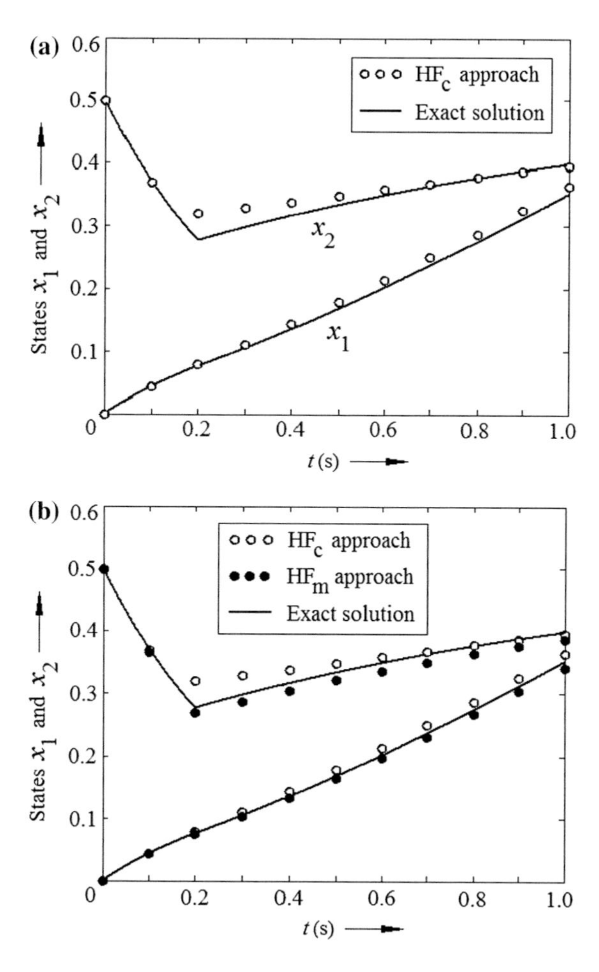

Fig. 8.17 Comparison of exact samples of the states x_1 and x_2 of the non-homogeneous system of Example 8.9 with the samples obtained using **a** the HF_c approach and **b** the HF_m approach, for $m = 10$ and $T = 1$ s

8.6 Conclusion

In this chapter, we have analysed the state space model of a non-homogeneous system and solved for the states in hybrid function platform. The method of analysis is attractive in the sense that it offers a simple recursive solution in a generalized form. This recursive equation has been used for the analysis of homogeneous systems by putting the condition $\mathbf{B} = 0$. Different types of numerical examples have been treated using the derived recursive matrix equations for homogeneous as well as non-homogeneous systems, and the results are compared with the exact solutions of the system states with error estimates. It is found that the HF method is a strong tool for such analysis. This fact is reflected through various tables and curves.

As an interesting example [6], we have taken up a set of simultaneous differential equations, which are no different from the well known homogeneous state equation, having oscillatory solution. This example has been treated for various step sizes, i.e., h = 0.1, 0.01, 0.001 and 0.0001 s and the maximum absolute errors incurred in HF domain analysis have been compared with other improved fifth order Runge-Kutta methods, such as MERKDP510, MERKDP512 and MERKDP514 suggested by Dormand and Prince [7].

Apart from deriving recursive equations for solving the system states, yet another recursive matrix equation has been derived for solving for the outputs equations for any non-homogeneous as well as homogeneous systems. As before, HF domain solutions are compared with the exact outputs and found to be reliably close.

References

1. Ogata, K.: Modern Control Engineering 5th ed. Prentice Hall of India, Upper Saddle River (2011)
2. Ogata, K.: System Dynamics, 4th ed. Pearson Education, New York City (2004)
3. Roychoudhury, S., Deb, A., Sarkar, G.: Analysis and synthesis of homogeneous/non-homogeneous control systems via orthogonal hybrid functions (HF) under states space environment. J. Inf. Optim. Sci. **35**(5 & 6), 431–482 (2014)
4. Deb, A., Sarkar, G., Ganguly, A., Biswas, A.: Approximation, integration and differentiation of time functions using a set of orthogonal hybrid functions (HF) and their application to solution of first order differential equations. Appl. Math. Comput. **218**(9), 4731–4759 (2012)
5. Jiang, J.H., Schaufelberger, W.: Block Pulse Functions and their Application in Control System, LNCIS, vol. 179. Springer, Berlin (1992)
6. Simos, T.E.: Modified Runge-Kutta methods for the numerical solution of ODEs with oscillating solutions. Appl. Math. Comput. **84**, 131–143 (1997)
7. Dormand, J.R., Prince, P.J.: New Runge-Kutta algorithms for numerical simulations in dynamical astronomy. Celest. Mech. **18**, 223–232 (1978)

Chapter 9
Time Varying System Analysis: State Space Approach

Abstract In this chapter, time varying system analysis is presented using state space approach in hybrid function domain. Both homogeneous and non-homogeneous systems are treated along with numerical examples. States and outputs of the systems are solved. Illustration has been provided with the support of five examples, six figures and nine tables.

As the heading implies, this chapter is dedicated for linear time varying (LTV) control system analysis.

A time varying system means, its parameters do vary with time. In the following, we intend to analyse two types of LTV control system in hybrid function platform, namely non-homogeneous system and homogeneous system [1, 2].

As discussed in Chap. 8, analysing a control system means, we determine the behavior of each system state of a system over a common time frame, knowing the system parameters and the input signal or forcing function. Like linear time invariant (LTI) systems, the output of a linear time varying system may be any one of the states or a linear combination of two or many states. Therefore, knowing all the states of the system, we can assess its performance.

In this chapter, the hybrid function set is employed for the analysis of time varying non-homogeneous as well as homogeneous systems described by their state space models [2].

First we take up the problem of analysis of a time varying non-homogeneous system in HF domain. After putting the specific condition of zero forcing function, we can easily arrive at the result of analysis of a time varying homogeneous system.

In practice, applications of time varying systems may be found in aircraft control, like during its takeoff, cruise and landing. Also, the aircraft has to adapt itself to the continuous decrease of its fuel leading to loss of weight.

The human vocal tract is another example. It is a time variant system due to time dependent nature of the shape of the vocal organs.

The last example is the discrete wavelet transform, used in modern signal processing. It is often used in its time variant form. And of course there are many more such examples.

A. Deb et al., *Analysis and Identification of Time-Invariant Systems, Time-Varying Systems, and Multi-Delay Systems using Orthogonal Hybrid Functions*,
Studies in Systems, Decision and Control 46, DOI 10.1007/978-3-319-26684-8_9

9.1 Analysis of Non-homogeneous Time Varying State Equation [3]

Consider the following n-state non-homogeneous equation of a time varying control system

$$\dot{\mathbf{x}}(t) = \mathbf{A}(t)\mathbf{x}(t) + \mathbf{B}(t)u(t) \tag{9.1}$$

where,

$\mathbf{A}(t)$ is the time varying system matrix of order n,
$\mathbf{B}(t)$ is the time varying input vector of order $(n \times 1)$,
$\mathbf{x}(t)$ is the state vector with the initial conditions $\mathbf{x}(0)$, of order $(n \times 1)$,
and $u(t)$ is the forcing function.
Integrating Eq. (9.1) we have

$$\mathbf{x}(t) - \mathbf{x}(0) = \int_0^t \mathbf{A}(\tau)\mathbf{x}(\tau)\mathrm{d}\tau + \int_0^t \mathbf{B}(\tau)u(\tau)\mathrm{d}\tau \tag{9.2}$$

Expanding $\mathbf{x}(t)$, $\mathbf{x}(0)$ and $u(t)$ via an m-set hybrid function, we get

$$\mathbf{x}(t) = \begin{bmatrix} x_1(t) \\ x_2(t) \\ \vdots \\ x_n(t) \end{bmatrix} \approx \begin{bmatrix} x_{10} & x_{11} & \cdots & x_{1(m-1)} \\ x_{20} & x_{21} & \cdots & x_{2(m-1)} \\ \vdots & \vdots & & \vdots \\ x_{n0} & x_{n1} & \cdots & x_{n(m-1)} \end{bmatrix} \mathbf{S}_{(m)} + \begin{bmatrix} (x_{11}-x_{10}) & (x_{12}-x_{11}) & \cdots & (x_{1m}-x_{1(m-1)}) \\ (x_{21}-x_{20}) & (x_{22}-x_{21}) & \cdots & (x_{2m}-x_{2(m-1)}) \\ \vdots & \vdots & & \vdots \\ (x_{n1}-x_{n0}) & (x_{n2}-x_{n1}) & \cdots & (x_{nm}-x_{n(m-1)}) \end{bmatrix} \mathbf{T}_{(m)} \tag{9.3}$$

$$\mathbf{x}(0) = \begin{bmatrix} x_1(0) \\ x_2(0) \\ \vdots \\ x_n(0) \end{bmatrix} \approx \begin{bmatrix} x_{10} & x_{10} & \cdots & x_{10} \\ x_{20} & x_{20} & \cdots & x_{20} \\ \vdots & \vdots & & \vdots \\ x_{n0} & x_{n0} & \cdots & x_{n0} \end{bmatrix} \mathbf{S}_{(m)} \tag{9.4}$$

$$
\begin{aligned}
u(t) \approx & \begin{bmatrix} u_0 & u_1 & u_2 & \ldots & u_{m-1} \end{bmatrix} \mathbf{S}_{(m)} \\
& + \begin{bmatrix} (u_1 - u_0) & (u_2 - u_1) & (u_3 - u_2) & \cdots & (u_m - u_{m-1}) \end{bmatrix} \mathbf{T}_{(m)}
\end{aligned}
\tag{9.5}
$$

We now follow the rule of multiplication of two time functions, as given in Chap. 2, Sect. 2.6.3 and Eq. (2.30), to expand $\mathbf{A}(t)\mathbf{x}(t)$ and $\mathbf{B}(t)u(t)$ via an m-set hybrid functions.

We have

$$
\begin{aligned}
\mathbf{A}(t)\mathbf{x}(t) &= \begin{bmatrix} a_{11}(t) & a_{12}(t) & \ldots & a_{1k}(t) & \ldots & a_{1n}(t) \\ a_{21}(t) & a_{22}(t) & \ldots & a_{2k}(t) & \ldots & a_{2n}(t) \\ \vdots & \vdots & & \vdots & & \vdots \\ a_{n1}(t) & a_{n2}(t) & \ldots & a_{nk}(t) & \ldots & a_{nn}(t) \end{bmatrix} \begin{bmatrix} x_1(t) \\ x_2(t) \\ \vdots \\ x_k(t) \\ \vdots \\ x_n(t) \end{bmatrix} \\
&= \begin{bmatrix} \sum_{j=1}^{n} a_{1j}(t)x_j(t) \\ \vdots \\ \sum_{j=1}^{n} a_{kj}(t)x_j(t) \\ \vdots \\ \sum_{j=1}^{n} a_{nj}(t)x_j(t) \end{bmatrix}
\end{aligned}
\tag{9.6}
$$

where, $a_{kj}(t)$ is the kjth element of the square matrix $\mathbf{A}(t)$.

Following (2.30), the kth term of the first element of the RHS column matrix of (9.6) can be represented in HF domain as

$$
\begin{aligned}
a_{1k}(t)x_k(t) &\approx \bar{a}_{1k}(t)\bar{x}_k(t) \\
&= \begin{bmatrix} a_{1k0}x_{k0} & a_{1k1}x_{k1} & a_{1k2}x_{k2} & \ldots & a_{1k(m-1)}x_{k(m-1)} \end{bmatrix} \mathbf{S}_{(m)} \\
&\quad + \begin{bmatrix} (a_{1k1}x_{k1} - a_{1k0}x_{k0}) & (a_{1k2}x_{k2} - a_{1k1}x_{k1}) & \ldots & \left(a_{1km}x_{km} - a_{1k(m-1)}x_{k(m-1)}\right) \end{bmatrix} \mathbf{T}_{(m)}
\end{aligned}
$$

We can express the first element of the resulting matrix of (9.6) as

$$
\begin{aligned}
&\sum_{j=1}^{n} a_{1j}(t)x_j(t) \\
&= \left[\,(a_{110}x_{10} + \cdots + a_{1n0}x_{n0}) \quad (a_{111}x_{11} + \cdots + a_{1n1}x_{n1})\right. \\
&\qquad \left.\cdots \quad \left(a_{11(m-1)}x_{1(m-1)} + \cdots + a_{1n(m-1)}x_{n(m-1)}\right)\,\right]\mathbf{S}_{(m)} \\
&\quad + \left[\{(a_{111}x_{11} - a_{110}x_{10}) + \cdots + (a_{1n1}x_{n1} - a_{1n0}x_{n0})\}\right. \\
&\qquad \{(a_{112}x_{12} - a_{111}x_{11}) + \cdots + (a_{1n2}x_{n2} - a_{1n1}x_{n1})\} \\
&\quad \left.\ldots\left\{\left(a_{11m}x_{1m} - a_{11(m-1)}x_{1(m-1)}\right) + \cdots + \left(a_{1nm}x_{nm} - a_{1n(m-1)}x_{n(m-1)}\right)\right\}\right]\mathbf{T}_{(m)} \\
&= \left[\sum_{j=1}^{n} a_{1j0}x_{j0} \quad \sum_{j=1}^{n} a_{1j1}x_{j1} \quad \cdots \quad \sum_{j=1}^{n} a_{1j(m-1)}x_{j(m-1)}\right]\mathbf{S}_{(m)} \\
&\quad + \left[\left(\sum_{j=1}^{n} a_{1j1}x_{j1} - \sum_{j=1}^{n} a_{1j0}x_{j0}\right) \quad \left(\sum_{j=1}^{n} a_{1j2}x_{j2} - \sum_{j=1}^{n} a_{1j1}x_{j1}\right)\right. \\
&\qquad \left.\cdots \quad \left(\sum_{j=1}^{n} a_{1jm}x_{jm} - \sum_{j=1}^{n} a_{1j(m-1)}x_{j(m-1)}\right)\right]\mathbf{T}_{(m)}
\end{aligned}
$$

$$
\begin{aligned}
\mathbf{A}(t)\mathbf{x}(t) \approx \bar{\mathbf{A}}(t)\bar{\mathbf{x}}(t) &= \begin{bmatrix} \sum_{j=1}^{n} a_{1j0}x_{j0} & \sum_{j=1}^{n} a_{1j1}x_{j1} & \cdots & \sum_{j=1}^{n} a_{1j(m-1)}x_{j(m-1)} \\ \sum_{j=1}^{n} a_{2j0}x_{j0} & \sum_{j=1}^{n} a_{2j1}x_{j1} & \cdots & \sum_{j=1}^{n} a_{2j(m-1)}x_{j(m-1)} \\ \vdots & \vdots & & \vdots \\ \sum_{j=1}^{n} a_{nj0}x_{j0} & \sum_{j=1}^{n} a_{nj1}x_{j1} & \cdots & \sum_{j=1}^{n} a_{nj(m-1)}x_{j(m-1)} \end{bmatrix}\mathbf{S}_{(m)} \\
&+ \begin{bmatrix} \left(\sum_{j=1}^{n} a_{1j1}x_{j1} - \sum_{j=1}^{n} a_{1j0}x_{j0}\right) & \left(\sum_{j=1}^{n} a_{1j2}x_{j2} - \sum_{j=1}^{n} a_{1j1}x_{j1}\right) & \cdots & \left(\sum_{j=1}^{n} a_{1jm}x_{jm} - \sum_{j=1}^{n} a_{1j(m-1)}x_{j(m-1)}\right) \\ \left(\sum_{j=1}^{n} a_{2j1}x_{j1} - \sum_{j=1}^{n} a_{2j0}x_{j0}\right) & \left(\sum_{j=1}^{n} a_{2j2}x_{j2} - \sum_{j=1}^{n} a_{2j1}x_{j1}\right) & \cdots & \left(\sum_{j=1}^{n} a_{2jm}x_{jm} - \sum_{j=1}^{n} a_{2j(m-1)}x_{j(m-1)}\right) \\ \vdots & \vdots & & \vdots \\ \left(\sum_{j=1}^{n} a_{nj1}x_{j1} - \sum_{j=1}^{n} a_{nj0}x_{j0}\right) & \left(\sum_{j=1}^{n} a_{nj2}x_{j2} - \sum_{j=1}^{n} a_{nj1}x_{j1}\right) & \cdots & \left(\sum_{j=1}^{n} a_{njm}x_{jm} - \sum_{j=1}^{n} a_{nj(m-1)}x_{j(m-1)}\right) \end{bmatrix}\mathbf{T}_{(m)} \\
&\triangleq \mathbf{A}_{\mathrm{XS}}^{\mathrm{T}}\mathbf{S}_{(m)} + \mathbf{A}_{\mathrm{XT}}^{\mathrm{T}}\mathbf{T}_{(m)}
\end{aligned}
\tag{9.7}
$$

Similarly, the product term $\mathbf{B}(t)u(t)$ in (9.2) can also be expressed in HF domain. We may write

$$
\mathbf{B}(t)u(t) = \begin{bmatrix} b_1(t) \\ b_2(t) \\ \vdots \\ b_i(t) \\ \vdots \\ b_n(t) \end{bmatrix} u(t) = \begin{bmatrix} b_1(t)u(t) \\ b_2(t)u(t) \\ \vdots \\ b_i(t)u(t) \\ \vdots \\ b_n(t)u(t) \end{bmatrix}
$$

where, $b_i(t)$ is the ith element of the matrix $\mathbf{B}(t)$.

Then

$$\mathbf{B}(t)u(t) \approx \begin{bmatrix} b_{10}u_0 & b_{11}u_1 & \cdots & b_{1(m-1)}u_{(m-1)} \\ b_{20}u_0 & b_{21}u_1 & \cdots & b_{2(m-1)}u_{(m-1)} \\ \vdots & \vdots & & \vdots \\ b_{n0}u_0 & b_{n1}u_1 & \cdots & b_{n(m-1)}u_{(m-1)} \end{bmatrix} \mathbf{S}_{(m)}$$

$$+ \begin{bmatrix} (b_{11}u_1 - b_{10}u_0) & (b_{12}u_2 - b_{11}u_1) & \cdots & \left(b_{1m}u_m - b_{1(m-1)}u_{(m-1)}\right) \\ (b_{21}u_1 - b_{20}u_0) & (b_{22}u_2 - b_{21}u_1) & \cdots & \left(b_{2m}u_m - b_{2(m-1)}u_{(m-1)}\right) \\ \vdots & \vdots & & \vdots \\ (b_{n1}u_1 - b_{n0}u_0) & (b_{n2}u_2 - b_{n1}u_1) & \cdots & \left(b_{nm}u_m - b_{n(m-1)}u_{(m-1)}\right) \end{bmatrix} \mathbf{T}_{(m)}$$

$$\triangleq \mathbf{B}_{\mathrm{US}}^{\mathrm{T}}\mathbf{S}_{(m)} + \mathbf{B}_{\mathrm{UT}}^{\mathrm{T}}\mathbf{T}_{(m)}$$

Therefore, considering the first term of the RHS of Eq. (9.2), we have

$$\int_0^t \mathbf{A}(\tau)\mathbf{x}(\tau)\,\mathrm{d}\tau \approx \left[\mathbf{A}_{\mathrm{XS}}^{\mathrm{T}} + \frac{1}{2}\mathbf{A}_{\mathrm{XT}}^{\mathrm{T}}\right]\mathbf{P1ss}_{(m)}\mathbf{S}_{(m)} + h\left[\mathbf{A}_{\mathrm{XS}}^{\mathrm{T}} + \frac{1}{2}\mathbf{A}_{\mathrm{XT}}^{\mathrm{T}}\right]\mathbf{T}_{(m)}$$

$$= \frac{1}{2}\begin{bmatrix} \sum_{j=1}^{n}\left(a_{1j0}x_{j0} + a_{1j1}x_{j1}\right) & \sum_{j=1}^{n}\left(a_{1j1}x_{j1} + a_{1j2}x_{j2}\right) & \cdots & \sum_{j=1}^{n}\left(a_{1j(m-1)}x_{j(m-1)} + a_{1jm}x_{jm}\right) \\ \sum_{j=1}^{n}\left(a_{2j0}x_{j0} + a_{2j1}x_{j1}\right) & \sum_{j=1}^{n}\left(a_{2j1}x_{j1} + a_{2j2}x_{j2}\right) & \cdots & \sum_{j=1}^{n}\left(a_{2j(m-1)}x_{j(m-1)} + a_{2jm}x_{jm}\right) \\ \vdots & \vdots & & \vdots \\ \sum_{j=1}^{n}\left(a_{nj0}x_{j0} + a_{nj1}x_{j1}\right) & \sum_{j=1}^{n}\left(a_{nj1}x_{j1} + a_{nj2}x_{j2}\right) & \cdots & \sum_{j=1}^{n}\left(a_{nj(m-1)}x_{j(m-1)} + a_{njm}x_{jm}\right) \end{bmatrix}\mathbf{P1ss}_{(m)}\mathbf{S}_{(m)}$$

$$+ \frac{h}{2}\begin{bmatrix} \sum_{j=1}^{n}\left(a_{1j0}x_{j0} + a_{1j1}x_{j1}\right) & \sum_{j=1}^{n}\left(a_{1j1}x_{j1} + a_{1j2}x_{j2}\right) & \cdots & \sum_{j=1}^{n}\left(a_{1j(m-1)}x_{j(m-1)} + a_{1jm}x_{jm}\right) \\ \sum_{j=1}^{n}\left(a_{2j0}x_{j0} + a_{2j1}x_{j1}\right) & \sum_{j=1}^{n}\left(a_{2j1}x_{j1} + a_{2j2}x_{j2}\right) & \cdots & \sum_{j=1}^{n}\left(a_{2j(m-1)}x_{j(m-1)} + a_{2jm}x_{jm}\right) \\ \vdots & \vdots & & \vdots \\ \sum_{j=1}^{n}\left(a_{nj0}x_{j0} + a_{nj1}x_{j1}\right) & \sum_{j=1}^{n}\left(a_{nj1}x_{j1} + a_{nj2}x_{j2}\right) & \cdots & \sum_{j=1}^{n}\left(a_{nj(m-1)}x_{j(m-1)} + a_{njm}x_{jm}\right) \end{bmatrix}\mathbf{T}_{(m)} \tag{9.8}$$

Similarly, the second term of the RHS of Eq. (9.2) can be expressed as

$$\int_0^t \mathbf{B}(\tau)u(\tau)\mathrm{d}\tau \approx \left[\mathbf{B}_{\mathrm{US}}^{\mathrm{T}} + \frac{1}{2}\mathbf{B}_{\mathrm{UT}}^{\mathrm{T}}\right]\mathbf{P1ss}_{(m)}\mathbf{S}_{(m)} + h\left[\mathbf{B}_{\mathrm{US}}^{\mathrm{T}} + \frac{1}{2}\mathbf{B}_{\mathrm{UT}}^{\mathrm{T}}\right]\mathbf{T}_{(m)}$$

$$= \frac{1}{2}\begin{bmatrix} (b_{10}u_0 + b_{11}u_1) & (b_{11}u_1 + b_{12}u_2) & \cdots & (b_{1(m-1)}u_{(m-1)} + b_{1m}u_m) \\ (b_{20}u_0 + b_{21}u_1) & (b_{21}u_1 + b_{22}u_2) & \cdots & (b_{2(m-1)}u_{(m-1)} + b_{2m}u_m) \\ \vdots & \vdots & & \vdots \\ (b_{n0}u_0 + b_{n1}u_1) & (b_{n1}u_1 + b_{n2}u_2) & \cdots & (b_{n(m-1)}u_{(m-1)} + b_{nm}u_m) \end{bmatrix}\mathbf{P1ss}_{(m)}\mathbf{S}_{(m)}$$

$$+ \frac{h}{2}\begin{bmatrix} (b_{10}u_0 + b_{11}u_1) & (b_{11}u_1 + b_{12}u_2) & \cdots & (b_{1(m-1)}u_{(m-1)} + b_{1m}u_m) \\ (b_{20}u_0 + b_{21}u_1) & (b_{21}u_1 + b_{22}u_2) & \cdots & (b_{2(m-1)}u_{(m-1)} + b_{2m}u_m) \\ \vdots & \vdots & & \vdots \\ (b_{n0}u_0 + b_{n1}u_1) & (b_{n1}u_1 + b_{n2}u_2) & \cdots & (b_{n(m-1)}u_{(m-1)} + b_{nm}u_m) \end{bmatrix}\mathbf{T}_{(m)} \tag{9.9}$$

Therefore, substituting **P1ss** from (4.9) in Eqs. (9.8) and (9.9), the RHS of Eq. (9.2) can be written as

$$\int_0^t \mathbf{A}(\tau)\mathbf{x}(\tau)\mathrm{d}\tau + \int_0^t \mathbf{B}(\tau)u(\tau)\mathrm{d}\tau,$$

$$= \frac{h}{2}\begin{bmatrix} 0 & \sum_{j=1}^{n}\left(a_{1j0}x_{j0} + a_{1j1}x_{j1}\right) + (b_{10}u_0 + b_{11}u_1) \\ 0 & \sum_{j=1}^{n}\left(a_{2j0}x_{j0} + a_{2j1}x_{j1}\right) + (b_{20}u_0 + b_{21}u_1) \\ \vdots & \vdots \\ 0 & \sum_{j=1}^{n}\left(a_{nj0}x_{j0} + a_{nj1}x_{j1}\right) + (b_{n0}u_0 + b_{n1}u_1) \end{bmatrix}$$

$$\begin{matrix} \sum_{j=1}^{n}\left\{\left(a_{1j0}x_{j0} + a_{1j1}x_{j1}\right) + \left(a_{1j1}x_{j1} + a_{1j2}x_{j2}\right)\right\} + (b_{10}u_0 + b_{11}u_1) + (b_{11}u_1 + b_{12}u_2) \cdots \\ \sum_{j=1}^{n}\left\{\left(a_{2j0}x_{j0} + a_{2j1}x_{j1}\right) + \left(a_{2j1}x_{j1} + a_{2j2}x_{j2}\right)\right\} + (b_{20}u_0 + b_{21}u_1) + (b_{21}u_1 + b_{22}u_2) \cdots \\ \vdots \\ \sum_{j=1}^{n}\left\{\left(a_{nj0}x_{j0} + a_{nj1}x_{j1}\right) + \left(a_{nj1}x_{j1} + a_{nj2}x_{j2}\right)\right\} + (b_{n0}u_0 + b_{n1}u_1) + (b_{n1}u_1 + b_{n2}u_2) \cdots \end{matrix}$$

$$\begin{bmatrix} \sum_{j=1}^{n}\left\{\left(a_{1j0}x_{j0}+a_{1j1}x_{j1}\right)+\cdots+\left(a_{1j(m-2)}x_{j(m-2)}+a_{1j(m-1)}x_{j(m-1)}\right)\right\}+\left\{\left(b_{10}u_0+b_{11}u_1\right)+\cdots+\left(b_{1(m-2)}u_{(m-2)}+b_{1(m-1)}u_{(m-1)}\right)\right\} \\ \sum_{j=1}^{n}\left\{\left(a_{2j0}x_{j0}+a_{2j1}x_{j1}\right)+\cdots+\left(a_{2j(m-2)}x_{j(m-2)}+a_{2j(m-1)}x_{j(m-1)}\right)\right\}+\left\{\left(b_{20}u_0+b_{21}u_1\right)+\cdots+\left(b_{2(m-2)}u_{(m-2)}+b_{2(m-1)}u_{(m-1)}\right)\right\} \\ \vdots \\ \sum_{j=1}^{n}\left\{\left(a_{nj0}x_{j0}+a_{nj1}x_{j1}\right)+\cdots+\left(a_{nj(m-2)}x_{j(m-2)}+a_{nj(m-1)}x_{j(m-1)}\right)\right\}+\left\{\left(b_{n0}u_0+b_{n1}u_1\right)+\cdots+\left(b_{n(m-2)}u_{(m-2)}+b_{n(m-1)}u_{(m-1)}\right)\right\} \end{bmatrix}\mathbf{S}_{(m)}$$

$$+\frac{h}{2}\begin{bmatrix} \sum_{j=1}^{n}\left(a_{1j0}x_{j0}+a_{1j1}x_{j1}\right)+\left(b_{10}u_0+b_{11}u_1\right) & \cdots & \sum_{j=1}^{n}\left(a_{1j(m-1)}x_{j(m-1)}+a_{1jm}x_{jm}\right)+\left(b_{1(m-1)}u_{(m-1)}+b_{1m}u_m\right) \\ \sum_{j=1}^{n}\left(a_{2j0}x_{j0}+a_{2j1}x_{j1}\right)+\left(b_{20}u_0+b_{21}u_1\right) & \cdots & \sum_{j=1}^{n}\left(a_{2j(m-1)}x_{j(m-1)}+a_{2jm}x_{jm}\right)+\left(b_{2(m-1)}u_{(m-1)}+b_{2m}u_m\right) \\ \vdots & & \vdots \\ \sum_{j=1}^{n}\left(a_{nj0}x_{j0}+a_{nj1}x_{j1}\right)+\left(b_{n0}u_0+b_{n1}u_1\right) & \cdots & \sum_{j=1}^{n}\left(a_{nj(m-1)}x_{j(m-1)}+a_{njm}x_{jm}\right)+\left(b_{n(m-1)}u_{(m-1)}+b_{nm}u_m\right) \end{bmatrix}\mathbf{T}_{(m)} \tag{9.10}$$

Using Eqs. (9.3) and (9.4), the LHS of Eq. (9.2) can be written as

$$\mathbf{x}(t)-\mathbf{x}(0)\approx\begin{bmatrix} 0 & (x_{11}-x_{10}) & (x_{12}-x_{10}) & \ldots & \left(x_{1(m-1)}-x_{10}\right) \\ 0 & (x_{21}-x_{20}) & (x_{22}-x_{20}) & \ldots & \left(x_{2(m-1)}-x_{20}\right) \\ \vdots & \vdots & \vdots & & \vdots \\ 0 & (x_{n1}-x_{n0}) & (x_{n2}-x_{n0}) & \ldots & \left(x_{n(m-1)}-x_{n0}\right) \end{bmatrix}\mathbf{S}_{(m)}$$
$$+\begin{bmatrix} (x_{11}-x_{10}) & (x_{12}-x_{11}) & \ldots & \left(x_{1m}-x_{1(m-1)}\right) \\ (x_{21}-x_{20}) & (x_{22}-x_{21}) & \ldots & \left(x_{2m}-x_{2(m-1)}\right) \\ \vdots & \vdots & & \vdots \\ (x_{n1}-x_{n0}) & (x_{n2}-x_{n1}) & \ldots & \left(x_{nm}-x_{n(m-1)}\right) \end{bmatrix}\mathbf{T}_{(m)} \tag{9.11}$$

Equating the SHF components of second column of Eqs. (9.10) and (9.11), we get,

$$\left.\begin{aligned} (x_{11}-x_{10}) &= \tfrac{h}{2}\left[\sum_{j=1}^{n}\left(a_{1j0}x_{j0}+a_{1j1}x_{j1}\right)+\left(b_{10}u_0+b_{11}u_1\right)\right] \\ (x_{21}-x_{20}) &= \tfrac{h}{2}\left[\sum_{j=1}^{n}\left(a_{2j0}x_{j0}+a_{2j1}x_{j1}\right)+\left(b_{20}u_0+b_{21}u_1\right)\right] \\ &\vdots \\ (x_{n1}-x_{n0}) &= \tfrac{h}{2}\left[\sum_{j=1}^{n}\left(a_{nj0}x_{j0}+a_{nj1}x_{j1}\right)+\left(b_{n0}u_0+b_{n1}u_1\right)\right] \end{aligned}\right\} \tag{9.12}$$

From these n number of equations, we can write

$$\left.\begin{aligned}
&\left(1-\frac{h}{2}a_{111}\right)x_{11}-\frac{h}{2}a_{121}x_{21}-\cdots-\frac{h}{2}a_{1n1}x_{n1}\\
&\quad=\left(1+\frac{h}{2}a_{110}\right)x_{10}+\frac{h}{2}a_{120}x_{20}+\cdots+\frac{h}{2}a_{1n0}x_{n0}+\frac{h}{2}(b_{10}u_0+b_{11}u_1)\\
&\quad-\frac{h}{2}a_{211}x_{11}+\left(1-\frac{h}{2}a_{221}\right)x_{21}-\cdots-\frac{h}{2}a_{2n1}x_{n1}\\
&\quad=\frac{h}{2}a_{210}x_{10}+\left(1+\frac{h}{2}a_{220}\right)x_{20}+\cdots+\frac{h}{2}a_{2n0}x_{n0}+\frac{h}{2}(b_{20}u_0+b_{21}u_1)\\
&\qquad\qquad\vdots\\
&\quad-\frac{h}{2}a_{n11}x_{11}-\frac{h}{2}a_{n21}x_{21}\cdots+\left(1-\frac{h}{2}a_{nn1}\right)x_{n1}\\
&\quad=\frac{h}{2}a_{n10}x_{10}+\frac{h}{2}a_{n20}x_{20}+\cdots+\left(1+\frac{h}{2}a_{nn0}\right)x_{n0}++\frac{h}{2}(b_{n0}u_0+b_{n1}u_1)
\end{aligned}\right\}\tag{9.13}$$

Writing in matrix form, we have

$$\begin{bmatrix}\left(1-\frac{h}{2}a_{111}\right) & -\frac{h}{2}a_{121} & \cdots & -\frac{h}{2}a_{1n1}\\ -\frac{h}{2}a_{211} & \left(1-\frac{h}{2}a_{221}\right) & \cdots & -\frac{h}{2}a_{2n1}\\ \vdots & \vdots & & \vdots\\ -\frac{h}{2}a_{n11} & -\frac{h}{2}a_{n21} & \cdots & \left(1-\frac{h}{2}_{nn1}\right)\end{bmatrix}\begin{bmatrix}x_{11}\\x_{21}\\\vdots\\x_{n1}\end{bmatrix}$$
$$=\begin{bmatrix}\left(1+\frac{h}{2}a_{110}\right) & \frac{h}{2}a_{120} & \cdots & \frac{h}{2}a_{1n0}\\ \frac{h}{2}a_{210} & \left(1+\frac{h}{2}a_{220}\right) & \cdots & \frac{h}{2}a_{2n0}\\ \vdots & \vdots & & \vdots\\ \frac{h}{2}a_{n10} & \frac{h}{2}a_{n20} & \cdots & \left(1+\frac{h}{2}a_{nn0}\right)\end{bmatrix}\begin{bmatrix}x_{10}\\x_{20}\\\vdots\\x_{n0}\end{bmatrix}+\frac{h}{2}\begin{bmatrix}b_{10}\\b_{20}\\\vdots\\b_{n0}\end{bmatrix}u_0+\frac{h}{2}\begin{bmatrix}b_{11}\\b_{21}\\\vdots\\b_{n1}\end{bmatrix}u_1\tag{9.14}$$

Similarly, comparing the SHF coefficients of third column of Eqs. (9.10) and (9.11), we have

$$\left.\begin{aligned}
(x_{12}-x_{10})&=\frac{h}{2}\left[\sum_{j=1}^{n}a_{1j0}x_{j0}+2\sum_{j=1}^{n}a_{1j1}x_{j1}+\sum_{j=1}^{n}a_{1j2}x_{j2}+(b_{10}u_0+b_{11}u_1)+(b_{11}u_1+b_{12}u_2)\right]\\
(x_{22}-x_{20})&=\frac{h}{2}\left[\sum_{j=1}^{n}a_{2j0}x_{j0}+2\sum_{j=1}^{n}a_{2j1}x_{j1}+\sum_{j=1}^{n}a_{2j2}x_{j2}+(b_{20}u_0+b_{21}u_1)+(b_{21}u_1+b_{22}u_2)\right]\\
&\qquad\vdots\\
(x_{n2}-x_{n0})&=\frac{h}{2}\left[\sum_{j=1}^{n}a_{nj0}x_{j0}+2\sum_{j=1}^{n}a_{nj1}x_{j1}+\sum_{j=1}^{n}a_{nj2}x_{j2}+(b_{n0}u_0+b_{n1}u_1)+(b_{n1}u_1+b_{n2}u_2)\right]
\end{aligned}\right\}\tag{9.15}$$

From the set of Eqs. (9.15), we can write

$$\left.\begin{aligned}
&\left(1-\frac{h}{2}a_{112}\right)x_{12}-\frac{h}{2}a_{122}x_{22}-\cdots-\frac{h}{2}a_{1n2}x_{n2}\\
&\quad=\left(1+\frac{h}{2}a_{111}\right)x_{11}+\frac{h}{2}a_{121}x_{21}+\cdots+\frac{h}{2}a_{1n1}x_{n1}+\frac{h}{2}(b_{11}u_1+b_{12}u_2)\\
&\quad-\frac{h}{2}a_{212}x_{12}+\left(1-\frac{h}{2}a_{222}\right)x_{22}-\cdots-\frac{h}{2}a_{2n2}x_{n2}\\
&\quad=\frac{h}{2}a_{211}x_{11}+\left(1+\frac{h}{2}a_{221}\right)x_{21}+\cdots+\frac{h}{2}a_{2n1}x_{n1}+\frac{h}{2}(b_{21}u_1+b_{22}u_2)\\
&\qquad\vdots\\
&\quad-\frac{h}{2}a_{n12}x_{12}-\frac{h}{2}a_{n22}x_{22}\cdots+\left(1-\frac{h}{2}a_{nn2}\right)x_{n2}\\
&=\frac{h}{2}a_{n11}x_{11}+\frac{h}{2}a_{n21}x_{21}+\cdots+\left(1+\frac{h}{2}a_{nn1}\right)x_{n1}++\frac{h}{2}(b_{n1}u_1+b_{n2}u_2)
\end{aligned}\right\} \tag{9.16}$$

Writing these equations in matrix form, we get

$$\begin{bmatrix}\left(1-\frac{h}{2}a_{112}\right) & -\frac{h}{2}a_{122} & \cdots & -\frac{h}{2}a_{1n2}\\ -\frac{h}{2}a_{212} & \left(1-\frac{h}{2}a_{222}\right) & \cdots & -\frac{h}{2}a_{2n2}\\ \vdots & \vdots & & \vdots\\ -\frac{h}{2}a_{n12} & -\frac{h}{2}a_{n22} & \cdots & \left(1-\frac{h}{2}a_{nn2}\right)\end{bmatrix}\begin{bmatrix}x_{12}\\x_{22}\\ \vdots\\ x_{n2}\end{bmatrix}$$
$$=\begin{bmatrix}\left(1+\frac{h}{2}a_{111}\right) & \frac{h}{2}a_{121} & \cdots & \frac{h}{2}a_{1n1}\\ \frac{h}{2}a_{211} & \left(1+\frac{h}{2}a_{221}\right) & \cdots & \frac{h}{2}a_{2n1}\\ \vdots & \vdots & & \vdots\\ \frac{h}{2}a_{n11} & \frac{h}{2}a_{n21} & \cdots & \left(1+\frac{h}{2}a_{nn1}\right)\end{bmatrix}\begin{bmatrix}x_{11}\\x_{21}\\ \vdots\\ x_{n1}\end{bmatrix}+\frac{h}{2}\begin{bmatrix}b_{11}\\b_{21}\\ \vdots\\ b_{n1}\end{bmatrix}u_1+\frac{h}{2}\begin{bmatrix}b_{12}\\b_{22}\\ \vdots\\ b_{n2}\end{bmatrix}u_2 \tag{9.17}$$

Equations (9.14) and (9.17), give a recursive solution for the states, for m number of subintervals.

That is,

$$\begin{bmatrix} x_{1(k+1)} \\ x_{2(k+1)} \\ \vdots \\ x_{n(k+1)} \end{bmatrix} = \begin{bmatrix} \left(\frac{2}{h} - a_{11(k+1)}\right) & -a_{12(k+1)} & \cdots & -a_{1n(k+1)} \\ -a_{21(k+1)} & \left(\frac{2}{h} - a_{22(k+1)}\right) & \cdots & -a_{2n(k+1)} \\ \vdots & \vdots & & \vdots \\ -a_{n1(k+1)} & -a_{n2(k+1)} & \cdots & \left(\frac{2}{h} - a_{nn(k+1)}\right) \end{bmatrix}^{-1}$$
$$\left(\begin{bmatrix} \left(\frac{2}{h} + a_{11k}\right) & a_{12k} & \cdots & a_{1nk} \\ a_{21k} & \left(\frac{2}{h} + a_{22k}\right) & \cdots & a_{2nk} \\ \vdots & \vdots & & \vdots \\ a_{n1k} & a_{n2k} & \cdots & \left(\frac{2}{h} + a_{nnk}\right) \end{bmatrix} \begin{bmatrix} x_{1k} \\ x_{2k} \\ \vdots \\ x_{nk} \end{bmatrix} + \begin{bmatrix} b_{1k} \\ b_{2k} \\ \vdots \\ b_{nk} \end{bmatrix} u_k + \begin{bmatrix} b_{1(k+1)} \\ b_{2(k+1)} \\ \vdots \\ b_{n(k+1)} \end{bmatrix} u_{(k+1)} \right) \tag{9.18}$$

for $k = 0, 1, 2, \ldots, (m - 1)$.

9.1.1 Numerical Examples

Example 9.1 (vide Appendix B, Program no. 26) Consider the non-homogeneous system $\dot{\mathbf{x}} = \mathbf{A}\mathbf{x} + \mathbf{B}u$, where $\mathbf{A} = \begin{bmatrix} 0 & 0 \\ t & 0 \end{bmatrix}$, $\mathbf{B} = \begin{bmatrix} 1 \\ 0 \end{bmatrix}$, $\mathbf{x}(0) = \begin{bmatrix} 1 \\ 1 \end{bmatrix}$ and $u = u(t)$, a unit step function.

The solution of the equation is $x_1(t) = 1 + t$ and $x_2(t) = 1 + \frac{t^2}{2} + \frac{t^3}{3}$.

Figure 9.1 graphically shows the comparison of the results obtained in HF domain using Eq. (9.18) with the exact curve. It is noted that the HF domain solutions (black dots) are right upon the exact curves. Table 9.1a, b show the comparison of the exact samples with HF domain solutions for the states of the system of Example 9.1 for $m = 4$ and $T = 1$ s.

Example 9.2 Consider the non-homogeneous system $\dot{\mathbf{x}} = \mathbf{A}\mathbf{x} + \mathbf{B}u$, where $\mathbf{A} = \begin{bmatrix} 0 & 1 \\ 0 & t \end{bmatrix}$, $\mathbf{B} = \begin{bmatrix} 0 \\ 1 \end{bmatrix}$, $\mathbf{x}(0) = \begin{bmatrix} 0 \\ 1 \end{bmatrix}$ and $u = u(t)$ a unit step function, having the solution

$$x_1(t) = t + \frac{1}{2}t^2 + \frac{1}{6}t^3 + \frac{1}{12}t^4 + \frac{1}{40}t^5 + \frac{1}{90}t^6 + \cdots$$
$$\text{and } x_2(t) = 1 + t + \frac{1}{2}t^2 + \frac{1}{3}t^3 + \frac{1}{8}t^4 + \frac{1}{15}t^5 + \frac{1}{48}t^6 + \frac{1}{105}t^7 + \cdots$$

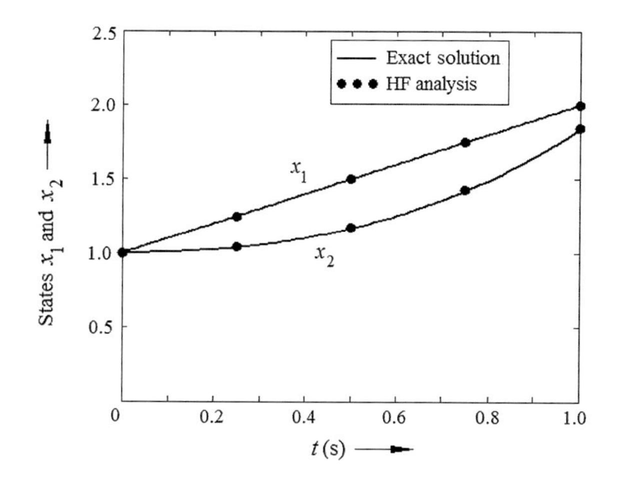

Fig. 9.1 Comparison of the HF domain solution of the states x_1 and x_2 of Example 9.1 with the exact solutions for $m = 4$ and $T = 1$ s (vide Appendix B, Program no. 26)

Table 9.1 Solution of the non-homogeneous system of Example 9.1 in HF domain compared with the exact samples along with percentage error at different sample points for (a) state x_1 and (b) state x_2, with $m = 4$, $T = 1$ s (vide Appendix B, Program no. 26)

t(s)	Samples of the exact solution s_d	Samples from HF analysis using Eq. (9.18) s_h	% Error $\varepsilon = \frac{s_d - s_h}{s_d} \times 100$
(a) System state x_1 ($m = 4$)			
0	1.00000000	1.00000000	0.00000000
$\frac{1}{4}$	1.25000000	1.25000000	0.00000000
$\frac{2}{4}$	1.50000000	1.50000000	0.00000000
$\frac{3}{4}$	1.75000000	1.75000000	0.00000000
$\frac{4}{4}$	2.00000000	2.00000000	0.00000000
(b) System state x_2 ($m = 4$)			
0	1.00000000	1.00000000	0.00000000
$\frac{1}{4}$	1.03645833	1.03906250	−0.25125660
$\frac{2}{4}$	1.16666667	1.17187500	−0.44642828
$\frac{3}{4}$	1.42187500	1.42968750	−0.54945055
$\frac{4}{4}$	1.83333333	1.84375000	−0.56818200

Table 9.2a and b show the comparison of the exact samples with HF domain solutions for the states of the system of Example 9.2 for $m = 4$ and $T = 1$ s.

Results obtained via direct expansion and using Eq. (9.18) are plotted in Fig. 9.2. It is noted that the HF domain solutions (black dots) are right upon the exact curves.

Table 9.2 Solution of the non-homogeneous system of Example 9.2 in HF domain compared with the exact samples along with percentage error at different sample points for (a) state x_1 and (b) state x_2 with $m = 4$, $T = 1$ s

t (s)	Samples of the exact solution s_d	Samples from HF analysis using Eq. (9.18) s_h	% Error $\varepsilon = \frac{s_d - s_h}{s_d} \times 100$
(a) System state x_1 ($m = 4$)			
0	0.00000000	0.00000000	–
$\frac{1}{4}$	0.28421139	0.28629032	−0.73147315
$\frac{2}{4}$	0.65228950	0.65833333	−0.92655638
$\frac{3}{4}$	1.13917694	1.15065814	−1.00785046
$\frac{4}{4}$	1.80486111	1.81990969	−0.83378050
(b) System state x_2 ($m = 4$)			
0	1.00000000	1.00000000	0. 00000000
$\frac{1}{4}$	1.28701739	1.29032258	−0.25681005
$\frac{2}{4}$	1.67696243	1.68602151	−0.54020768
$\frac{3}{4}$	2.23222525	2.25257694	−0.91172205
$\frac{4}{4}$	3.05535714	3.10143546	−1.50811568

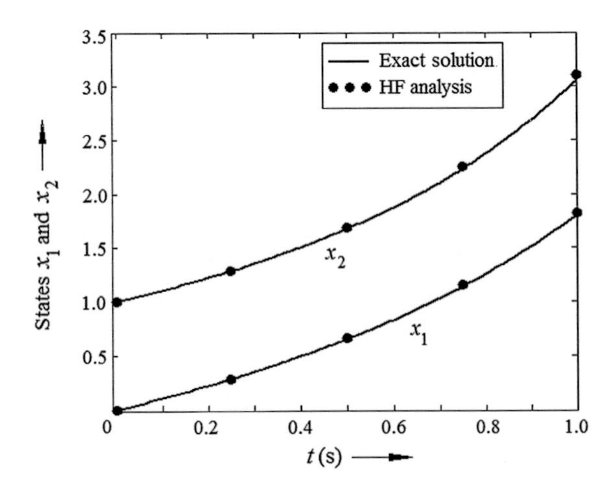

Fig. 9.2 Comparison of the HF domain solution of the states of Example 9.2 with the exact solutions for $m = 4$ and $T = 1$ s

9.2 Determination of Output of a Non-homogeneous Time Varying System

Consider the output of a time varying non-homogeneous system described by

$$\mathbf{y}(t) = \mathbf{C}(t)\mathbf{x}(t) + \mathbf{D}(t)\,u(t)$$

where,

$\mathbf{x}(t)$ is the *state vector* given by $\mathbf{x}(t) = \begin{bmatrix} x_1 & x_2 & \cdots & \cdots & x_n \end{bmatrix}^T$,
$\mathbf{y}(t)$ is the *output vector*, is expressed by $\mathbf{y}(t) \triangleq \begin{bmatrix} y_1 & y_2 & \cdots & \cdots & y_v \end{bmatrix}^T$,
$u(t)$ is the *input vector*, is expressed by $u(t) \triangleq \begin{bmatrix} u_1 & u_2 & \cdots & \cdots & u_r \end{bmatrix}^T$,
$\mathbf{C}(t)$ is the time varying *output matrix* and
$\mathbf{D}(t)$ is the time varying *direct transmission matrix.*

As we have shown in Sect. 9.1, we can easily expand the products $\mathbf{C}(t)\mathbf{x}(t)$ and $\mathbf{D}(t)\,u(t)$ in HF domain, vide Eqs. (9.6) and (9.7). Since, $\mathbf{C}(t)$, $\mathbf{D}(t)$, $x(t)$ and $u(t)$ are already known, $y(t)$ can easily be determined.

9.3 Analysis of Homogeneous Time Varying State Equation [3]

If we consider the time varying input vector $\mathbf{B}(t)$ as zero, the resulting expression from Eq. (9.18) provides the recursive solution for an n-state homogeneous time varying system.

That is,

$$\begin{bmatrix} x_{1(k+1)} \\ x_{2(k+1)} \\ \vdots \\ x_{n(k+1)} \end{bmatrix} = \begin{bmatrix} \left(\frac{2}{h} - a_{11(k+1)}\right) & -a_{12(k+1)} & \cdots & -a_{1n(k+1)} \\ -a_{21(k+1)} & \left(\frac{2}{h} - a_{22(k+1)}\right) & \cdots & -a_{2n(k+1)} \\ \vdots & \vdots & & \vdots \\ -a_{n1(k+1)} & -a_{n2(k+1)} & \cdots & \left(\frac{2}{h} - a_{nn(k+1)}\right) \end{bmatrix}^{-1} \begin{bmatrix} \left(\frac{2}{h} + a_{11k}\right) & a_{12k} & \cdots & a_{1nk} \\ a_{21k} & \left(\frac{2}{h} + a_{22k}\right) & \cdots & a_{2nk} \\ \vdots & \vdots & & \vdots \\ a_{n1k} & a_{n2k} & \cdots & \left(\frac{2}{h} + a_{nnk}\right) \end{bmatrix} \begin{bmatrix} x_{1k} \\ x_{2k} \\ \vdots \\ x_{nk} \end{bmatrix} \tag{9.19}$$

for $k = 0, 1, 2, \ldots, (m-1)$

For a time-invariant system, $\mathbf{A}(t) = \mathbf{A}$ and $\mathbf{B}(t) = \mathbf{B}$. For such a system Eq. (9.18) can be modified as

$$\begin{bmatrix} x_{1(k+1)} \\ x_{2(k+1)} \\ \vdots \\ x_{n(k+1)} \end{bmatrix} = \begin{bmatrix} \left(\frac{2}{h} - a_{11}\right) & -a_{12} & \cdots & -a_{1n} \\ -a_{21} & \left(\frac{2}{h} - a_{22}\right) & \cdots & -a_{2n} \\ \vdots & \vdots & & \vdots \\ -a_{n1} & -a_{n2} & \cdots & \left(\frac{2}{h} - a_{nn}\right) \end{bmatrix}^{-1}$$
$$\left(\begin{bmatrix} \left(\frac{2}{h} + a_{11}\right) & a_{12} & \cdots & a_{1n} \\ a_{21} & \left(\frac{2}{h} + a_{22}\right) & \cdots & a_{2n} \\ \vdots & \vdots & & \vdots \\ a_{n1} & a_{n2} & \cdots & \left(\frac{2}{h} + a_{nn}\right) \end{bmatrix} \begin{bmatrix} x_{1k} \\ x_{2k} \\ \vdots \\ x_{nk} \end{bmatrix} + \begin{bmatrix} b_1 \\ b_2 \\ \vdots \\ b_n \end{bmatrix} u_k + \begin{bmatrix} b_1 \\ b_2 \\ \vdots \\ b_n \end{bmatrix} u_{(k+1)} \right) \tag{9.20}$$

This (9.20) can be written as

$$\mathbf{x}_{k+1} = \left[\frac{2}{h}\mathbf{I} - \mathbf{A}\right]^{-1} \left(\left[\frac{2}{h}\mathbf{I} + \mathbf{A}\right] \mathbf{x}_k + \mathbf{B}(u_k + u_{k+1}) \right), \quad \text{for } k = 0, 1, 2, \cdots, (m-2) \tag{9.21}$$

where,

$$\mathbf{x}_k = \begin{bmatrix} x_{1k} \\ x_{2k} \\ \vdots \\ x_{nk} \end{bmatrix} \quad \text{and} \quad \mathbf{x}_{(k+1)} = \begin{bmatrix} x_{1(k+1)} \\ x_{2(k+1)} \\ \vdots \\ x_{n(k+1)} \end{bmatrix}$$

In (9.21), if we set the vector **B** as zero, we get the recursive solution for the states of a homogeneous linear time-invariant (LTI) system.

9.3.1 Numerical Examples

Example 9.3 Consider the first order homogeneous system $\dot{x} = Ax$ where $A = 2t$ and $x(0) = 1$. having the solution $x(t) = \exp(t^2)$.

It is noted from Table 9.3a, b that an increasing m from 4 to 10 improves the accuracy to such a degree that the HF domain solution almost coincides with the exact samples of the states. Figures 9.3 and 9.4 prove this point graphically.

Example 9.4 [4] Consider the homogeneous system $\dot{\mathbf{x}} = \mathbf{A}\mathbf{x}$, where $\mathbf{A} = \begin{bmatrix} 0 & 0 \\ t & 0 \end{bmatrix}$ and $\mathbf{x}(0) = \begin{bmatrix} 1 \\ 1 \end{bmatrix}$ having the solution $x_1(t) = 1$ and $x_2(t) = \frac{1}{2}t^2 + 1$.

Table 9.3 Solution of the homogeneous system of Example 9.3 in HF domain compared with the exact samples along with percentage error at different sample points for (a) $m = 4$, $T = 1$ s and (b) $m = 10$, $T = 1$ s

t(s)	Samples of the exact solution s_d	Samples from HF analysis using Eq. (9.19) s_h	% Error $\varepsilon = \frac{s_d - s_h}{s_d} \times 100$
(a) System state x ($m = 4$)			
0	1.00000000	1.00000000	0.00000000
$\frac{1}{4}$	1.06449446	1.06666667	−0.20406024
$\frac{2}{4}$	1.28402542	1.29523810	−0.87324439
$\frac{3}{4}$	1.75505466	1.79340659	−2.18522709
$\frac{4}{4}$	2.71828183	2.83956044	−4.46159072
(b) System state x ($m = 10$)			
0	1.00000000	1.00000000	0.00000000
$\frac{1}{10}$	1.01005017	1.01010101	−0.00503341
$\frac{2}{10}$	1.04081077	1.04102247	−0.02033991
$\frac{3}{10}$	1.09417428	1.09468342	−0.04653189
$\frac{4}{10}$	1.17351087	1.17450409	−0.08463663
$\frac{5}{10}$	1.28402542	1.28577290	−0.13609388
$\frac{6}{10}$	1.43332941	1.43623568	−0.20276358
$\frac{7}{10}$	1.63231622	1.63699981	−0.28692909
$\frac{8}{10}$	1.89648088	1.90390195	−0.39130740
$\frac{9}{10}$	2.24790799	2.25957594	−0.51905817
$\frac{10}{10}$	2.71828183	2.73659753	−0.67379695

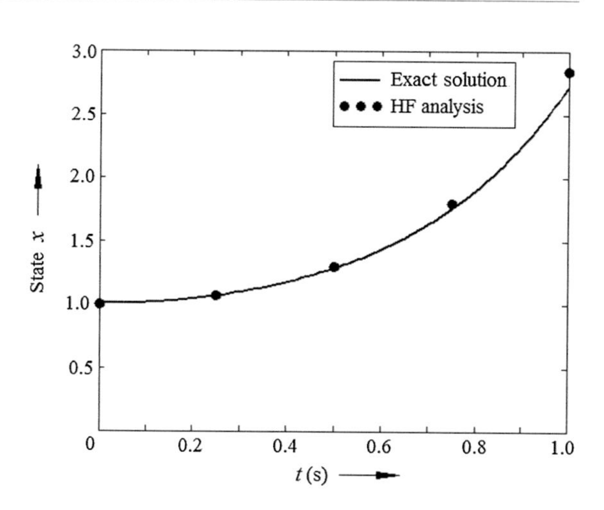

Fig. 9.3 Comparison of the HF domain solution of state x of Example 9.3 with the exact solutions for $m = 4$ and $T = 1$ s

We see that the system state x_2, obtained via direct expansion and the proposed HF domain technique, are the same leading to zero error, vide Table 9.4.

Figure 9.5 shows the results obtained via HF analysis with the exact solutions, for $m = 4$ and $T = 1$ s.

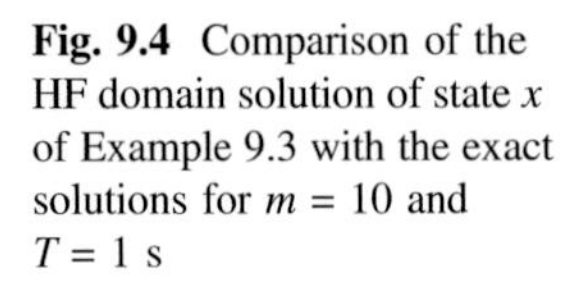

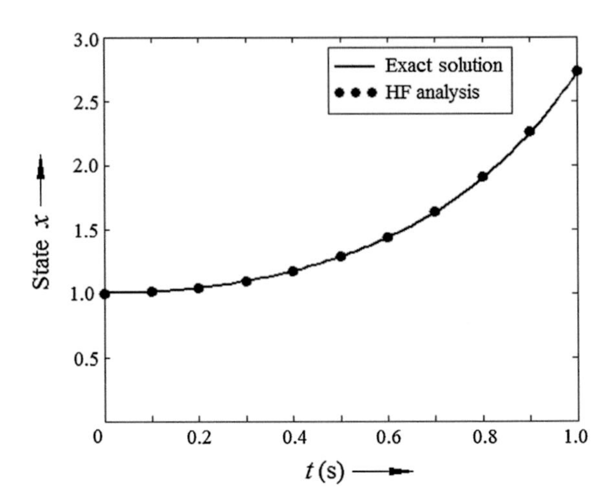

Fig. 9.4 Comparison of the HF domain solution of state x of Example 9.3 with the exact solutions for $m = 10$ and $T = 1$ s

Table 9.4 HF domain solution of the state x_2 of the system of Example 9.4 compared with the exact samples along with percentage error at different sample points, for $m = 4$ and $T = 1$ s

t(s)	System state x_2 ($m = 4$)		
	Samples of the exact solution s_d	Samples from HF analysis using Eq. (9.19) s_h	% Error $\varepsilon = \frac{s_d - s_h}{s_d} \times 100$
0	1.00000000	1.00000000	0.00000000
$\frac{1}{4}$	1.03125000	1.03125000	0.00000000
$\frac{2}{4}$	1.12500000	1.12500000	0.00000000
$\frac{3}{4}$	1.28125000	1.28125000	0.00000000
$\frac{4}{4}$	1.50000000	1.50000000	0.00000000

Example 9.5 [5] (vide Appendix B, Program no. 27) Consider the homogeneous system $\dot{\mathbf{x}} = \mathbf{A}\mathbf{x}$, where $\mathbf{A} = \begin{bmatrix} \cos(t) & \sin(t) \\ -\sin(t) & \cos(t) \end{bmatrix}$ and $\mathbf{x}(0) = \begin{bmatrix} 1 \\ 2 \end{bmatrix}$ having the solution

$$x_1(t) = (\cos(1 - \cos t) + 2\sin(1 - \cos t))e^{\sin t}$$
$$\text{and}\, x_2(t) = (-\sin(1 - \cos t) + 2\cos(1 - \cos t))e^{\sin t}$$

Analysis of the given system produces the following results:

Tables 9.5a, b show the comparison of the exact samples with HF domain solutions for the states of the system of Example 9.5 for $m = 8$ and $T = 1$ s. Results obtained via direct expansion and using Eq. (9.19) are plotted in Fig. 9.6. It is noted that the HF domain solutions (black dots) are right upon the exact curves.

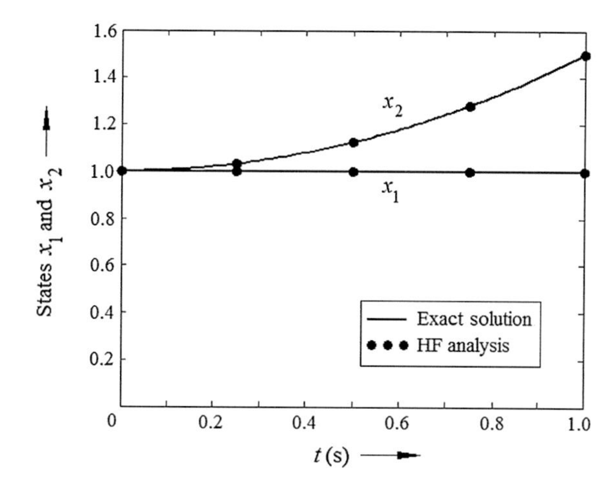

Fig. 9.5 Comparison of the HF domain solution of the states x_1 and x_2 of Example 9.4 with the exact solutions for $m = 4$ and $T = 1$ s

Table 9.5 HF domain solution of the (a) state x_1 and (b) state x_2 of the homogeneous system of Example 9.5 compared with the exact samples along with percentage error at different sample points, with $m = 8$, $T = 1$ s (vide Appendix B, Program no. 27)

t(s)	Samples of the exact solution s_d	Samples from HF analysis using Eq. (9.19), s_h	% Error $\varepsilon = \frac{s_d - s_h}{s_d} \times 100$
(a) System state x_1 (m = 8)			
0	1.00000000	1.00000000	0.00000000
$\frac{1}{8}$	1.15042193	1.15148482	−0.09239132
$\frac{2}{8}$	1.35969222	1.36188071	−0.16095481
$\frac{3}{8}$	1.63917010	1.64229144	−0.19042197
$\frac{4}{8}$	1.99751628	2.00098980	−0.17389195
$\frac{5}{8}$	2.43785591	2.44061935	−0.11335535
$\frac{6}{8}$	2.95465368	2.95517291	−0.01757329
$\frac{7}{8}$	3.53102052	3.52746220	0.10077313
$\frac{8}{8}$	4.13741725	4.12800452	0.22750256
(b) System state x_2 (m = 8)			
0	2.00000000	2.00000000	0. 00000000
$\frac{1}{8}$	2.25665267	2.25592250	0.03235633
$\frac{2}{8}$	2.52034777	2.51828652	0.08178435
$\frac{3}{8}$	2.77758241	2.77342604	0.14963984
$\frac{4}{8}$	3.00888958	3.00181231	0.23521202
$\frac{5}{8}$	3.18903687	3.17834568	0.33524824
$\frac{6}{8}$	3.28860794	3.27401580	0.44371784
$\frac{7}{8}$	3.27727645	3.25919411	0.55174900
$\frac{8}{8}$	3.12867401	3.10841539	0.64751457

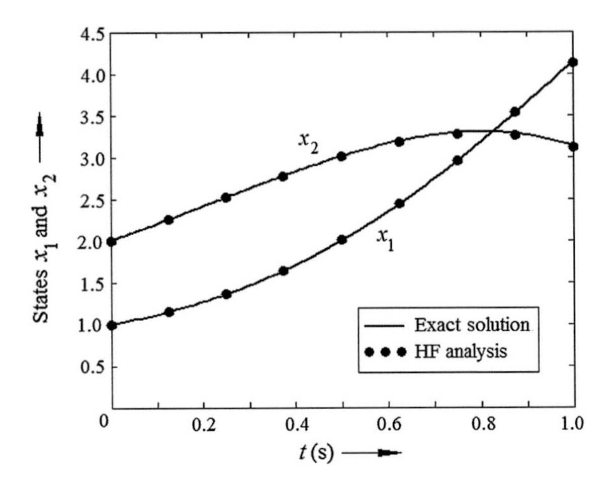

Fig. 9.6 Comparison of the HF domain solution of the states of Example 9.5 with the exact solutions for $m = 8$ and $T = 1$ s (vide Appendix B, Program no. 27)

9.4 Determination of Output of a Homogeneous Time Varying System

Consider the output of a time varying homogeneous system described by

$$\mathbf{y}(t) = \mathbf{C}(t)\mathbf{x}(t)$$

where,

$\mathbf{x}(t)$ is the *state vector* given by $\mathbf{x}(t) = [x_1 \quad x_2 \quad \ldots \quad \ldots \quad x_n]^{\mathrm{T}}$,
$\mathbf{y}(t)$ is the *output vector*, is expressed by $\mathbf{y}(t) \triangleq [y_1 \quad y_2 \quad \ldots \quad \ldots \quad y_v]^{\mathrm{T}}$ and
$\mathbf{C}(t)$ is the time varying *output matrix*.
We can easily solve for $y(t)$ in a fashion similar to that of Sect. 9.2.

9.5 Conclusion

In this chapter, we have analysed the state space model of a time varying non-homogeneous system and solved for the states in hybrid function platform. The proficiency of the method has been illustrated by suitable examples. Also, by putting $\mathbf{B}(t) = 0$, the same method has been applied successfully to analyze homogeneous time varying systems as well. With slight modification, we arrive from Eqs. (9.19) to (9.21) which is suitable for the analysis of time-invariant non-homogeneous systems. Further, by setting $\mathbf{B} = 0$, Eq. (9.21) becomes suitable for the analysis of linear time-invariant homogeneous system.

Using the above mentioned recursive matrix equations, different types of numerical examples have been treated for homogeneous as well as non-homogeneous systems, and the results are compared with the exact solutions along with error estimates for each sample point. It is found that the HF method is a strong as well as convenient tool for such analysis. This fact is reflected through various tables and curves.

For the analysis of output equations, since it has been treated in Chap. 8, we avoid repeating the same here. However, it has already been shown that the HF domain solutions are reliably close.

References

1. Ogata, K.: Modern Control Engineering, 5th edn. Prentice Hall of India (2011)
2. Fogiel, M. (ed.): The Automatic Control Systems/ Robotics Problem Solver. Research & Education Association, New Jersey (2000)
3. Roychoudhury, S., Deb, A., Sarkar, G.: Analysis and synthesis of time-varying systems via orthogonal hybrid functions (HF) in states space environment. Int. J. Dyn. Control. **3**(4), 389–402 (2015)
4. Rao, G.P.: Piecewise Constant Orthogonal Functions and Their Application in Systems and Control, LNC1S, vol. 55. Springer, Berlin (1983)
5. Jiang, J.H., Schaufelberger, W.: Block Pulse Functions and Their Application in Control System, LNCIS, vol. 179. Springer, Berlin (1992)

Chapter 10
Multi-delay System Analysis: State Space Approach

Abstract This chapter is devoted to multi-delay system analysis using state space approach in hybrid function domain. First, the theory of HF domain approximation of functions with time delay is presented. This is followed by integration of functions with time delay. Then analysis of both homogeneous and non-homogeneous systems are given along with numerical examples. States of the systems are solved. Illustration of the theory has been provided with the support of eight examples, nineteen figures and four tables.

In this chapter, we deal with multi-delay control systems. In practice we can find time delays in different electrical systems, industrial processes, mechanical systems, population growth, economic growth and chemical processes. We intend to analyse two types of control systems in hybrid function platform, namely non-homogeneous system and homogeneous system.

Analysing a time delay control system means, knowing the system parameters, the nature of input signal or forcing function and the presence of delay in the system, we determine the behavior of each system states of a time delay system over a common time frame, to assess the performance of the system.

In this chapter, the hybrid function set is employed for the analysis of time delay non-homogeneous as well as homogeneous systems described in state space platform. Here we have converted the multi-delay differential equation to an algebraic form using the orthogonal hybrid function (HF) set.

First we take up the problem of analysis of a delayed non-homogeneous system in HF domain, because after putting the specific condition of zero forcing function, we can arrive at the result of analysis of a delayed homogeneous system.

10.1 HF Domain Approximation of Function with Time Delay

Let us consider a time function $f(t)$ which consists of single delay or multiple delays along the time scale. In case of HF approximation of the delayed time function, following situations may arise:

A. Deb et al., *Analysis and Identification of Time-Invariant Systems, Time-Varying Systems, and Multi-Delay Systems using Orthogonal Hybrid Functions*,
Studies in Systems, Decision and Control 46, DOI 10.1007/978-3-319-26684-8_10

- The function $f(t)$ consists of multiple delays with one delay equal to τ (say) and the rest its integral multiples. Then we can select the sampling interval h judiciously as equal to the delay τ, or τ may be integral multiple of h when h is smaller than ν. Then we can approximate the delayed time function in HF domain comfortably.
- The function $f(t)$ consists of multiple delays with one delay equal to τ (say) while the other delays are *not* integral multiples of τ. This situation may be handled in a different way. In this case the sampling interval h has to be chosen in such a way that all the delays are integral multiples of h. So we find the highest common factor (HCF) of the delays to determine a suitable value of the sampling interval h. That is, if $h_n = \text{highest common factor}(\tau_1, \tau_2, \ldots, \tau_n)$, the sampling interval $h = \frac{h_n}{k}$, where, k is an integer greater than zero.

Figure 10.1a shows equidistant samples of a time function $f(t)$ while Fig. 10.1b illustrates its delayed version $f(t-\tau)$ with $\tau = 2h$.

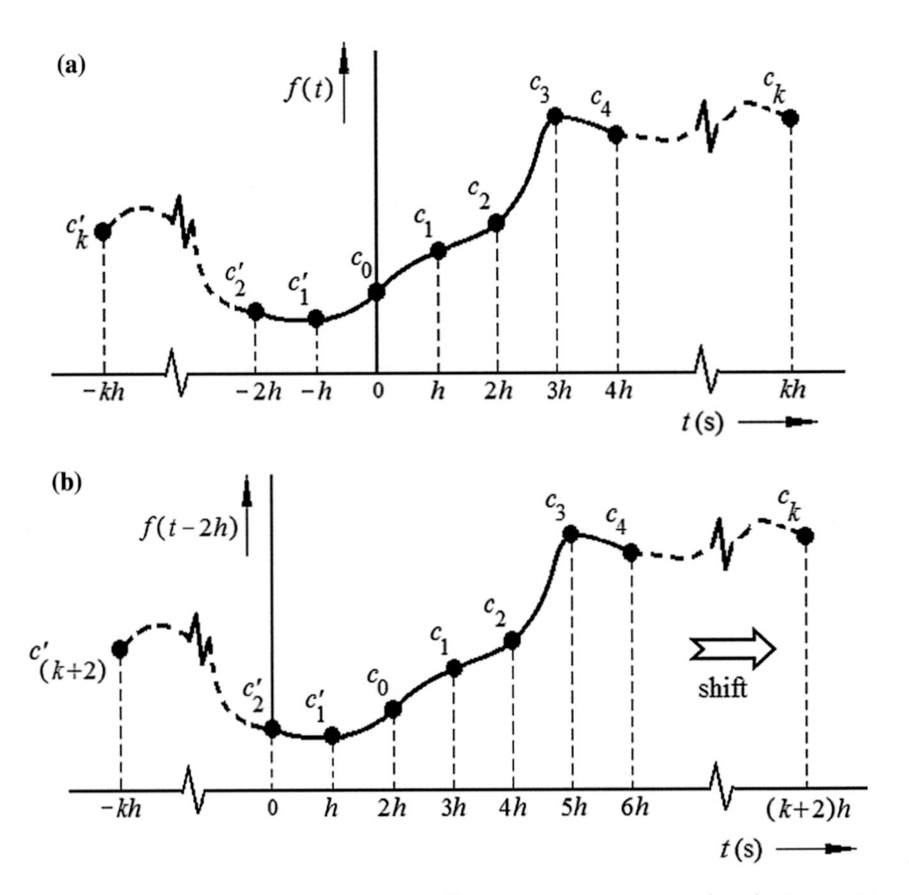

Fig. 10.1 Equidistant samples of a a function $f(t)$ and b its delayed version $f(t-\tau)$ with $\tau = 2h$

Let us now recall the delay matrix Q, defined as

$$\mathbf{Q}_{(m)} \triangleq [\![0 \quad 1 \quad 0 \quad 0 \quad \cdots \quad 0 \quad 0]\!]_{(m \times m)}$$

$\mathrm{Q}_{(m)}$ has the property

$$\mathbf{Q}^k_{(m)} = \left[\left[\underbrace{0 \quad 0 \quad \ldots 0 \quad 0}_{k \text{ terms}} \quad 1 \quad \underbrace{0 \quad 0 \quad \ldots \quad 0 \quad 0}_{(m-k-1) \text{ terms}}\right]\right]_{(m \times m)}$$

and $\mathbf{Q}^k_{(m)} = \mathbf{0}_{(m)}$

If we post-multiply any row matrix **A** by a **Q** matrix of proper dimension, the **A** matrix will be shifted towards *right* by one element, and a zero will be introduced as the first element of **AQ**.

However, if we post-multiply **AQ** by **Q** again, the result will be shifted further towards *right* by another element, and the vacant place at the start of $\mathbf{AQ}^2$ will again be taken up by another zero. If we continue such post multiplication with **Q**, at each step of multiplication the result will be shifted towards *right* by one element.

That means, if

$$\mathbf{A} = [a_{11} \quad a_{12} \quad \cdots \quad a_{1m}]_{(1\times m)}, \text{ then}$$

$$\mathbf{AQ}^k = \left[\underbrace{0 \quad 0 \quad \ldots \quad 0 \quad 0}_{k \text{ terms}} \quad a_{11} \quad a_{12} \quad \ldots \quad a_{1(m-k)}\right]_{(1\times m)} \tag{10.1}$$

We find that, transpose of **Q**, namely $\mathbf{Q}^{\mathrm{T}}$, behaves in a manner contrary to that of **Q**. That is, if we post-multiply a row matrix **A** by the $\mathbf{Q}^{\mathrm{T}}$ matrix of proper dimension, the result will be the **A** matrix shifted towards *left* by one element, and a zero will be introduced as the last element of $\mathbf{AQ}^{\mathrm{T}}$. Thus, we can call $\mathbf{Q}^{\mathrm{T}}$ as the *advance* matrix.

Like repeated multiplication by **Q**, if we proceed similarly with $\mathbf{Q}^{\mathrm{T}}$, at each stage of multiplication, the resulting row matrix will be shifted towards *left* by one element.

That means

$$\mathbf{A}\left[\mathbf{Q}^{\mathrm{T}}\right]^k = \left[\underbrace{a_{1(m-k)} \quad a_{1(m-k+1)} \quad \cdots \quad a_{1(m-1)} \quad a_{1m}}_{(m-k)\text{terms}} \quad \underbrace{0 \quad 0 \quad \ldots \quad 0 \quad 0}_{k \text{ terms}}\right]_{(1\times m)} \tag{10.2}$$

In approximating time functions with delays, $\mathbf{Q}$ as well as $\mathbf{Q}^{\mathrm{T}}$ have vital roles to play.

For convenience in representation, let us define

$$\mathbf{Q}^{T} \triangleq \mathbf{Q}_{\mathrm{t}}$$

Referring to Fig. 10.1a, we can represent the function $f(t)$ in HF domain with a sampling period of h in the interval $t \in [0, mh)$. That is

$$\begin{aligned} f(t) \approx \bar{f}(t) &= [c_0 \quad c_1 \quad c_2 \quad \dots \quad c_{m-1}]\mathbf{S}_{(m)}(t) \\ &\quad + [(c_1 - c_0) \quad (c_2 - c_1) \quad (c_3 - c_2) \quad \dots \quad (c_m - c_{m-1})]\mathbf{T}_{(m)}(t) \\ &\triangleq \mathbf{C}_{\mathrm{S}}^{\mathrm{T}}\mathbf{S}_{(m)} + \mathbf{C}_{\mathrm{T}}^{\mathrm{T}}\mathbf{T}_{(m)} \end{aligned}$$

where $\bar{f}(t)$ is the piecewise linear approximation of $f(t)$ in HF domain.

Consider that we know m number of samples of the function $f(t)$ for time $t < 0$. That is, we want to work with a row matrix of dimension $(1 \times m)$. Thus, for $t < 0$, the function $f(t)$ may be represented via sample-and-hold functions as

$$f(t)|_{t<0} \approx [c'_m \quad c'_{m-1} \quad \dots \quad c'_3 \quad c'_2 \quad c'_1]\mathbf{S}_{(m)}(t) \tag{10.3}$$

Now, consider the samples of the function for $t \geq 0$. Again, if we consider m number of samples and represent the function $f(t)$ via sample-and-hold functions, we have

$$f(t)|_{t>0} \approx [c_0 \quad c_1 \quad c_2 \quad \dots \quad c_{m-1}]\mathbf{S}_{(m)}(t) \tag{10.4}$$

If we work with m samples in total within the time interval $t \in [-kh, (m-k)h]$ then we can express this particular section of the function in SHF domain as

$$f(t)|_{-kh \le t \le (m-k)h} \approx [c'_k \quad c'_{k-1} \quad c'_{k-2} \quad \dots \quad c'_1 \quad c_0 \quad c_1 \quad \dots \quad c_{m-k-1}]\mathbf{S}_{(m)}(t) \tag{10.5}$$

Using Eq. (10.5), we can handle a function delayed by kh (k being an integer). Also, by employing the $\mathbf{Q}$ and $\mathbf{Q}_{\mathrm{t}}$ matrices, we can arrive at (10.5) from the generalized Eqs. (10.3) and (10.4). That is

$$\begin{aligned} f(t)|_{-kh \le t \le (m-k)h} &\approx [c_0 \quad c_1 \quad c_2 \quad \dots \quad c_{m-1}]\mathbf{Q}^k\mathbf{S}_m(t) \\ &\quad + [c'_m \quad c'_{m-1} \quad \dots \quad c'_2 \quad c'_1]\mathbf{Q}_{\mathrm{t}}^k\mathbf{S}_m(t) \\ &= [c'_k \quad c'_{k-1} \quad c'_{k-2} \quad \dots \quad c'_1 \quad c_0 \quad c_1 \quad \dots \quad c_{m-k-1}]\mathbf{S}_{(m)}(t) \end{aligned}$$

Now, representation of the delayed function $f(t - kh)$ in HF domain in an interval $t \in [-kh, (m-k)h]$ is given by

$$
\begin{aligned}
f(t-kh) \approx \bar{f}(t-kh) &= \left[c'_k \quad c'_{k-1} \quad c'_{k-2} \quad \ldots \quad c'_1 \quad c_0 \quad c_1 \quad \ldots \quad c_{m-k-1} \right] \mathbf{S}_{(m)}(t) \\
&\quad + \left[\left(c'_{k-1} - c'_k\right) \quad \left(c'_{k-2} - c'_{k-1}\right) \quad \ldots \quad \left(c_0 - c'_1\right) \quad \left(c_1 - c_0\right) \right. \\
&\quad \left. \ldots \left(c_{m-k} - c_{m-k-1}\right) \right] \mathbf{T}_{(m)}(t) \\
&= \left[c_0 \quad c_1 \quad c_2 \quad \ldots \quad c_{m-1} \right] \mathbf{Q}^k \mathbf{S}_m(t) \\
&\quad + \left[c'_m \quad c'_{m-1} \quad \ldots \quad c'_2 \quad c'_1 \right] \mathbf{Q}_{\mathrm{t}}^k \mathbf{S}_m(t) \\
&\quad + \left[(c_1 - c_0) \quad (c_2 - c_1) \quad (c_3 - c_2) \quad \ldots \quad (c_m - c_{m-1}) \right] \mathbf{Q}^k \mathbf{T}_m(t) \\
&\quad + \left[(c'_{m-1} - c'_m) \quad (c'_{m-2} - c'_{m-1}) \quad \ldots \quad (c'_1 - c'_2) \quad (c_0 - c'_1) \right] \mathbf{Q}_t^k \mathbf{T}_m(t) \\
&\triangleq \left[\mathbf{C}_{\mathrm{S}}^{\mathrm{T}} \mathbf{Q}^k + \mathbf{C}_{\mathrm{S}\tau}^{\mathrm{T}} \mathbf{Q}_{\mathrm{t}}^k \right] \mathbf{S}_{(m)} + \left[\mathbf{C}_{\mathrm{T}}^{\mathrm{T}} \mathbf{Q}^k + \mathbf{C}_{\mathrm{T}\tau}^{\mathrm{T}} \mathbf{Q}_{\mathrm{t}}^k \right] \mathbf{T}_{(m)}
\end{aligned}
\tag{10.6}
$$

where, use is made of Eq. (10.5), $\mathbf{Q}^k$ is termed as the *delay* matrix of degree k and $\mathbf{Q}_{\mathrm{t}}^k$ is termed as the *advance* matrix of degree k.

Here $\mathbf{C}_{\mathrm{S}\tau}^{\mathrm{T}}$ is the sample-and-hold coefficient vector for initial values of the state, representing the coefficients belong to time scale $-kh \leq t < 0$. Whereas $\mathbf{C}_{\mathrm{T}\tau}^{\mathrm{T}}$ is the triangular function coefficient vector for initial values of the state, representing the coefficients belong to time scale $-kh \leq t < 0$.

10.1.1 Numerical Examples

Example 10.1 Consider a function $f_1(t) = t$ having a delay at t = 0.4 s. Using Eq. (10.6), approximations of this function is graphically shown in Fig. 10.2, for m = 10 and T = 1 s.

Here the delay time $\tau = kh = 0.4$ s, and time period T = 1 s. That means k = 4.

Using Eq. (10.1), the approximation of function $f_1(t - kh) = (t - 0.4)$ can be represented by

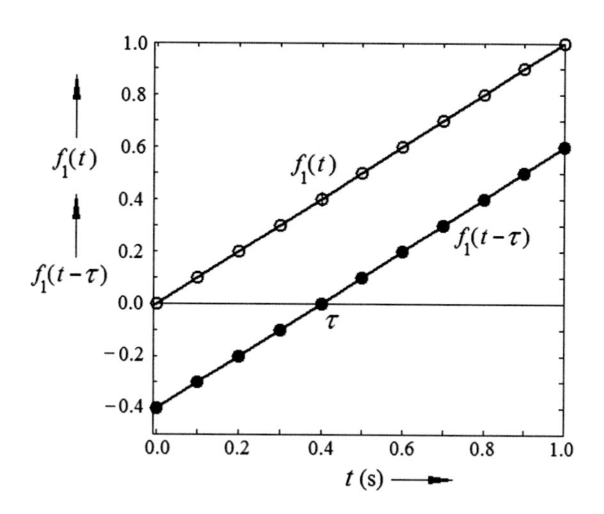

Fig. 10.2 Graphical representation of HF domain approximation with exact function, of function $f_1(t)$ of Example 10.1, for m = 10 and T = 1 s

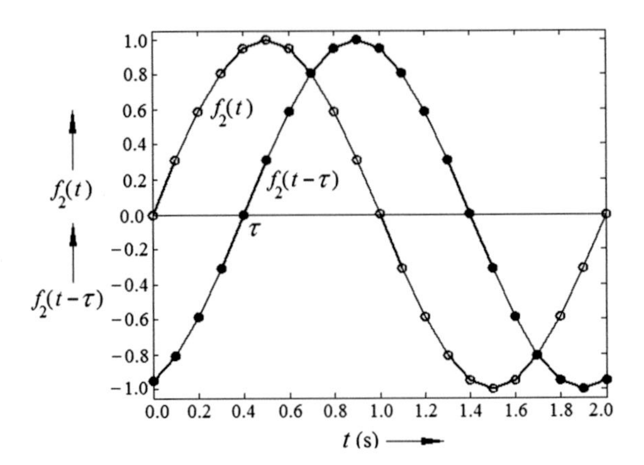

Fig. 10.3 Pictorial representation of HF domain approximation of $f_2(t) = sin[\pi(t-0.4)]$ of Example 10.2 along with the exact function for m = 20 and T = 2 s

$$(t-0.4) \approx [-0.4 \quad -0.3 \quad -0.2 \quad -0.1 \quad 0 \quad 0.1 \quad 0.2 \quad 0.3 \quad 0.4 \quad 0.5]\mathbf{S}_{(10)}(t) + [0.1 \quad 0.1 \quad 0.1 \quad 0.1 \quad 0.1 \quad 0.1 \quad 0.1 \quad 0.1 \quad 0.1 \quad 0.1]\mathbf{T}_{(10)}(t) \tag{10.7}$$

In Fig. 10.2, we see that the samples of the delayed time function, obtained via HF domain approximation, coincide with exact samples of the function $f_1(t-kh)$ due to its linear nature.

Example 10.2 Consider a function $f_2(t) = sin(\pi t)$ over a time period of T = 2 s, having a delay of τ = 0.4 s. For m = 20 and T = 1 s, h = 0.1 and $k = \frac{\tau}{h}$.

Using Eq. (10.6), the approximation of this function is graphically shown in Fig. 10.3 as in Example 10.2.

10.2 Integration of Functions with Time Delay

Let us consider a time delayed square integrable function $f(t-kh)$ which can be expanded in HF domain, using Eq. (10.6). Now, if we consider all the initial values of the function for $-kh \le t < 0$ to be zero, then integrating Eq. (10.6) with respect to t, we have

$$\int f(t-kh)\mathrm{d}t \approx \int \mathbf{C}_\mathrm{S}^\mathrm{T}\mathbf{Q}^k\mathbf{S}_{(m)}\mathrm{d}t + \int \mathbf{C}_\mathrm{T}^\mathrm{T}\mathbf{Q}^k\mathbf{T}_{(m)}\mathrm{d}t \tag{10.8}$$

Using Eqs. (10.8), (4.9) and (4.18), the integration of a time delayed function with all zero initial values, can be expressed as

$$\int f(t-kh)\mathrm{d}t \approx \left[\mathbf{C}_\mathrm{S}^\mathrm{T} + \frac{1}{2}\mathbf{C}_\mathrm{T}^\mathrm{T}\right]\mathbf{Q}^k\mathbf{P1ss}_{(m)}\mathbf{S}_{(m)} + h\left[\mathbf{C}_\mathrm{S}^\mathrm{T} + \frac{1}{2}\mathbf{C}_\mathrm{T}^\mathrm{T}\right]\mathbf{Q}^k\mathbf{T}_{(m)} \tag{10.9}$$

But in reality, most of the practical systems have some non-zero initial values of their states. In that case, the integration of a delayed time function can be represented by

$$\int f(t-kh)\,\mathrm{d}t \approx \int \left[\mathbf{C}_\mathrm{S}^\mathrm{T}\mathbf{Q}^k + \mathbf{C}_{\mathrm{S}\tau}^\mathrm{T}\mathbf{Q}_\mathrm{t}^k\right]\mathbf{S}_{(m)}\mathrm{d}t + \int \left[\mathbf{C}_\mathrm{T}^\mathrm{T}\mathbf{Q}^k + \mathbf{C}_{\mathrm{T}\tau}^\mathrm{T}\mathbf{Q}_\mathrm{t}^k\right]\mathbf{T}_{(m)}\mathrm{d}t \tag{10.10}$$

Using Eqs. (10.10), (4.9) and (4.18), integration of a time delayed function with initial values, may be expressed as

$$\int f(t-kh)\mathrm{d}t \approx \left[\left(\mathbf{C}_\mathrm{S}^\mathrm{T}\mathbf{Q}^k + \mathbf{C}_{\mathrm{S}\tau}^\mathrm{T}\mathbf{Q}_\mathrm{t}^k\right) + \frac{1}{2}\left(\mathbf{C}_\mathrm{T}^\mathrm{T}\mathbf{Q}^k + \mathbf{C}_{\mathrm{T}\tau}^\mathrm{T}\mathbf{Q}_\mathrm{t}^k\right)\right]\mathbf{P1ss}_{(m)}\mathbf{S}_{(m)} + h\left[\left(\mathbf{C}_\mathrm{S}^\mathrm{T}\mathbf{Q}^k + \mathbf{C}_{\mathrm{S}\tau}^\mathrm{T}\mathbf{Q}_\mathrm{t}^k\right) + \frac{1}{2}\left(\mathbf{C}_\mathrm{T}^\mathrm{T}\mathbf{Q}^k + \mathbf{C}_{\mathrm{T}\tau}^\mathrm{T}\mathbf{Q}_\mathrm{t}^k\right)\right]\mathbf{T}_{(m)} \tag{10.11}$$

Using Eq. (10.11), integration of some basic time delayed functions are illustrated below.

10.2.1 Numerical Examples

Example 10.3 Consider the function $f_1(t) = t$ having a delay at t = 0.4 s, from Example 10.1. Using Eq. (10.11), integration of the function in HF domain is plotted with the exact integration in Fig. 10.4, for m = 10 and T = 1 s.

Here we note that the samples obtained via HF domain integration approach coincide with the exact integration curve.

Example 10.4: Consider the function $f_2(t) = \sin(\pi t)$ having a delay of τ = 0.4 s. Using Eq. (10.11), integration of the function in HF domain is plotted along with the exact curve in Fig. 10.5, for m = 20 and T = 2 s.

As in Example 10.2, here, k = 4, the delay time $\tau = kh = 0.4$ s, and time period T = 2 s.

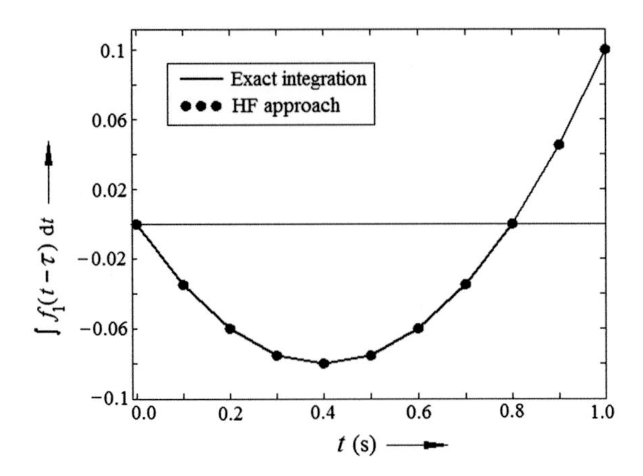

Fig. 10.4 Graphical representation of HF domain integration of the function $f_1(t) = (t - 0.4)$ of Example 10.1 along with its exact integration, for $m = 10$ and $T = 1$ s

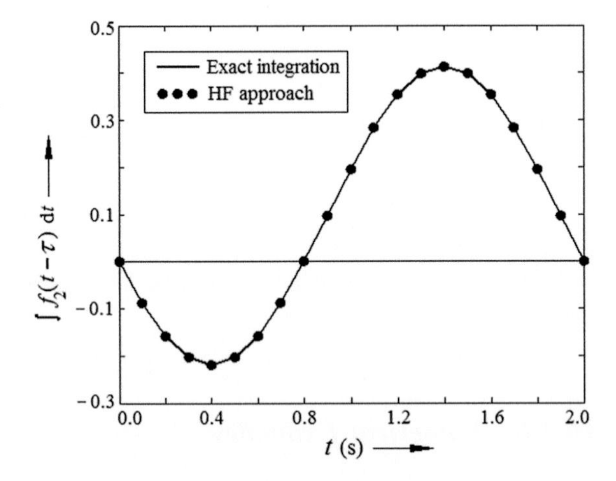

Fig. 10.5 Graphical representation of HF domain integration, of the function $f_2(t)$ of Example 10.2, along with its exact integration, for $m = 10$ and $T = 1$ s

10.3 Analysis of Non-homogeneous State Equations with Delay

Consider the non-homogeneous delayed state equation as,

$$\dot{\mathbf{x}}(t) = \mathbf{A}\mathbf{x}(t) + \mathbf{B}u(t) + \sum_{k=1}^{M} \mathbf{A}_k \mathbf{x}(t - kh) + \sum_{\lambda=1}^{N} \mathbf{B}_\lambda u(t - \lambda h) \tag{10.12}$$

where, $\mathbf{A}$ is an $(n \times n)$ system matrix given by

$$\mathbf{A} \triangleq \begin{bmatrix} a_{11} & a_{12} & a_{13} & \cdots & a_{1n} \\ a_{21} & a_{22} & a_{23} & \cdots & a_{2n} \\ \vdots & \vdots & \vdots & & \vdots \\ \vdots & \vdots & \vdots & & \vdots \\ a_{n1} & a_{n2} & a_{n3} & \cdots & a_{nn} \end{bmatrix}$$

$\mathbf{A}_k$ is an $(n \times n)$ system matrix associated with the delayed state vector,
$\mathbf{B}$ is the $(n \times 1)$ input vector given by $\mathbf{B} = [b_1 \quad b_2 \quad \cdots \quad \cdots \quad b_n]^T$
$\mathbf{B}_k$ is the $(n \times 1)$ input vector associated with the delayed input,
$\mathbf{x}(t)$ is the state vector given by $\mathbf{x}(t) = [x_1 \quad x_2 \quad \ldots \quad \ldots \quad x_n]^{\mathrm{T}}$ with the initial conditions

$\mathbf{x}(0) = [x_1(0) \quad x_2(0) \quad \ldots \quad \ldots \quad x_n(0)]^{\mathrm{T}}$ where, $[\ldots]^{\mathrm{T}}$denotes transpose,

$\mathbf{x}(t - kh)$ is the state vector with a delay of $t = kh$, k being an integer,
u is the forcing function,
$u(t - \lambda h)$ is the forcing function with a delay of $t = \lambda h$, λ being an integer,

and M and N are upper limits of k and λ respectively.

For simplicity, let us consider the following multi-delay system

$$\begin{aligned} \dot{\mathbf{x}}(t) &= \mathbf{A}\mathbf{x}(t) + \mathbf{B}u(t) + \mathbf{A}_1\mathbf{x}(t - k_1 h) + \mathbf{A}_2\mathbf{x}(t - k_2 h) \\ &\quad + \mathbf{B}_1 u(t - \lambda_1 h) + \mathbf{B}_2 u(t - \lambda_2 h) \end{aligned} \tag{10.13}$$

Integrating Eq. (10.13), we get

$$\begin{aligned} \mathbf{x}(t) - \mathbf{x}(0) &= \mathbf{A}\int \mathbf{x}(t)\mathrm{d}t + \mathbf{B}\int u(t)\mathrm{d}t + \mathbf{A}_1\int \mathbf{x}(t - k_1 h)\mathrm{d}t + \mathbf{A}_2\int \mathbf{x}(t - k_2 h)\mathrm{d}t \\ &\quad + \mathbf{B}_1\int u(t - \lambda_1 h)\mathrm{d}t + \mathbf{B}_2\int u(t - \lambda_2 h)\mathrm{d}t \end{aligned} \tag{10.14}$$

Expanding $\mathbf{x}(t)$, $\mathbf{x}(0)$, $\mathbf{x}(t-k_1h)$, $\mathbf{x}(t-k_2h)$, $u(t)$, $u(t-\lambda_1 h)$ and $u(t-\lambda_2 h)$ via an m-set hybrid function, we get

$$\begin{aligned}
\mathbf{x}(t) &\triangleq \mathbf{C}_{\mathrm{Sx}}\mathbf{S}_{(m)} + \mathbf{C}_{\mathrm{Tx}}\mathbf{T}_{(m)}, \mathbf{x}(0) \triangleq \mathbf{C}_{\mathrm{Sx0}}\mathbf{S}_{(m)} + \mathbf{C}_{\mathrm{Tx0}}\mathbf{T}_{(m)} \\
\mathbf{x}(t-k_1h) &\triangleq \left[\mathbf{C}_{\mathrm{Sx}}\mathbf{Q}^{k1} + \mathbf{C}_{\mathrm{Sx}\tau1}\mathbf{Q}_{\mathrm{t}}^{k1}\right]\mathbf{S}_{(m)} + \left[\mathbf{C}_{\mathrm{Tx}}\mathbf{Q}^{k1} + \mathbf{C}_{\mathrm{Tx}\tau1}\mathbf{Q}_{\mathrm{t}}^{k1}\right]\mathbf{T}_{(m)} \\
\mathbf{x}(t-k_2h) &\triangleq \left[\mathbf{C}_{\mathrm{Sx}}\mathbf{Q}^{k2} + \mathbf{C}_{\mathrm{Sx}\tau2}\mathbf{Q}_{\mathrm{t}}^{k2}\right]\mathbf{S}_{(m)} + \left[\mathbf{C}_{\mathrm{Tx}}\mathbf{Q}^{k2} + \mathbf{C}_{\mathrm{Tx}\tau2}\mathbf{Q}_{\mathrm{t}}^{k2}\right]\mathbf{T}_{(m)} \\
u(t) &\triangleq \mathbf{C}_{\mathrm{Su}}^{\mathrm{T}}\mathbf{S}_{(m)} + \mathbf{C}_{\mathrm{Tu}}^{\mathrm{T}}\mathbf{T}_{(m)} \\
u(t-\beta_1h) &\triangleq \left[\mathbf{C}_{\mathrm{Su}}^{\mathrm{T}}\mathbf{Q}^{\beta1} + \mathbf{C}_{\mathrm{Su}\tau1}^{\mathrm{T}}\mathbf{Q}_{\mathrm{t}}^{\beta1}\right]\mathbf{S}_{(m)} + \left[\mathbf{C}_{\mathrm{Tu}}^{\mathrm{T}}\mathbf{Q}^{\beta1} + \mathbf{C}_{\mathrm{Tu}\tau1}^{\mathrm{T}}\mathbf{Q}_{\mathrm{t}}^{\beta1}\right]\mathbf{T}_{(m)} \\
u(t-\beta_2h) &\triangleq \left[\mathbf{C}_{\mathrm{Su}}^{\mathrm{T}}\mathbf{Q}^{\beta2} + \mathbf{C}_{\mathrm{Su}\tau2}^{\mathrm{T}}\mathbf{Q}_{\mathrm{t}}^{\beta2}\right]\mathbf{S}_{(m)} + \left[\mathbf{C}_{\mathrm{Tu}}^{\mathrm{T}}\mathbf{Q}^{\beta2} + \mathbf{C}_{\mathrm{Tu}\tau2}^{\mathrm{T}}\mathbf{Q}_{\mathrm{t}}^{\beta2}\right]\mathbf{T}_{(m)}
\end{aligned}$$

where,

$$\mathbf{C}_{\mathrm{Sx}} = \begin{bmatrix} \mathbf{C}_{\mathrm{Sx1}}^{\mathrm{T}} \\ \mathbf{C}_{\mathrm{Sx2}}^{\mathrm{T}} \\ \vdots \\ \vdots \\ \mathbf{C}_{\mathrm{Sx}i}^{\mathrm{T}} \\ \vdots \\ \vdots \\ \mathbf{C}_{\mathrm{Sx}n}^{\mathrm{T}} \end{bmatrix}_{(n\times m)}, \mathbf{C}_{\mathrm{Tx}} = \begin{bmatrix} \mathbf{C}_{\mathrm{Tx1}}^{\mathrm{T}} \\ \mathbf{C}_{\mathrm{Tx2}}^{\mathrm{T}} \\ \vdots \\ \vdots \\ \mathbf{C}_{\mathrm{Tx}i}^{\mathrm{T}} \\ \vdots \\ \vdots \\ \mathbf{C}_{\mathrm{Tx}n}^{\mathrm{T}} \end{bmatrix}_{(n\times m)}, \mathbf{C}_{\mathrm{Sx0}} = \begin{bmatrix} \mathbf{C}_{\mathrm{Sx01}}^{\mathrm{T}} \\ \mathbf{C}_{\mathrm{Sx02}}^{\mathrm{T}} \\ \vdots \\ \vdots \\ \mathbf{C}_{\mathrm{Sx0}i}^{\mathrm{T}} \\ \vdots \\ \vdots \\ \mathbf{C}_{\mathrm{Sx0}n}^{\mathrm{T}} \end{bmatrix}_{(n\times m)}, \mathbf{C}_{\mathrm{Tx0}} = \begin{bmatrix} \mathbf{C}_{\mathrm{Tx01}}^{\mathrm{T}} \\ \mathbf{C}_{\mathrm{Tx02}}^{\mathrm{T}} \\ \vdots \\ \vdots \\ \mathbf{C}_{\mathrm{Tx0}i}^{\mathrm{T}} \\ \vdots \\ \vdots \\ \mathbf{C}_{\mathrm{Tx0}n}^{\mathrm{T}} \end{bmatrix}_{(n\times m)}$$

$$\begin{aligned}
\mathbf{C}_{\mathrm{Sx}i} &= \begin{bmatrix} c_{\mathrm{Sx}i1} & c_{\mathrm{Sx}i2} & c_{\mathrm{Sx}i3} & \cdots & \cdots & c_{\mathrm{Sx}im} \end{bmatrix}^{\mathrm{T}} \\
\mathbf{C}_{\mathrm{Tx}i} &= \begin{bmatrix} c_{\mathrm{Tx}i1} & c_{\mathrm{Tx}i2} & c_{\mathrm{Tx}i3} & \cdots & \cdots & c_{\mathrm{Tx}im} \end{bmatrix}^{\mathrm{T}} \\
\mathbf{C}_{\mathrm{Sx0}i} &= \begin{bmatrix} c_{\mathrm{Sx0}i1} & c_{\mathrm{Sx0}i2} & c_{\mathrm{Sx0}i3} & \cdots & \cdots & c_{\mathrm{Sx0}im} \end{bmatrix}^{\mathrm{T}} \\
\mathbf{C}_{\mathrm{Tx0}i} &= \begin{bmatrix} c_{\mathrm{Tx0}i1} & c_{\mathrm{Tx0}i2} & c_{\mathrm{Tx0}i3} & \cdots & \cdots & c_{\mathrm{Tx0}im} \end{bmatrix}^{\mathrm{T}} \\
\mathbf{C}_{\mathrm{su}} &= \begin{bmatrix} c_{\mathrm{su},1} & c_{\mathrm{su},2} & c_{\mathrm{su},3} & \cdots & \cdots & c_{\mathrm{su},m} \end{bmatrix}^{\mathrm{T}} \\
\mathbf{C}_{\mathrm{Tu}} &= \begin{bmatrix} c_{\mathrm{Tu},1} & c_{\mathrm{Tu},2} & c_{\mathrm{Tu},3} & \cdots & \cdots & c_{\mathrm{Tu},m} \end{bmatrix}^{\mathrm{T}}
\end{aligned}$$

Therefore LHS of Eq. (10.14), can be expressed as

$$\mathbf{x}(t) - \mathbf{x}(0) \approx (\mathbf{C}_{\mathrm{Sx}} - \mathbf{C}_{\mathrm{Sx0}})\mathbf{S}_{(m)} + (\mathbf{C}_{\mathrm{Tx}} - \mathbf{C}_{\mathrm{Tx0}})\mathbf{T}_{(m)} \tag{10.15}$$

The initial values of all the states being constants, they always essentially represent step functions. Hence, HF domain expansions of the initial values will always yield null coefficient matrices for the **T** vectors.

Therefore in HF domain, integration of the above terms can be represented by

$$\int \mathbf{x}(t)\mathrm{d}t \approx \left[\mathbf{C}_{\mathrm{Sx}} + \tfrac{1}{2}\mathbf{C}_{\mathrm{Tx}}\right]\mathbf{P1ss}\ \mathbf{S}_{(m)} + h\left[\mathbf{C}_{\mathrm{Sx}} + \tfrac{1}{2}\mathbf{C}_{\mathrm{Tx}}\right]\mathbf{T}_{(m)}$$

$$\begin{aligned}\int \mathbf{x}(t-k_1h)\mathrm{d}t \approx & \left[\left(\mathbf{C}_{\mathrm{Sx}}\mathbf{Q}^{k1} + \mathbf{C}_{\mathrm{Sx}\tau 1}\mathbf{Q}_{\mathrm{t}}^{k1}\right) + \tfrac{1}{2}\left(\mathbf{C}_{\mathrm{Tx}}\mathbf{Q}^{k1} + \mathbf{C}_{\mathrm{Tx}\tau 1}\mathbf{Q}_{\mathrm{t}}^{k1}\right)\right]\mathbf{P1ss}\,\mathbf{S}_{(m)} \\ & + h\left[\left(\mathbf{C}_{\mathrm{Sx}}\mathbf{Q}^{k1} + \mathbf{C}_{\mathrm{Sx}\tau 1}\mathbf{Q}_{\mathrm{t}}^{k1}\right) + \tfrac{1}{2}\left(\mathbf{C}_{\mathrm{Tx}}\mathbf{Q}^{k1} + \mathbf{C}_{\mathrm{Tx}\tau 1}\mathbf{Q}_{\mathrm{t}}^{k1}\right)\right]\mathbf{T}_{(m)}\end{aligned}$$

$$\begin{aligned}\int \mathbf{x}(t-k_2h)\mathrm{d}t \approx & \left[\left(\mathbf{C}_{\mathrm{Sx}}\mathbf{Q}^{k2} + \mathbf{C}_{\mathrm{Sx}\tau 2}\mathbf{Q}_{\mathrm{t}}^{k2}\right) + \tfrac{1}{2}\left(\mathbf{C}_{\mathrm{Tx}}\mathbf{Q}^{k2} + \mathbf{C}_{\mathrm{Tx}\tau 2}\mathbf{Q}_{\mathrm{t}}^{k2}\right)\right]\mathbf{P1ss}\,\mathbf{S}_{(m)} \\ & + h\left[\left(\mathbf{C}_{\mathrm{Sx}}\mathbf{Q}^{k2} + \mathbf{C}_{\mathrm{Sx}\tau 2}\mathbf{Q}_{\mathrm{t}}^{k2}\right) + \tfrac{1}{2}\left(\mathbf{C}_{\mathrm{Tx}}\mathbf{Q}^{k2} + \mathbf{C}_{\mathrm{Tx}\tau 2}\mathbf{Q}_{\mathrm{t}}^{k2}\right)\right]\mathbf{T}_{(m)}\end{aligned}$$

$$\int u(t)\mathrm{d}t \approx \left[\mathbf{C}_{\mathrm{Su}}^{\mathrm{T}} + \tfrac{1}{2}\mathbf{C}_{\mathrm{Tu}}^{\mathrm{T}}\right]\mathbf{P1ss}\ \mathbf{S}_{(m)} + h\left[\mathbf{C}_{\mathrm{Su}}^{\mathrm{T}} + \tfrac{1}{2}\mathbf{C}_{\mathrm{Tu}}^{\mathrm{T}}\right]\mathbf{T}_{(m)}$$

$$\begin{aligned}\int u(t-\beta_1h)\mathrm{d}t \approx & \left[\left(\mathbf{C}_{\mathrm{Su}}^{\mathrm{T}}\mathbf{Q}^{\beta 1} + \mathbf{C}_{\mathrm{Su}\tau 1}^{\mathrm{T}}\mathbf{Q}_{\mathrm{t}}^{\beta 1}\right) + \tfrac{1}{2}\left(\mathbf{C}_{\mathrm{Tu}}^{\mathrm{T}}\mathbf{Q}^{\beta 1} + \mathbf{C}_{\mathrm{Tu}\tau 1}^{\mathrm{T}}\mathbf{Q}_{\mathrm{t}}^{\beta 1}\right)\right]\mathbf{P1ss}\,\mathbf{S}_{(m)} \\ & + h\left[\left(\mathbf{C}_{\mathrm{Su}}^{\mathrm{T}}\mathbf{Q}^{\beta 1} + \mathbf{C}_{\mathrm{Su}\tau 1}^{\mathrm{T}}\mathbf{Q}_{\mathrm{t}}^{\beta 1}\right) + \tfrac{1}{2}\left(\mathbf{C}_{\mathrm{Tu}}^{\mathrm{T}}\mathbf{Q}^{\beta 1} + \mathbf{C}_{\mathrm{Tu}\tau 1}^{\mathrm{T}}\mathbf{Q}_{\mathrm{t}}^{\beta 1}\right)\right]\mathbf{T}_{(m)}\end{aligned}$$

$$\begin{aligned}\int u(t-\beta_2h)\mathrm{d}t \approx & \left[\left(\mathbf{C}_{\mathrm{Su}}^{\mathrm{T}}\mathbf{Q}^{\beta 2} + \mathbf{C}_{\mathrm{Su}\tau 2}^{\mathrm{T}}\mathbf{Q}_{\mathrm{t}}^{\beta 2}\right) + \tfrac{1}{2}\left(\mathbf{C}_{\mathrm{Tu}}^{\mathrm{T}}\mathbf{Q}^{\beta 2} + \mathbf{C}_{\mathrm{Tu}\tau 2}^{\mathrm{T}}\mathbf{Q}_{\mathrm{t}}^{\beta 2}\right)\right]\mathbf{P1ss}\,\mathbf{S}_{(m)} \\ & + h\left[\left(\mathbf{C}_{\mathrm{Su}}^{\mathrm{T}}\mathbf{Q}^{\beta 2} + \mathbf{C}_{\mathrm{Su}\tau 2}^{\mathrm{T}}\mathbf{Q}_{\mathrm{t}}^{\beta 2}\right) + \tfrac{1}{2}\left(\mathbf{C}_{\mathrm{Tu}}^{\mathrm{T}}\mathbf{Q}^{\beta 2} + \mathbf{C}_{\mathrm{Tu}\tau 2}^{\mathrm{T}}\mathbf{Q}_{\mathrm{t}}^{\beta 2}\right)\right]\mathbf{T}_{(m)}\end{aligned}$$

Substituting the above terms in RHS of Eq. (10.14), we get

$$\begin{aligned}& \mathbf{A}\int \mathbf{x}(t)\mathrm{d}t + \mathbf{B}\int u(t)\mathrm{d}t + \mathbf{A}_1\int \mathbf{x}(t-k_1h)\mathrm{d}t + \mathbf{A}_2\int \mathbf{x}(t-k_2h)\mathrm{d}t \\ & \quad + \mathbf{B}_1\int u(t-\beta_1h)\,\mathrm{d}t + \mathbf{B}_2\int u(t-\beta_2h)\,\mathrm{d}t \\ & \approx \mathbf{A}\left[\left(\mathbf{C}_{\mathrm{Sx}} + \tfrac{1}{2}\mathbf{C}_{\mathrm{Tx}}\right)\mathbf{P1ss}\ \mathbf{S}_{(m)} + h\left(\mathbf{C}_{\mathrm{Sx}} + \tfrac{1}{2}\mathbf{C}_{\mathrm{Tx}}\right)\mathbf{T}_{(m)}\right] \\ & \quad + \mathbf{B}\left[\left(\mathbf{C}_{\mathrm{Su}}^{\mathrm{T}} + \tfrac{1}{2}\mathbf{C}_{\mathrm{Tu}}^{\mathrm{T}}\right)\mathbf{P1ss}\ \mathbf{S}_{(m)} + h\left(\mathbf{C}_{\mathrm{Su}}^{\mathrm{T}} + \tfrac{1}{2}\mathbf{C}_{\mathrm{Tu}}^{\mathrm{T}}\right)\mathbf{T}_{(m)}\right] \\ & \quad + \mathbf{A}_1\left[\left(\mathbf{C}_{\mathrm{Sx}}\mathbf{Q}^{k1} + \mathbf{C}_{\mathrm{Sx}\tau 1}\mathbf{Q}_{\mathrm{t}}^{k1}\right) + \tfrac{1}{2}\left(\mathbf{C}_{\mathrm{Tx}}\mathbf{Q}^{k1} + \mathbf{C}_{\mathrm{Tx}\tau 1}\mathbf{Q}_{\mathrm{t}}^{k1}\right)\right]\mathbf{P1ss}\,\mathbf{S}_{(m)} \\ & \quad + \mathbf{A}_1h\left[\left(\mathbf{C}_{\mathrm{Sx}}\mathbf{Q}^{k1} + \mathbf{C}_{\mathrm{Sx}\tau 1}\mathbf{Q}_{\mathrm{t}}^{k1}\right) + \tfrac{1}{2}\left(\mathbf{C}_{\mathrm{Tx}}\mathbf{Q}^{k1} + \mathbf{C}_{\mathrm{Tx}\tau 1}\mathbf{Q}_{\mathrm{t}}^{k1}\right)\right]\mathbf{T}_{(m)} \\ & \quad + \mathbf{A}_2\left[\left(\mathbf{C}_{\mathrm{Sx}}\mathbf{Q}^{k2} + \mathbf{C}_{\mathrm{Sx}\tau 2}\mathbf{Q}_{\mathrm{t}}^{k2}\right) + \tfrac{1}{2}\left(\mathbf{C}_{\mathrm{Tx}}\mathbf{Q}^{k2} + \mathbf{C}_{\mathrm{Tx}\tau 2}\mathbf{Q}_{\mathrm{t}}^{k2}\right)\right]\mathbf{P1ss}\,\mathbf{S}_{(m)} \\ & \quad + \mathbf{A}_2h\left[\left(\mathbf{C}_{\mathrm{Sx}}\mathbf{Q}^{k2} + \mathbf{C}_{\mathrm{Sx}\tau 2}\mathbf{Q}_{\mathrm{t}}^{k2}\right) + \tfrac{1}{2}\left(\mathbf{C}_{\mathrm{Tx}}\mathbf{Q}^{k2} + \mathbf{C}_{\mathrm{Tx}\tau 2}\mathbf{Q}_{\mathrm{t}}^{k2}\right)\right]\mathbf{T}_{(m)} \\ & \quad + \mathbf{B}_1\left[\left(\mathbf{C}_{\mathrm{Su}}^{\mathrm{T}}\mathbf{Q}^{\beta 1} + \mathbf{C}_{\mathrm{Su}\tau 1}^{\mathrm{T}}\mathbf{Q}_{\mathrm{t}}^{\beta 1}\right) + \tfrac{1}{2}\left(\mathbf{C}_{\mathrm{Tu}}^{\mathrm{T}}\mathbf{Q}^{\beta 1} + \mathbf{C}_{\mathrm{Tu}\tau 1}^{\mathrm{T}}\mathbf{Q}_{\mathrm{t}}^{\beta 1}\right)\right]\mathbf{P1ss}\,\mathbf{S}_{(m)} \\ & \quad + \mathbf{B}_1h\left[\left(\mathbf{C}_{\mathrm{Su}}^{\mathrm{T}}\mathbf{Q}^{\beta 1} + \mathbf{C}_{\mathrm{Su}\tau 1}^{\mathrm{T}}\mathbf{Q}_{\mathrm{t}}^{\beta 1}\right) + \tfrac{1}{2}\left(\mathbf{C}_{\mathrm{Tu}}^{\mathrm{T}}\mathbf{Q}^{\beta 1} + \mathbf{C}_{\mathrm{Tu}\tau 1}^{\mathrm{T}}\mathbf{Q}_{\mathrm{t}}^{\beta 1}\right)\right]\mathbf{T}_{(m)} \\ & \quad + \mathbf{B}_2\left[\left(\mathbf{C}_{\mathrm{Su}}^{\mathrm{T}}\mathbf{Q}^{\beta 2} + \mathbf{C}_{\mathrm{Su}\tau 2}^{\mathrm{T}}\mathbf{Q}_{\mathrm{t}}^{\beta 2}\right) + \tfrac{1}{2}\left(\mathbf{C}_{\mathrm{Tu}}^{\mathrm{T}}\mathbf{Q}^{\beta 2} + \mathbf{C}_{\mathrm{Tu}\tau 2}^{\mathrm{T}}\mathbf{Q}_{\mathrm{t}}^{\beta 2}\right)\right]\mathbf{P1ss}\,\mathbf{S}_{(m)} \\ & \quad + \mathbf{B}_2h\left[\left(\mathbf{C}_{\mathrm{Su}}^{\mathrm{T}}\mathbf{Q}^{\beta 2} + \mathbf{C}_{\mathrm{Su}\tau 2}^{\mathrm{T}}\mathbf{Q}_{\mathrm{t}}^{\beta 2}\right) + \tfrac{1}{2}\left(\mathbf{C}_{\mathrm{Tu}}^{\mathrm{T}}\mathbf{Q}^{\beta 2} + \mathbf{C}_{\mathrm{Tu}\tau 2}^{\mathrm{T}}\mathbf{Q}_{\mathrm{t}}^{\beta 2}\right)\right]\mathbf{T}_{(m)}\end{aligned} \tag{10.16}$$

Now equating the SHF part of Eqs. (10.15) and (10.16), we get

$$\begin{aligned}\mathbf{C}_{\mathrm{Sx}}-\mathbf{C}_{\mathrm{Sx0}}=&\,\mathbf{A}\left(\mathbf{C}_{\mathrm{Sx}}+\tfrac{1}{2}\mathbf{C}_{\mathrm{Tx}}\right)\mathbf{P1ss}+\mathbf{B}\left(\mathbf{C}_{\mathrm{Su}}^{\mathrm{T}}+\tfrac{1}{2}\mathbf{C}_{\mathrm{Tu}}^{\mathrm{T}}\right)\mathbf{P1ss}\\&+\mathbf{A}_1\left[\left(\mathbf{C}_{\mathrm{Sx}}\mathbf{Q}^{k1}+\mathbf{C}_{\mathrm{Sx\tau1}}\mathbf{Q}_{\mathrm{t}}^{k1}\right)+\tfrac{1}{2}\left(\mathbf{C}_{\mathrm{Tx}}\mathbf{Q}^{k1}+\mathbf{C}_{\mathrm{Tx\tau1}}\mathbf{Q}_{\mathrm{t}}^{k1}\right)\right]\mathbf{P1ss}\\&+\mathbf{A}_2\left[\left(\mathbf{C}_{\mathrm{Sx}}\mathbf{Q}^{k2}+\mathbf{C}_{\mathrm{Sx\tau2}}\mathbf{Q}_{\mathrm{t}}^{k2}\right)+\tfrac{1}{2}\left(\mathbf{C}_{\mathrm{Tx}}\mathbf{Q}^{k2}+\mathbf{C}_{\mathrm{Tx\tau2}}\mathbf{Q}_{\mathrm{t}}^{k2}\right)\right]\mathbf{P1ss}\\&+\mathbf{B}_1\left[\left(\mathbf{C}_{\mathrm{Su}}^{\mathrm{T}}\mathbf{Q}^{\beta1}+\mathbf{C}_{\mathrm{Su\tau1}}^{\mathrm{T}}\mathbf{Q}_{\mathrm{t}}^{\beta1}\right)+\tfrac{1}{2}\left(\mathbf{C}_{\mathrm{Tu}}^{\mathrm{T}}\mathbf{Q}^{\beta1}+\mathbf{C}_{\mathrm{Tu\tau1}}^{\mathrm{T}}\mathbf{Q}_{\mathrm{t}}^{\beta1}\right)\right]\mathbf{P1ss}\\&+\mathbf{B}_2\left[\left(\mathbf{C}_{\mathrm{Su}}^{\mathrm{T}}\mathbf{Q}^{\beta2}+\mathbf{C}_{\mathrm{Su\tau2}}^{\mathrm{T}}\mathbf{Q}_{\mathrm{t}}^{\beta2}\right)+\tfrac{1}{2}\left(\mathbf{C}_{\mathrm{Tu}}^{\mathrm{T}}\mathbf{Q}^{\beta2}+\mathbf{C}_{\mathrm{Tu\tau2}}^{\mathrm{T}}\mathbf{Q}_{\mathrm{t}}^{\beta2}\right)\right]\mathbf{P1ss}\end{aligned}\tag{10.17}$$

Similarly, equating the TF part of Eqs. (10.15) and (10.16), we have

$$\begin{aligned}\mathbf{C}_{\mathrm{Tx}}=&\,\mathbf{A}h\left(\mathbf{C}_{\mathrm{Sx}}+\tfrac{1}{2}\mathbf{C}_{\mathrm{Tx}}\right)+\mathbf{B}h\left(\mathbf{C}_{\mathrm{Su}}^{\mathrm{T}}+\tfrac{1}{2}\mathbf{C}_{\mathrm{Tu}}^{\mathrm{T}}\right)\\&+\mathbf{A}_1h\left[\left(\mathbf{C}_{\mathrm{Sx}}\mathbf{Q}^{k1}+\mathbf{C}_{\mathrm{Sx\tau1}}\mathbf{Q}_{\mathrm{t}}^{k1}\right)+\tfrac{1}{2}\left(\mathbf{C}_{\mathrm{Tx}}\mathbf{Q}^{k1}+\mathbf{C}_{\mathrm{Tx\tau1}}\mathbf{Q}_{\mathrm{t}}^{k1}\right)\right]\\&+\mathbf{A}_2h\left[\left(\mathbf{C}_{\mathrm{Sx}}\mathbf{Q}^{k2}+\mathbf{C}_{\mathrm{Sx\tau2}}\mathbf{Q}_{\mathrm{t}}^{k2}\right)+\tfrac{1}{2}\left(\mathbf{C}_{\mathrm{Tx}}\mathbf{Q}^{k2}+\mathbf{C}_{\mathrm{Tx\tau2}}\mathbf{Q}_{\mathrm{t}}^{k2}\right)\right]\\&+\mathbf{B}_1h\left[\left(\mathbf{C}_{\mathrm{Su}}^{\mathrm{T}}\mathbf{Q}^{\beta1}+\mathbf{C}_{\mathrm{Su\tau1}}^{\mathrm{T}}\mathbf{Q}_{\mathrm{t}}^{\beta1}\right)+\tfrac{1}{2}\left(\mathbf{C}_{\mathrm{Tu}}^{\mathrm{T}}\mathbf{Q}^{\beta1}+\mathbf{C}_{\mathrm{Tu\tau1}}^{\mathrm{T}}\mathbf{Q}_{\mathrm{t}}^{\beta1}\right)\right]\\&+\mathbf{B}_2h\left[\left(\mathbf{C}_{\mathrm{Su}}^{\mathrm{T}}\mathbf{Q}^{\beta2}+\mathbf{C}_{\mathrm{Su\tau2}}^{\mathrm{T}}\mathbf{Q}_{\mathrm{t}}^{\beta2}\right)+\tfrac{1}{2}\left(\mathbf{C}_{\mathrm{Tu}}^{\mathrm{T}}\mathbf{Q}^{\beta2}+\mathbf{C}_{\mathrm{Tu\tau2}}^{\mathrm{T}}\mathbf{Q}_{\mathrm{t}}^{\beta2}\right)\right]\end{aligned}\tag{10.18}$$

Therefore, from Eqs. (10.17) and (10.18), we can write

$$\mathbf{C}_{\mathrm{Tx}}\,\mathbf{P1ss}=h\left(\mathbf{C}_{\mathrm{Sx}}-\mathbf{C}_{\mathrm{Sx0}}\right)$$

Substituting this relation in Eq. (10.17), we have

$$\begin{aligned}\mathbf{C}_{\mathrm{Sx}}-\mathbf{C}_{\mathrm{Sx0}}=&\,\mathbf{A}\mathbf{C}_{\mathrm{Sx}}\mathbf{P1ss}+\mathbf{B}\left(\mathbf{C}_{\mathrm{Su}}^{\mathrm{T}}+\tfrac{1}{2}\mathbf{C}_{\mathrm{Tu}}^{\mathrm{T}}\right)\mathbf{P1ss}+\tfrac{h}{2}\mathbf{A}(\mathbf{C}_{\mathrm{Sx}}-\mathbf{C}_{\mathrm{Sx0}})\\&+\mathbf{A}_1\mathbf{C}_{\mathrm{Sx}}\mathbf{Q}^{k1}\mathbf{P1ss}+\tfrac{h}{2}\mathbf{A}_1(\mathbf{C}_{\mathrm{Sx}}-\mathbf{C}_{\mathrm{Sx0}})\mathbf{Q}^{k1}+\mathbf{A}_1\left(\mathbf{C}_{\mathrm{Sx\tau1}}+\tfrac{1}{2}\mathbf{C}_{\mathrm{Tx\tau1}}\right)\mathbf{Q}_{\mathrm{t}}^{k1}\mathbf{P1ss}\\&+\mathbf{A}_2\mathbf{C}_{\mathrm{Sx}}\mathbf{Q}^{k2}\mathbf{P1ss}+\tfrac{h}{2}\mathbf{A}_2(\mathbf{C}_{\mathrm{Sx}}-\mathbf{C}_{\mathrm{Sx0}})\mathbf{Q}^{k2}+\mathbf{A}_2\left(\mathbf{C}_{\mathrm{Sx\tau2}}+\tfrac{1}{2}\mathbf{C}_{\mathrm{Tx\tau2}}\right)\mathbf{Q}_{\mathrm{t}}^{k2}\mathbf{P1ss}\\&+\mathbf{B}_1\left[\left(\mathbf{C}_{\mathrm{Su}}^{\mathrm{T}}\mathbf{Q}^{\beta1}+\mathbf{C}_{\mathrm{Su\tau1}}^{\mathrm{T}}\mathbf{Q}_{\mathrm{t}}^{\beta1}\right)+\tfrac{1}{2}\left(\mathbf{C}_{\mathrm{Tu}}^{\mathrm{T}}\mathbf{Q}^{\beta1}+\mathbf{C}_{\mathrm{Tu\tau1}}^{\mathrm{T}}\mathbf{Q}_{\mathrm{t}}^{\beta1}\right)\right]\mathbf{P1ss}\\&+\mathbf{B}_2\left[\left(\mathbf{C}_{\mathrm{Su}}^{\mathrm{T}}\mathbf{Q}^{\beta2}+\mathbf{C}_{\mathrm{Su\tau2}}^{\mathrm{T}}\mathbf{Q}_{\mathrm{t}}^{\beta2}\right)+\tfrac{1}{2}\left(\mathbf{C}_{\mathrm{Tu}}^{\mathrm{T}}\mathbf{Q}^{\beta2}+\mathbf{C}_{\mathrm{Tu\tau2}}^{\mathrm{T}}\mathbf{Q}_{\mathrm{t}}^{\beta2}\right)\right]\mathbf{P1ss}\end{aligned}\tag{10.19}$$

$$\begin{aligned}
&\text{or, } \mathbf{C}_{\mathrm{Sx}} - \mathbf{A}\mathbf{C}_{\mathrm{Sx}}\mathbf{P1ss} - \frac{h}{2}\mathbf{A}\mathbf{C}_{\mathrm{Sx}} - \mathbf{A}_1\mathbf{C}_{\mathrm{Sx}}\mathbf{Q}^{k1}\mathbf{P1ss} - \frac{h}{2}\mathbf{A}_1\mathbf{C}_{\mathrm{Sx}}\mathbf{Q}^{k1} \\
&\quad - \mathbf{A}_2\mathbf{C}_{\mathrm{Sx}}\mathbf{Q}^{k2}\mathbf{P1ss} - \frac{h}{2}\mathbf{A}_2\mathbf{C}_{\mathrm{Sx}}\mathbf{Q}^{k2} \\
&= \mathbf{C}_{\mathrm{Sx0}} - \frac{h}{2}\mathbf{A}\mathbf{C}_{\mathrm{Sx0}} - \frac{h}{2}\mathbf{A}_1\mathbf{C}_{\mathrm{Sx0}}\mathbf{Q}^{k1} + \mathbf{A}_1\left(\mathbf{C}_{\mathrm{Sx}\tau1} + \frac{1}{2}\mathbf{C}_{\mathrm{Tx}\tau1}\right)\mathbf{Q}_{\mathrm{t}}^{k1}\mathbf{P1ss} \\
&\quad - \frac{h}{2}\mathbf{A}_2\mathbf{C}_{\mathrm{Sx0}}\mathbf{Q}^{k2} + \mathbf{A}_2\left(\mathbf{C}_{\mathrm{Sx}\tau2} + \frac{1}{2}\mathbf{C}_{\mathrm{Tx}\tau2}\right)\mathbf{Q}_{\mathrm{t}}^{k2}\mathbf{P1ss} + \mathbf{B}\left(\mathbf{C}_{\mathrm{Su}}^{\mathrm{T}} + \frac{1}{2}\mathbf{C}_{\mathrm{Tu}}^{\mathrm{T}}\right)\mathbf{P1ss} \\
&\quad + \mathbf{B}_1\left[\left(\mathbf{C}_{\mathrm{Su}}^{\mathrm{T}}\mathbf{Q}^{\beta1} + \mathbf{C}_{\mathrm{Su}\tau1}^{\mathrm{T}}\mathbf{Q}_{\mathrm{t}}^{\beta1}\right) + \frac{1}{2}\left(\mathbf{C}_{\mathrm{Tu}}^{\mathrm{T}}\mathbf{Q}^{\beta1} + \mathbf{C}_{\mathrm{Tu}\tau1}^{\mathrm{T}}\mathbf{Q}_{\mathrm{t}}^{\beta1}\right)\right]\mathbf{P1ss} \\
&\quad + \mathbf{B}_2\left[\left(\mathbf{C}_{\mathrm{Su}}^{\mathrm{T}}\mathbf{Q}^{\beta2} + \mathbf{C}_{\mathrm{Su}\tau2}^{\mathrm{T}}\mathbf{Q}_{\mathrm{t}}^{\beta2}\right) + \frac{1}{2}\left(\mathbf{C}_{\mathrm{Tu}}^{\mathrm{T}}\mathbf{Q}^{\beta2} + \mathbf{C}_{\mathrm{Tu}\tau2}^{\mathrm{T}}\mathbf{Q}_{\mathrm{t}}^{\beta2}\right)\right]\mathbf{P1ss}
\end{aligned} \tag{10.20}$$

Replacing **P1ss** in Eq. (10.20), following (8.12), we get

$$\begin{aligned}
&\mathbf{C}_{\mathrm{Sx}} - \mathbf{A}\mathbf{C}_{\mathrm{Sx}}\mathbf{P} - \mathbf{A}_1\mathbf{C}_{\mathrm{Sx}}\mathbf{Q}^{k1}\mathbf{P} - \mathbf{A}_2\mathbf{C}_{\mathrm{Sx}}\mathbf{Q}^{k2}\mathbf{P} \\
&= \left[\mathbf{I} - \frac{h}{2}\mathbf{A} - \frac{h}{2}\mathbf{A}_1\mathbf{Q}^{k1} - \frac{h}{2}\mathbf{A}_2\mathbf{Q}^{k2}\right]\mathbf{C}_{\mathrm{Sx0}} + \mathbf{A}_1\left(\mathbf{C}_{\mathrm{Sx}\tau1} + \frac{1}{2}\mathbf{C}_{\mathrm{Tx}\tau1}\right)\mathbf{Q}_{\mathrm{t}}^{k1}\mathbf{P1ss} \\
&\quad + \mathbf{A}_2\left(\mathbf{C}_{\mathrm{Sx}\tau2} + \frac{1}{2}\mathbf{C}_{\mathrm{Tx}\tau2}\right)\mathbf{Q}_{\mathrm{t}}^{k2}\mathbf{P1ss} + \mathbf{B}\left(\mathbf{C}_{\mathrm{Su}}^{\mathrm{T}} + \frac{1}{2}\mathbf{C}_{\mathrm{Tu}}^{\mathrm{T}}\right)\mathbf{P1ss} \\
&\quad + \mathbf{B}_1\left[\left(\mathbf{C}_{\mathrm{Su}}^{\mathrm{T}}\mathbf{Q}^{\beta1} + \mathbf{C}_{\mathrm{Su}\tau1}^{\mathrm{T}}\mathbf{Q}_{\mathrm{t}}^{\beta1}\right) + \frac{1}{2}\left(\mathbf{C}_{\mathrm{Tu}}^{\mathrm{T}}\mathbf{Q}^{\beta1} + \mathbf{C}_{\mathrm{Tu}\tau1}^{\mathrm{T}}\mathbf{Q}_{\mathrm{t}}^{\beta1}\right)\right]\mathbf{P1ss} \\
&\quad + \mathbf{B}_2\left[\left(\mathbf{C}_{\mathrm{Su}}^{\mathrm{T}}\mathbf{Q}^{\beta2} + \mathbf{C}_{\mathrm{Su}\tau2}^{\mathrm{T}}\mathbf{Q}_{\mathrm{t}}^{\beta2}\right) + \frac{1}{2}\left(\mathbf{C}_{\mathrm{Tu}}^{\mathrm{T}}\mathbf{Q}^{\beta2} + \mathbf{C}_{\mathrm{Tu}\tau2}^{\mathrm{T}}\mathbf{Q}_{\mathrm{t}}^{\beta2}\right)\right]\mathbf{P1ss}
\end{aligned} \tag{10.21}$$

Now, the $(i + 1)$th column of $\mathbf{AC}_{\mathrm{Sx}}\mathbf{P}$ is

$$[\mathbf{AC}_{\mathrm{Sx}}\mathbf{P}]_{i+1} = \begin{bmatrix} a_{11} & a_{12} & \cdots & a_{1n} \\ a_{21} & a_{22} & \cdots & a_{2n} \\ \vdots & \vdots & & \vdots \\ a_{n1} & a_{n2} & \cdots & a_{nn} \end{bmatrix} \begin{bmatrix} c_{\mathrm{Sx11}} & c_{\mathrm{Sx12}} & \cdots & c_{\mathrm{Sx1}m} \\ c_{\mathrm{Sx21}} & c_{\mathrm{Sx22}} & \cdots & c_{\mathrm{Sx2}m} \\ \vdots & \vdots & & \vdots \\ c_{\mathrm{Sx}n1} & c_{\mathrm{Sx}n2} & \cdots & c_{\mathrm{Sx}nm} \end{bmatrix} h \overset{(i+1)-\text{th column}}{\overset{\downarrow}{\begin{bmatrix} 1 \\ \vdots \\ 1 \\ \frac{1}{2} \\ 0 \\ \vdots \\ 0 \end{bmatrix}}} \begin{matrix} \\ \\ \\ \leftarrow (i+1)\text{th element} \\ \leftarrow (i+2)\text{th element} \\ \\ \\ \end{matrix}$$

$$= h \begin{bmatrix} a_{11} & a_{12} & \cdots & a_{1n} \\ a_{21} & a_{22} & \cdots & a_{2n} \\ \vdots & \vdots & & \vdots \\ a_{n1} & a_{n2} & \cdots & a_{nn} \end{bmatrix} \begin{bmatrix} c_{\mathrm{Sx11}} + c_{\mathrm{Sx12}} + \cdots + c_{\mathrm{Sx1}i} + \frac{1}{2}c_{\mathrm{Sx1}(i+1)} \\ c_{\mathrm{Sx21}} + c_{\mathrm{Sx22}} + \cdots + c_{\mathrm{Sx2}i} + \frac{1}{2}c_{\mathrm{Sx2}(i+1)} \\ \vdots \\ c_{\mathrm{Sx}n1} + c_{\mathrm{Sx}n2} + \cdots + c_{\mathrm{Sx}ni} + \frac{1}{2}c_{\mathrm{Sx}n\,(i+1)} \end{bmatrix}$$

$$= h\begin{bmatrix} a_{11}\left(c_{Sx11}+c_{Sx12}+\cdots+c_{Sx1i}+\frac{1}{2}c_{Sx1(i+1)}\right)+a_{12}\left(c_{Sx21}+c_{Sx22}+\cdots+c_{Sx2i}+\frac{1}{2}c_{Sx2(i+1)}\right) \\ +\cdots+a_{1n}\left(c_{Sxn1}+c_{Sxn2}+\cdots+c_{Sxni}+\frac{1}{2}c_{Sxn(i+1)}\right) \\ a_{21}\left(c_{Sx11}+c_{Sx12}+\cdots+c_{Sx1i}+\frac{1}{2}c_{Sx1(i+1)}\right)+a_{22}\left(c_{Sx21}+c_{Sx22}+\cdots+c_{Sx2i}+\frac{1}{2}c_{Sx2(i+1)}\right) \\ +\cdots+a_{2n}\left(c_{Sxn1}+c_{Sxn2}+\cdots+c_{Sxni}+\frac{1}{2}c_{Sxn(i+1)}\right) \\ \vdots \\ a_{n1}\left(c_{Sx11}+c_{Sx12}+\cdots+c_{Sx1i}+\frac{1}{2}c_{Sx1(i+1)}\right)+a_{n2}\left(c_{Sx21}+c_{Sx22}+\cdots+c_{Sx2i}+\frac{1}{2}c_{Sx2(i+1)}\right) \\ +\cdots+a_{nn}\left(c_{Sxn1}+c_{Sxn2}+\cdots+c_{Sxni}+\frac{1}{2}c_{Sxn(i+1)}\right) \end{bmatrix} \tag{10.22}$$

Similarly, the $(i + 1)$th column of $\mathbf{A}_1\mathbf{C}_{Sx}\mathbf{Q}^{k1}\mathbf{P}$ is

$$\left[\mathbf{A}_1\mathbf{C}_{Sx}\mathbf{Q}^{k1}\mathbf{P}\right]_{i+1}$$

$$= h\begin{bmatrix} a_{(1)11} & a_{(1)12} & \cdots & a_{(1)1n} \\ a_{(1)21} & a_{(1)22} & \cdots & a_{(1)2n} \\ \vdots & \vdots & & \vdots \\ a_{(1)n1} & a_{(1)n2} & \cdots & a_{(1)nn} \end{bmatrix}\begin{bmatrix} c_{Sx11} & c_{Sx12} & \cdots & c_{Sx1m} \\ c_{Sx21} & c_{Sx22} & \cdots & c_{Sx2m} \\ \vdots & \vdots & & \vdots \\ c_{Sxn1} & c_{Sxn2} & \cdots & c_{Sxnm} \end{bmatrix}\overset{\substack{(i+1)-\text{thcolumn} \\ \downarrow}}{\begin{bmatrix} 1 \\ \vdots \\ 1 \\ \frac{1}{2} \\ 0 \\ \vdots \\ 0 \end{bmatrix}} \begin{matrix} \\ \\ \\ \leftarrow \begin{matrix}(i-k1+1) \\ \text{th element}\end{matrix} \\ \\ \\ \\ \end{matrix}$$

$$= h\begin{bmatrix} a_{(1)11} & a_{(1)12} & \cdots & a_{(1)1n} \\ a_{(1)21} & a_{(1)22} & \cdots & a_{(1)2n} \\ \vdots & \vdots & & \vdots \\ a_{(1)n1} & a_{(1)n2} & \cdots & a_{(1)nn} \end{bmatrix}\begin{bmatrix} c_{Sx11}+c_{Sx12}+\cdots+c_{Sx1(i-k1)}+\frac{1}{2}c_{Sx1(i-k1+1)} \\ c_{Sx21}+c_{Sx22}+\cdots+c_{Sx2(i-k1)}+\frac{1}{2}c_{Sx2(i-k1+1)} \\ \vdots \\ c_{Sxn1}+c_{Sxn2}+\cdots+c_{Sxn(i-k1)}+\frac{1}{2}c_{Sxn(i-k1+1)} \end{bmatrix}$$

$$= h\begin{bmatrix} a_{(1)11}\left(c_{Sx11}+c_{Sx12}+\cdots+c_{Sx1(i-k1)}+\frac{1}{2}c_{Sx1(i-k1+1)}\right) \\ +a_{(1)12}\left(c_{Sx21}+c_{Sx22}+\cdots+c_{Sx2(i-k1)}+\frac{1}{2}c_{Sx2(i-k1+1)}\right)+\cdots+ \\ +a_{(1)1n}\left(c_{Sxn1}+c_{Sxn2}+\cdots+c_{Sxn(i-k1)}+\frac{1}{2}c_{Sxn(i-k1+1)}\right) \\ a_{(1)21}\left(c_{Sx11}+c_{Sx12}+\cdots+c_{Sx1(i-k1)}+\frac{1}{2}c_{Sx1(i-k1+1)}\right) \\ +a_{(1)22}\left(c_{Sx21}+c_{Sx22}+\cdots+c_{Sx2(i-k1)}+\frac{1}{2}c_{Sx2(i-k1+1)}\right)+\cdots+ \\ +a_{(1)2n}\left(c_{Sxn1}+c_{Sxn2}+\cdots+c_{Sxn(i-k1)}+\frac{1}{2}c_{Sxn(i-k1+1)}\right) \\ \vdots \\ a_{(1)n1}\left(c_{Sx11}+c_{Sx12}+\cdots+c_{Sx1(i-k1)}+\frac{1}{2}c_{Sx1(i-k1+1)}\right) \\ +a_{(1)n2}\left(c_{Sx21}+c_{Sx22}+\cdots+c_{Sx2(i-k1)}+\frac{1}{2}c_{Sx2(i-k1+1)}\right)+\cdots+ \\ +a_{(1)nn}\left(c_{Sxn1}+c_{Sxn2}+\cdots+c_{Sxn(i-k1)}+\frac{1}{2}c_{Sxn(i-k1+1)}\right) \end{bmatrix} \quad \text{for } (i+1)>k1 \tag{10.23}$$

Similarly, the $(i + 1)$th column of $\mathbf{A}_2\mathbf{C}_{\mathrm{Sx}}\mathbf{Q}^{k2}\mathbf{P}$ is given by

$$\left[\mathbf{A}_2\mathbf{C}_{\mathrm{Sx}}\mathbf{Q}^{k2}\mathbf{P}\right]_{i+1}$$

$$= h\begin{bmatrix} a_{(2)11} & a_{(2)12} & \cdots & a_{(2)1n} \\ a_{(2)21} & a_{(2)22} & \cdots & a_{(2)2n} \\ \vdots & \vdots & & \vdots \\ a_{(2)n1} & a_{(2)n2} & \cdots & a_{(2)nn} \end{bmatrix}\begin{bmatrix} c_{\mathrm{Sx}11} + c_{\mathrm{Sx}12} + \cdots + c_{\mathrm{Sx}1(i-k2)} + \frac{1}{2}c_{\mathrm{Sx}1(i-k2+1)} \\ c_{\mathrm{Sx}21} + c_{\mathrm{Sx}22} + \cdots + c_{\mathrm{Sx}2(i-k2)} + \frac{1}{2}c_{\mathrm{Sx}2(i-k2+1)} \\ \vdots \\ c_{\mathrm{Sx}n1} + c_{\mathrm{Sx}n2} + \cdots + c_{\mathrm{Sx}n(i-k2)} + \frac{1}{2}c_{\mathrm{Sx}n(i-k2+1)} \end{bmatrix}$$

$$= h\begin{bmatrix} a_{(2)11}\left(c_{\mathrm{Sx}11} + c_{\mathrm{Sx}12} + \cdots + c_{\mathrm{Sx}1(i-k2)} + \frac{1}{2}c_{\mathrm{Sx}1(i-k2+1)}\right) \\ \quad + a_{(2)12}\left(c_{\mathrm{Sx}21} + c_{\mathrm{Sx}22} + \cdots + c_{\mathrm{Sx}2(i-k2)} + \frac{1}{2}c_{\mathrm{Sx}2(i-k2+1)}\right) + \cdots + \\ \quad + a_{(2)1n}\left(c_{\mathrm{Sx}n1} + c_{\mathrm{Sx}n2} + \cdots + c_{\mathrm{Sx}n(i-k2)} + \frac{1}{2}c_{\mathrm{Sx}n(i-k2+1)}\right) \\ a_{(2)21}\left(c_{\mathrm{Sx}11} + c_{\mathrm{Sx}12} + \cdots + c_{\mathrm{Sx}1(i-k2)} + \frac{1}{2}c_{\mathrm{Sx}1(i-k2+1)}\right) \\ \quad + a_{(2)22}\left(c_{\mathrm{Sx}21} + c_{\mathrm{Sx}22} + \cdots + c_{\mathrm{Sx}2(i-k2)} + \frac{1}{2}c_{\mathrm{Sx}2(i-k2+1)}\right) + \cdots + \\ \quad + a_{(2)2n}\left(c_{\mathrm{Sx}n1} + c_{\mathrm{Sx}n2} + \cdots + c_{\mathrm{Sx}n(i-k2)} + \frac{1}{2}c_{\mathrm{Sx}n(i-k2+1)}\right) \\ \vdots \\ a_{(2)n1}\left(c_{\mathrm{Sx}11} + c_{\mathrm{Sx}12} + \cdots + c_{\mathrm{Sx}1(i-k2)} + \frac{1}{2}c_{\mathrm{Sx}1(i-k2+1)}\right) \\ \quad + a_{(2)n2}\left(c_{\mathrm{Sx}21} + c_{\mathrm{Sx}22} + \cdots + c_{\mathrm{Sx}2(i-k2)} + \frac{1}{2}c_{\mathrm{Sx}2(i-k2+1)}\right) + \cdots + \\ \quad + a_{(2)nn}\left(c_{\mathrm{Sx}n1} + c_{\mathrm{Sx}n2} + \cdots + c_{\mathrm{Sx}n(i-k2)} + \frac{1}{2}c_{\mathrm{Sx}n(i-k2+1)}\right) \end{bmatrix} \quad \text{for } (i+1) > k2 \tag{10.24}$$

Now, the ith column of $\mathbf{A}\mathbf{C}_{\mathrm{Sx}}\mathbf{P}$ is

$$= h\begin{bmatrix} a_{11}\left(c_{\mathrm{Sx}11} + c_{\mathrm{Sx}12} + \cdots + c_{\mathrm{Sx}1(i-1)} + \frac{1}{2}c_{\mathrm{Sx}1i}\right) + a_{12}\left(c_{\mathrm{Sx}21} + c_{\mathrm{Sx}22} + \cdots + c_{\mathrm{Sx}2(i-1)} + \frac{1}{2}c_{\mathrm{Sx}2i}\right) \\ + \cdots + a_{1n}\left(c_{\mathrm{Sx}n1} + c_{\mathrm{Sx}n2} + \cdots + c_{\mathrm{Sx}n(i-1)} + \frac{1}{2}c_{\mathrm{Sx}ni}\right) \\ a_{21}\left(c_{\mathrm{Sx}11} + c_{\mathrm{Sx}12} + \cdots + c_{\mathrm{Sx}1(i-1)} + \frac{1}{2}c_{\mathrm{Sx}1i}\right) + a_{22}\left(c_{\mathrm{Sx}21} + c_{\mathrm{Sx}22} + \cdots + c_{\mathrm{Sx}2(i-1)} + \frac{1}{2}c_{\mathrm{Sx}2i}\right) \\ + \cdots + a_{2n}\left(c_{\mathrm{Sx}n1} + c_{\mathrm{Sx}n2} + \cdots + c_{\mathrm{Sx}n(i-1)} + \frac{1}{2}c_{\mathrm{Sx}ni}\right) \\ \vdots \\ a_{n1}\left(c_{\mathrm{Sx}11} + c_{\mathrm{Sx}12} + \cdots + c_{\mathrm{Sx}1(i-1)} + \frac{1}{2}c_{\mathrm{Sx}1i}\right) + a_{n2}\left(c_{\mathrm{Sx}21} + c_{\mathrm{Sx}22} + \cdots + c_{\mathrm{Sx}2(i-1)} + \frac{1}{2}c_{\mathrm{Sx}2i}\right) \\ + \cdots + a_{nn}\left(c_{\mathrm{Sx}n1} + c_{\mathrm{Sx}n2} + \cdots + c_{\mathrm{Sx}n(i-1)} + \frac{1}{2}c_{\mathrm{Sx}ni}\right) \end{bmatrix} \tag{10.25}$$

The ith column of $\mathbf{A}_1\mathbf{C}_{Sx}\mathbf{Q}^{k1}\mathbf{P}$ is

$$
= h\begin{bmatrix}
a_{(1)11}\left(c_{Sx11}+c_{Sx12}+\cdots+c_{Sx1(i-k1-1)}+\frac{1}{2}c_{Sx1(i-k1)}\right) \\
\quad +a_{(1)12}\left(c_{Sx21}+c_{Sx22}+\cdots+c_{Sx2(i-k1-1)}+\frac{1}{2}c_{Sx2(i-k1)}\right)+\cdots+ \\
\quad\quad +a_{(1)1n}\left(c_{Sxn1}+c_{Sxn2}+\cdots+c_{Sxn(i-k1-1)}+\frac{1}{2}c_{Sxn(i-k1)}\right) \\
a_{(1)21}\left(c_{Sx11}+c_{Sx12}+\cdots+c_{Sx1(i-k1-1)}+\frac{1}{2}c_{Sx1(i-k1)}\right) \\
\quad +a_{(1)22}\left(c_{Sx21}+c_{Sx22}+\cdots+c_{Sx2(i-k1-1)}+\frac{1}{2}c_{Sx2(i-k1)}\right)+\cdots+ \\
\quad\quad +a_{(1)2n}\left(c_{Sxn1}+c_{Sxn2}+\cdots+c_{Sxn(i-k1-1)}+\frac{1}{2}c_{Sxn(i-k1)}\right) \\
\vdots \\
a_{(1)n1}\left(c_{Sx11}+c_{Sx12}+\cdots+c_{Sx1(i-k1-1)}+\frac{1}{2}c_{Sx1(i-k1)}\right) \\
\quad +a_{(1)n2}\left(c_{Sx21}+c_{Sx22}+\cdots+c_{Sx2(i-k1-1)}+\frac{1}{2}c_{Sx2(i-k1)}\right)+\cdots+ \\
\quad\quad +a_{(1)nn}\left(c_{Sxn1}+c_{Sxn2}+\cdots+c_{Sxn(i-k1-1)}+\frac{1}{2}c_{Sxn(i-k1)}\right)
\end{bmatrix}
$$

for $i > k1$

(10.26)

The ith column of $\mathbf{A}_2\mathbf{C}_{Sx}\mathbf{Q}^{k2}\mathbf{P}$ is

$$
= h\begin{bmatrix}
a_{(2)11}\left(c_{Sx11}+c_{Sx12}+\cdots+c_{Sx1(i-k2-1)}+\frac{1}{2}c_{Sx1(i-k2)}\right) \\
\quad +a_{(2)12}\left(c_{Sx21}+c_{Sx22}+\cdots+c_{Sx2(i-k2-1)}+\frac{1}{2}c_{Sx2(i-k2)}\right)+\cdots+ \\
\quad\quad +a_{(2)1n}\left(c_{Sxn1}+c_{Sxn2}+\cdots+c_{Sxn(i-k2-1)}+\frac{1}{2}c_{Sxn(i-k2)}\right) \\
a_{(2)21}\left(c_{Sx11}+c_{Sx12}+\cdots+c_{Sx1(i-k2-1)}+\frac{1}{2}c_{Sx1(i-k2)}\right) \\
\quad +a_{(2)22}\left(c_{Sx21}+c_{Sx22}+\cdots+c_{Sx2(i-k2-1)}+\frac{1}{2}c_{Sx2(i-k2)}\right)+\cdots+ \\
\quad\quad +a_{(2)2n}\left(c_{Sxn1}+c_{Sxn2}+\cdots+c_{Sxn(i-k2-1)}+\frac{1}{2}c_{Sxn(i-k2)}\right) \\
\vdots \\
a_{(2)n1}\left(c_{Sx11}+c_{Sx12}+\cdots+c_{Sx1(i-k2-1)}+\frac{1}{2}c_{Sx1(i-k2)}\right) \\
\quad +a_{(2)n2}\left(c_{Sx21}+c_{Sx22}+\cdots+c_{Sx2(i-k2-1)}+\frac{1}{2}c_{Sx2(i-k2)}\right)+\cdots+ \\
\quad\quad +a_{(2)nn}\left(c_{Sxn1}+c_{Sxn2}+\cdots+c_{Sxn(i-k2-1)}+\frac{1}{2}c_{Sxn(i-k2)}\right)
\end{bmatrix}
$$

for $i > k2$

(10.27)

After subtracting the ith column of LHS of Eq. (10.21) from its $(i + 1)$th column, using the relations from (10.22) to (10.27), we get

$$
\begin{aligned}
&\left([\mathbf{C}_{\mathrm{Sx}}]_{i+1}-[\mathbf{A}\mathbf{C}_{\mathrm{Sx}}\mathbf{P}]_{i+1}-\left[\mathbf{A}_1\mathbf{C}_{\mathrm{Sx}}\mathbf{Q}^{k1}\mathbf{P}\right]_{i+1}-\left[\mathbf{A}_2\mathbf{C}_{\mathrm{Sx}}\mathbf{Q}^{k2}\mathbf{P}\right]_{i+1}\right)\\
&\quad-\left([\mathbf{C}_{\mathrm{Sx}}]_{i}-[\mathbf{A}\mathbf{C}_{\mathrm{Sx}}\mathbf{P}]_{i}-\left[\mathbf{A}_1\mathbf{C}_{\mathrm{Sx}}\mathbf{Q}^{k1}\mathbf{P}\right]_{i}-\left[\mathbf{A}_2\mathbf{C}_{\mathrm{Sx}}\mathbf{Q}^{k2}\mathbf{P}\right]_{i}\right)\\
&=\left([\mathbf{C}_{\mathrm{Sx}}]_{i+1}-[\mathbf{C}_{\mathrm{Sx}}]_{i}\right)-\left([\mathbf{A}\mathbf{C}_{\mathrm{Sx}}\mathbf{P}]_{i+1}-[\mathbf{A}\mathbf{C}_{\mathrm{Sx}}\mathbf{P}]_{i}\right)\\
&\quad-\left(\left[\mathbf{A}_1\mathbf{C}_{\mathrm{Sx}}\mathbf{Q}^{k1}\mathbf{P}\right]_{i+1}-\left[\mathbf{A}_1\mathbf{C}_{\mathrm{Sx}}\mathbf{Q}^{k1}\mathbf{P}\right]_{i}\right)-\left(\left[\mathbf{A}_2\mathbf{C}_{\mathrm{Sx}}\mathbf{Q}^{k2}\mathbf{P}\right]_{i+1}-\left[\mathbf{A}_2\mathbf{C}_{\mathrm{Sx}}\mathbf{Q}^{k2}\mathbf{P}\right]_{i}\right)\\
&=\begin{bmatrix} c_{\mathrm{Sx}1(i+1)}\\ c_{\mathrm{Sx}2(i+1)}\\ \vdots\\ c_{\mathrm{Sx}n(i+1)}\end{bmatrix}-\begin{bmatrix} c_{\mathrm{Sx}1i}\\ c_{\mathrm{Sx}2i}\\ \vdots\\ c_{\mathrm{Sx}ni}\end{bmatrix}-\frac{h}{2}\mathbf{A}\begin{bmatrix} c_{\mathrm{Sx}1i}\\ c_{\mathrm{Sx}2i}\\ \vdots\\ c_{\mathrm{Sx}ni}\end{bmatrix}-\frac{h}{2}\mathbf{A}\begin{bmatrix} c_{\mathrm{Sx}1(i+1)}\\ c_{\mathrm{Sx}2(i+1)}\\ \vdots\\ c_{\mathrm{Sx}n(i+1)}\end{bmatrix}-\frac{h}{2}\mathbf{A}_1\begin{bmatrix} c_{\mathrm{Sx}1(i-k1)}\\ c_{\mathrm{Sx}2(i-k1)}\\ \vdots\\ c_{\mathrm{Sx}n(i-k1)}\end{bmatrix}\\
&\quad-\frac{h}{2}\mathbf{A}_1\begin{bmatrix} c_{\mathrm{Sx}1(i-k1+1)}\\ c_{\mathrm{Sx}2(i-k1+1)}\\ \vdots\\ c_{\mathrm{Sx}n(i-k1+1)}\end{bmatrix}-\frac{h}{2}\mathbf{A}_2\begin{bmatrix} c_{\mathrm{Sx}1(i-k2)}\\ c_{\mathrm{Sx}2(i-k2)}\\ \vdots\\ c_{\mathrm{Sx}n(i-k2)}\end{bmatrix}-\frac{h}{2}\mathbf{A}_2\begin{bmatrix} c_{\mathrm{Sx}1(i-k2+1)}\\ c_{\mathrm{Sx}2(i-k2+1)}\\ \vdots\\ c_{\mathrm{Sx}n(i-k2+1)}\end{bmatrix}
\end{aligned}
\tag{10.28}
$$

Similar to Eq. (10.28), we subtract the ith column of RHS of Eq. (10.21) from its $(i+1)$th column and express the result of subtraction in parts. First, we subtract the first terms of the two columns in the RHS of Eq. (10.21) and get

$$
\begin{aligned}
&\left(\left[\mathbf{I}-\frac{h}{2}\mathbf{A}-\frac{h}{2}\mathbf{A}_1\mathbf{Q}^{k1}-\frac{h}{2}\mathbf{A}_2\mathbf{Q}^{k2}\right]\mathbf{C}_{\mathrm{Sx}0}\right)_{i+1}\\
&\quad-\left(\left[\mathbf{I}-\frac{h}{2}\mathbf{A}-\frac{h}{2}\mathbf{A}_1\mathbf{Q}^{k1}-\frac{h}{2}\mathbf{A}_2\mathbf{Q}^{k2}\right]\mathbf{C}_{\mathrm{Sx}0}\right)_{i}=0
\end{aligned}
$$

Subtractions of the remaining terms, we have

$$
\begin{aligned}
&\left[\mathbf{A}_1\left(\mathbf{C}_{\mathrm{Sx}\tau1}+\frac{1}{2}\mathbf{C}_{\mathrm{Tx}\tau1}\right)\mathbf{Q}_{\mathrm{t}}^{k1}\mathbf{P1ss}\right]_{i+1}-\left[\mathbf{A}_1\left(\mathbf{C}_{\mathrm{Sx}\tau1}+\frac{1}{2}\mathbf{C}_{\mathrm{Tx}\tau1}\right)\mathbf{Q}_{\mathrm{t}}^{k1}\mathbf{P1ss}\right]_{i}\\
&=h\,\mathbf{A}_1\left(\mathbf{C}_{\mathrm{Sx}\tau1}+\frac{1}{2}\mathbf{C}_{\mathrm{Tx}\tau1}\right)\left[\mathbf{Q}_{\mathrm{t}}^{k1}\right]_{i}
\end{aligned}
$$

Similarly,

$$
\begin{aligned}
&\left[\mathbf{A}_2\left(\mathbf{C}_{\mathrm{Sx}\tau2}+\frac{1}{2}\mathbf{C}_{\mathrm{Tx}\tau2}\right)\mathbf{Q}_{\mathrm{t}}^{k2}\mathbf{P1ss}\right]_{i+1}-\left[\mathbf{A}_2\left(\mathbf{C}_{\mathrm{Sx}\tau2}+\frac{1}{2}\mathbf{C}_{\mathrm{Tx}\tau2}\right)\mathbf{Q}_{\mathrm{t}}^{k2}\mathbf{P1ss}\right]_{i}\\
&=h\,\mathbf{A}_2\left(\mathbf{C}_{\mathrm{Sx}\tau2}+\frac{1}{2}\mathbf{C}_{\mathrm{Tx}\tau2}\right)\left[\mathbf{Q}_{\mathrm{t}}^{k2}\right]_{i}
\end{aligned}
$$

Now

$$\left[\mathbf{B}\left(\mathbf{C}_{\mathrm{Su}}^{\mathrm{T}}+\frac{1}{2}\mathbf{C}_{\mathrm{Tu}}^{\mathrm{T}}\right)\mathbf{P1ss}\right]_{i+1}-\left[\mathbf{B}\left(\mathbf{C}_{\mathrm{Su}}^{\mathrm{T}}+\frac{1}{2}\mathbf{C}_{\mathrm{Tu}}^{\mathrm{T}}\right)\mathbf{P1ss}\right]_{i}=h\,\mathbf{B}\left(\mathbf{C}_{\mathrm{Su}}^{\mathrm{T}}+\frac{1}{2}\mathbf{C}_{\mathrm{Tu}}^{\mathrm{T}}\right)_{i}$$

Similarly

$$\begin{aligned}&\left(\mathbf{B}_1\left[\left(\mathbf{C}_{\mathrm{Su}}^{\mathrm{T}}\mathbf{Q}^{\beta 1}+\mathbf{C}_{\mathrm{Su}\tau 1}^{\mathrm{T}}\mathbf{Q}_{\mathrm{t}}^{\beta 1}\right)+\frac{1}{2}\left(\mathbf{C}_{\mathrm{Tu}}^{\mathrm{T}}\mathbf{Q}^{\beta 1}+\mathbf{C}_{\mathrm{Tu}\tau 1}^{\mathrm{T}}\mathbf{Q}_{\mathrm{t}}^{\beta 1}\right)\right]\mathbf{P1ss}\right)_{i+1}\\&\qquad-\left(\mathbf{B}_1\left[\left(\mathbf{C}_{\mathrm{Su}}^{\mathrm{T}}\mathbf{Q}^{\beta 1}+\mathbf{C}_{\mathrm{Su}\tau 1}^{\mathrm{T}}\mathbf{Q}_{\mathrm{t}}^{\beta 1}\right)+\frac{1}{2}\left(\mathbf{C}_{\mathrm{Tu}}^{\mathrm{T}}\mathbf{Q}^{\beta 1}+\mathbf{C}_{\mathrm{Tu}\tau 1}^{\mathrm{T}}\mathbf{Q}_{\mathrm{t}}^{\beta 1}\right)\right]\mathbf{P1ss}\right)_{i}\\&=h\,\mathbf{B}_1\left(\mathbf{C}_{\mathrm{Su}}^{\mathrm{T}}+\frac{1}{2}\mathbf{C}_{\mathrm{Tu}}^{\mathrm{T}}\right)\left[\mathbf{Q}^{\beta 1}\right]_i+h\,\mathbf{B}_1\left(\mathbf{C}_{\mathrm{Su}\tau 1}^{\mathrm{T}}+\frac{1}{2}\mathbf{C}_{\mathrm{Tu}\tau 1}^{\mathrm{T}}\right)\left[\mathbf{Q}_{\mathrm{t}}^{\beta 1}\right]_i\end{aligned}$$

and

$$\begin{aligned}&\left(\mathbf{B}_2\left[\left(\mathbf{C}_{\mathrm{Su}}^{\mathrm{T}}\mathbf{Q}^{\beta 2}+\mathbf{C}_{\mathrm{Su}\tau 2}^{\mathrm{T}}\mathbf{Q}_{\mathrm{t}}^{\beta 2}\right)+\frac{1}{2}\left(\mathbf{C}_{\mathrm{Tu}}^{\mathrm{T}}\mathbf{Q}^{\beta 2}+\mathbf{C}_{\mathrm{Tu}\tau 2}^{\mathrm{T}}\mathbf{Q}_{\mathrm{t}}^{\beta 2}\right)\right]\mathbf{P1ss}\right)_{i+1}\\&\quad-\left(\mathbf{B}_2\left[\left(\mathbf{C}_{\mathrm{Su}}^{\mathrm{T}}\mathbf{Q}^{\beta 2}+\mathbf{C}_{\mathrm{Su}\tau 2}^{\mathrm{T}}\mathbf{Q}_{\mathrm{t}}^{\beta 2}\right)+\frac{1}{2}\left(\mathbf{C}_{\mathrm{Tu}}^{\mathrm{T}}\mathbf{Q}^{\beta 2}+\mathbf{C}_{\mathrm{Tu}\tau 2}^{\mathrm{T}}\mathbf{Q}_{\mathrm{t}}^{\beta 2}\right)\right]\mathbf{P1ss}\right)_{i}\\&\quad=h\,\mathbf{B}_2\left(\mathbf{C}_{\mathrm{Su}}^{\mathrm{T}}+\frac{1}{2}\mathbf{C}_{\mathrm{Tu}}^{\mathrm{T}}\right)\left[\mathbf{Q}^{\beta 2}\right]_i+h\,\mathbf{B}_2\left(\mathbf{C}_{\mathrm{Su}\tau 2}^{\mathrm{T}}+\frac{1}{2}\mathbf{C}_{\mathrm{Tu}\tau 2}^{\mathrm{T}}\right)\left[\mathbf{Q}_{\mathrm{t}}^{\beta 2}\right]_i\end{aligned}$$

Therefore, finally the subtraction of ith columns of Eq. (10.21) from respective $(i + 1)$th columns can be expressed as

$$\begin{aligned}&\begin{bmatrix}c_{\mathrm{Sx}1(i+1)}\\c_{\mathrm{Sx}2(i+1)}\\\vdots\\c_{\mathrm{Sx}n(i+1)}\end{bmatrix}-\begin{bmatrix}c_{\mathrm{Sx}1i}\\c_{\mathrm{Sx}2i}\\\vdots\\c_{\mathrm{Sx}ni}\end{bmatrix}-\frac{h}{2}\mathbf{A}\begin{bmatrix}c_{\mathrm{Sx}1i}\\c_{\mathrm{Sx}2i}\\\vdots\\c_{\mathrm{Sx}ni}\end{bmatrix}-\frac{h}{2}\mathbf{A}\begin{bmatrix}c_{\mathrm{Sx}1(i+1)}\\c_{\mathrm{Sx}2(i+1)}\\\vdots\\c_{\mathrm{Sx}n(i+1)}\end{bmatrix}-\frac{h}{2}\mathbf{A}_1\begin{bmatrix}c_{\mathrm{Sx}1(i-k1)}\\c_{\mathrm{Sx}2(i-k1)}\\\vdots\\c_{\mathrm{Sx}n(i-k1)}\end{bmatrix}\\&-\frac{h}{2}\mathbf{A}_1\begin{bmatrix}c_{\mathrm{Sx}1(i-k1+1)}\\c_{\mathrm{Sx}2(i-k1+1)}\\\vdots\\c_{\mathrm{Sx}n(i-k1+1)}\end{bmatrix}-\frac{h}{2}\mathbf{A}_2\begin{bmatrix}c_{\mathrm{Sx}1(i-k2)}\\c_{\mathrm{Sx}2(i-k2)}\\\vdots\\c_{\mathrm{Sx}n(i-k2)}\end{bmatrix}-\frac{h}{2}\mathbf{A}_2\begin{bmatrix}c_{\mathrm{Sx}1(i-k2+1)}\\c_{\mathrm{Sx}2(i-k2+1)}\\\vdots\\c_{\mathrm{Sx}n(i-k2+1)}\end{bmatrix}\\&=h\mathbf{A}_1\left(\mathbf{C}_{\mathrm{Sx}\tau 1}+\frac{1}{2}\mathbf{C}_{\mathrm{Tx}\tau 1}\right)\left[\mathbf{Q}_{\mathrm{t}}^{k1}\right]_i+h\mathbf{A}_2\left(\mathbf{C}_{\mathrm{Sx}\tau 2}+\frac{1}{2}\mathbf{C}_{\mathrm{Tx}\tau 2}\right)\left[\mathbf{Q}_{\mathrm{t}}^{k2}\right]_i+h\,\mathbf{B}\left(\mathbf{C}_{\mathrm{Su}}^{\mathrm{T}}+\frac{1}{2}\mathbf{C}_{\mathrm{Tu}}^{\mathrm{T}}\right)_i\\&\quad+h\,\mathbf{B}_1\left(\mathbf{C}_{\mathrm{Su}}^{\mathrm{T}}+\frac{1}{2}\mathbf{C}_{\mathrm{Tu}}^{\mathrm{T}}\right)\left[\mathbf{Q}^{\beta 1}\right]_i+h\,\mathbf{B}_1\left(\mathbf{C}_{\mathrm{Su}\tau 1}^{\mathrm{T}}+\frac{1}{2}\mathbf{C}_{\mathrm{Tu}\tau 1}^{\mathrm{T}}\right)\left[\mathbf{Q}_{\mathrm{t}}^{\beta 1}\right]_i\\&\quad+h\,\mathbf{B}_2\left(\mathbf{C}_{\mathrm{Su}}^{\mathrm{T}}+\frac{1}{2}\mathbf{C}_{\mathrm{Tu}}^{\mathrm{T}}\right)\left[\mathbf{Q}^{\beta 2}\right]_i+h\,\mathbf{B}_2\left(\mathbf{C}_{\mathrm{Su}\tau 2}^{\mathrm{T}}+\frac{1}{2}\mathbf{C}_{\mathrm{Tu}\tau 2}^{\mathrm{T}}\right)\left[\mathbf{Q}_{\mathrm{t}}^{\beta 2}\right]_i\\&=h\mathbf{A}_1\left(\mathbf{C}_{\mathrm{Sx}\tau 1}+\frac{1}{2}\mathbf{C}_{\mathrm{Tx}\tau 1}\right)_{i+k1}+h\mathbf{A}_2\left(\mathbf{C}_{\mathrm{Sx}\tau 2}+\frac{1}{2}\mathbf{C}_{\mathrm{Tx}\tau 2}\right)_{i+k2}+h\,\mathbf{B}\left(\mathbf{C}_{\mathrm{Su}}^{\mathrm{T}}+\frac{1}{2}\mathbf{C}_{\mathrm{Tu}}^{\mathrm{T}}\right)_i\\&\quad+h\,\mathbf{B}_1\left(\mathbf{C}_{\mathrm{Su}}^{\mathrm{T}}+\frac{1}{2}\mathbf{C}_{\mathrm{Tu}}^{\mathrm{T}}\right)_{i-\beta 1}+h\,\mathbf{B}_1\left(\mathbf{C}_{\mathrm{Su}\tau 1}^{\mathrm{T}}+\frac{1}{2}\mathbf{C}_{\mathrm{Tu}\tau 1}^{\mathrm{T}}\right)_{i+\beta 1}\\&\quad+\mathbf{B}_2\,h\left(\mathbf{C}_{\mathrm{Su}}^{\mathrm{T}}+\frac{1}{2}\mathbf{C}_{\mathrm{Tu}}^{\mathrm{T}}\right)_{i-\beta 2}+\mathbf{B}_2\,h\left(\mathbf{C}_{\mathrm{Su}\tau 2}^{\mathrm{T}}+\frac{1}{2}\mathbf{C}_{\mathrm{Tu}\tau 2}^{\mathrm{T}}\right)_{i+\beta 2}\end{aligned}\tag{10.29}$$

Following Eq. (10.12), the generalized form of Eq. (10.29) can be written as

$$\left[\mathbf{I}-\frac{h}{2}\mathbf{A}\right]\begin{bmatrix} c_{\mathrm{Sx}1(i+1)} \\ c_{\mathrm{Sx}2(i+1)} \\ \vdots \\ c_{\mathrm{Sx}n(i+1)} \end{bmatrix} = \left[\mathbf{I}+\frac{h}{2}\mathbf{A}\right]\begin{bmatrix} c_{\mathrm{Sx}1i} \\ c_{\mathrm{Sx}2i} \\ \vdots \\ c_{\mathrm{Sx}ni} \end{bmatrix} + \frac{h}{2}\sum_{k=1}^{Nx}[\mathbf{A}_k][\mathbf{C}_{\mathrm{Sx}}]_{i-k} + \frac{h}{2}\sum_{k=1}^{Nx}[\mathbf{A}_k][\mathbf{C}_{\mathrm{Sx}}]_{i-k+1}$$
$$+ h\sum_{k=1}^{Nx}[\mathbf{A}_k]\left[\mathbf{C}_{\mathrm{Sx}\tau k}+\frac{1}{2}\mathbf{C}_{\mathrm{Tx}\tau k}\right]_{i+k} + h\,\mathbf{B}\left(\mathbf{C}_{\mathrm{Su}}^{\mathrm{T}}+\frac{1}{2}\mathbf{C}_{\mathrm{Tu}}^{\mathrm{T}}\right)_i$$
$$+ h\sum_{\lambda=1}^{Nu}[\mathbf{B}_\lambda]\left[\mathbf{C}_{\mathrm{Su}}^{\mathrm{T}}+\frac{1}{2}\mathbf{C}_{\mathrm{Tu}}^{\mathrm{T}}\right]_{i-\lambda} + h\sum_{\lambda=1}^{Nu}[\mathbf{B}_\lambda]\left[\mathbf{C}_{\mathrm{Su}\tau\lambda}^{\mathrm{T}}+\frac{1}{2}\mathbf{C}_{\mathrm{Tu}\tau\lambda}^{\mathrm{T}}\right]_{i+\lambda} \tag{10.30}$$

From Eq. (10.30), using matrix inversion, we have

$$\begin{bmatrix} c_{\mathrm{Sx}1(i+1)} \\ c_{\mathrm{Sx}2(i+1)} \\ \vdots \\ c_{\mathrm{Sx}n(i+1)} \end{bmatrix} = \left[\mathbf{I}-\frac{h}{2}\mathbf{A}\right]^{-1}\left\{\left[\mathbf{I}+\frac{h}{2}\mathbf{A}\right]\begin{bmatrix} c_{\mathrm{Sx}1i} \\ c_{\mathrm{Sx}2i} \\ \vdots \\ c_{\mathrm{Sx}ni} \end{bmatrix} + \frac{h}{2}\sum_{k=1}^{Nx}[\mathbf{A}_k][\mathbf{C}_{\mathrm{Sx}}]_{i-k} + \frac{h}{2}\sum_{k=1}^{Nx}[\mathbf{A}_k][\mathbf{C}_{\mathrm{Sx}}]_{i-k+1}\right.$$
$$+ h\sum_{k=1}^{Nx}[\mathbf{A}_k]\left[\mathbf{C}_{\mathrm{Sx}\tau k}+\frac{1}{2}\mathbf{C}_{\mathrm{Tx}\tau k}\right]_{i+k} + h\,\mathbf{B}\left(\mathbf{C}_{\mathrm{Su}}^{\mathrm{T}}+\frac{1}{2}\mathbf{C}_{\mathrm{Tu}}^{\mathrm{T}}\right)_i$$
$$\left.+ h\sum_{\lambda=1}^{Nu}[\mathbf{B}_\lambda]\left[\mathbf{C}_{\mathrm{Su}}^{\mathrm{T}}+\frac{1}{2}\mathbf{C}_{\mathrm{Tu}}^{\mathrm{T}}\right]_{i-\lambda} + h\sum_{\lambda=1}^{Nu}[\mathbf{B}_\lambda]\left[\mathbf{C}_{\mathrm{Su}\tau\lambda}^{\mathrm{T}}+\frac{1}{2}\mathbf{C}_{\mathrm{Tu}\tau\lambda}^{\mathrm{T}}\right]_{i+\lambda}\right\}$$

or,

$$\begin{bmatrix} c_{\mathrm{Sx}1(i+1)} \\ c_{\mathrm{Sx}2(i+1)} \\ \vdots \\ c_{\mathrm{Sx}n(i+1)} \end{bmatrix} = \left[\frac{2}{h}\mathbf{I}-\mathbf{A}\right]^{-1}\left\{\left[\frac{2}{h}\mathbf{I}+\mathbf{A}\right]\begin{bmatrix} c_{\mathrm{Sx}1i} \\ c_{\mathrm{Sx}2i} \\ \vdots \\ c_{\mathrm{Sx}ni} \end{bmatrix} + \sum_{k=1}^{Nx}[\mathbf{A}_k][\mathbf{C}_{\mathrm{Sx}}]_{i-k} + \sum_{k=1}^{Nx}[\mathbf{A}_k][\mathbf{C}_{\mathrm{Sx}}]_{i-k+1}\right.$$
$$+ 2\sum_{k=1}^{Nx}[\mathbf{A}_k]\left[\mathbf{C}_{\mathrm{Sx}\tau k}+\frac{1}{2}\mathbf{C}_{\mathrm{Tx}\tau k}\right]_{i+k} + 2\,\mathbf{B}\left(\mathbf{C}_{\mathrm{Su}}^{\mathrm{T}}+\frac{1}{2}\mathbf{C}_{\mathrm{Tu}}^{\mathrm{T}}\right)_i$$
$$\left.+ 2\sum_{\lambda=1}^{Nu}[\mathbf{B}_\lambda]\left[\mathbf{C}_{\mathrm{Su}}^{\mathrm{T}}+\frac{1}{2}\mathbf{C}_{\mathrm{Tu}}^{\mathrm{T}}\right]_{i-\lambda} + 2\sum_{\lambda=1}^{Nu}[\mathbf{B}_\lambda]\left[\mathbf{C}_{\mathrm{Su}\tau\lambda}^{\mathrm{T}}+\frac{1}{2}\mathbf{C}_{\mathrm{Tu}\tau\lambda}^{\mathrm{T}}\right]_{i+\lambda}\right\} \tag{10.31}$$

The inverse in (10.31) can always be made to exist by judicious choice of h.

Equation (10.31) provides a simple recursive solution of the states of a multi-delay non-homogeneous system, or, in other words, time samples of the states, with a sampling period of h knowing the system matrix $\mathbf{A}$, the input matrix $\mathbf{B}$, the input signal u, the delay matrices for states and for input, and the initial values of the states.

For a time-invariant system, all the delay matrices are zero. For such a system Eq. (10.31) can be modified as

$$\begin{bmatrix} c_{\mathrm{Sx1}(i+1)} \\ c_{\mathrm{Sx2}(i+1)} \\ \vdots \\ c_{\mathrm{Sxn}(i+1)} \end{bmatrix} = \left[\frac{2}{h}\mathbf{I}-\mathbf{A}\right]^{-1}\left[\frac{2}{h}\mathbf{I}+\mathbf{A}\right]\begin{bmatrix} c_{\mathrm{Sx1}i} \\ c_{\mathrm{Sx2}i} \\ \vdots \\ c_{\mathrm{Sx}ni} \end{bmatrix} + 2\left[\frac{2}{h}\mathbf{I}-\mathbf{A}\right]^{-1}\mathbf{B}\left[\mathbf{C}_{\mathrm{Su}}^{\mathrm{T}}+\frac{1}{2}\mathbf{C}_{\mathrm{Tu}}^{\mathrm{T}}\right]_i$$

like Eq. (8.21).

10.3.1 Numerical Examples

Example 10.5 (vide Appendix B, Program no. 28)

Consider the non-homogeneous time-delay system

$$\dot{x}(t) = x(t-1) + u(t), \quad 0 \le t \le 1$$
$$x(t) = 1, \quad \text{for} \quad -1 \le t \le 0$$
$$\text{and} \quad u(t) = \begin{cases} -2.1 + 1.05t, & 0 \le t \le 1 \\ -1.05, & 1 \le t \le 2 \end{cases}$$

having the solution

$$x(t) = \begin{cases} 1 - 1.1t + 0.525t^2, & 0 \le t \le 1 \\ -0.25 + 1.575t - 1.075t^2 + 0.175t^3, & 1 \le t \le 2 \end{cases}$$

The exact solution of the state $x(t)$ along with samples computed using the HF approach, are shown in Fig. 10.6, for $m = 4$ and $m = 8$ with $T = 2$ s.

For quantitative analysis of this HF based approach, we have compared sample-wise computational error with the exact samples, in Table 10.1. From Table 10.1 it is noted that, instead of using a small number of segments ($m = 4$), the samples obtained using HF domain analysis are reasonably close to their respective exact samples.

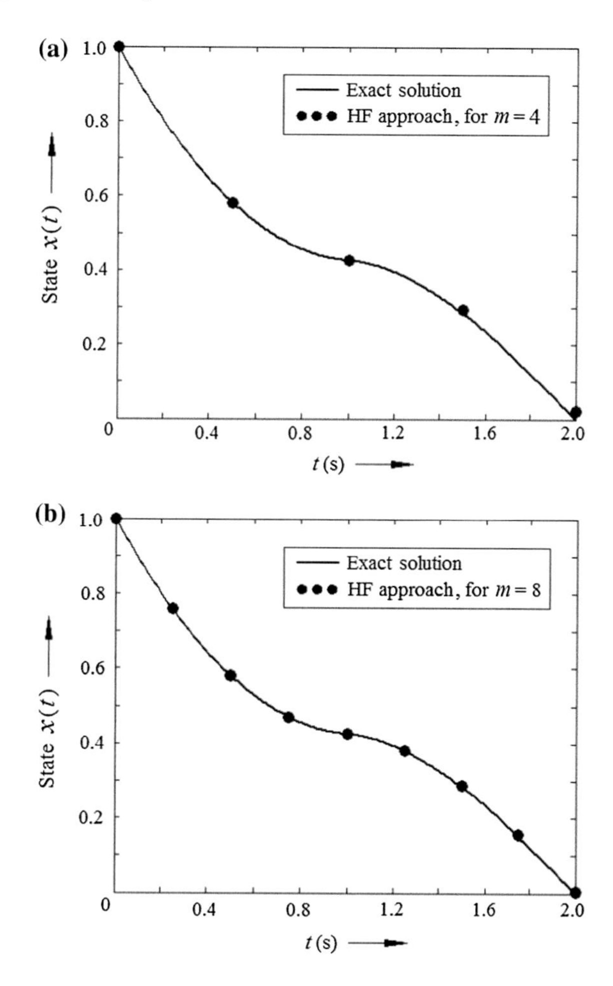

Fig. 10.6 Comparison of the exact samples of the state $x(t)$ of the non-homogeneous time-delay system of Example 10.5, with the samples obtained in HF domain, for a $m = 4$ and b $m = 8$, with $T = 2$ s (vide Appendix B, Program no. 28)

Example 10.6 Consider the non-homogeneous time-delay system

$$\dot{x}(t) = -x(t) - 2x\left(t - \frac{1}{4}\right) + 2u\left(t - \frac{1}{4}\right), \quad 0 \leq t \leq 1$$

$$x(t) = u(t) = 0, \quad \text{for } -\frac{1}{4} \leq t \leq 0$$

$$\text{and } u(t) = 1 \quad \text{for } t \geq 0$$

Table 10.1 Solution of state $x(t)$ of the non- homogeneous time delay system of Example 10.5 with comparison of exact samples and corresponding samples obtained via HF domain with percentage errors at different sample points for (a) m = 4 and (b) m = 8, with T = 2 s (vide Appendix B, Program no. 28)

t (s)	Exact samples of the state x_d	Samples from HF analysis, using Eq. (10.31), x_h	% Error $\varepsilon = \frac{x_d - x_h}{x_d} \times 100$
(a) System state $x(t)$			
0	1.00000000	1.00000000	0.00000000
$\frac{2}{4}$	0.58125000	0.58125000	0.00000000
$\frac{4}{4}$	0.42500000	0.42500000	0.00000000
$\frac{6}{4}$	0.28437500	0.29531250	−3.84615385
$\frac{8}{4}$	0.00000000	0.02187500	–
(b) System state $x(t)$			
0	1.00000000	0.00000000	0.00000000
$\frac{2}{8}$	0.75781250	0.75781250	0.00000000
$\frac{4}{8}$	0.58125000	0.58125000	0.00000000
$\frac{6}{8}$	0.47031250	0.47031250	0.00000000
$\frac{8}{8}$	0.42500000	0.42500000	0.00000000
$\frac{10}{8}$	0.38085938	0.38222656	−0.35897239
$\frac{12}{8}$	0.28437500	0.28710938	−0.96154022
$\frac{14}{8}$	0.15195313	0.15605469	−2.69922706
$\frac{16}{8}$	0.00000000	0.00546875	–

having the solution

$$x(t) = \begin{cases} 0, \quad 0 \le t \le \frac{1}{4} \\ 2 - 2\exp\left(-\left(t - \frac{1}{4}\right)\right), \quad \frac{1}{4} \le t \le \frac{1}{2} \\ -2 - 2\exp\left(-\left(t - \frac{1}{4}\right)\right) + (2 + 4t)\exp\left(-\left(t - \frac{1}{2}\right)\right), \quad \frac{1}{2} \le t \le \frac{3}{4} \\ 6 - 2\exp\left(-\left(t - \frac{1}{4}\right)\right) + (2 + 4t)\exp\left(-\left(t - \frac{1}{2}\right)\right) - \left(\frac{17}{4} + 2t + 4t^2\right)\exp\left(-\left(t - \frac{3}{4}\right)\right), \quad \frac{3}{4} \le t \le 1 \end{cases}$$

The exact solution of the state $x(t)$ along with results computed using the HF approach, are shown in Fig. 10.7, for m = 10 and T = 1 s.

Example 10.7 Consider a homogeneous time-delay system

$$\begin{aligned} \dot{x}(t) &= x(t - 0.35) + x(t - 0.7) + u(t), \\ x(t) &= 0, \quad \text{for } t \le 0, \\ u(t) &= 1 \end{aligned}$$

The exact solution is

$$x(t) = \begin{cases} t, \quad 0 \le t \le 0.35 \\ t + \frac{1}{2}(t - 0.35)^2, \quad 0.35 \le t \le 0.7 \\ t + \frac{1}{2}(t - 0.35)^2 + \frac{1}{2}(t - 0.7)^2 + \frac{1}{6}(t - 0.7)^3, \quad 0.7 \le t \le 1.05 \end{cases}$$

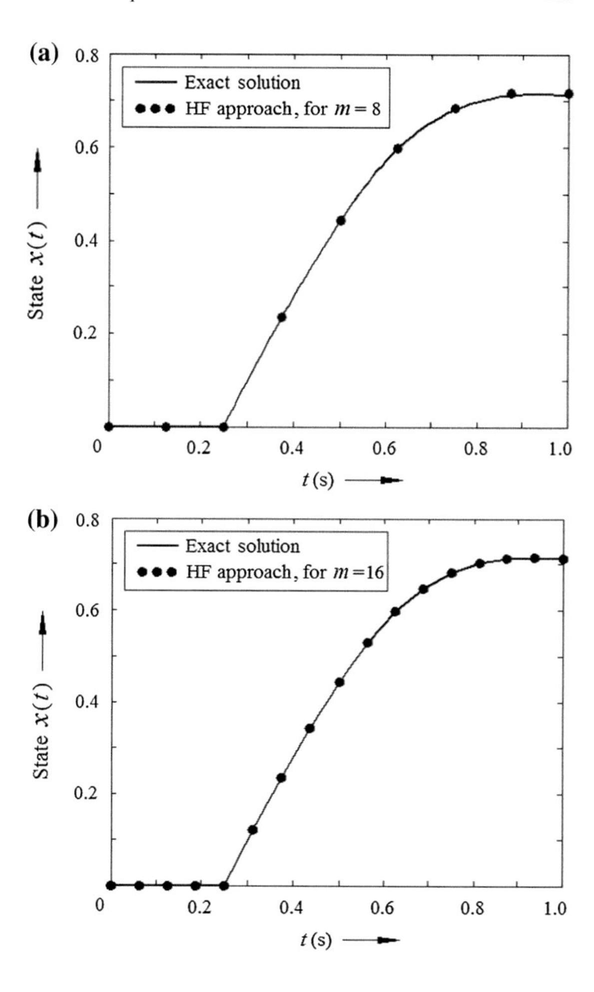

Fig. 10.7 Comparison of exact samples of the state $x(t)$ of the non-homogeneous time-delay system of Example 10.6, with the samples obtained in HF domain, for **a** $m = 8$ and **b** $m = 16$, with $T = 1$ s

The exact solution of the state $x(t)$ along with the results obtained via HF approach, are shown in Fig. 10.8a and b, for $m = 3$ and 6, for $T = 1$ s. Figure 10.9 graphically compares the exact solution of the state with the results obtained using Walsh series ($m = 4$), Taylor series (4th order) and HF domain ($m = 6$) analyses. It seems that the HF based analysis results are more close than the results obtained via Walsh series or Taylor series. To show the differences in the three results, Fig. 10.10 is presented where a magnified view of some portion of Fig. 10.9. is

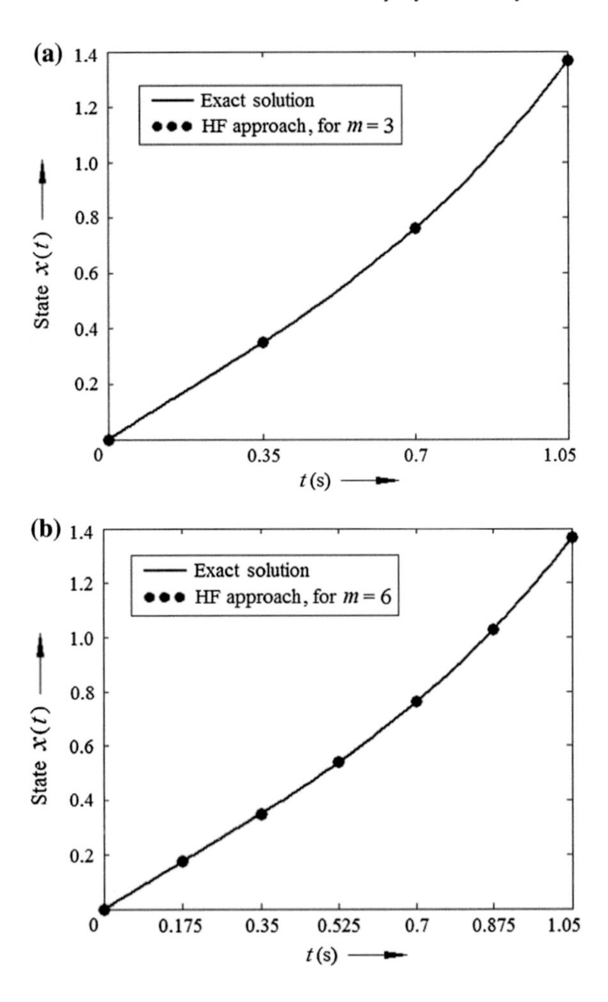

Fig. 10.8 Comparison of exact samples of the state $x(t)$ of the non-homogeneous time-delay system of Example 10.7, with the samples obtained in HF domain, for **a** $m = 3$ and **b** $m = 6$, with $T = 1.05$ s

shown for better clarity. In Fig. 10.10, it is noted that the samples obtained via HF domain analysis fall directly on the exact curve, while this is not the case for Taylor series as well as Walsh series analyses.

To assess this attribute quantitatively, we present Table 10.2 where the MISE's [1, 2] of three different analyses are compared.

It is noted from Table 10.2 that the MISE for Taylor series analysis [3] is much less compared to Walsh series analysis [4]. But the HF analysis produces even less

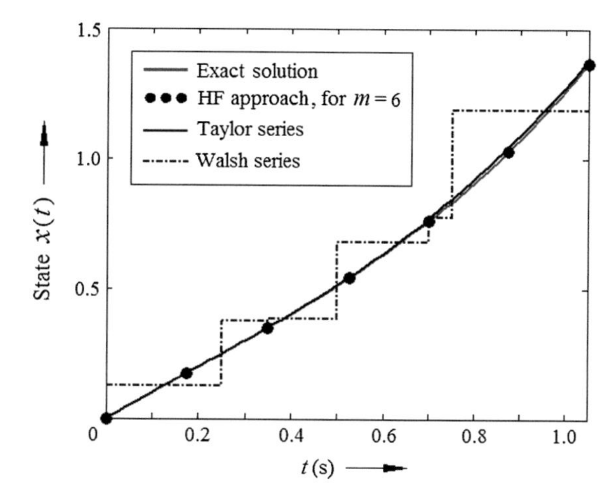

Fig. 10.9 Comparison of the exact samples of the state $x(t)$ of the non-homogeneous time-delay system of Example 10.7, for $T = 1.05$ s, with a the samples obtained via HF domain analysis for $m = 6$, b Taylor series analysis of 4th order [3] and c Walsh series analysis [4] with $m = 4$

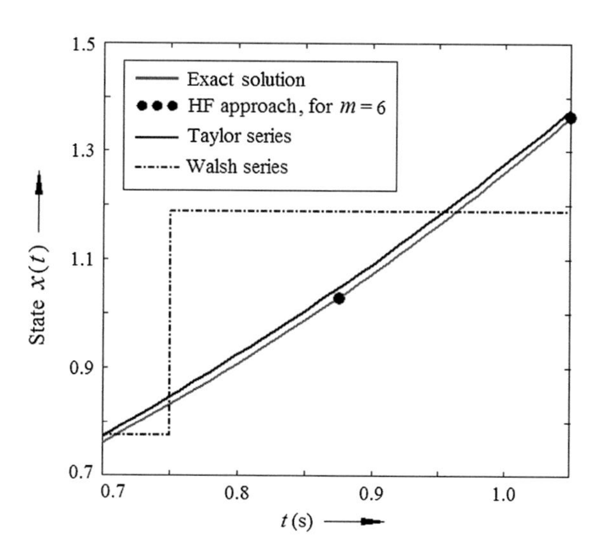

Fig. 10.10 Magnified view of Fig. 10.9, of Example 10.7

Table 10.2 Computation of MISE for the analysis of state $x(t)$ of Example 10.7 for $T = 1.05$ s, for (a) Walsh series analysis [4] with $m = 4$ (b) Taylor series analysis of 4th order [3] and (c) HF domain analysis for $m = 6$

Walsh series analysis with $m = 4$ ($MISE_{Walsh}$)	4th order Taylor series analysis ($MISE_{Taylor}$)	HF domain analysis with $m = 6$ ($MISE_{HF}$)
0.01385083	9.04236472e-05	1.67268017e-05

Table 10.3 Computation of percentage increase in MISE for Walsh analysis ($m = 4$) and Taylor analysis (4th order) with respect to HF domain analysis ($m = 6$), for the state $x(t)$ of Example 10.7, with $T = 1.05$ s

Percentage increase in MISE $\frac{\mathrm{MISE}_{\mathrm{Walsh}} - \mathrm{MISE}_{\mathrm{HF}}}{\mathrm{MISE}_{\mathrm{HF}}} \times 100$	Percentage increase in MISE $\frac{\mathrm{MISE}_{\mathrm{Taylor}} - \mathrm{MISE}_{\mathrm{HF}}}{\mathrm{MISE}_{\mathrm{HF}}} \times 100$
82,706.2069	440.591376

error. Also, a fourth order Taylor series involves much more computation burden compared to the HF analysis. Therefore, the HF domain analysis proves to be more efficient and it produces best results compared to the other two methods.

This is further evident from Table 10.3 where the MISE's of Walsh series and Taylor series analyses are expressed as percentages of the MISE of HF based analysis.

The numerical figures are self explanatory, because, in case of Walsh series analysis, the increase in percentage error is 82,706.2069 % and that for the Taylor series analysis is 440.591376 %.

10.4 Analysis of Homogeneous State Equations with Delay

For a homogeneous system, B and the input delay matrices are zero and Eq. (10.26) will be reduced to

$$\begin{bmatrix} c_{Sx1(i+1)} \\ c_{Sx2(i+1)} \\ \vdots \\ c_{Sxn(i+1)} \end{bmatrix} = \left[\mathbf{I} - \frac{h}{2}\mathbf{A}\right]^{-1} \left\{ \left[\mathbf{I} + \frac{h}{2}\mathbf{A}\right] \begin{bmatrix} c_{Sx1i} \\ c_{Sx2i} \\ \vdots \\ c_{Sxni} \end{bmatrix} + \frac{h}{2}\sum_{k=1}^{Nx} [\mathbf{A}_k][\mathbf{C}_{\mathrm{Sx}}]_{i-k} + \frac{h}{2}\sum_{k=1}^{Nx} [\mathbf{A}_k][\mathbf{C}_{\mathrm{Sx}}]_{i-k+1} \right.$$
$$\left. + h\sum_{k=1}^{Nx} [\mathbf{A}_k]\left[\mathbf{C}_{\mathrm{Sx}\tau k} + \frac{1}{2}\mathbf{C}_{\mathrm{Tx}\tau k}\right]_{i+k} \right\} \tag{10.32}$$

The inverse in (10.32) can always be made to exist by judicious choice of h.

Equation (10.32) provides a simple recursive solution of the states of a multi-delay homogeneous system, or, in other words, time samples of the states, with a sampling period of h knowing the system matrix $\mathbf{A}$, the delay matrices for states, and the initial values of the states.

10.4.1 Numerical Examples

Example 10.8 Consider a homogeneous time-delay system

$$\begin{aligned} \dot{x}(t) &= 4x\left(t-\frac{1}{4}\right), \quad 0 \leq t \leq 1 \\ x(0) &= 1 \\ x(t) &= 0, \quad \text{for } -\frac{1}{4} \leq t < 0 \end{aligned}$$

having the solution

$$x(t) = \begin{cases} 1, \quad 0 \leq t \leq \frac{1}{4} \\ 1 + 4\left(t - \frac{1}{4}\right), \quad \frac{1}{4} \leq t \leq \frac{1}{2} \\ 1 + 4\left(t - \frac{1}{4}\right) + 8\left(t - \frac{1}{2}\right)^2, \quad \frac{1}{2} \leq t \leq \frac{3}{4} \\ 1 + 4\left(t - \frac{1}{4}\right) + 8\left(t - \frac{1}{2}\right)^2 + \frac{32}{3}\left(t - \frac{3}{4}\right)^3, \quad \frac{3}{4} \leq t \leq 1 \end{cases}$$

The exact solution of the state $x(t)$ along with results computed using the HF approach, are shown in Fig. 10.11, for $m = 4$, 12, 20 with $T = 1$ s.

In Example 10.8, the initial values of the state $x(t)$ are zero for time $t < 0$ second, and the state jump to one at $t = 0$ s. If we do not approximate the initial values in HF domain, considering this jump, will analysis the state of the delay system with error. Though we can reduce this error using increasing number of segments, m.

In Example 10.8, it is noted that a jump is involved in the initial values of the state at $t = 0$. To obtain even better results in HF domain, we utilize the concept of jump discontinuity, as illustrated in Chap. 3. Hence we can refine the results in system analysis, using minimum number of sub-intervals. This is evident from Fig. 10.12a and b.

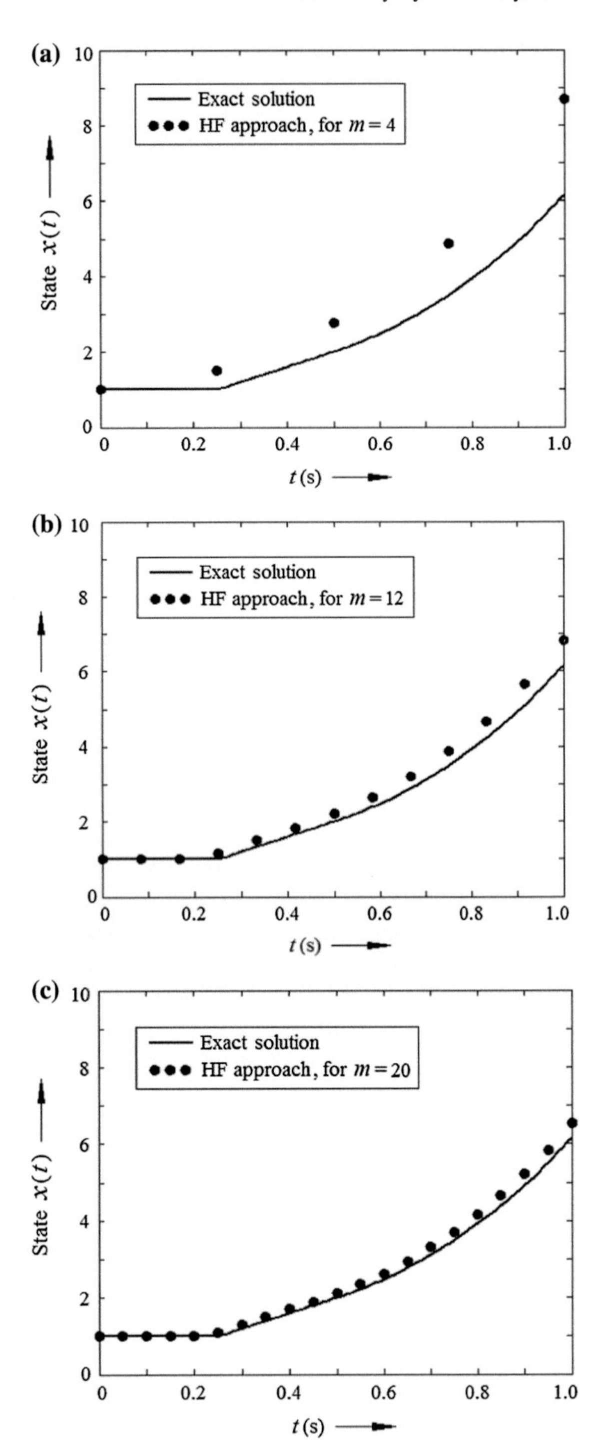

Fig. 10.11 Comparison of exact samples of the state $x(t)$ of the homogeneous time-delay system of Example 10.8, with the samples obtained in HF domain, for **a** $m = 4$, **b** $m = 12$ and **c** $m = 20$, with $T = 1$ s

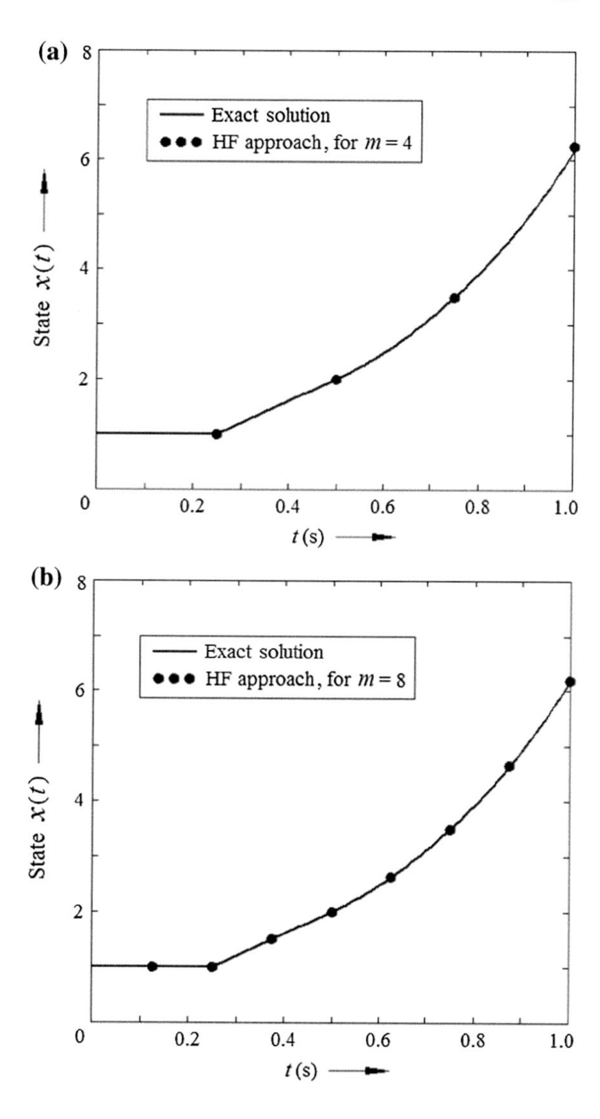

Fig. 10.12 Comparison of exact samples of the state $x(t)$ of the homogeneous time-delay system of Example 10.8, with the samples obtained in HF domain, considering the concept of jump discontinuity, for a $m = 4$, and b $m = 8$, with $T = 1$ s

10.5 Conclusion

In this chapter, we have analysed the state space models of non-homogeneous multi-delay systems in hybrid function platform. In the generalized form of solution, vide Eq. (10.31), it is noted that the structure of the solution is recursive in manner. Thus, the computation is simple as well as attractive.

The same Eq. (10.31) has been used for the analysis of homogeneous systems by simply putting B = 0.

Different types of numerical examples have been treated using the derived matrix equations for homogeneous as well as non-homogeneous systems, and the results are compared with the exact solutions of the system states with respective error estimates. These facts are reflected in Figs. 10.6, 10.7, 10.8, 10.9, 10.10, 10.11, 10.12 and Table 10.1, 10.2 and 10.3 both qualitatively and quantitatively.

References

1. Rao, G.P.: Piecewise Constant Orthogonal Functions and Their Application in Systems and Control, LNC1S, vol. 55. Springer, Berlin (1983)
2. Jiang, J.H., Schaufelberger, W.: Block Pulse Functions and Their Application in Control System, LNCIS, vol. 179. Springer, Berlin (1992)
3. Chung, H.Y., Sun, Y.Y.: Taylor series analysis of multi-delay system. J. Franklin Instt. **324**(1), 65–72 (1987)
4. Chen, W.L.: Walsh series analysis of multi-delay systems. J. Franklin Inst. **313**(4), 207–217 (1982)

Chapter 11
Time Invariant System Analysis: Method of Convolution

Abstract This chapter presents analysis of both open loop as well as closed loop systems based upon the idea of convolution in HF domain. Three numerical examples have been treated, and for clarity, thirteen figures and six tables have been presented.

As the heading implies, this chapter is dedicated for linear time invariant (LTI) control system [1] analysis using convolution method. The convolution operation in HF domain was discussed in detail in Chap. 7 where the key equation giving the samples of the result of convolution was Eq. (7.21).

In any control system block diagram, we find many blocks through which a signal passes. Since the block diagram is usually drawn in Laplace domain, we frequently have products of two functions described in s-domain. Such a product in s-domain means convolution in time domain. Thus, analysis of control systems involves convolution. The principles of convolution in HF domain are now employed for analysis of time invariant non-homogeneous system, both open loop as well as closed loop. As mentioned earlier, HF domain analysis is always based upon function samples and it provides attractive computational advantages.

11.1 Analysis of an Open Loop System

An input $r(t)$ is applied to a causal SISO system, as shown in Fig. 11.1, at $t = 0$. If the impulse response of the plant is $g(t)$, we obtain the output $y(t)$ simply by convolution of $r(t)$ and $g(t)$. Knowing $r(t)$ and $g(t)$, we can employ the generalized Eq. (7.21) for computing the samples of the output $y(t)$ so that the result is obtained in HF domain [2–4].

A. Deb et al., *Analysis and Identification of Time-Invariant Systems, Time-Varying Systems, and Multi-Delay Systems using Orthogonal Hybrid Functions*,
Studies in Systems, Decision and Control 46, DOI 10.1007/978-3-319-26684-8_11

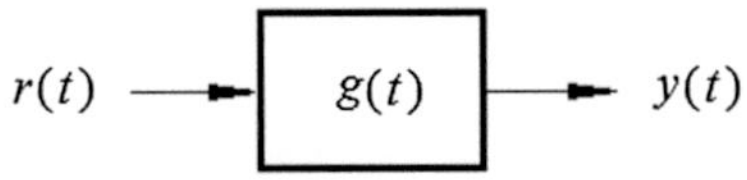

Fig. 11.1 An open loop SISO control system

11.1.1 Numerical Examples

Example 11.1 Consider a linear open loop system, shown in Fig. 11.2, with its impulse response given by $g_1(t) = \exp(-t)$.

Taking $T = 1$ s, $m = 4$ and $h = T/m = 0.25$ s, we analyze the system in HF domain for a step input. Then the actual output is $y_1(t) = 1 - \exp(-t)$

In HF domain, for $m = 4$ and $T = 1$ s, $u(t)$ and $g_1(t)$ are given by

$$u(t) = [1 \quad 1 \quad 1 \quad 1]\mathbf{S}_{(4)} + [0 \quad 0 \quad 0 \quad 0]\mathbf{T}_{(4)}$$

and

$$\begin{aligned} g_1(t) = & [1.00000000 \quad 0.77880078 \quad 0.60653066 \quad 0.47236655]\mathbf{S}_{(4)} \\ & + [-0.22119922 \quad -0.17227012 \quad -0.13416411 \quad -0.10448711]\mathbf{T}_{(4)} \end{aligned}$$

Using Eq. (7.18) or (7.21), convolution of $u(t)$ and $g_1(t)$ in hybrid function domain yields the output as

$$\begin{aligned} y_{1c}(t) = & [0.00000000 \quad 0.22235010 \quad 0.39551653 \quad 0.53037868]\mathbf{S}_{(4)} \\ & + [0.22235010 \quad 0.17316643 \quad 0.13486215 \quad 0.10503075]\mathbf{T}_{(4)} \end{aligned}$$

Direct expansion of the output $y_{1d}(t)$, in HF domain, for $m = 4$ and $T = 1$ s, is

$$\begin{aligned} y_{1d}(t) = & [0.00000000 \quad 0.22119922 \quad 0.39346934 \quad 0.52763345]\mathbf{S}_{(4)} \\ & + [0.22119922 \quad 0.17227012 \quad 0.13416411 \quad 0.10448711]\mathbf{T}_{(4)} \end{aligned}$$

We compute percentage errors at different sample points of the function $y_{1c}(t)$ and compare the same with respective reference samples of the function $y_{1d}(t)$ for $m = 4$ and $m = 10$. These are presented in Tables 11.1 and 11.2. Also, percentage errors at different sample points for different values of m are plotted in Fig. 11.3 to

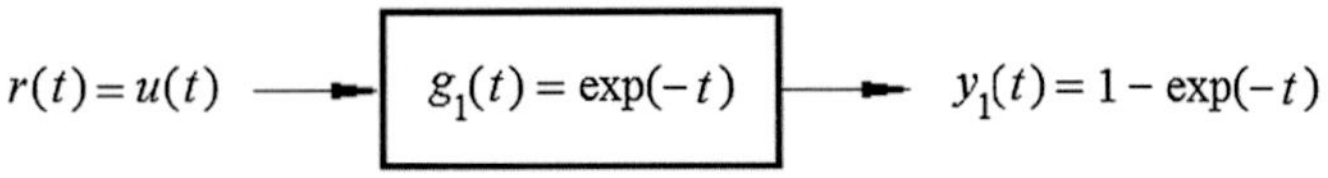

Fig. 11.2 System having an impulse response $\exp(-t)$ with a unit step input

Table 11.1 Percentage error of different samples of $y_{1c}(t)$ for Example 11.1, with $T = 1$ s, $m = 4$

t (s)	Via direct expansion in HF domain (y_{1d})	Via convolution in HF domain, using (7.21) (y_{1c})	% Error $\varepsilon = \frac{(y_{1d}-y_{1c})}{y_{1d}} \times 100$
0	0.00000000	0.00000000	–
$\frac{1}{4}$	0.22119922	0.22235010	−0.52029160
$\frac{2}{4}$	0.39346934	0.39551653	−0.52029160
$\frac{3}{4}$	0.52763345	0.53037868	−0.52029160
$\frac{4}{4}$	0.63212056	0.63540943	−0.52029160

Table 11.2 Percentage error of different samples of $y_{1c}(t)$ for Example 11.1, with $T = 1$ s, $m = 10$

t (s)	Via direct expansion in HF domain (y_{1d})	Via convolution in HF Domain, using (7.21) (y_{1c})	% Error $\varepsilon = \frac{(y_{1d}-y_{1c})}{y_{1d}} \times 100$
0	0.00000000	0.00000000	–
$\frac{1}{10}$	0.09516258	0.09524187	−0.08331945
$\frac{2}{10}$	0.18126925	0.18142028	−0.08331945
$\frac{3}{10}$	0.25918178	0.25939773	−0.08331945
$\frac{4}{10}$	0.32967995	0.32995464	−0.08331945
$\frac{5}{10}$	0.39346934	0.39379718	−0.08331945
$\frac{6}{10}$	0.45118836	0.45156429	−0.08331945
$\frac{7}{10}$	0.50341470	0.50383414	−0.08331945
$\frac{8}{10}$	0.55067104	0.55112985	−0.08331945
$\frac{9}{10}$	0.59343034	0.59392478	−0.08331945
$\frac{10}{10}$	0.63212056	0.63264724	−0.08331945

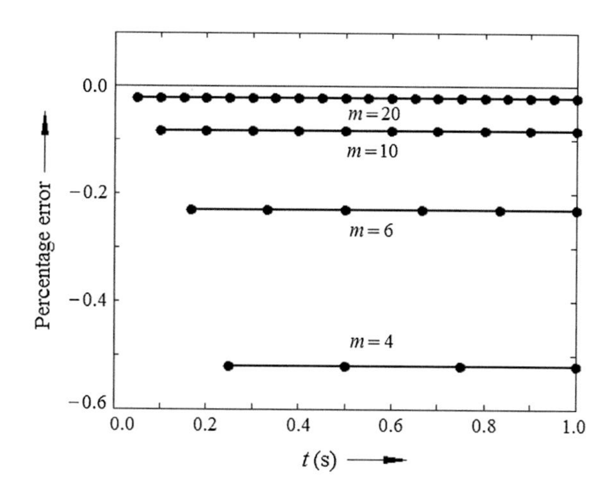

Fig. 11.3 Percentage error at different sample points for different values of m, with $T = 1$ s for Example 11.1

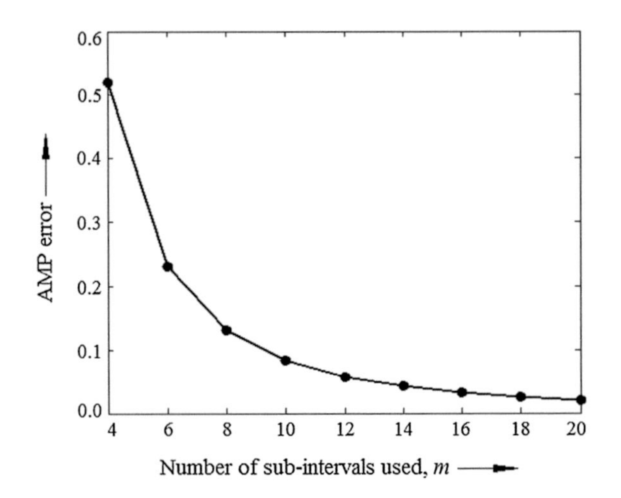

Fig. 11.4 AMP Error for different values of m, with $T = 1$ s for Example 11.1

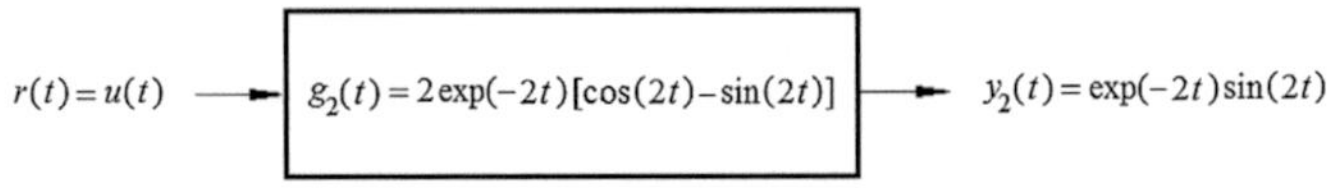

$r(t) = u(t)$ → $g_2(t) = 2\exp(-2t)[\cos(2t) - \sin(2t)]$ → $y_2(t) = \exp(-2t)\sin(2t)$

Fig. 11.5 Open loop SISO system with the impulse response $2\exp(-2t)[\cos(2t) - \sin(2t)]$ and a unit step input

show that with increasing m, error reduces at a very fast rate. Figure 11.4 shows the AMP error for different values of m. The curve resembles a rectangular hyperbola. It is seen that for $m = 20$, average absolute percentage error is less than 0.05 %.

Example 11.2 (vide Appendix B, Program no. 29) Consider the open loop system, shown in Fig. 11.5, with an impulse response given by

$$g_2(t) = 2\exp(-2t)[\cos(2t) - \sin(2t)]$$

Taking $T = 1$ s, $m = 4$ and we analyze the system in HF domain for a step input. The exact solution is $y_2(t) = \exp(-2t)\sin(2t)$.

In HF domain, for $m = 4$ and $T = 1$ s, $u(t)$ and $g_2(t)$ are given by

$$u(t) = [1 \quad 1 \quad 1 \quad 1]\mathbf{S}_{(4)} + [0 \quad 0 \quad 0 \quad 0]\mathbf{T}_{(4)}$$

and

$$\begin{aligned} g_2(t) = {} & [2.00000000 \quad 0.48298888 \quad -0.22158753 \quad -0.41357523]\mathbf{S}_{(4)} \\ & + [-1.51701112 \quad -0.70457641 \quad -0.19198770 \quad 0.05481648]\mathbf{T}_{(4)} \end{aligned}$$

Convolution of $u(t)$ and $g_2(t)$ in HF domain yields the output $y_2(t)$ as

$$y_{2c}(t) = [0.00000000 \quad 0.31037361 \quad 0.34304878 \quad 0.26365344]\mathbf{S}_{(4)} + [0.31037361 \quad 0.03267517 \quad -0.07939534 \quad -0.09654175]\mathbf{T}_{(4)}$$

Direct expansion of the output $y_{2d}(t)$, in HF domain, for $m = 4$ and $T = 1$ s, is

$$y_{2d}(t) = [0.00000000 \quad 0.29078628 \quad 0.30955988 \quad 0.22257122]\mathbf{S}_{(4)} + [0.29078629 \quad 0.01877359 \quad -0.08698866 \quad -0.09951119]\mathbf{T}_{(4)}$$

Now we compare the corresponding sample points of $y_{2c}(t)$ and $y_{2d}(t)$, of example 2, and compute percentage errors at different sample points with reference to the samples of $y_{2d}(t)$ for $m = 4$ and $m = 10$. The results of comparison are presented in tabular form in Tables 11.3 and 11.4.

Table 11.3 Percentage error of different samples of $y_{2c}(t)$ for Example 11.2, with $T = 1$ s, $m = 4$ (vide Appendix B, Program no. 29)

t (s)	Via direct expansion in HF domain (y_{2d})	Via convolution in HF domain, using (7.21) (y_{2c})	% Error $\varepsilon = \frac{(y_{2d}-y_{2c})}{y_{2d}} \times 100$
0	0.00000000	0.00000000	–
$\frac{1}{4}$	0.29078628	0.31037361	−6.73598553
$\frac{2}{4}$	0.30955988	0.34304878	−10.81823151
$\frac{3}{4}$	0.22257122	0.26365344	−18.45801075
$\frac{4}{4}$	0.12306002	0.16711169	−35.79689134

Table 11.4 Percentage errors of samples of $y_{2c}(t)$ for Example 11.2, with $T = 1$ s, $m = 10$ (vide Appendix B, Program no. 29)

t (s)	Via direct expansion in HF domain (y_{2d})	Via convolution in HF domain, using (7.21) (y_{2c})	% Error $\varepsilon = \frac{(y_{2d}-y_{2c})}{y_{2d}} \times 100$
0	0.00000000	0.00000000	–
$\frac{1}{10}$	0.16265669	0.16397540	−0.81072892
$\frac{2}{10}$	0.26103492	0.26358786	−0.97800816
$\frac{3}{10}$	0.30988236	0.31353208	−1.17777597
$\frac{4}{10}$	0.32232887	0.32691139	−1.42168985
$\frac{5}{10}$	0.30955988	0.31490417	−1.72641797
$\frac{6}{10}$	0.28072478	0.28666632	−2.11650235
$\frac{7}{10}$	0.24300891	0.24939830	−2.62928134
$\frac{8}{10}$	0.20181043	0.20851818	−3.32378529
$\frac{9}{10}$	0.16097593	0.16789439	−4.29781955
$\frac{10}{10}$	0.12306002	0.13010323	−5.72338986

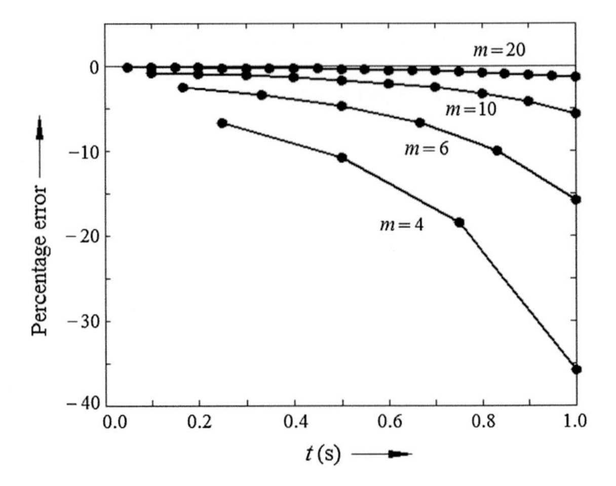

Fig. 11.6 Percentage error at different sample points for different values of m, with $T = 1$ s for Example 11.2 (vide Appendix B, Program no. 29)

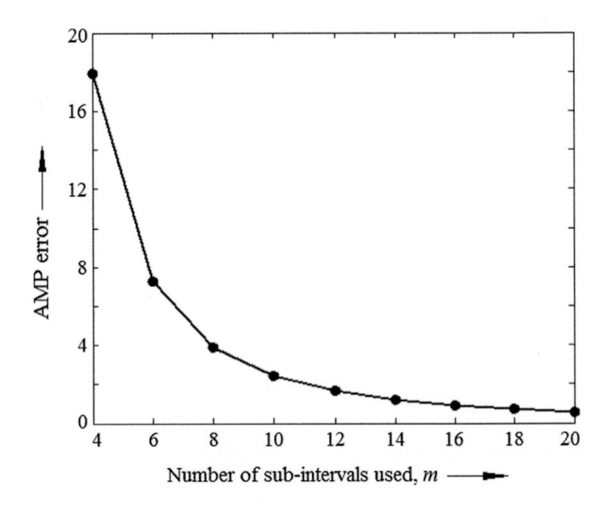

Fig. 11.7 AMP Error for different values of m, with $T = 1$ s for Example 11.2 (vide Appendix B, Program no. 29)

Also, percentage errors at different sample points for different values of m ($m = 4, 6, 10, 20$) are plotted in Fig. 11.6 to show that with increasing m, error reduces quite rapidly.

Figure 11.7 shows the AMP error for several values of m. It is seen that for $m = 20$, AMP error is less than 1.0 %.

11.2 Analysis of a Closed Loop System

Consider a single-input-single-output (SISO) time-invariant system [1].

An input $r(t)$ is applied to the system at $t = 0$. The block diagram of the system using time variables is shown in Fig. 11.8. Application of $r(t)$ to the system $g(t)$ with feedback $h(t)$ produces the corresponding output $y(t)$ for $t \geq 0$.

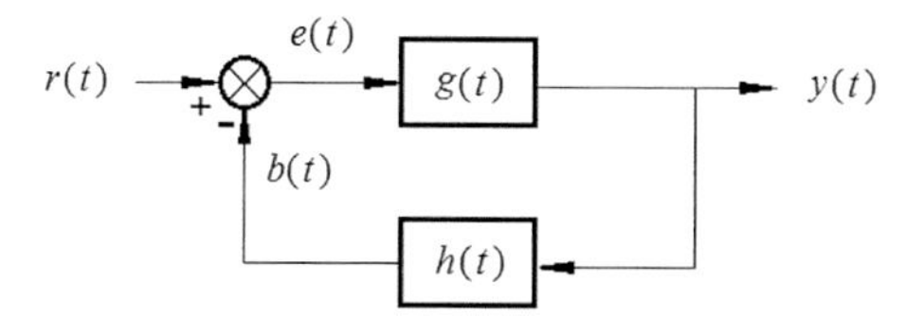

Fig. 11.8 Block diagram of a closed loop control system

Considering $r(t)$, $g(t)$, $y(t)$ and $h(t)$ to be bounded (i.e. the system is BIBO stable) and absolutely integrable over $t \in [0, T)$, all these functions may be expanded via HF series. For $m = 4$, we can write and

$$\left.\begin{aligned} r(t) &\triangleq \mathbf{R}_S^T \mathbf{S}_{(4)} + \mathbf{R}_T^T \mathbf{T}_{(4)} \\ g(t) &\triangleq \mathbf{G}_S^T \mathbf{S}_{(4)} + \mathbf{G}_T^T \mathbf{T}_{(4)} \\ y(t) &\triangleq \mathbf{Y}_S^T \mathbf{S}_{(4)} + \mathbf{Y}_T^T \mathbf{T}_{(4)} \\ \text{and}\, h(t) &\triangleq \mathbf{H}_S^T \mathbf{S}_{(4)} + \mathbf{H}_T^T \mathbf{T}_{(4)} \end{aligned}\right\} \tag{11.1}$$

Where

$$\begin{array}{ll} \mathbf{R}_S^T = [\,r_0 \quad r_1 \quad r_2 \quad r_3\,] & \mathbf{R}_T^T = [\,(r_1 - r_0) \quad (r_2 - r_1) \quad (r_3 - r_2) \quad (r_4 - r_3)\,] \\ \mathbf{G}_S^T = [\,g_0 \quad g_1 \quad g_2 \quad g_3\,] & \mathbf{G}_T^T = [\,(g_1 - g_0) \quad (g_2 - g_1) \quad (g_3 - g_2) \quad (g_4 - g_3)\,] \\ \mathbf{Y}_S^T = [\,y_0 \quad y_1 \quad y_2 \quad y_3\,] & \mathbf{Y}_T^T = [\,(y_1 - y_0) \quad (y_2 - y_1) \quad (y_3 - y_2) \quad (y_4 - y_3)\,] \\ \mathbf{H}_S^T = [\,h_0 \quad h_1 \quad h_2 \quad h_3\,] & \mathbf{H}_T^T = [\,(h_1 - h_0) \quad (h_2 - h_1) \quad (h_3 - h_2) \quad (h_4 - h_3)\,] \end{array}$$

Output of the feedback system is $b(t) = y(t) * h(t)$.
Following Eq. (7.17), we can write

$$b(t) = y(t) * h(t)$$

$$\begin{aligned} b(t) &= \frac{h}{6}[\,y_0 \quad y_1 \quad y_2 \quad y_3\,] \begin{bmatrix} 0 & H_0 & H_1 & H_2 \\ 0 & H_4 & H_5 & H_6 \\ 0 & 0 & H_4 & H_5 \\ 0 & 0 & 0 & H_4 \end{bmatrix} \mathbf{S}_{(4)} \\ &+ \frac{h}{6}\left\{ [\,y_0 \quad y_1 \quad y_2 \quad y_3\,] \begin{bmatrix} H_0 & (H_1 - H_0) & (H_2 - H_1) & (H_3 - H_2) \\ 0 & H_0 & (H_1 - H_0) & (H_2 - H_1) \\ 0 & 0 & H_0 & (H_1 - H_0) \\ 0 & 0 & 0 & H_0 \end{bmatrix} \right. \\ &\left. + [\,y_1 \quad y_2 \quad y_3 \quad y_4\,] \begin{bmatrix} H_4 & H_8 & H_9 & H_{10} \\ 0 & H_4 & H_8 & H_9 \\ 0 & 0 & H_4 & H_8 \\ 0 & 0 & 0 & H_4 \end{bmatrix} \right\} \mathbf{T}_{(4)} \end{aligned} \tag{11.2}$$

where

$$\left.\begin{array}{ll} H_0 \triangleq 2h_1 + h_0 & H_1 \triangleq 2h_2 + h_1 \\ H_2 \triangleq 2h_3 + h_2 & H_3 \triangleq 2h_4 + h_3 \\ H_4 \triangleq h_1 + 2h_0 & H_5 \triangleq h_2 + 4h_1 + h_0 \\ H_6 \triangleq h_3 + 4h_2 + h_1 & H_7 \triangleq h_4 + 4h_3 + h_2 \\ H_8 \triangleq h_2 + h_1 - 2h_0 & H_9 \triangleq h_3 + h_2 - 2h_1 \\ H_{10} \triangleq h_4 + h_3 - 2h_2 & \end{array}\right\} \tag{11.3}$$

After simplification, we have

$$\begin{aligned} b(t) = & \frac{h}{6}[0 \quad (H_0y_0 + H_4y_1) \quad (H_1y_0 + H_5y_1 + H_4y_2) \quad (H_2y_0 + H_6y_1 + H_5y_2 + H_4y_3)]\mathbf{S}_{(4)} \\ & + \frac{h}{6}[\{H_0y_0 + H_4y_1\} \quad \{(H_1 - H_0)y_0 + (H_0 + H_8)y_1 + H_4y_2\} \\ & \{(H_2 - H_1)y_0 + (H_1 - H_0 + H_9)y_1 + (H_0 + H_8)y_2 + H_4y_3\} \\ & \{(H_3 - H_2)y_0 + (H_2 - H_1 + H_{10})y_1 + (H_1 - H_0 + H_9)y_2 + (H_0 + H_8)y_3 + H_4y_4\}]\mathbf{T}_{(4)} \end{aligned} \tag{11.4}$$

Now the error signal of the system is

$$\begin{aligned} e(t) &= r(t) - b(t) \\ &= \Big[r_0 \quad r_1 - \tfrac{h}{6}(H_0y_0 + H_4y_1) \quad r_2 - \tfrac{h}{6}(H_1y_0 + H_5y_1 + H_4y_2) \\ &\qquad r_3 - \tfrac{h}{6}(H_2y_0 + H_6y_1 + H_5y_2 + H_4y_3)\Big]\mathbf{S}_{(4)} \\ &\quad + \Big[(r_1 - r_0) - \tfrac{h}{6}(H_0y_0 + H_4y_1) \quad (r_2 - r_1) - \tfrac{h}{6}\{(H_1 - H_0)y_0 + (H_0 + H_8)y_1 + H_4y_2\} \\ &\qquad (r_3 - r_2) - \tfrac{h}{6}\{(H_2 - H_1)y_0 + (H_1 - H_0 + H_9)y_1 + (H_0 + H_8)y_2 + H_4y_3\} \\ &\qquad (r_4 - r_3) - \tfrac{h}{6}\{(H_3 - H_2)y_0 + (H_2 - H_1 + H_{10})y_1 + (H_1 - H_0 + H_9)y_2 \\ &\quad + (H_0 + H_8)y_3 + H_4y_4]\}\mathbf{T}_{(4)} \end{aligned} \tag{11.5}$$

Again, direct expansion of the error signal $e(t)$ in HF domain is

$$e(t) \triangleq [e_0 \quad e_1 \quad e_2 \quad e_3]\mathbf{S}_{(4)} + [(e_1 - e_0) \quad (e_2 - e_1) \quad (e_3 - e_2) \quad (e_4 - e_3)]\mathbf{T}_{(4)} \tag{11.6}$$

Comparing Eqs. (11.5) and (11.6), HF coefficients of error $e(t)$ are

$$\begin{aligned}
e_0 &= r_0 \\
e_1 &= r_1 - \frac{h}{6}(H_0 y_0 + H_4 y_1) \\
e_2 &= r_2 - \frac{h}{6}(H_1 y_0 + H_5 y_1 + H_4 y_2) \\
e_3 &= r_3 - \frac{h}{6}(H_2 y_0 + H_6 y_1 + H_5 y_2 + H_4 y_3) \\
e_4 &= r_4 - \frac{h}{6}\{H_3 y_0 + (H_2 - H_1 + H_{10} + H_6) y_1 + (H_1 - H_0 + H_9 + H_5) y_2 \\
&\quad + (H_0 + H_8 + H_4) y_3 + H_4 y_4\} \\
&= r_4 - \frac{h}{6}(H_3 y_0 + H_7 y_1 + H_6 y_2 + H_5 y_3 + H_4 y_4)
\end{aligned} \tag{11.7}$$

Hence, the output $y(t)$ of the system is

$$y(t) = e(t) * g(t)$$

Thus, following Eq. (7.17), we can write

$$\begin{aligned}
y(t) \approx{}& \frac{h}{6}\begin{bmatrix} e_0 & e_1 & e_2 & e_3 \end{bmatrix} \begin{bmatrix} 0 & G_0 & G_1 & G_2 \\ 0 & G_4 & G_5 & G_6 \\ 0 & 0 & G_4 & G_5 \\ 0 & 0 & 0 & G_4 \end{bmatrix} \mathbf{S}_{(4)} \\
&+ \frac{h}{6}\left\{ \begin{bmatrix} e_0 & e_1 & e_2 & e_3 \end{bmatrix} \begin{bmatrix} G_0 & (G_1 - G_0) & (G_2 - G_1) & (G_3 - G_2) \\ 0 & G_0 & (G_1 - G_0) & (G_2 - G_1) \\ 0 & 0 & G_0 & (G_1 - G_0) \\ 0 & 0 & 0 & G_0 \end{bmatrix} \right. \\
&\left. + \frac{h}{6}\begin{bmatrix} e_1 & e_2 & e_3 & e_4 \end{bmatrix} \begin{bmatrix} G_4 & G_8 & G_9 & G_{10} \\ 0 & G_4 & G_8 & G_9 \\ 0 & 0 & G_4 & G_8 \\ 0 & 0 & 0 & G_4 \end{bmatrix} \right\} \mathbf{T}_{(4)}
\end{aligned} \tag{11.8}$$

where

$$\left.\begin{array}{ll}
G_0 \triangleq 2g_1 + g_0 & G_1 \triangleq 2g_2 + g_1 \\
G_2 \triangleq 2g_3 + g_2 & G_3 \triangleq 2g_4 + g_3 \\
G_4 \triangleq g_1 + 2g_0 & G_5 \triangleq g_2 + 4g_1 + g_0 \\
G_6 \triangleq g_3 + 4g_2 + g_1 & G_7 \triangleq g_4 + 4g_3 + g_2 \\
G_8 \triangleq g_2 + g_1 - 2g_0 & G_9 \triangleq g_3 + g_2 - 2g_1 \\
G_{10} \triangleq g_4 + g_3 - 2g_2 &
\end{array}\right\} \tag{11.9}$$

and

$$\left.\begin{array}{ll} E_0 \triangleq 2e_1 + e_0 & E_1 \triangleq 2e_2 + e_1 \\ E_2 \triangleq 2e_3 + e_2 & E_3 \triangleq 2e_4 + e_3 \\ E_4 \triangleq e_1 + 2e_0 & E_5 \triangleq e_2 + 4e_1 + e_0 \\ E_6 \triangleq e_3 + 4e_2 + e_1 & E_7 \triangleq e_4 + 4e_3 + e_2 \\ E_8 \triangleq e_2 + e_1 - 2e_0 & E_9 \triangleq e_3 + e_2 - 2e_1 \\ E_{10} \triangleq e_4 + e_3 - 2e_2 & \end{array}\right\} \tag{11.10}$$

Hence,

$$\begin{aligned} y(t) \approx \frac{h}{6}[0 \quad (G_0e_0 + G_4e_1) \quad (G_1e_0 + G_5e_1 + G_4e_2) \\ (G_2e_0 + G_6e_1 + G_5e_2 + G_4e_3)]\mathbf{S}_{(4)} \\ + \frac{h}{6}[\{G_0e_0 + G_4e_1\} \quad \{(G_1 - G_0)e_0 + (G_0 + G_8)e_1 + G_4e_2\} \\ \{(G_2 - G_1)e_0 + (G_1 - G_0 + G_9)e_1 + (G_0 + G_8)e_2 + G_4e_3\} \\ \{(G_3 - G_2)e_0 + (G_2 - G_1 + G_{10})e_1 + (G_1 - G_0 + G_9)e_2 + (G_0 + G_8)e_3 + G_4e_4\}]\mathbf{T}_{(4)} \end{aligned}$$

$$\begin{aligned} \text{or,} \quad y(t) \approx \frac{h}{6}[0 \quad (G_0e_0 + G_4e_1) \quad (G_1e_0 + G_5e_1 + G_4e_2) \\ (G_2e_0 + G_6e_1 + G_5e_2 + G_4e_3)]\mathbf{S}_{(4)} \\ \frac{h}{6}[\{G_0e_0 + G_4e_1\} \quad \{(G_1 - G_0)e_0 + (G_5 - G_4)e_1 + G_4e_2\} \\ \{(G_2 - G_1)e_0 + (G_6 - G_5)e_1 + (G_5 - G_4)e_2 + G_4e_3\} \\ \{(G_3 - G_2)e_0 + (G_7 - G_6)e_1 + (G_6 - G_5)e_2 + (G_5 - G_4)e_3 + G_4e_4\}]\mathbf{T}_{(4)} \end{aligned} \tag{11.11}$$

Since convolution operation is commutative, using Eq. (11.10) instead of (11.9), an alternative expression for the output $y(t)$ can be

$$\begin{aligned} y(t) \approx \frac{h}{6}[0 \quad (E_0g_0 + E_4g_1) \quad (E_1g_0 + E_5g_1 + E_4g_2) \\ \times (E_2g_0 + E_6g_1 + E_5g_2 + E_4g_3)]\mathbf{S}_{(4)} \\ + \frac{h}{6}[\{E_0g_0 + E_4g_1\} \quad \{(E_1 - E_0)g_0 + (E_5 - E_4)g_1 + E_4g_2\} \\ \times \{(E_2 - E_1)g_0 + (E_6 - E_5)g_1 + (E_5 - E_4)g_2 + E_4g_3\} \\ \times \{(E_3 - E_2)g_0 + (E_7 - E_6)g_1 + (E_6 - E_5)g_2 + (E_5 - E_4)g_3 + E_4g_4\}]\mathbf{T}_{(4)} \end{aligned} \tag{11.12}$$

Again, direct expansion of the output $y(t)$ of the system in HF domain is

$$\begin{aligned} y(t) \triangleq [y_0 \quad y_1 \quad y_2 \quad y_3]\mathbf{S}_{(4)} \\ + [(y_1 - y_0) \quad (y_2 - y_1) \quad (y_3 - y_2) \quad (y_4 - y_3)]\mathbf{T}_{(4)} \end{aligned} \tag{11.13}$$

We equate respective coefficients of $y(t)$ from Eqs. (11.11) and (11.13), and use Eqs. (11.3), (11.7), (11.9) and (11.10) to determine the five output coefficients

(*i*)
$$y_0 = 0 \tag{11.14}$$

(*ii*)
$$\begin{aligned} y_1 &\approx \frac{h}{6}(G_0 e_0 + G_4 e_1) \\ &= \frac{h}{6}\left[G_0 r_0 + G_4 r_1 - \frac{h}{6}(G_4 H_0) y_0 - \frac{h}{6}(G_4 H_4) y_1\right] \end{aligned}$$

Solving for y_1

$$y_1 = \frac{h}{6}\frac{1}{\left\{1 + \frac{h^2}{36} G_4 H_4\right\}}\left[G_0 r_0 + G_4 r_1 - \frac{h}{6}(G_4 H_0) y_0\right] \tag{11.15}$$

(*iii*)
$$\begin{aligned} y_2 &\approx \frac{h}{6}(G_1 e_0 + G_5 e_1 + G_4 e_2) \\ &= \frac{h}{6}\left[G_1 r_0 + \left\{r_1 - \frac{h}{6}(H_0 y_0 + H_4 y_1) G_5\right\} + \left\{r_2 - \frac{h}{6}(H_1 y_0 + H_5 y_1 + H_4 y_2) G_4\right\}\right] \end{aligned}$$

Solving for y_2

$$\begin{aligned} y_2 = \frac{h}{6}\frac{1}{\left\{1 + \frac{h^2}{36} G_4 H_4\right\}}\bigg[& G_1 r_0 + G_5 r_1 + G_4 r_2 - \frac{h}{6}(G_5 H_0 + G_4 H_1) y_0 \\ & - \frac{h}{6}(G_5 H_4 + G_4 H_5) y_1\bigg] \end{aligned} \tag{11.16}$$

(*iv*)
$$\begin{aligned} y_3 &\approx \frac{h}{6}(G_2 e_0 + G_6 e_1 + G_5 e_2 + G_4 e_3) \\ &= \frac{h}{6}\left[G_2 r_0 + \left\{r_1 - \frac{h}{6}(H_0 y_0 + H_4 y_1) G_6\right\} + \left\{r_2 - \frac{h}{6}(H_1 y_0 + H_5 y_1 + H_4 y_2) G_5\right\}\right. \\ &\quad \left. + \left\{r_3 - \frac{h}{6}(H_2 y_0 + H_6 y_1 + H_5 y_2 + H_4 y_3) G_4\right\}\right] \end{aligned}$$

Solving for y_3

$$y_3 = \frac{h}{6}\frac{1}{\left\{1+\frac{h^2}{36}G_4H_4\right\}}\left[G_2r_0+G_6r_1+G_5r_2-\frac{h}{6}(G_6H_0+G_5H_1+G_4H_2)y_0 - \frac{h}{6}(G_6H_4+G_5H_5+G_4H_6)y_1-\frac{h}{6}(G_5H_4+G_4H_5)y_2\right] \tag{11.17}$$

(v)

$$\begin{aligned} y_4 &\approx \frac{h}{6}[(G_3-G_2)e_0+(G_7-G_6)e_1+(G_6-G_5)e_2+(G_5-G_4)e_3+G_4e_4] \\ &\quad + \frac{h}{6}[G_2e_0+G_6e_1+G_5e_2+G_4e_3] \\ &= \frac{h}{6}[G_3e_0+G_7e_1+G_6e_2+G_5e_3+G_4e_4] \\ &= \frac{h}{6}\left[G_3r_0+G_7r_1+G_6r_2+G_5r_3+G_4r_4+\frac{h}{6}(G_7H_0+G_6H_1+G_5H_2+G_4H_3)y_0 \right. \\ &\quad -\frac{h}{6}(G_7H_4+G_6H_5+G_5H_6+G_4H_7)y_1-\frac{h}{6}(G_6H_4+G_5H_5+G_4H_6)y_2 \\ &\quad \left. -\frac{h}{6}(G_5H_4+G_4H_5)y_3-\frac{h}{6}G_4H_4\right] \end{aligned}$$

Solving for y_4

$$y_4 = \frac{h}{6}\frac{1}{\left\{1+\frac{h^2}{36}G_4H_4\right\}}\left[G_3r_0+G_7r_1+G_6r_2+G_5r_3+G_4r_4-\frac{h}{6}(G_7H_0+G_6H_1+G_5H_2+G_4H_3)y_0 - \frac{h}{6}(G_7H_4+G_6H_5+G_5H_6+G_4H_7)y_1-\frac{h}{6}(G_6H_4+G_5H_5+G_4H_6)y_2 - \frac{h}{6}(G_5H_4+G_4H_5)y_3\right] \tag{11.18}$$

To write down the generalized form of the output coefficients, first we express $G_i's$ and $H_i's$ by their following general forms:

$$G_i = \begin{cases} 2g_{(i+1)}+g_i & \text{for } i=0 \text{ to } (m-1) \\ g_1+2g_0 & \text{for } i=m \\ g_{(i-m+1)}+4g_{(i-m)}+g_{(i-m-1)} & \text{for } i=(m+1) \text{ to } (2m-1) \\ g_{(i-2m+2)}+g_{(i-2m+1)}-2g_{(i-2m)} & \text{for } i=2m \text{ to } (3m-2) \end{cases} \tag{11.19}$$

$$H_i = \begin{cases} 2h_{(i+1)} + h_i & \text{for} \quad i = 0 \text{ to } (m-1) \\ h_1 + 2h_0 & \text{for} \quad i = m \\ h_{(i-m+1)} + 4h_{(i-m)} + h_{(i-m-1)} & \text{for} \quad i = (m+1) \text{ to } (2m-1) \\ h_{(i-2m+2)} + h_{(i-2m+1)} - 2h_{(i-2m)} & \text{for} \quad i = 2m \text{ to } (3m-2) \end{cases} \tag{11.20}$$

It is noted that $G_i's$ and $H_i's$ have the same general form.

Now, we use Eqs. (11.14)–(11.20) to write down the generalized form of the output coefficients as

$$y_i = \frac{\frac{h}{6}\left[G_{(i-1)}r_0 + \sum\limits_{p=1}^{i} G_{(m+i-p)}r_p - \frac{h}{6}\sum\limits_{p=1}^{i} G_{(m+i-p)}H_{(p-1)}y_0 - \frac{h}{6}\sum\limits_{p=2}^{i} K_{(i-p+2)}y_{(p-1)}\right]}{1 + \frac{h^2}{36}G_m H_m} \tag{11.21}$$

where $K_j = \sum\limits_{p=1}^{j} G_{(m+j-p)}H_{(m+p-1)}$ and $j = i - p + 2$.

11.2.1 Numerical Examples

To analyse a closed loop system, we use Eqs. (11.14)–(11.21) to determine the output coefficients in HF domain. The results are compared with direct expression of the output of the system in HF domain.

Example 11.3 (vide Appendix B, Program no. 30) Consider a closed loop system with input $r(t)$, plant $g_3(t)$, feedback $h_3(t)$ and output $y_3(t)$, shown in Fig. 11.9.

Taking $T = 1$ s and $m = 4$, we analyse the system in HF domain for a step input $u(t)$ and feedback $h_3(t) = 4u(t)$ with its impulse response given by $g_3(t) = 2\exp(-4t)$. The exact solution of the system is $y_3(t) = \exp(-2t)\sin(2t)$.

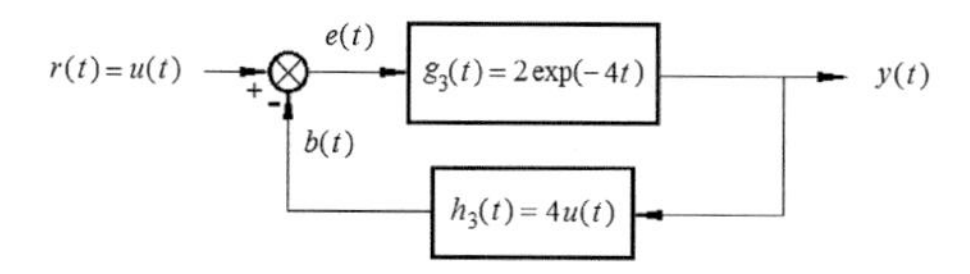

Fig. 11.9 A closed loop system with unit step input

Using Eqs. (11.19)–(11.21), the output coefficients of closed loop system $y_{3c}(t)$ in HF domain, for $m = 4$ and $T = 1$ s, is

$$\begin{aligned} y_{3c}(t) = & \left[0.00000000 \quad 0.31126040 \quad 0.33909062 \quad 0.24469593\right]\mathbf{S}_{(4)} \\ & + \left[0.31126040 \quad 0.02783022 \quad -0.09439469 \quad -0.11398545\right]\mathbf{T}_{(4)} \end{aligned}$$

Direct expansion of the output $y_{3d}(t)$, in HF domain, for $m = 4$ and $T = 1$ s, is

$$\begin{aligned} y_{3d}(t) = & \left[0.00000000 \quad 0.29078629 \quad 0.30955988 \quad 0.22257122\right]\mathbf{S}_{(4)} \\ & + \left[0.29078629 \quad 0.01877358 \quad -0.08698865 \quad -0.09951120\right]\mathbf{T}_{(4)} \end{aligned}$$

Now we compare the corresponding samples of $y_{3c}(t)$ and $y_{3d}(t)$ and compute percentage errors at different sample points with reference to the samples of $y_3(t)$

Table 11.5 Percentage error at different sample points of $y_{3c}(t)$ for Example 11.3, with $m = 4$, $T = 1$ s and (b) $m = 10$, $T = 1$ s (vide Appendix B, Program no. 30)

t (s)	Via direct expansion in HF domain (y_{3d})	Via convolution in HF domain, using (11.19)–(11.21) (y_{3c})	% Error $\varepsilon = \frac{(y_{3d} - y_{3c})}{y_{3d}} \times 100$
(a)			
0	0.00000000	0.00000000	–
$\frac{1}{4}$	0.29078629	0.31126040	−7.04094744
$\frac{2}{4}$	0.30955988	0.33909062	−9.53959161
$\frac{3}{4}$	0.22257122	0.24469593	−9.94050884
$\frac{4}{4}$	0.12306002	0.13071048	−6.21685153
(b)			
0	0.00000000	0.00000000	–
$\frac{1}{10}$	0.16265669	0.16411049	−0.89378245
$\frac{2}{10}$	0.26103492	0.26393607	−1.11140316
$\frac{3}{10}$	0.30988236	0.31388535	−1.29177880
$\frac{4}{10}$	0.32232887	0.32694277	−1.43142705
$\frac{5}{10}$	0.30955988	0.31428062	−1.52498673
$\frac{6}{10}$	0.28072478	0.28511656	−1.56444413
$\frac{7}{10}$	0.24300891	0.24674586	−1.53778446
$\frac{8}{10}$	0.20181043	0.20468938	−1.42656376
$\frac{9}{10}$	0.16097593	0.16290972	−1.20128964
$\frac{10}{10}$	0.12306002	0.12405920	−0.81193829

for $m = 4$ and $m = 10$. The results of computation are presented in tabular form in Table 11.5a, b. Figure 11.10a, b shows the graphical comparison of improvement in HF domain convolution with increasing m, for $m = 4$ and $m = 10$. Also, percentage errors at different sample points for different values of m ($m = 4$, 6, 10 and 20) are plotted in Fig. 11.11 to show that with increasing m, error reduces at a very fast rate.

Figure 11.12 shows AMP error for several different values of m. It is seen that for $m = 20$, AMP error is less than 1.0 %.

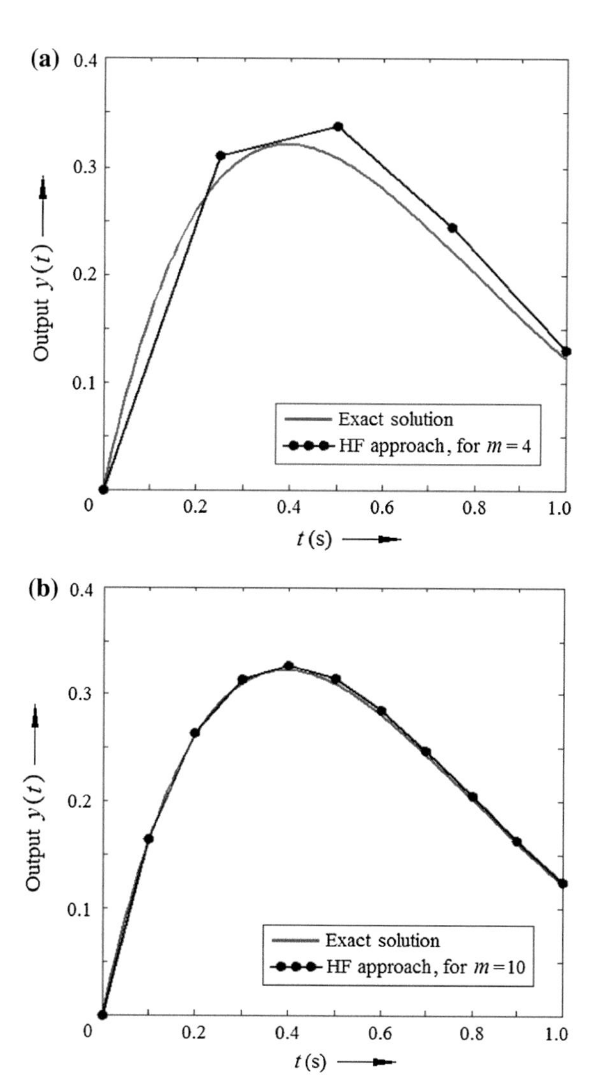

Fig. 11.10 Output of the closed loop system of Example 11.3, obtained via convolution in HF domain for **a** $m = 4$, $T = 1$ s and **b** $m = 10$, $T = 1$ s (vide Appendix B, Program no. 30)

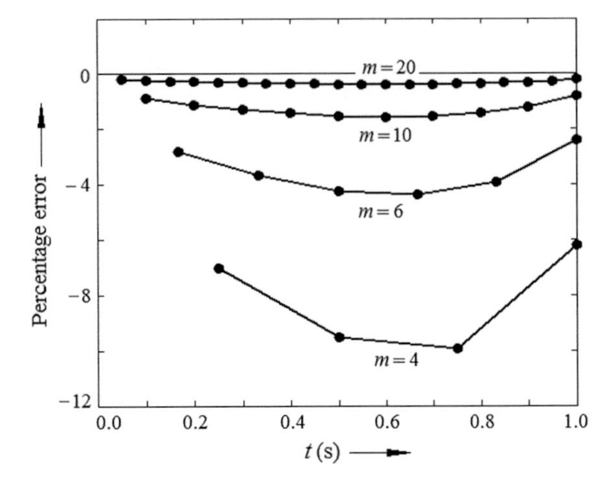

Fig. 11.11 Percentage error at different sample points for different values of m, with $T = 1$ s for Example 11.3 (vide Appendix B, Program no. 30)

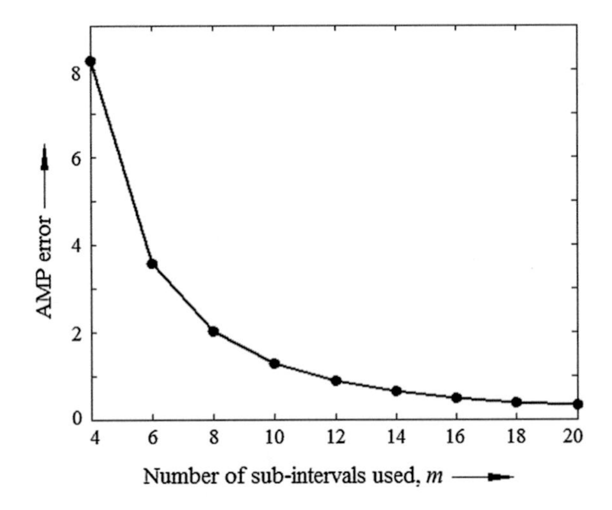

Fig. 11.12 AMP error for different values of m, with $T = 1$ s for Example 11.3 (vide Appendix B, Program no. 30)

11.3 Conclusion

In this chapter, we have studied the analysis of open loop as well as closed loop linear systems in hybrid function domain, employing the rules of HF domain convolution. As a foundation, convolution of basic component functions of the HF set was discussed in Chap. 7 and relevant sub-results were derived. These sub-results were subsequently used to determine the convolution of two time functions.

Using this theory, output of a linear open loop system is determined. For the open loop system, two examples are treated and related tables are presented for $m = 4$ and 10, m being the number of component functions used in HF set. For these two cases, percentage errors at different sample points are also computed. These are tabulated in Tables 11.1, 11.2, 11.3 and 11.4. It is observed that with increasing m, the error is rapidly reduced. This fact is also supported by Figs. 11.3, 11.4, 11.6, 11.7, 11.11 and 11.12.

For analyzing a closed loop control system the same technique has been employed and Eq. (11.8) has been derived. In this equation, the output is obtained from algebraic operations of some special matrices determined from the convolution operation in HF domain. However, apart from Eq. (11.8), an interesting *recursive* equation has been derived in the form of Eq. (11.21). A few relevant examples have been treated for $m = 4$ and 10, and the results are found to be in good agreement with the exact solutions. Percentage errors at different sample points are shown in Table 11.5a, b. As observed before, with increasing m, the error is reduced rapidly and this fact is apparent from Figs. 11.11 and 11.12.

References

1. Ogata, K.: Modern Control Engineering (5th Ed.). Prentice Hall of India (2011)
2. Deb, A., Sarkar, G., Bhattacharjee, M., Sen, S.K.: A new set of piecewise constant orthogonal functions for the analysis of linear SISO systems with sample-and-hold. J. Franklin Instt. **335B** (2), 333–358 (1998)
3. Deb, A., Sarkar, G., Sengupta, A.: Triangular Orthogonal Functions for the Analysis of Continuous Time Systems. Anthem Press, London (2011)
4. Biswas, A.: Analysis and synthesis of continuous control systems using a set of orthogonal hybrid functions. Ph. D. Dissertation, University of Calcutta (2015)

Chapter 12
System Identification Using State Space Approach: Time Invariant Systems

Abstract In this chapter, HF domain identification of time invariant systems in state space is presented. Both homogeneous and non-homogeneous systems are treated. State and output matrices of the systems are identified. A non-homogeneous system with a jump discontinuity at input is also identified. Illustration has been provided with the support of six examples, eleven figures and eleven tables.

In this chapter, we deal with linear time invariant (LTI) control systems [1]. We intend to identify two types of LTI systems in hybrid function platform, namely non-homogeneous system and homogeneous system [1, 2].

Identifying a control system [3] means, knowing the input signal or forcing function and the output of the system, we determine the system parameters. For a time invariant system, these parameters do not vary with time.

In a homogeneous system, though no external signal is applied, the initial conditions of system states may exist. While in a non-homogeneous system, we deal with a situation where the initial conditions and external input signals exist simultaneously.

Here, we solve the problem of system identification [4, 5], both for non-homogeneous systems and homogeneous systems, described as state space models, by using the concept of HF domain.

First, we take up the problem of identification [6] of a non-homogeneous system, because after putting the specific condition of zero forcing function in the solution, we can arrive at the result of identification of a homogeneous system quite easily.

12.1 Identification of a Non-homogeneous System [3]

For solving the identification problem, we refer to Eq. (8.20). That is

A. Deb et al., *Analysis and Identification of Time-Invariant Systems, Time-Varying Systems, and Multi-Delay Systems using Orthogonal Hybrid Functions*,
Studies in Systems, Decision and Control 46, DOI 10.1007/978-3-319-26684-8_12

$$\left[\frac{2}{h}\mathbf{I}-\mathbf{A}\right]\begin{bmatrix} c_{\mathrm{Sx}1(i+1)} \\ c_{\mathrm{Sx}2(i+1)} \\ \vdots \\ c_{\mathrm{Sx}n(i+1)} \end{bmatrix} - \left[\frac{2}{h}\mathbf{I}+\mathbf{A}\right]\begin{bmatrix} c_{\mathrm{Sx}1i} \\ c_{\mathrm{Sx}2i} \\ \vdots \\ c_{\mathrm{Sx}ni} \end{bmatrix} = 2\mathbf{B}\left(\mathbf{C}_{\mathrm{Su}}^{\mathrm{T}}+\frac{1}{2}\mathbf{C}_{\mathrm{Tu}}^{\mathrm{T}}\right)_i$$

$$\text{or,}\mathbf{A}\begin{bmatrix} c_{\mathrm{Sx}1i}+c_{\mathrm{Sx}1(i+1)} \\ c_{\mathrm{Sx}2i}+c_{\mathrm{Sx}2(i+1)} \\ \vdots \\ c_{\mathrm{Sx}ni}+c_{\mathrm{Sx}n(i+1)} \end{bmatrix} = \frac{2}{h}\begin{bmatrix} c_{\mathrm{Sx}1(i+1)}-c_{\mathrm{Sx}1i} \\ c_{\mathrm{Sx}2(i+1)}-c_{\mathrm{Sx}2i} \\ \vdots \\ c_{\mathrm{Sx}n(i+1)}-c_{\mathrm{Sx}ni} \end{bmatrix} - 2\mathbf{B}\left(\mathbf{C}_{\mathrm{Su}}^{\mathrm{T}}+\frac{1}{2}\mathbf{C}_{\mathrm{Tu}}^{\mathrm{T}}\right)_i \tag{12.1}$$

It is noted that, if we try to solve **A** from Eq. (12.1) we meet with one immediate difficulty. The coefficient matrix of **A** is a column matrix, and so no question of inversion arises.

To tackle this problem, we proceed to construct a coefficient matrix which is square and can be inverted. To achieve this end, we can write down $(n - 1)$ more equations similar to Eq. (12.1), by varying the index i from i to $(i + n - 1)$ by incrementing i by 1. For example, by incrementing i by 1, we can write down the following equation:

$$\mathbf{A}\begin{bmatrix} c_{\mathrm{Sx}1(i+1)}+c_{\mathrm{Sx}1(i+2)} \\ c_{\mathrm{Sx}2(i+1)}+c_{\mathrm{Sx}2(i+2)} \\ \vdots \\ c_{\mathrm{Sx}n(i+1)}+c_{\mathrm{Sx}n(i+2)} \end{bmatrix} = \frac{2}{h}\begin{bmatrix} c_{\mathrm{Sx}1(i+2)}-c_{\mathrm{Sx}1(i+1)} \\ c_{\mathrm{Sx}2(i+2)}-c_{\mathrm{Sx}2(i+1)} \\ \vdots \\ c_{\mathrm{Sx}n(i+2)}-c_{\mathrm{Sx}n(i+1)} \end{bmatrix} - 2\mathbf{B}\left(\mathbf{C}_{\mathrm{Su}}^{\mathrm{T}}+\frac{1}{2}\mathbf{C}_{\mathrm{Tu}}^{\mathrm{T}}\right)_{i+1} \tag{12.2}$$

From these n numbers of equations, we rearrange them in a manner to produce the following equation:

$$\begin{aligned} &\mathbf{A}\begin{bmatrix} c_{\mathrm{Sx}1i}+c_{\mathrm{Sx}1(i+1)} & c_{\mathrm{Sx}1(i+1)}+c_{\mathrm{Sx}1(i+2)} & \cdots & c_{\mathrm{Sx}1(i+n-1)}+c_{\mathrm{Sx}1(i+n)} \\ c_{\mathrm{Sx}2i}+c_{\mathrm{Sx}2(i+1)} & c_{\mathrm{Sx}2(i+1)}+c_{\mathrm{Sx}2(i+2)} & \cdots & c_{\mathrm{Sx}2(i+n-1)}+c_{\mathrm{Sx}2(i+n)} \\ \vdots & \vdots & & \vdots \\ c_{\mathrm{Sx}ni}+c_{\mathrm{Sx}n(i+1)} & c_{\mathrm{Sx}n(i+1)}+c_{\mathrm{Sx}n(i+2)} & \cdots & c_{\mathrm{Sx}n(i+n-1)}+c_{\mathrm{Sx}n(i+n)} \end{bmatrix}_{n\times n} \\ &= \frac{2}{h}\begin{bmatrix} c_{\mathrm{Sx}1(i+1)}-c_{\mathrm{Sx}1i} & c_{\mathrm{Sx}1(i+2)}-c_{\mathrm{Sx}1(i+1)} & \cdots & c_{\mathrm{Sx}1(i+n)}-c_{\mathrm{Sx}1(i+n-1)} \\ c_{\mathrm{Sx}2(i+1)}-c_{\mathrm{Sx}2i} & c_{\mathrm{Sx}2(i+2)}-c_{\mathrm{Sx}2(i+1)} & \cdots & c_{\mathrm{Sx}2(i+n)}-c_{\mathrm{Sx}2(i+n-1)} \\ \vdots & \vdots & & \vdots \\ c_{\mathrm{Sx}n(i+1)}-c_{\mathrm{Sx}ni} & c_{\mathrm{Sx}n(i+2)}-c_{\mathrm{Sx}n(i+1)} & \cdots & c_{\mathrm{Sx}n(i+n)}-c_{\mathrm{Sx}n(i+n-1)} \end{bmatrix}_{n\times n} \\ &- 2\mathbf{B}\left[\left(\mathbf{C}_{\mathrm{Su}}^{\mathrm{T}}+\frac{1}{2}\mathbf{C}_{\mathrm{Tu}}^{\mathrm{T}}\right)_i \left(\mathbf{C}_{\mathrm{Su}}^{\mathrm{T}}+\frac{1}{2}\mathbf{C}_{\mathrm{Tu}}^{\mathrm{T}}\right)_{i+1} \cdots \left(\mathbf{C}_{\mathrm{Su}}^{\mathrm{T}}+\tfrac{1}{2}\mathbf{C}_{\mathrm{Tu}}^{\mathrm{T}}\right)_{i+n-1}\right]_{1\times n} \end{aligned} \tag{12.3}$$

Now, in (12.3), we have the coefficient matrix of **A** which is square of order n and invertible.

It is evident from (12.3) that for solving n elements of the matrix $\mathbf{A}$, we need $(n + 1)$ samples of the states and subsequently use them to form an $(n \times n)$ coefficient matrix. That is, if $\mathbf{A}$ has a dimension (2×2), say, we need at least 3 samples of the states. Also, if we have m samples of the states, we normally use a set of consecutive samples from the states and the inputs, starting from any ith sample. Hence, if any system has the order n, the number of sub-intervals to be chosen should be greater than n to take advantage of Eq. (12.3). That is, $m \geq n$.

It is noted that the pattern of the coefficient matrix of $\mathbf{A}$ in (12.3) indicates that each column of the coefficient matrix uses a pair of consecutive samples. However, if we choose the samples of the states and the inputs erratically, to obtain reliable results, such erratic choice should not violate the indicated pattern requirement. Thus, it is a necessity to consider $2n$ number of samples, instead of $(n + 1)$ number of samples, to identify a system of order n.

Now, from Eq. (12.3), we have identified the system matrix $\mathbf{A}$ as,

$$\mathbf{A} = \left\{ \frac{2}{h} \begin{bmatrix} c_{Sx1(i+1)} - c_{Sx1i} & c_{Sx1(i+2)} - c_{Sx1(i+1)} & \cdots & c_{Sx1(i+n)} - c_{Sx1(i+n-1)} \\ c_{Sx2(i+1)} - c_{Sx2i} & c_{Sx2(i+2)} - c_{Sx2(i+1)} & \cdots & c_{Sx2(i+n)} - c_{Sx2(i+n-1)} \\ \vdots & \vdots & & \vdots \\ c_{Sxn(i+1)} - c_{Sxni} & c_{Sxn(i+2)} - c_{Sxn(i+1)} & \cdots & c_{Sxn(i+n)} - c_{Sxn(i+n-1)} \end{bmatrix} \right.$$
$$\left. -2\mathbf{B}\left[\left(\mathbf{C}_{Su}^{T} + \frac{1}{2}\mathbf{C}_{Tu}^{T}\right)_i \left(\mathbf{C}_{Su}^{T} + \frac{1}{2}\mathbf{C}_{Tu}^{T}\right)_{i+1} \cdots \left(\mathbf{C}_{Su}^{T} + \tfrac{1}{2}\mathbf{C}_{Tu}^{T}\right)_{i+n-1} \right] \right\} \tag{12.4}$$
$$\times \begin{bmatrix} c_{Sx1i} + c_{Sx1(i+1)} & c_{Sx1(i+1)} + c_{Sx1(i+2)} & \cdots & c_{Sx1(i+n-1)} + c_{Sx1(i+n)} \\ c_{Sx2i} + c_{Sx2(i+1)} & c_{Sx2(i+1)} + c_{Sx2(i+2)} & \cdots & c_{Sx2(i+n-1)} + c_{Sx2(i+n)} \\ \vdots & \vdots & & \vdots \\ c_{Sxni} + c_{Sxn(i+1)} & c_{Sxn(i+1)} + c_{Sxn(i+2)} & \cdots & c_{Sxn(i+n-1)} + c_{Sxn(i+n)} \end{bmatrix}^{-1}$$

12.1.1 Numerical Examples

Example 12.1 [1] (vide Appendix B, Program no. 31) Consider the system $\dot{\mathbf{x}}(t) = \mathbf{A}\mathbf{x}(t) + \mathbf{B}u(t)$ where $\mathbf{A} = \begin{bmatrix} 0 & 1 \\ -2 & -3 \end{bmatrix}$ $\mathbf{B} = \begin{bmatrix} 0 \\ 1 \end{bmatrix}$ and $\mathbf{x}_0 = \begin{bmatrix} 0 \\ 0.5 \end{bmatrix}$ with a unit step forcing function.

This system is identified using Eq. (12.4). The identified elements of system for increasing number of segments m, are tabulated in Table 12.1.

Table 12.1 Identification of the non-homogeneous system matrix **A** of Example 12.1 for different values of m = 4, 10, 12 15, 20, 25 tabulated with the actual elements of **A** (vide Appendix B, Program no. 31)

Elements of system matrix A	Exact values	HF domain solution for					
		$m = 4$	$m = 10$	$m = 12$	$m = 15$	$m = 20$	$m = 25$
a_{11}	0	0.0000	−0.0000	−0.0000	−0.0000	−0.0000	−0.0000
a_{12}	1	0.9948	0.9992	0.9994	0.9996	0.9998	0.9999
a_{21}	−2	−2.0000	−2.0000	−2.0000	−2.0000	−2.0000	−2.0000
a_{22}	−3	−2.9948	−2.9992	−2.9994	−2.9996	−2.9998	−2.9999

Table 12.2 Hybrid function based system identification for Example 12.1 with m = 10

Elements of system matrix A	Exact values (E)	HF domain solution for (m = 10) (H)	% Error $\varepsilon = \frac{E-H}{E} \times 100$
a_{11}	0	0.0000	–
a_{12}	1	0.9992	0.0800
a_{21}	−2	−2.0000	0.0000
a_{22}	−3	−2.9992	0.0266

The results are compared with the actual elements of **A** and corresponding percentage errors are computed

Table 12.2 shows the comparison of the actual elements of the system matrix with respective computed elements in HF domain along with percentage errors for m = 10 and T = 1 s.

Results obtained for Example 12.1, using Eq. (12.4), are plotted in Fig. 12.1. It is noted that, with an increase in m from 4 to 25, the HF domain solution (black dots) improves rapidly.

Example 12.2 Consider the system $\dot{\mathbf{x}}(t) = \mathbf{A}\mathbf{x}(t) + \mathbf{B}u(t)$ where $\mathbf{A} = \begin{bmatrix} 0 & 1 & 0 \\ 0 & 0 & 1 \\ -6 & -11 & -6 \end{bmatrix}$, $\mathbf{B} = \begin{bmatrix} 0 \\ 0 \\ 1 \end{bmatrix}$ and $\mathbf{x}_0 = \begin{bmatrix} 1 \\ 0 \\ 0 \end{bmatrix}$ with a unit step forcing function.

This system is identified using Eq. (12.4). The results are tabulated in Table 12.3.

Table 12.4 shows the comparison of the actual elements of the system matrix with respective computed elements in HF domain along with percentage errors for m = 12 and T = 1 s.

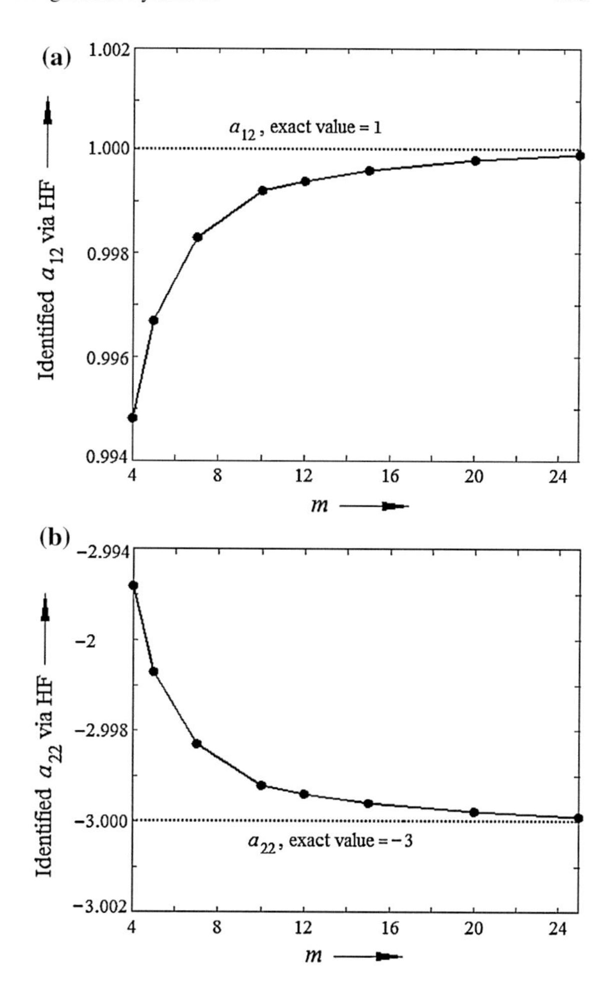

Fig. 12.1 Hybrid function domain system identification for increasing m, for the elements **a** a_{12} and **b** a_{22} (vide Appendix B, Program no. 31)

Table 12.3 Identification of the non-homogeneous system matrix **A** of Example 12.2 for different values of m = 4, 10, 12 15, 20, 25 tabulated with the actual elements of **A**

Elements of system matrix A	Exact values	HF domain solution for					
		$m = 4$	$m = 10$	$m = 12$	$m = 15$	$m = 20$	$m = 25$
a_{11}	0	0.0222	0.0041	0.0028	0.0018	0.0010	0.0007
a_{12}	1	1.0518	1.0091	1.0063	1.0041	1.0023	1.0015
a_{13}	0	0.0282	0.0049	0.0034	0.0022	0.0012	0.0008
a_{21}	0	−0.1423	−0.0246	−0.0172	−0.0110	−0.0062	−0.0040
a_{22}	0	−0.2978	−0.0501	−0.0348	−0.0223	−0.0126	−0.0080
a_{23}	1	0.8813	0.9795	0.9857	0.9908	0.9948	0.9967
a_{31}	−6	−5.3961	−5.8971	−5.9283	−5.9540	−5.9740	−5.9834
a_{32}	−11	−9.7937	−10.8000	−10.8610	−10.9110	−10.9500	−10.9680
a_{33}	−6	−5.5705	−5.9260	−5.9484	−5.9669	−5.9813	−5.9880

Table 12.4 Hybrid function based system identification for Example 12.2 with $m = 12$

Elements of system matrix A	Exact values (E)	HF domain solution for (m = 12) (H)	% Error $\varepsilon = \frac{E-H}{E} \times 100$
a_{11}	0	0.0028	–
a_{12}	1	1.0063	−0.6300
a_{13}	0	0.0034	–
a_{21}	0	−0.0172	–
a_{22}	0	−0.0348	–
a_{23}	1	0.9857	1.4300
a_{31}	−6	−5.9283	1.1950
a_{32}	−11	−10.8610	1.2636
a_{33}	−6	−5.9484	0.8600

The elements obtained via HF domain are compared with the actual elements of **A** and corresponding percentage errors are computed

Results obtained for Example 12.2, using Eq. (12.4) is plotted in Fig. 12.2a–e. It is noted that, with an increase in m the HF domain solutions (black dots) has improved much more from 4 to 24.

12.2 Identification of Output Matrix of a Non-homogeneous System [3]

Referring to Eq. (8.31), we can write,

$$\begin{aligned} \mathbf{y}_S &= \mathbf{C}\mathbf{C}_{Sx} + \mathbf{D}\mathbf{C}_{Su} \\ \text{or, } \mathbf{C}\mathbf{C}_{Sx} &= \mathbf{y}_S - \mathbf{D}\mathbf{C}_{Su} \\ \text{or, } \mathbf{C} &= [\mathbf{y}_S - \mathbf{D}\mathbf{C}_{Su}]\mathbf{C}_{Sx}^{-1} \end{aligned} \tag{12.5}$$

Similarly from Eq. (8.32), we have

$$\begin{aligned} \mathbf{y}_T &= \mathbf{C}\,\mathbf{C}_{Tx} + \mathbf{D}\mathbf{C}_{Tu} \\ \text{or, } \mathbf{C} &= [\mathbf{y}_T - \mathbf{D}\mathbf{C}_{Tu}]\mathbf{C}_{Tx}^{-1} \end{aligned} \tag{12.6}$$

For identification of the output matrix **C**, we can use either Eq. (12.5) or (12.6). If we use Eq. (12.5), the dimension of the matrix $\mathbf{C}_{Sx}$ should be $n \times n$, so that it is invertible. That is, for n states, we need to expand each state in HF domain and consider n consecutive samples of each state to form the square matrix $\mathbf{C}_{Sx}$.

If we use Eq. (12.6), for identification of the output matrix **C**, we need (n + 1) samples of the output y and of each state. This will help to form n number of respective triangular function coefficients of the variables when they are expanded in HF domain. This is because, as before, the dimension of the matrix $\mathbf{C}_{Tx}$ should

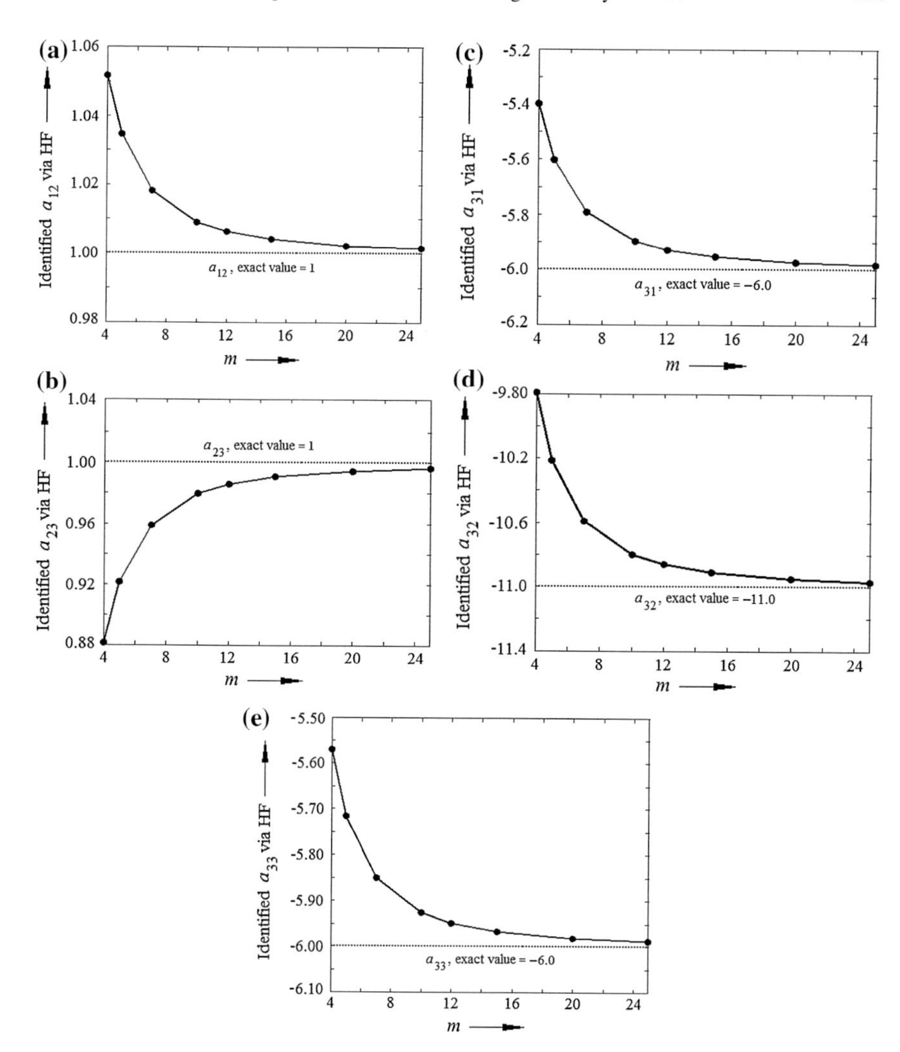

Fig. 12.2 Hybrid function based identification for Example 12.2 for the elements **a** a_{12}, **b** a_{23}, **c** a_{31}, **d** a_{32} and **e** a_{33} of **A**

be such that it is invertible. Meeting this condition will lead to the identification of the output matrix quite easily.

12.2.1 Numerical Examples

Example 12.3 Consider a system of Example 12.1 having two states and let its output curve be given by Fig. 12.3. Using three samples from the output curve and Eq. (12.5) or (12.6), we identify the output matrix as

$$\mathbf{C} = \begin{bmatrix} 1 & 0 \end{bmatrix}.$$

Example 12.4 (vide Appendix B, Program no. 32) Consider a system of Example 12.2 having three states and the output curve of Fig. 12.4. From the given output curve, we consider four samples, and then using Eq. (12.5) or (12.6), we may identify the output matrix as

$$\mathbf{C} = \begin{bmatrix} 4 & 5 & 1 \end{bmatrix}.$$

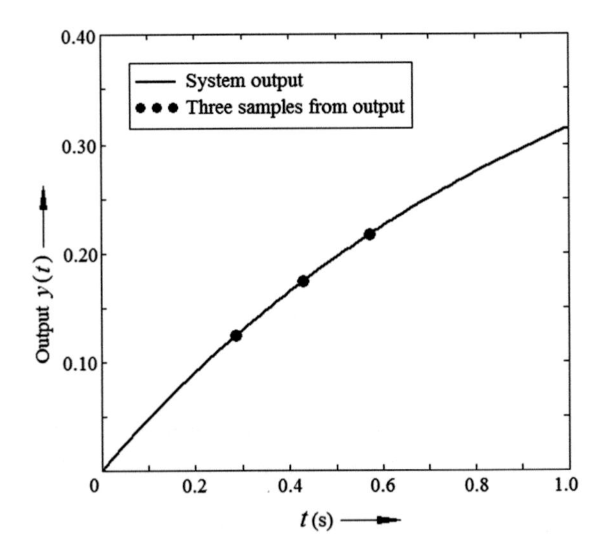

Fig. 12.3 Hybrid function based identification of system output matrix **C** using three samples from the exact solution of Example 12.3

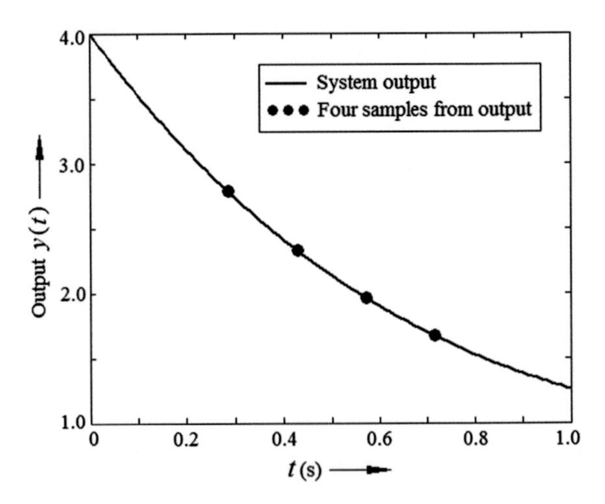

Fig. 12.4 Hybrid function based identification of system output matrix **C** using four samples from the exact solution of Example 12.4 (vide Appendix B, Program no. 32)

12.3 Identification of a Homogeneous System [3]

Similarly if we like to identify the system matrix **A** for an $n \times n$ homogeneous system, Eq. (12.4) will be modified to

$$\mathbf{A} = \frac{2}{h}\begin{bmatrix} c_{\mathrm{Sx}1(i+1)} - c_{\mathrm{Sx}1i} & c_{\mathrm{Sx}1(i+2)} - c_{\mathrm{Sx}1(i+1)} & \cdots & c_{\mathrm{Sx}1(i+n)} - c_{\mathrm{Sx}1(i+n-1)} \\ c_{\mathrm{Sx}2(i+1)} - c_{\mathrm{Sx}2i} & c_{\mathrm{Sx}2(i+2)} - c_{\mathrm{Sx}2(i+1)} & \cdots & c_{\mathrm{Sx}2(i+n)} - c_{\mathrm{Sx}2\,(i+n-1)} \\ \vdots & \vdots & & \vdots \\ c_{\mathrm{Sx}n(i+1)} - c_{\mathrm{Sx}ni} & c_{\mathrm{Sx}n(i+2)} - c_{\mathrm{Sx}n(i+1)} & \cdots & c_{\mathrm{Sx}n(i+n)} - c_{\mathrm{Sx}n\,(i+n-1)} \end{bmatrix}$$
$$\times \begin{bmatrix} c_{\mathrm{Sx}1i} + c_{\mathrm{Sx}1(i+1)} & c_{\mathrm{Sx}1(i+1)} + c_{\mathrm{Sx}1(i+2)} & \cdots & c_{\mathrm{Sx}1(i+n-1)} + c_{\mathrm{Sx}1(i+n)} \\ c_{\mathrm{Sx}2i} + c_{\mathrm{Sx}2(i+1)} & c_{\mathrm{Sx}2(i+1)} + c_{\mathrm{Sx}2(i+2)} & \cdots & c_{\mathrm{Sx}2(i+n-1)} + c_{\mathrm{Sx}2(i+n)} \\ \vdots & \vdots & & \vdots \\ c_{\mathrm{Sx}ni} + c_{\mathrm{Sx}n(i+1)} & c_{\mathrm{Sx}n(i+1)} + c_{\mathrm{Sx}n(i+2)} & \cdots & c_{\mathrm{Sx}n(i+n-1)} + c_{\mathrm{Sx}n\,(i+n)} \end{bmatrix}^{-1} \tag{12.7}$$

12.4 Identification of Output Matrix of a Homogeneous System [3]

Referring to Eq. (12.5), we can identify the output matrix of a homogeneous system by substituting a condition of zero to the direct coupling matrix **D**, and we get

$$\mathbf{C} = \mathbf{y}_{\mathrm{S}}\mathbf{C}_{\mathrm{Sx}}^{-1} \tag{12.8}$$

Similarly from Eq. (12.6), we have

$$\mathbf{C} = \mathbf{y}_{\mathrm{T}}\,\mathbf{C}_{\mathrm{Tx}}^{-1} \tag{12.9}$$

12.5 Identification of a Non-homogeneous System with Jump Discontinuity at Input

Equation (12.4) is derived in HF domain and is based upon the conventional HF technique. That is, it is essentially the $\mathrm{HF_c}$ based approach and it has one inconvenience: it comes up with utterly erroneous results for identification, shown later, if the samples are chosen from the zones where the jumps have occurred. And for system analysis as well, vide Figs. 8.16 and 8.17, and Table 8.16, we ended up with unacceptable errors in the results due to jump discontinuities in the inputs.

But we can modify Eq. (12.4) to make it suitable for the $\mathrm{HF_m}$ based approach, so that it can come up with good results in spite of the jump discontinuities. The modification is quite simple in the sense that in the RHS of Eq. (12.4), all the triangular function coefficient matrices associated with the matrix $\mathbf{B}$ have to be modified as discussed in Chap. 3, Eq. (3.12). That is, all the $\mathbf{C}_{\mathrm{Tu}}^{\mathrm{T}}$'s in (12.4) are to be replaced by $\mathbf{C'}_{\mathrm{Tu}}^{\mathrm{T}}$, where $\mathbf{C'}_{\mathrm{Tu}}^{\mathrm{T}} \triangleq \mathbf{C}_{\mathrm{Tu}}^{\mathrm{T}} \mathbf{J}_{k(m)}$. This modification will come up with good results for system identification, shown in the following, even if we choose the samples of the states from the region containing the jump. It will also be shown that if we take the samples of the inputs from a region excluding the jump portion, and corresponding samples of the states, the results of identification through any of the $\mathrm{HF_c}$ or $\mathrm{HF_m}$ approach will have no difference at all.

The $\mathrm{HF_m}$ approach shows its usefulness only when the samples are selected from the jump region. Thus, with the $\mathrm{HF_m}$ approach we can be careless about the chosen region where from the involved samples are selected. But, as delineated, we have to be careful in selecting the samples while using the $\mathrm{HF_c}$ based approach because it is no different from the conventional HF domain analysis.

From (12.4), in $\mathrm{HF_c}$ approach, we write

$$\mathbf{A} = \left\{ \frac{2}{h} \begin{bmatrix} c_{\mathrm{Sx}1(i+1)} - c_{\mathrm{Sx}1i} & \cdots & c_{\mathrm{Sx}1(i+n)} - c_{\mathrm{Sx}1(i+n-1)} \\ \vdots & \vdots & \vdots \\ c_{\mathrm{Sx}n(i+1)} - c_{\mathrm{Sx}ni} & \cdots & c_{\mathrm{Sx}n(i+n)} - c_{\mathrm{Sx}n(i+n-1)} \end{bmatrix} - 2\mathbf{B}\left[\left(\mathbf{C}_{\mathrm{Su}}^{\mathrm{T}} + \frac{1}{2}\mathbf{C}_{\mathrm{Tu}}^{\mathrm{T}} \right)_i \cdots \left(\mathbf{C}_{\mathrm{Su}}^{\mathrm{T}} + \tfrac{1}{2}\mathbf{C}_{\mathrm{Tu}}^{\mathrm{T}} \right)_{i+n-1} \right] \right\} \times \begin{bmatrix} c_{\mathrm{Sx}1i} + c_{\mathrm{Sx}1(i+1)} & \cdots & c_{\mathrm{Sx}1(i+n-1)} + c_{\mathrm{Sx}1(i+n)} \\ \vdots & \vdots & \vdots \\ c_{\mathrm{Sx}ni} + c_{\mathrm{Sx}n(i+1)} & \cdots & c_{\mathrm{Sx}n(i+n-1)} + c_{\mathrm{Sx}n(i+n)} \end{bmatrix}^{-1} \tag{12.10}$$

Whereas from (12.4), in $\mathrm{HF_m}$ approach, we have

$$\mathbf{A} = \left\{ \frac{2}{h} \begin{bmatrix} c_{\mathrm{Sx}1(i+1)} - c_{\mathrm{Sx}1i} & \cdots & c_{\mathrm{Sx}1(i+n)} - c_{\mathrm{Sx}1(i+n-1)} \\ \vdots & \vdots & \vdots \\ c_{\mathrm{Sx}n(i+1)} - c_{\mathrm{Sx}ni} & \cdots & c_{\mathrm{Sx}n(i+n)} - c_{\mathrm{Sx}n(i+n-1)} \end{bmatrix} - 2\mathbf{B}\left[\left(\mathbf{C}_{\mathrm{Su}}^{\mathrm{T}} + \frac{1}{2}\mathbf{C'}_{\mathrm{Tu}}^{\mathrm{T}} \right)_i \cdots \left(\mathbf{C}_{\mathrm{Su}}^{\mathrm{T}} + \tfrac{1}{2}\mathbf{C'}_{\mathrm{Tu}}^{\mathrm{T}} \right)_{i+n-1} \right] \right\} \times \begin{bmatrix} c_{\mathrm{Sx}1i} + c_{\mathrm{Sx}1(i+1)} & \cdots & c_{\mathrm{Sx}1(i+n-1)} + c_{\mathrm{Sx}1(i+n)} \\ \vdots & \vdots & \vdots \\ c_{\mathrm{Sx}ni} + c_{\mathrm{Sx}n(i+1)} & \cdots & c_{\mathrm{Sx}n(i+n-1)} + c_{\mathrm{Sx}n(i+n)} \end{bmatrix}^{-1} \tag{12.11}$$

Similarly, if we want to identify the system matrix **A** for an $n \times n$ homogeneous system, we simply put **B** = 0 in Eq. (12.11) and get

$$\mathbf{A} = \frac{2}{h}\begin{bmatrix} c_{Sx1(i+1)} - c_{Sx1i} & \cdots & c_{Sx1(i+n)} - c_{Sx1(i+n-1)} \\ \vdots & & \vdots \\ c_{Sxn(i+1)} - c_{Sxni} & \cdots & c_{Sxn(i+n)} - c_{Sxn(i+n-1)} \end{bmatrix}\begin{bmatrix} c_{Sx1i} + c_{Sx1(i+1)} & \cdots & c_{Sx1(i+n-1)} + c_{Sx1(i+n)} \\ \vdots & & \vdots \\ c_{Sxni} + c_{Sxn(i+1)} & \cdots & c_{Sxn(i+n-1)} + c_{Sxn(i+n)} \end{bmatrix}^{-1} \tag{12.12}$$

12.5.1 Numerical Examples

Example 12.5 Consider the non-homogeneous system $\dot{\mathbf{x}}(t) = \mathbf{A}\mathbf{x}(t) + \mathbf{B}u(t) + \mathbf{B}u(t-a)$ where $\mathbf{A} = \begin{bmatrix} 0 & 1 \\ -2 & -3 \end{bmatrix}$, $\mathbf{B} = \begin{bmatrix} 0 \\ 1 \end{bmatrix}$, $\mathbf{x}(0) = \begin{bmatrix} 0 \\ 0.5 \end{bmatrix}$, $u(t)$ is a unit step function and $u(t-a)$ is a delayed unit step function.

The system has the solution

$$x_1(t) = 1 - \frac{1}{2}\exp(-t) - \left[\exp(-(t-a)) - \frac{1}{2}\exp(-2(t-a))\right]u(t-a)$$

$$x_2(t) = \frac{1}{2}\exp(-t) + [\exp(-(t-a)) - \exp(-2(t-a))]u(t-a)$$

Knowing the states and inputs, this system can be identified in HF domain.

It may be noted that the input to the system has a jump discontinuity. Though we can represent this function using conventional HF set, the approximation is not exact due to the jump. However if we employ the $\mathrm{HF_m}$ based approximation technique, the input function may be represented in an exact manner. So we use both the approximation techniques and take help of Eqs. (12.10) and (12.11) to identify the system. Here, we take $a = 0.5$ s. It is expected that the $\mathrm{HF_m}$ approach will bring out much better result compared to the $\mathrm{HF_c}$ approach.

The results obtained via the $\mathrm{HF_c}$ approach and $\mathrm{HF_m}$ approach, for five different values of m, are tabulated in Table 12.5a, b respectively, where the samples of the inputs and corresponding states are purposely selected from the jump region. It is noted that for the $\mathrm{HF_c}$ approach the effort ends in a fiasco, producing results which are unreliable. But the $\mathrm{HF_m}$ approach identifies the system with much less error.

However, if the samples are chosen from a region excluding the sub-interval containing the jump point, the results of identification derived via any of the approaches yields the same results. Table 12.6 presents these results.

From the data presented in Table 12.5b, we show a typical comparison of the actual elements of the system matrix with respective elements computed using the $\mathrm{HF_m}$ based approach for $m = 10$, in Table 12.7.

Percentage errors are computed to figure out the efficiency of the method through quantitative estimates. To have an idea about the behavior of error with increasing m, for both $\mathrm{HF_c}$ and $\mathrm{HF_m}$ approach, we compute percentage errors of all

Table 12.5 Identification of the system of Example 12.5 using the (a) HF_c approach and the (b) HF_m approach for different values of m = 8, 10, 20, 40 and 50 for T = 1 s, with the samples chosen from the jump region

Elements of system matrix A	Exact values	$m = 8$	$m = 10$	$m = 20$	$m = 40$	$m = 50$
(a) HF_c based approach						
a_{11}	0	0.0000	0.0000	0.0000	0.0000	0.0000
a_{12}	1	0.9987	0.9992	0.9998	0.9999	1.0000
a_{21}	−2	−10.5104	−12.5083	−22.5042	−42.5021	−52.5017
a_{22}	−3	0.1006	1.3859	7.8492	20.8118	27.2967
(b) HF_m based approach						
a_{11}	0	0.0000	0.0000	0.0000	0.0000	0.0000
a_{12}	1	0.9987	0.9992	0.9998	0.9999	1.0000
a_{21}	−2	−2.0000	−2.0000	−2.0000	−2.0000	−2.0000
a_{22}	−3	−2.9987	−2.9992	−2.9998	−2.9999	−3.0000

The results are compared with the actual elements of **A**. It is noted that the results obtained via the HF_c approach are simply unreliable while that obtained via the HF_m approach are reasonably accurate

Table 12.6 Identification of the system of Example 12.5, considering the samples chosen from a region excluding the sub-interval containing the jump point

Elements of system matrix A	Exact values	HF_c or HF_m based approach				
		$m = 8$	$m = 10$	$m = 20$	$m = 40$	$m = 50$
a_{11}	0	0.0000	0.0000	0.0000	0.0000	0.0000
a_{12}	1	0.9987	0.9992	0.9998	0.9999	1.0000
a_{21}	−2	−2.0000	−2.0000	−2.0000	−2.0000	−2.0000
a_{22}	−3	−2.9987	−2.9992	−2.9998	−2.9999	−3.0000

It is noted that in such a case both the approaches yields the same results. These are compared with the actual elements of the system matrix **A** for different values of m = 8, 10, 20, 40 and 50 with T = 1 s

Table 12.7 Identification of the system of Example 12.5 with comparison of actual elements of **A** with corresponding elements computed via HF_m approach, for m = 20, T = 1 s

Elements of system matrix A	Exact values (E)	Values obtained via HF_m approach for m = 20 (H)	% Error $\varepsilon = \frac{E-H}{E} \times 100$
a_{11}	0	0.0000	–
a_{12}	1	0.9998	0.0200
a_{21}	−2	−2.0000	0.0000
a_{22}	−3	−2.9998	0.0067

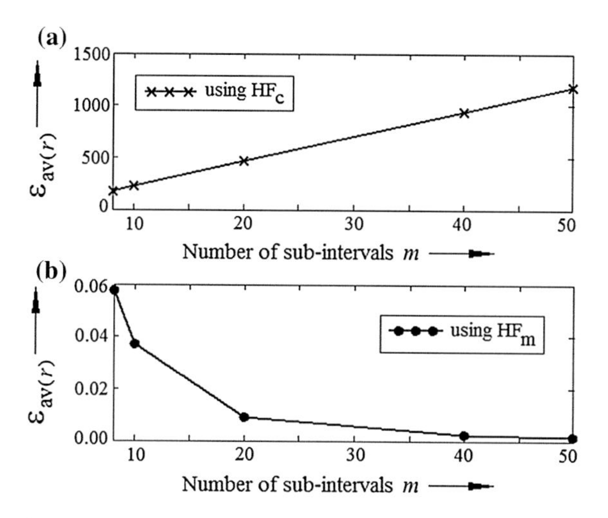

Fig. 12.5 Error in system identification for increasing m (m = 8, 10, 20, 40 and 50) using the **a** HF$_c$ approach and the **b** HF$_m$ approach for Example 12.5, with $T = 1$ s. It is noted that the error is large and increases with increasing m for the HF$_c$ approach, while for the HF$_m$ approach, the error is comparatively much smaller and decreases steadily and rapidly with increasing m

the elements of a system matrix **A** of order n for any particular value of m and calculate AMP error, vide Eq. (4.45). Only in this case, number of elements in the denominator will be $r = n^2$.

Figure 12.5 is drawn to show the variation of $\varepsilon_{av(r)}$ with m for both the HF$_c$ and HF$_m$ approaches. It is noted that for the HF$_c$ approach, $\varepsilon_{av(r)}$ is large and increases with increasing m, but for the HF$_m$ approach, it is comparatively much smaller and decreases steadily and rapidly with increasing m.

Example 12.6 (vide Appendix B, Program no. 33) Consider the non-homogeneous system $\dot{\mathbf{x}}(t) = \mathbf{A}\mathbf{x}(t) + \mathbf{B}\,u_1(t) + \mathbf{B}\,u_2(t-a)$ where $\mathbf{A} = \begin{bmatrix} 0 & 1 \\ -2 & -3 \end{bmatrix}$, $\mathbf{B} = \begin{bmatrix} 0 \\ 1 \end{bmatrix}$, $\mathbf{x}(0) = \begin{bmatrix} 0 \\ 0.5 \end{bmatrix}$ $u_1(t)$ is a ramp function and $u_2(t-a)$ is a delayed unit step function, having the jump at $t = a$ s.
The system has the solution

$$x_1(t) = -\frac{1}{4} + \frac{1}{2}t + \frac{3}{2}\exp(-t) - \frac{3}{4}\exp(-2t) - \left[\exp(-(t-a)) - \frac{1}{2}\exp(-2(t-a))\right]u(t-a)$$

$$x_2(t) = \frac{1}{2} - \frac{3}{2}\exp(-t) + \frac{3}{2}\exp(-2t) + [\exp(-(t-a)) - \exp(-2(t-a))]u(t-a)$$

Knowing the states and inputs, this system can be identified in HF domain.

It may be noted that the input to the system has a jump discontinuity which can not be represented in an exact manner if we approximate this function using conventional HF set. Here, even if we employ the HF$_m$ based approximation technique, the input function can not be represented exactly, because of its nature. Anyway, we use both the approximation techniques two identify the system using Eqs. (12.10) and (12.11). Here, we take $a = 0.5$ s. Here also, we expect that the HF_m approach will produce better result compared to the HF_c approach. However, if we employ the combined HF_c and HF_m technique, the input function can be approximated in an exact fashion as was done for the function of Example 3.5, vide Eq. (3.18). For the combined HF_c and HF_m technique, we expect best results of all the above three approaches.

The inputs $u_1(t)$ and $u_2(t-a)$ to the system are basically a combination of a ramp function and a delayed step function. Such a function can be expanded in HF domain using a combination of HF_c and HF_m approaches, as was done in Eq. (3.18) in Sect. 3.5.2. In such a case, to solve the identification problem via Eq. (12.11), the $\mathbf{C}'^{T}_{Tu}$ matrix has to be modified accordingly.

Knowing the states and inputs, this system can be identified in HF domain using the Eqs. (12.10) and (12.11). That is, we identify the system using both the HF_c and HF_m based approaches.

The results obtained via the HF_c, HF_m and the combined approaches, for five different values of m, are tabulated in Table 12.8a, b, c respectively, where the samples of the inputs and corresponding states are purposely selected from the jump region. It is noted that for the HF_c approach the effort ends in a fiasco, producing

Table 12.8 Identification of the system of Example 12.6 using the (a) HF_c approach, (b) HF_m approach and (c) the combination of HF_c and HF_m approaches, for different values of m = 8, 10, 20, 40 and 50 for $T = 1$ s, with the samples chosen from the jump region

Elements of system matrix A	Exact values	$m = 8$	$m = 10$	$m = 20$	$m = 40$	$m = 50$
(a) HF_c based approach						
a_{11}	0	0.0022	0.0019	0.0008	0.0002	0.0002
a_{12}	1	0.9858	0.9905	0.9974	0.9993	0.9995
a_{21}	−2	−12.6762	−15.4476	−30.1201	−60.4888	−75.7946
a_{22}	−3	2.1893	4.7296	18.4059	46.9492	61.3603
(b) HF_m based approach						
a_{11}	0	0.0022	0.0019	0.0008	0.0002	0.0002
a_{12}	1	0.9858	0.9905	0.9974	0.9993	0.9995
a_{21}	−2	−0.6663	−0.6572	−0.5952	−0.5382	−0.5244
a_{22}	−3	−3.6110	−3.7482	−4.0637	−4.2470	−4.2861
(c) Combination of HF_c and HF_m based approaches						
a_{11}	0	0.0022	0.0019	0.0008	0.0002	0.0002
a_{12}	1	0.9858	0.9905	0.9974	0.9993	0.9995
a_{21}	−2	−2.0007	−2.0018	−2.0012	−2.0004	−2.0003
a_{22}	−3	−2.9665	−2.9775	−2.9937	−2.9983	−2.9989

The results are compared with the actual elements of **A**. It is noted that the elements computed via the HF_c approach are simply unrecognizable. So is the case for identification via the HF_m approach. But use of the combined approach (HF_c and HF_m) comes up with reasonably accurate results (vide Appendix B, Program no. 33)

results which are unrecognizable. The HF_m approach, though ends up with much less error compared to the HF_c approach, also fails to identify the system. However, since the combined approach can represent the input function exactly, it identifies the system with much less error.

Figure 12.6 is drawn to show the variation of AMP error with different values of m (m = 8, 10, 20, 40 and 50) for the HF_c, HF_m and the combined approaches. However, if the samples are chosen from a region excluding the sub-interval containing the jump point, the results of identification derived via any of the approaches yields the same results. Table 12.9 presents these results.

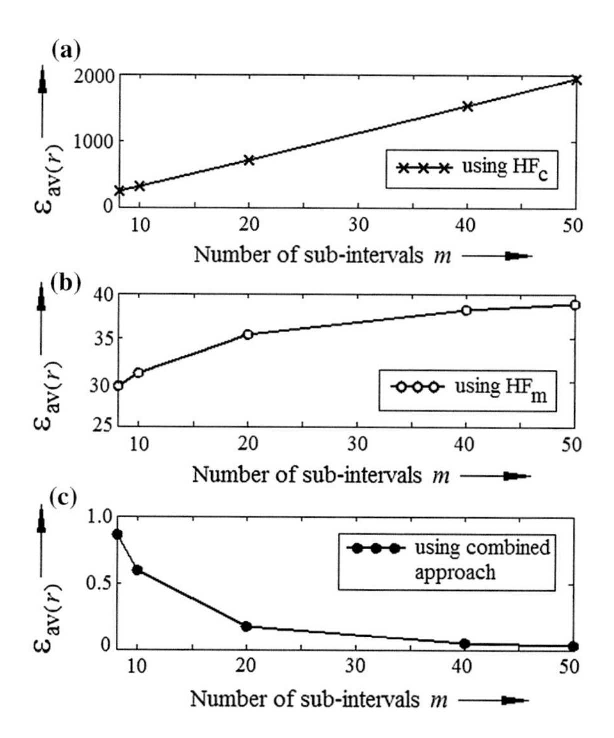

Fig. 12.6 Error in system identification for increasing m (m = 8, 10, 20, 40 and 50) using the **a** HF_c approach, **b** HF_m approach and **c** combined approach, for Example 12.6, with $T = 1$ s. It is noted that for the HF_c approach the error is large, and it increases linearly with increasing m. So is the case for error for the HF_m approach, though the rate of increase is much decreased. But for the combined approach (HF_c and HF_m), the error is very very less in comparison and decreases steadily and rapidly with increasing m (vide Appendix B, Program no. 33)

Table 12.9 Identification of the system of Example 12.6, considering the samples chosen from a region excluding the sub-interval containing the jump point

Elements of system matrix A	Exact values	HF_c or HF_m based approach				
		$m = 8$	$m = 10$	$m = 20$	$m = 40$	$m = 50$
a_{11}	0	0.2829	0.0493	0.0059	0.0012	0.0008
a_{12}	1	0.8195	0.9711	0.9972	0.9995	0.9997
a_{21}	−2	−2.5797	−2.1011	−2.0120	−2.0025	−2.0016
a_{22}	−3	−2.6303	−2.9409	−2.9942	−2.9989	−2.9993

It is noted that in such a case both the approaches yields the same results. These are compared with the actual elements of the system matrix **A** for different values of m = 8, 10, 20, 40 and 50 with T = 1 s (vide Appendix B, Program no. 33)

12.6 Conclusion

In this chapter the system identification problem has been tackled. For an $(n \times n)$ system, the system matrix **A** is determined considering $(n + 1)$ number of consecutive samples of the states, the input vector **B** and the input function. The output matrix **C** is determined from the knowledge of $(n + 1)$ number of consecutive samples of the output function y and those of the states.

Also in this chapter, we have used the modified approach of hybrid function domain, as discussed in Chap. 3, for identification of non-homogeneous systems involving jump discontinuity in the applied input, thus affecting the system states.

Equation (12.10) presents the equation for solving the elements of the systems matrix **A** via HF_c approach. Equation (12.11) is the *ultimate* equation for solving the elements of **A** using HF_m approach.

It is noted that both the equations work with the samples of the states and the inputs. And also, while using these samples for obtaining the elements of **A**, the samples of the states are always to be considered in pairs. This fact is evident from the structure of the Eqs. (12.10) and (12.11).

If the system matrix has a dimension of n, the number of sample pairs to be considered for determining the solution is n pairs. If consecutive pairs are chosen, then number of samples involved will be $(n + 1)$. But if we choose the pairs randomly from the sample sequence, then obviously, the number of samples will become $2n$. However, in both the cases, matching samples are to be chosen from the input functions.

The major difference between (12.10) and (12.11) is, while (12.11) can work effectively with samples chosen from anywhere within the time interval under consideration, Eq. (12.10) has restrictions. That is, if the samples are chosen from the *jump* portions of any of the functions, it will simply *fail* to compute the elements of **A**. Should the selection of samples be made from *non-jump* portions, both the Eqs. (12.10) and (12.11) produce the same result.

All these points are presented via Tables 12.5 and 12.6.

Figure 12.5a, b present the variation of AMP error with m for Example 12.5. It is noted that though the samples are chosen intentionally from the *jump portion*, the error is large and increases with m for the HF_c approach, but for the HF_m approach, the error is comparatively much smaller and decreases steadily and rapidly with m.

Similarly, for Example 12.6, the behavior of AMP error with m is illustrated using Tables 12.8 and 12.9, and Fig. 12.6, when the samples are chosen from the *jump portion*. It is noted that for the HF_c approach the error is large and increases linearly with increasing m. So is the case for error for the HF_m approach, though the rate of increase is much less. But for the combined approach (HF_c and HF_m), the error is not only much less compared to the other two methods, but decreases rapidly with increasing m.

The above results prove beyond any doubt that when handling stair case functions with jump discontinuities, the HF based *modified* approach is superior to the HF based *conventional* approach. However for jumps of different nature, the combined HF_c and HF_m approach is much more suitable than either of HF_c and HF_m approaches. This is because, good results are dependent upon proper approximation of jump functions at the input. These facts are useful for identification of control systems with jump discontinuity in the input functions.

References

1. Ogata, K.: Modern Control Engineering, 5th edn. Prentice Hall of India, New Delhi (2011)
2. Ogata, K.: System Dynamics, 4th edn. Pearson Education, Upper Saddle River (2004)
3. Roychoudhury, S., Deb, A., Sarkar, G.: Analysis and synthesis of homogeneous/non-homogeneous control systems via orthogonal hybrid functions (HF) under states space environment. J. Inf. Optim. Sci. **35**(5–6), 431–482 (2014)
4. Rao, G.P.: Piecewise Constant Orthogonal Functions and Their Application to Systems and Control, LNCIS Series, vol. 55. Springer, Berlin (1983)
5. Jiang, J.H., Schaufelberger, W.: Block Pulse Functions and Their Application in Control System, LNCIS, vol. 179. Springer, Berlin (1992)
6. Unbehauen, H., Rao, G.P.: Identification of Continuous Systems. North-Holland, Amsterdam (1987)

Chapter 13
System Identification Using State Space Approach: Time Varying Systems

Abstract This chapter discusses HF domain identification of time varying systems in state space. Both homogeneous and non-homogeneous systems are treated. Illustration has been provided with the support of three examples, ten figures and two tables.

As the name suggests, in this chapter, we deal with linear time varying (LTV) control systems [1]. We intend to identify two types of LTV systems in hybrid function platform, namely non-homogeneous system and homogeneous system [1, 2]. For a time varying system, these parameters do vary with time.

The HF domain identification [3] pivot upon the samples of involved functions. In today's digital world, this is an important advantage.

Here, we solve the problem of system identification [4, 5], both for non-homogeneous systems and homogeneous systems, described as state space models, by using the concept of HF domain. First, we take up the problem of identification of a non-homogeneous time varying system, because after putting the specific condition of zero forcing function in the solution, we can arrive at the result of identification of a homogeneous system quite easily.

13.1 Identification of a Non-homogeneous System [3]

Consider the system given by Eq. (9.1).

Its time varying system matrix $\mathbf{A}(t)$ is given by

$$\mathbf{A}(t) \triangleq \begin{bmatrix} a_{11}(t) & a_{12}(t) & \cdots & a_{1n}(t) \\ a_{21}(t) & a_{22}(t) & \cdots & a_{2n}(t) \\ \vdots & \vdots & & \vdots \\ a_{n1}(t) & a_{n2}(t) & \cdots & a_{nn}(t) \end{bmatrix}$$

A. Deb et al., *Analysis and Identification of Time-Invariant Systems, Time-Varying Systems, and Multi-Delay Systems using Orthogonal Hybrid Functions*,
Studies in Systems, Decision and Control 46, DOI 10.1007/978-3-319-26684-8_13

To identify the elements of $\mathbf{A}(t)$, we take help of the first equation of the equation set (9.12) and also the first equation of the equation set (9.15). Thus, we have

$$(x_{11} - x_{10}) = \frac{h}{2}\left[\sum_{j=1}^{n}\left(a_{1j0}\,x_{j0} + a_{1j1}\,x_{j1}\right) + \left(b_{10}\,u_0 + b_{11}\,u_1\right)\right] \tag{13.1}$$

and

$$\begin{aligned}(x_{12} - x_{10}) = \frac{h}{2}\Bigg[&\sum_{j=1}^{n} a_{1j0}\,x_{j0} + 2\sum_{j=1}^{n} a_{1j1}\,x_{j1} + \sum_{j=1}^{n} a_{1j2}\,x_{j2} + \left(b_{10}\,u_0 + b_{11}\,u_1\right)\\ &+ \left(b_{11}\,u_1 + b_{12}\,u_2\right)\Bigg]\end{aligned} \tag{13.2}$$

For a second order system, we make use of the generalized Eqs. (13.1) and (13.2), and solve for the off-diagonal elements $a_{12}(t)$ and $a_{21}(t)$ as

$$\begin{aligned}a_{12\,(k+1)} = \frac{1}{x_{2\,(k+1)}}\Bigg[&\left(\frac{2}{h} - a_{11\,(k+1)}\right)x_{1\,(k+1)} - \left(\frac{2}{h} + a_{11k}\right)x_{1k} - a_{12k}\,x_{2k}\\ &- \left(b_{1k}\,u_k + b_{1\,(k+1)}\,u_{(k+1)}\right)\Bigg]\end{aligned} \tag{13.3}$$

$$\begin{aligned}a_{21\,(k+1)} = \frac{1}{x_{1\,(k+1)}}\Bigg[&\left(\frac{2}{h} - a_{22\,(k+1)}\right)x_{2\,(k+1)} - \left(\frac{2}{h} + a_{22k}\right)x_{2k} - a_{21k}\,x_{1k}\\ &- \left(b_{2k}\,u_k + b_{2\,(k+1)}\,u_{(k+1)}\right)\Bigg]\end{aligned} \tag{13.4}$$

It can be shown that, for an nth order system, the off-diagonal elements can be represented by the following general form:

$$\begin{aligned}a_{ij\,(k+1)} = \frac{1}{x_{j\,(k+1)}}\Bigg[&\left(\frac{2}{h} - a_{ii\,(k+1)}\right)x_{i\,(k+1)} - \left(\frac{2}{h} + a_{iik}\right)x_{ik} - \sum_{\substack{l \neq j\\ l \neq i}}^{n} a_{il\,(k+1)}x_{l\,(k+1)} - \sum_{l \neq i}^{n} a_{ilk}x_{lk}\\ &- \left(b_{ik}u_k + b_{i\,(k+1)}u_{(k+1)}\right)\Bigg]\end{aligned} \tag{13.5}$$

From Eqs. (13.1) and (13.2), the diagonal elements $a_{11}(t)$ and $a_{22}(t)$ can be solved as

$$a_{11\,(k+1)} = \frac{2}{h} - \frac{1}{x_{1\,(k+1)}}\left[\left(\frac{2}{h} + a_{11k}\right)x_{1k} + a_{12k}\,x_{2k} + a_{12\,(k+1)}\,x_{2\,(k+1)} - \left(b_{1k}\,u_k + b_{1\,(k+1)}\,u_{(k+1)}\right)\right] \tag{13.6}$$

$$a_{22\,(k+1)} = \frac{2}{h} - \frac{1}{x_{2\,(k+1)}}\left[a_{21k}\,x_{1k} + \left(\frac{2}{h} + a_{22k}\right)x_{2k} + a_{21\,(k+1)}\,x_{1(k+1)} - \left(b_{2k}\,u_k + b_{2\,(k+1)}\,u_{(k+1)}\right)\right] \tag{13.7}$$

For an nth order system, the diagonal elements can be represented by the following general form:

$$a_{ii\,(k+1)} = \frac{2}{h} - \frac{1}{x_{i\,(k+1)}}\left[\left(\frac{2}{h} + a_{iik}\right)x_{ik} + \sum_{l\neq i}^{n} a_{il\,(k+1)}\,x_{l\,(k+1)} + \sum_{l\neq i}^{n} a_{ilk}\,x_{lk} + \left(b_{ik}\,u_k + b_{i\,(k+1)}\,u_{(k+1)}\right)\right] \tag{13.8}$$

Using above relevant equations, all unknown parameters of the system can be computed.

13.1.1 Numerical Examples

Example 13.1 (vide Appendix B, Program no. 34) Consider the non-homogeneous system $\dot{\mathbf{x}} = \mathbf{A}\mathbf{x} + \mathbf{B}u$ where $\mathbf{A} = \begin{bmatrix} 0 & 0 \\ t & 0 \end{bmatrix}$, $\mathbf{B} = \begin{bmatrix} 1 \\ 0 \end{bmatrix}$, $\mathbf{x}(0) = \begin{bmatrix} 1 \\ 1 \end{bmatrix}$ and $u = u(t)$, a unit step function.

The solution of the equation is

$$x_1(t) = 1 + t$$

and

$$x_2(t) = 1 + \frac{t^2}{2} + \frac{t^3}{3}$$

Identification of the given system parameter produces the following results:

Table 13.1 shows the comparison of the exact samples of $a_{21}(t)$ and the samples obtained in HF domain using Eq. (13.5) and the percentage errors of respective samples for m = 8, T = 1 s.

Table 13.1 Identification of the non-homogeneous system parameter $a_{21}(t)$ of Example 13.1 in HF domain compared with its exact samples along with percentage error at different sample points for m = 8 and T = 1 s (vide Appendix B, Program no. 34)

t (s)	System parameter $a_{21}(t)$ (m = 8)		
	Samples of the exact solution	Samples from HF domain synthesis	% Error
0	0.00000000	0.00000000	–
$\frac{1}{8}$	0.12500000	0.12037037	3.70370370
$\frac{2}{8}$	0.25000000	0.24999999	0.00000000
$\frac{3}{8}$	0.37500000	0.37121212	1.01010101
$\frac{4}{8}$	0.50000000	0.49999999	0.00000000
$\frac{5}{8}$	0.62500000	0.62179487	0.51282051
$\frac{6}{8}$	0.75000000	0.75000000	0.00000000
$\frac{7}{8}$	0.87500000	0.87222222	0.31746031
$\frac{8}{8}$	1.00000000	1.00000000	0.00000000

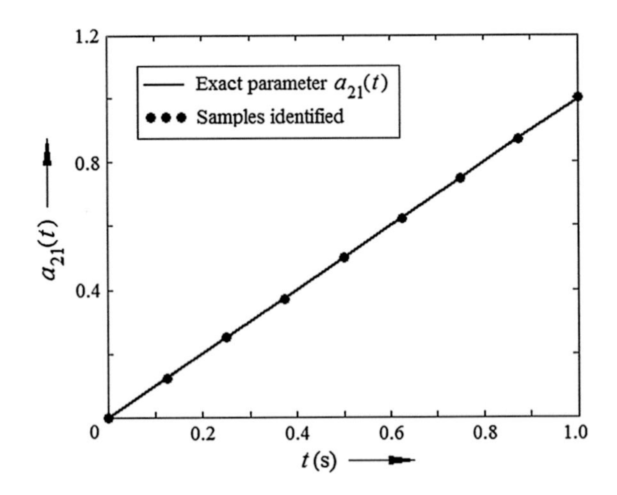

Fig. 13.1 Identification of $a_{21}(t)$ of the system matrix of Example 13.1 via HF domain for m = 8 and T = 1 s (vide Appendix B, Program no. 34)

Figure 13.1 compares graphically the exact curve for $a_{21}(t)$ with its hybrid function domain solution. Though the sample points obtained via HF domain technique seem to be reasonably close with the exact curve, a scrutiny of the error column of Table 13.1 reveals that there is a slight tendency of oscillation in the HF domain results. This possibly is due to numerical instability in the computation. Such oscillation turned out to be predominant for another case study presented in the following.

13.2 Identification of a Homogeneous System [3]

Similarly if we like to identify the time-varying system matrix $\mathbf{A}(t)$ for an $n \times n$ homogeneous system, Eqs. (13.5) and (13.8) will be modified to

$$a_{ij\,(k+1)} = \frac{1}{x_{j\,(k+1)}}\left[\left(\frac{2}{h} - a_{ii\,(k+1)}\right)x_{i\,(k+1)} - \left(\frac{2}{h} + a_{iik}\right)x_{ik} - \sum_{\substack{l \neq j \\ l \neq i}}^{n} a_{il\,(k+1)}x_{l\,(k+1)} \cdot - \sum_{l \neq i}^{n} a_{ilk}x_{lk}\right] \tag{13.9}$$

It can be shown that, for an nth order homogeneous time-varying system, the off-diagonal elements can be identified by the Eq. (13.9).

For an nth order homogeneous time-varying system, the diagonal elements can be represented by the following general form:

$$a_{ii\,(k+1)} = \frac{2}{h} - \frac{1}{x_{i\,(k+1)}}\left[\left(\frac{2}{h} + a_{iik}\right)x_{ik} + \sum_{l \neq i}^{n} a_{il\,(k+1)}\,x_{l\,(k+1)} + \sum_{l \neq i}^{n} a_{ilk}\,x_{lk}\right] \tag{13.10}$$

Using above relevant equations, all unknown parameters of the homogeneous system can be computed.

13.2.1 Numerical Examples

Example 13.2 (vide Appendix B, Program no. 35) Consider the time-varying homogeneous system $\dot{\mathbf{x}} = \mathbf{Ax}$, where, $\mathbf{A} = \begin{bmatrix} \cos(t) & \sin(t) \\ -\sin(t) & \cos(t) \end{bmatrix}$ and $\mathbf{x}(0) = \begin{bmatrix} 1 \\ 2 \end{bmatrix}$ having the solution

$$x_1(t) = (\cos(1 - \cos t) + 2\sin(1 - \cos t))e^{\sin t}$$

and

$$x_2(t) = (-\sin(1 - \cos t) + 2\cos(1 - \cos t))e^{\sin t}$$

Identification of the given system parameters produces results shown in Figs. 13.2 through 13.6.

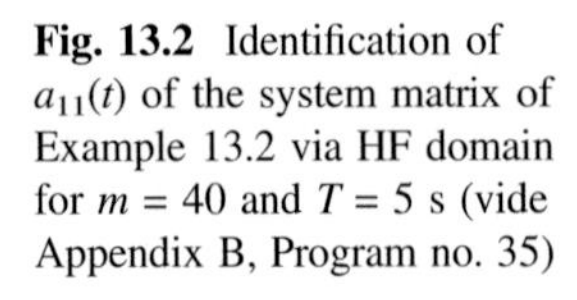

Fig. 13.2 Identification of $a_{11}(t)$ of the system matrix of Example 13.2 via HF domain for m = 40 and T = 5 s (vide Appendix B, Program no. 35)

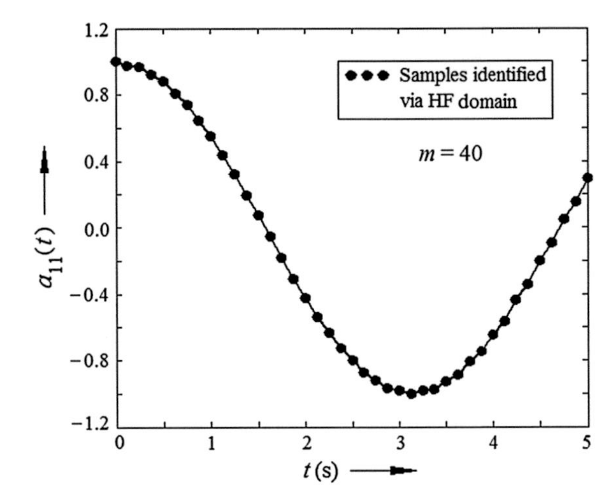

Fig. 13.3 Identification of $a_{21}(t)$ of the system matrix of Example 13.2 via HF domain for m = 40 and T = 6 s (vide Appendix B, Program no. 35)

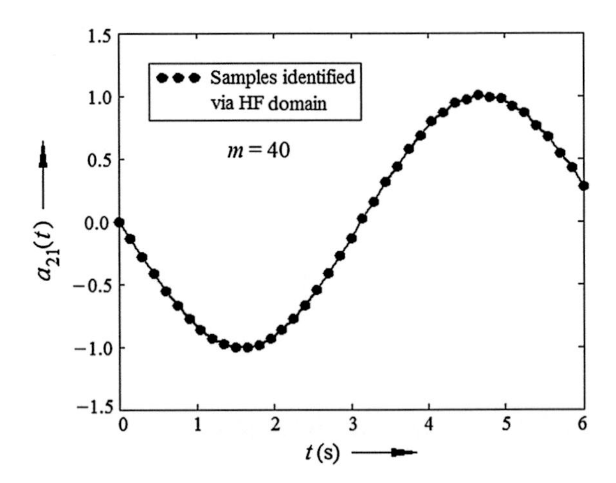

For the given homogeneous system, Figs. 13.2 and 13.3 illustrate the identified parameters $a_{11}(t)$ and $a_{21}(t)$ in HF domain for m = 40 and over a time period of 5 and 6 s respectively.

But interestingly, if instead of $a_{11}(t)$ and $a_{21}(t)$, we attempt to identify the elements $a_{12}(t)$ and $a_{22}(t)$, we are met with numerical instability. That is, knowing $a_{11}(t)$ and $a_{21}(t)$, the HF domain solution of $a_{12}(t)$ shows somewhat erratic results even for m = 50 over a 20 s interval. The nature of instability is illustrated in Fig. 13.4.

To investigate this phenomenon, we have computed the same result with another higher value of m, namely, m = 100, for a time period T = 20 s. Figure 13.5 shows the results. While numerical instability is encountered, no specific pattern of such instability emerged from the study.

Similarly, the effort to identify the parameter $a_{22}(t)$ ends up with the same fate. Figure 13.6 demonstrates this instability.

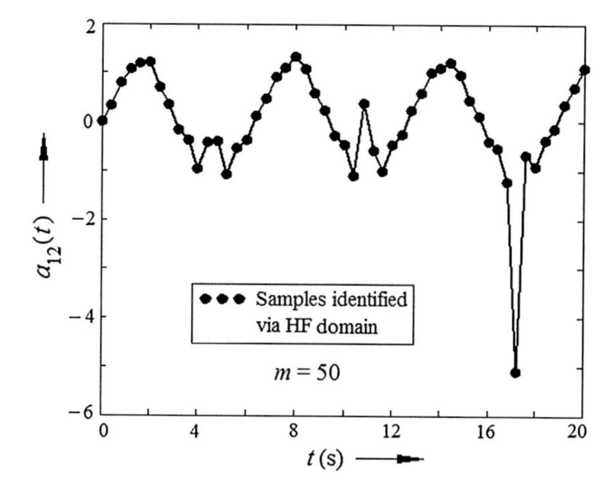

Fig. 13.4 Identification of $a_{12}(t)$ of system matrix of Example 13.2 via HF domain is met with numerical instability for $m = 50$ and $T = 20$ s

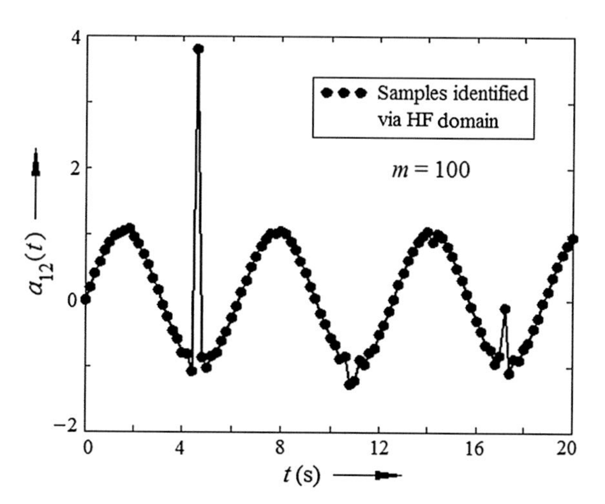

Fig. 13.5 Identification of $a_{12}(t)$ of system matrix of Example 13.2 again shows numerical instability for $m = 100$ and $T = 20$ s, though the pattern is different from the other two cases

Example 13.3 Consider the time-varying homogeneous system $\dot{\mathbf{x}} = \mathbf{A}\mathbf{x}$, where, $\mathbf{A} = \begin{bmatrix} 0 & 1 \\ 0 & t \end{bmatrix}$ and $\mathbf{x}(0) = \begin{bmatrix} 0 \\ 1 \end{bmatrix}$ having the solution

$$x_1(t) = \ln\left\{\left(\frac{2+t}{2-t}\right)^2 \exp(-t)\right\}$$

and

$$x_2(t) = \exp\left(\frac{t^2}{2}\right)$$

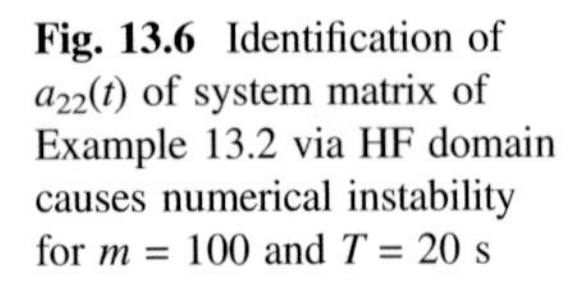

Fig. 13.6 Identification of $a_{22}(t)$ of system matrix of Example 13.2 via HF domain causes numerical instability for $m = 100$ and $T = 20$ s

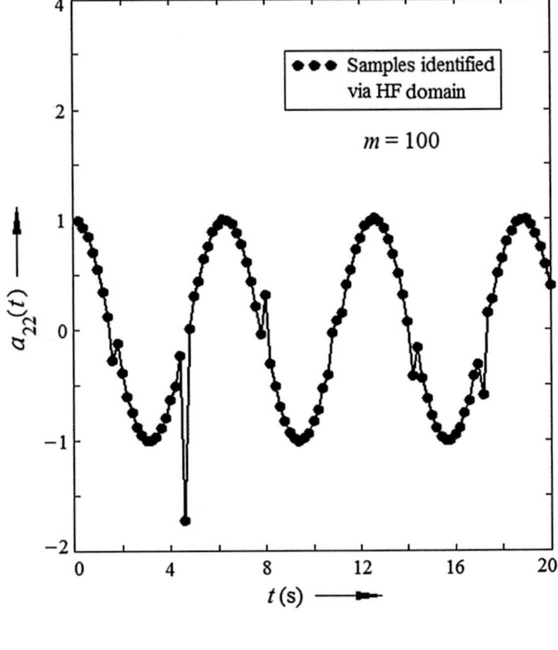

Fig. 13.7 Identification of $a_{12}(t)$ of the system matrix of Example 13.3 via HF domain, for $m = 10$ and $T = 1$ s

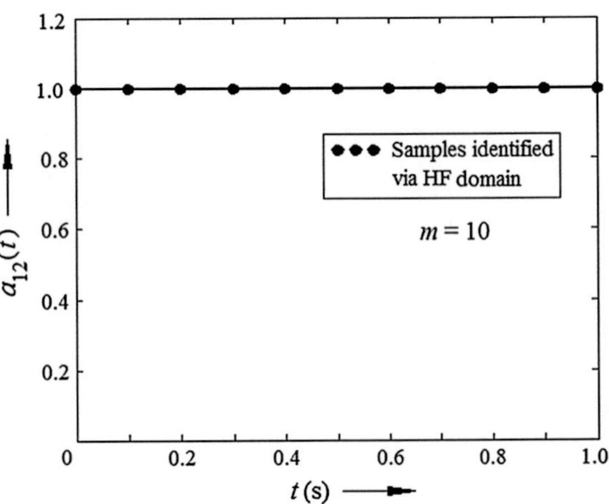

Identification of the given system parameters produces results shown in Figs. 13.7 through 13.9.

For the given homogeneous system, Figs. 13.7 and 13.8 illustrate the identified parameters $a_{12}(t)$ and $a_{22}(t)$ in HF domain for $m = 10$ and over a time period of 1 s. It seems that the results are readily acceptable.

To study the results more minutely, Table 13.2 is presented where the exact samples of $a_{12}(t)$ and $a_{22}(t)$ are compared with the samples computed via HF domain analysis using Eqs. (13.9) and (13.10), for $m = 10$, $T = 1$ s.

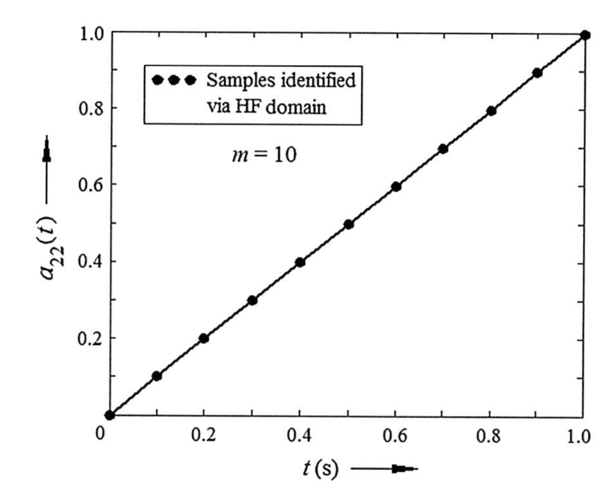

Fig. 13.8 Identification of $a_{22}(t)$ of the system matrix of Example 13.3 via HF domain, for $m = 10$ and $T = 1$ s

Table 13.2 Identification of the system parameters $a_{12}(t)$ and $a_{22}(t)$ of Example 13.3 in HF domain compared with its exact samples at different sample points for $m = 10$ and $T = 1$ s

t (s)	System parameter $a_{12}(t)$ ($m = 10$)		System parameter $a_{22}(t)$ ($m = 10$)	
	Samples of the exact solution s_d	Samples from HF domain synthesis s_h	Samples of the exact solution s_d	Samples from HF domain synthesis s_h
0	1.00000000	1.00000000	0.00000000	0.00000000
$\frac{1}{10}$	1.00000000	0.99833417	0.10000000	0.09975042
$\frac{2}{10}$	1.00000000	0.99995041	0.20000000	0.19949588
$\frac{3}{10}$	1.00000000	0.99830032	0.30000000	0.29923145
$\frac{4}{10}$	1.00000000	0.99980359	0.40000000	0.39895217
$\frac{5}{10}$	1.00000000	0.99822872	0.50000000	0.49865310
$\frac{6}{10}$	1.00000000	0.99956392	0.60000000	0.59832929
$\frac{7}{10}$	1.00000000	0.99810797	0.70000000	0.69797582
$\frac{8}{10}$	1.00000000	0.99922356	0.80000000	0.79758775
$\frac{9}{10}$	1.00000000	0.99788543	0.90000000	0.89716016
$\frac{10}{10}$	1.00000000	0.99868608	1.00000000	0.99668814

Interestingly, we find that though the sample points obtained for $a_{12}(t)$, via HF domain technique seem to be extremely close with the exact samples, a scrutiny of Table 13.2 reveals that there is a slight tendency of oscillation due to numerical instability in the HF domain results for $a_{12}(t)$. Noting this fact, such oscillation has been studied in more detail for three different values of m, namely, $m = 10$, 20 and 30. These results are depicted in Fig. 13.9, which seems to be oscillation free. But a magnified view of a portion of Fig. 13.9 tells another story which is self evident from Fig. 13.10. From the Fig. 13.10, it is noted that, the tendency of this oscillation can be reduced with increasing number of segments.

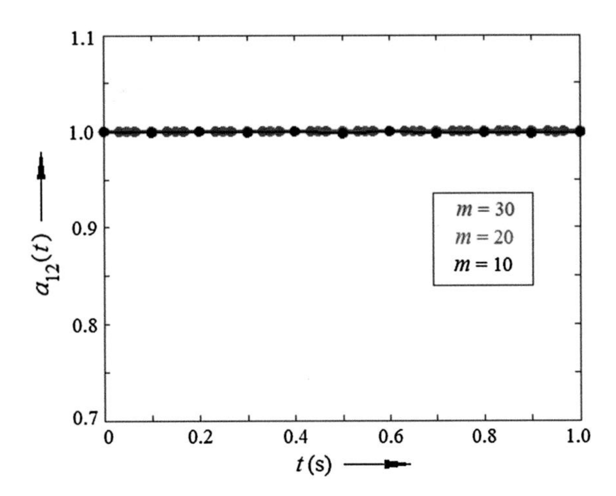

Fig. 13.9 Identification of $a_{12}(t)$ of the system matrix of Example 13.3 via HF domain, for three different values of m (10, 20 and 30) for $T = 1$ s

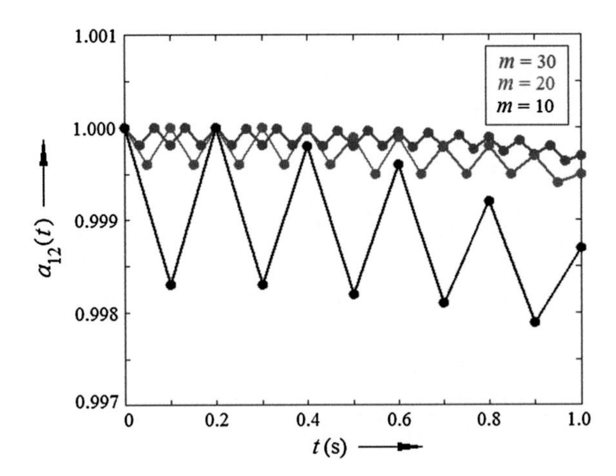

Fig. 13.10 Magnified view of Fig. 13.9 showing the identified parameter $a_{12}(t)$ of Example 13.3 via HF domain, for three different values of m (10, 20 and 30) for $T = 1$ s

13.3 Conclusion

In this chapter, we have presented a generalized method for identifying non-homogeneous as well as homogeneous time-varying systems. The proficiency of the method has been illustrated by suitable examples. Also, by putting $\mathbf{B}(t) = 0$, the same methods have been successfully applied to identify homogeneous time-varying systems as well.

Here the linear time-varying system identification problem has been solved for an n-state system. However, the limitation of the proposed method is, it can solve only a maximum of n number of system parameters of the system matrix $\mathbf{A}(t)$ out of n^2 parameters. Further, an essential requirement is, for solving the above mentioned n parameters, the remaining $(n^2 - n)$ parameters have to be known.

Some examples have been treated to establish the validity of the HF domain methods. However, for the case of system identification, numerical instability is encountered in Example 13.2. Such instability is vigorous and the same has been illustrated via Figs. 13.4, 13.5 and 13.6. Since no regular pattern has immerged from different unstable results, the reason for such instability needs to be explored in detail.

A scrutiny of Table 13.2 for Example 13.3 reveals that there is a slight tendency of oscillation for $a_{12}(t)$ in the HF domain results due to numerical instability. While Fig. 13.9 can not reveal such oscillation, its magnified view, shown in Fig. 13.10, can. It is noted that, such oscillation can be reduced with increased number of sub-intervals m.

References

1. Ogata, K.: Modern Control Engineering (5th Ed.). Prentice Hall of India (2011)
2. Fogiel, M. (Chief Ed.): The Automatic Control Systems/Robotics Problem Solver. Research & Education Association, New Jersey (2000)
3. Roychoudhury, S., Deb, A., Sarkar, G.: Analysis and synthesis of time-varying systems via orthogonal hybrid functions (HF) in states space environment. Int. J. Dyn. Control **3**(4), 389–402 (2015)
4. Rao, G.P.: Piecewise Constant Orthogonal Functions and their Application in Systems and Control, LNCIS, vol. 55. Springer, Berlin (1983)
5. Jiang, J.H., Schaufelberger, W.: Block Pulse Functions and their Application in Control System, LNCIS, vol. 179. Springer, Berlin (1992)

Chapter 14
Time Invariant System Identification: Via 'Deconvolution'

Abstract Control system identification using hybrid function domain 'deconvolution' technique is discussed in this chapter. Both open loop as well as closed loop systems have been identified. Two numerical examples have been treated, and for clarity, eight figures and four tables have been presented.

System identification [1–3] is a common problem encountered in the design of control systems. The known components, usually the plant under control, are assumed to be described satisfactorily by its respective models. Then, the problem of identification is the characterization of the assumed model based on some observations or measurements.

It is well known that one may set up more than one model for a dynamic system, and in control system design the choice of the most suitable one depends heavily on the design method [3] being used. While in classical design, a nonparametric model such as an impulse response function is more appropriate. Kwong and Chen [4], in their work, presented a method based upon block pulse function (BPF) to identify an unknown plant modeled by impulse-response.

In this chapter, as the name suggests, the orthogonal hybrid function (HF) set [1], a combination of sample-and-hold functions (SHF) and triangular functions (TF), is employed to identify an unknowing plant using method of deconvolution.

14.1 Control System Identification Via 'Deconvolution'

The basic equation which relates the input-output of a control system in Laplace domain is given by

$$\begin{aligned} \mathrm{C(s)} &= \mathrm{G(s)\,R(s)} \\ \text{or}\quad \mathrm{G(s)} &= \frac{\mathrm{C(s)}}{\mathrm{R(s)}} \end{aligned} \tag{14.1}$$

A. Deb et al., *Analysis and Identification of Time-Invariant Systems, Time-Varying Systems, and Multi-Delay Systems using Orthogonal Hybrid Functions*,
Studies in Systems, Decision and Control 46, DOI 10.1007/978-3-319-26684-8_14

where $\mathbf{C}(s)$ is the Laplace transformed output, $\mathbf{G}(s)$ is the transfer function of the system and $\mathbf{R}(s)$ is the Laplace transformed input to the system. The problem of identification is basically to determine $\mathbf{G}(s)$ of Eq. (14.1).

14.1.1 Open Loop Control System Identification [5]

Consider two time functions $r(t)$ and $g(t)$ expanded in hybrid function domain. Considering the convolution between the time functions $r(t)$ and $g(t)$, we determine the output $y(t)$ in HF domain using Eq. (7.15). For $m = 4$, $T = 1$ s, the convolution result can be written as

$$
\begin{aligned}
y(t) = {} & \frac{h}{6}\begin{bmatrix} g_0 & g_1 & g_2 & g_3 \end{bmatrix}\begin{bmatrix} 0 & (2r_1+r_0) & (2r_2+r_1) & (2r_3+r_2) \\ 0 & (r_1+2r_0) & (r_2+4r_1+r_0) & (r_3+4r_2+r_1) \\ 0 & 0 & (r_1+2r_0) & (r_2+4r_1+r_0) \\ 0 & 0 & 0 & (r_1+2r_0) \end{bmatrix}\mathbf{S}_{(4)} \\
& + \frac{h}{6}\left[g_0(2r_1+r_0)+g_1(r_1+2r_0) \quad g_0(2r_2-r_1-r_0)+g_1(r_2+3r_1-r_0)+g_2(r_1+2r_0)\right. \\
& \quad g_0(2r_3-r_2-r_1)+g_1(r_3+3r_2-3r_1-r_0)+g_2(r_2+3r_1-r_0)+g_3(r_1+2r_0) \\
& \quad g_0(2r_4-r_3-r_2)+g_1(r_4+3r_3-3r_2-r_1)+g_2(r_3+3r_2-3r_1-r_0) \\
& \quad \left.+g_3(r_2+3r_1-r_0)+g_4(r_1+2r_0)\right]\mathbf{T}_{(4)}
\end{aligned}
\tag{14.2}
$$

Comparing the SHF vectors of output $y(t)$ with Eq. (14.2), we get

$$
\begin{bmatrix} y_0 & y_1 & y_2 & y_3 \end{bmatrix} = \frac{h}{6}\begin{bmatrix} g_0 & g_1 & g_2 & g_3 \end{bmatrix}\begin{bmatrix} 0 & (2r_1+r_0) & (2r_2+r_1) & (2r_3+r_2) \\ 0 & (r_1+2r_0) & (r_2+4r_1+r_0) & (r_3+4r_2+r_1) \\ 0 & 0 & (r_1+2r_0) & (r_2+4r_1+r_0) \\ 0 & 0 & 0 & (r_1+2r_0) \end{bmatrix}
\tag{14.3}
$$

where, because the first column contains all zeros, the matrix is singular and thus poses restriction on its inversion. To avoid this problem, the leading element of the first column may be replaced by a very small positive number ε. But before introducing ε, the scaling factor $\frac{h}{6}$ is multiplied with each element of the square matrix on the RHS of (14.3). This is done to avoid any adverse effect of the scaling factor $\frac{h}{6}$ upon ε.

Then Eq. (14.3) becomes

$$
\begin{bmatrix} y_0 & y_1 & y_2 & y_3 \end{bmatrix} = \begin{bmatrix} g_0 & g_1 & g_2 & g_3 \end{bmatrix}\begin{bmatrix} \varepsilon & \frac{h}{6}(2r_1+r_0) & \frac{h}{6}(2r_2+r_1) & \frac{h}{6}(2r_3+r_2) \\ 0 & \frac{h}{6}(r_1+2r_0) & \frac{h}{6}(r_2+4r_1+r_0) & \frac{h}{6}(r_3+4r_2+r_1) \\ 0 & 0 & \frac{h}{6}(r_1+2r_0) & \frac{h}{6}(r_2+4r_1+r_0) \\ 0 & 0 & 0 & \frac{h}{6}(r_1+2r_0) \end{bmatrix}
\tag{14.4}
$$

Now let,

$$\left.\begin{array}{ll} R_0 \triangleq 2r_1 + r_0, & R_6 \triangleq r_3 + 4r_2 + r_1, \\ R_1 \triangleq 2r_2 + r_1, & R_7 \triangleq r_4 + 4r_3 + r_2, \\ R_2 \triangleq 2r_3 + r_2, & R_8 \triangleq r_2 + r_1 - 2r_0, \\ R_3 \triangleq 2r_4 + r_3, & R_9 \triangleq r_3 + r_2 - 2r_1, \\ R_4 \triangleq r_1 + 2r_0, & R_{10} \triangleq r_4 + r_3 - 2r_2. \\ R_5 \triangleq r_2 + 4r_1 + r_0, & \end{array}\right\} \tag{14.5}$$

Using Eqs. (14.5) and (14.4) can be written as

$$\begin{aligned} [y_0 \quad y_1 \quad y_2 \quad y_3] &= [g_0 \quad g_1 \quad g_2 \quad g_3] \begin{bmatrix} \varepsilon & \frac{h}{6}R_0 & \frac{h}{6}R_1 & \frac{h}{6}R_2 \\ 0 & \frac{h}{6}R_4 & \frac{h}{6}R_5 & \frac{h}{6}R_6 \\ 0 & 0 & \frac{h}{6}R_4 & \frac{h}{6}R_5 \\ 0 & 0 & 0 & \frac{h}{6}R_4 \end{bmatrix} \\ \text{or, } [y_0 \quad y_1 \quad y_2 \quad y_3] &= [g_0 \quad g_1 \quad g_2 \quad g_3]\mathbf{R} \end{aligned} \tag{14.6}$$

$$\text{where, } \mathbf{R} \triangleq \begin{bmatrix} \varepsilon & \frac{h}{6}R_0 & \frac{h}{6}R_1 & \frac{h}{6}R_2 \\ 0 & \frac{h}{6}R_4 & \frac{h}{6}R_5 & \frac{h}{6}R_6 \\ 0 & 0 & \frac{h}{6}R_4 & \frac{h}{6}R_5 \\ 0 & 0 & 0 & \frac{h}{6}R_4 \end{bmatrix}$$

Hence,

$$[g_0 \quad g_1 \quad g_2 \quad g_3] = [y_0 \quad y_1 \quad y_2 \quad y_3]\mathbf{W} \tag{14.7}$$

where, $\mathbf{R}^{-1} \triangleq \mathbf{W}$

The computation of Eq. (14.7) was carried out with different small values of ε to determine the first four samples of the impulse response function $g(t)$. It was noted that the values of the coefficients $g_0, g_1, g_2, \ldots$, etc. did not alter with different values of ε.

This was investigated theoretically and it was noted that the matrix **R** being upper triangular, its inverse **W** is also upper triangular in nature. For an upper triangular matrix, if the leading element is ε, in its inverse, only the elements of the first row contains expressions involving the factor ε.

Now, in the row matrix $[y_0 \quad y_1 \quad y_2 \quad y_3]$, the first element y_0 is zero. Usually, for realistic causal systems, this is so. If the matrix multiplication in Eq. (14.7) was executed using this value of y_0, the first element g_0 of plant impulse response would have turned out to be zero. To avoid this problem, like the leading element of **R**, the first element y_0 of the output was also replaced by the same small number ε, used in **R**. Hence, when **W** was pre-multiplied by the row matrix in (14.7), the result did not contain any term involving ε. That is, the choice of ε does not affect the result.

To determine the fifth sample of $g(t)$, we proceed as follows.

We equate the last elements of the TF parts of Eqs. (11.13) and (14.2) to get

$$y_4 - y_3 = \frac{h}{6}[g_0(2r_4 - r_3 - r_2) + g_1(r_4 + 3r_3 - 3r_2 - r_1) + g_2(r_3 + 3r_2 - 3r_1 - r_0) + g_3(r_2 + 3r_1 - r_0) + g_4(r_1 + 2r_0)]$$

Solving for coefficient g_4, we have

$$g_4 = \frac{[y_4 - y_3] - \frac{h}{6}[g_0(R_3 - R_2) + g_1(R_7 - R_6) + g_2(R_6 - R_5) + g_3(R_5 - R_4)]}{\frac{h}{6}R_4}$$

For an analysis involving m terms, the generalized expression for the last coefficient of the function $g(t)$ is

$$g_m = \frac{[y_m - y_{(m-1)}] - \frac{h}{6}\left[g_0\{R_{(m-1)} - R_{(m-2)}\} + \sum_{i=1}^{m-1} g_i\{R_{(2m-i)} - R_{(2m-i-1)}\}\right]}{\frac{h}{6}R_m} \tag{14.8}$$

From Eqs. (14.7) and (14.8), we can calculate all the coefficient of the impulse response of the plant.

14.1.1.1 Numerical Examples

Now, we consider an open loop system, whose input $u(t)$ and output $y(t)$ are known in HF domain. With this information, we identify the plant $g(t)$ in HF domain. For such identification in HF domain, we use Eqs. (14.7) and (14.8).

Example 14.1 (vide Appendix B, Program no. 36) Consider an open loop system with input $r_1(t) = u(t)$ and output $y_1(t) = 1 - \exp(-t)$. The plant $g_1(t) = \exp(-t)$ is computed in HF domain via deconvolution and compared with the HF domain direct expansion of the plant impulse response.

Let, time $T = 1$ s, $m = 4$ and 10, and $\varepsilon = 10^{-5}$.

Tables 14.1 and 14.2 present the quantitative results and these are graphically compared with the exact impulse response of the plant in Figs. 14.1 and 14.2, for $m = 4$ and $m = 10$. Figure 14.3 shows the plots of percentage errors at different sample points for various values of m when the exact samples of $g_1(t)$ are compared with the samples computed via HF domain technique. Figure 14.4 shows the plot of AMP error for different values of m. It is noted that the value of average absolute percentage error decreases drastically with increasing m.

Table 14.1 Percentage error at different sample points of the impulse response of the plant $g_1(t)$ of Example 14.1 for $T = 1$ s, $m = 4$ and $\varepsilon = 10^{-5}$ (vide Appendix B, Program no. 36)

t (s)	Samples obtained from direct expansion of $g_1(t)$ g_{1d}	Samples of $g_1(t)$ obtained via HF domain analysis using Eqs. (14.7) and (14.8) g_{1c}	% Error $\varepsilon = \frac{g_{1d}-g_{1c}}{g_{1d}} \times 100$
0	1.00000000	1.00000000	0.00000000
$\frac{1}{4}$	0.77880078	0.76959374	1.18220832
$\frac{2}{4}$	0.60653066	0.60856725	−0.33577721
$\frac{3}{4}$	0.47236655	0.46474560	1.61335479
$\frac{4}{4}$	0.36787944	0.37115129	−0.88938024

Table 14.2 Percentage error at different sample points of the impulse response of the plant $g_1(t)$ of Example 14.1 for $T = 1$ s, $m = 10$ and $\varepsilon = 10^{-5}$ (vide Appendix B, Program no. 36)

t (s)	Samples obtained from direct expansion of $g_1(t)$ g_{1d}	Samples of $g_1(t)$ obtained via HF domain analysis using Eqs. (14.7) and (14.8) g_{1c}	% Error $\varepsilon = \frac{g_{1d}-g_{1c}}{g_{1d}} \times 100$
0	1.00000000	1.00000000	0.00000000
$\frac{1}{10}$	0.90483742	0.90325164	0.17525566
$\frac{2}{10}$	0.81873075	0.81888166	−0.01843180
$\frac{3}{10}$	0.74081822	0.73936899	0.19562594
$\frac{4}{10}$	0.67032005	0.67059450	−0.04094445
$\frac{5}{10}$	0.60653066	0.60519322	0.22050627
$\frac{6}{10}$	0.54881164	0.54918725	−0.06844146
$\frac{7}{10}$	0.49658530	0.49533940	0.25089517
$\frac{8}{10}$	0.44932896	0.44978740	−0.10202639
$\frac{9}{10}$	0.40656966	0.40539869	0.28801225
$\frac{10}{10}$	0.36787944	0.36840568	−0.14304711

From the error columns of Tables 14.1 and 14.2, and from Fig. 14.3, it is noted that there is a slight oscillation in the result, and such oscillations are reduced with increasing m.

14.1.2 Closed Loop Control System Identification

Consider the closed loop system shown in Fig. 11.8. The impulse response of the plant is $g(t)$. Hence, we obtain the output $y(t)$ simply by convolution of $e(t)$ and $g(t)$. In line with Eq. (7.15), for $m = 4$, $T = 1$ s, convolution result can directly be written as

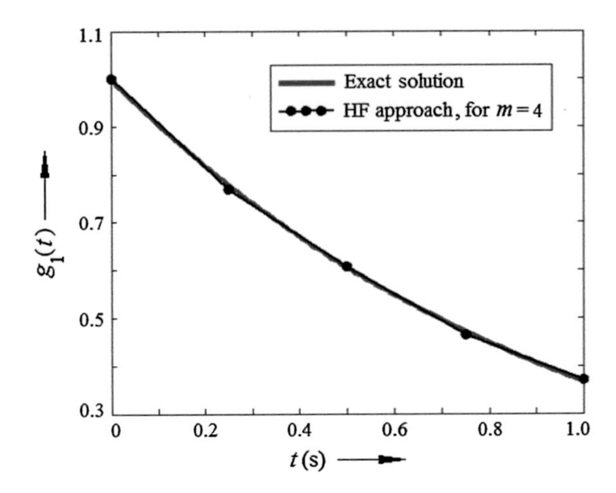

Fig. 14.1 Samples of the plant $g_1(t)$ of Example 14.1 identified using HF domain deconvolution are compared with the exact curve for $m = 4$ and $T = 1$ s (vide Appendix B, Program no. 36)

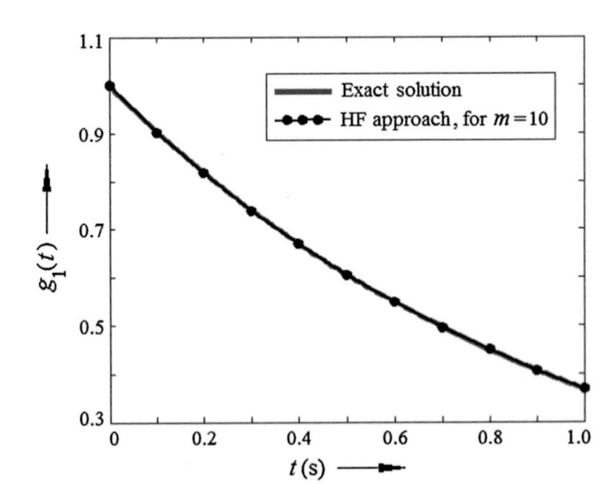

Fig. 14.2 Samples of the plant $g_1(t)$ of Example 14.1 identified using HF domain deconvolution are compared with the exact curve for $m = 10$ and $T = 1$ s (vide Appendix B, Program no. 36)

$$
\begin{aligned}
y(t) = \frac{h}{6}[g_0 \quad g_1 \quad g_2 \quad g_3]
\begin{bmatrix}
0 & (2e_1+e_0) & (2e_2+e_1) & (2e_3+e_2) \\
0 & (e_1+2e_0) & (e_2+4e_1+e_0) & (e_3+4e_2+e_1) \\
0 & 0 & (e_1+2e_0) & (e_2+4e_1+e_0) \\
0 & 0 & 0 & (e_1+2e_0)
\end{bmatrix}\mathbf{S}_{(4)} \\
+\frac{h}{6}[\{g_0(2e_1+e_0)+g_1(e_1+2e_0)\} \\
\{g_0(2e_2-e_1-e_0)+g_1(e_2+3e_1-e_0)+g_2(e_1+2e_0)\} \\
\{g_0(2e_3-e_2-e_1)+g_1(e_3+3e_2-3e_1-e_0)+g_2(e_2+3e_1-e_0)+g_3(e_1+2e_0)\} \\
\{g_0(2e_4-e_3-e_2)+g_1(e_4+3e_3-3e_2-e_1)+g_2(e_3+3e_2-3e_1-e_0) \\
+g_3(e_2+3e_1-e_0)+g_4(e_1+2e_0)\}]\mathbf{T}_{(4)}
\end{aligned}
\tag{14.9}
$$

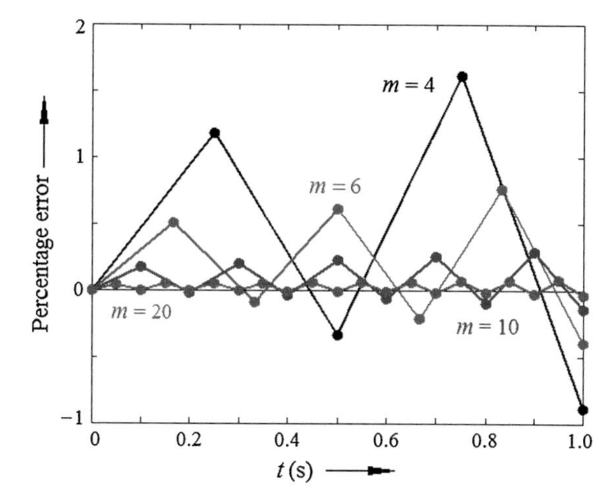

Fig. 14.3 Percentage error at different sample points computed via HF domain for different values of m with $T = 1$ s and $\varepsilon = 10^{-5}$ for Example 14.1. It is observed that percentage error decreases with increasing m

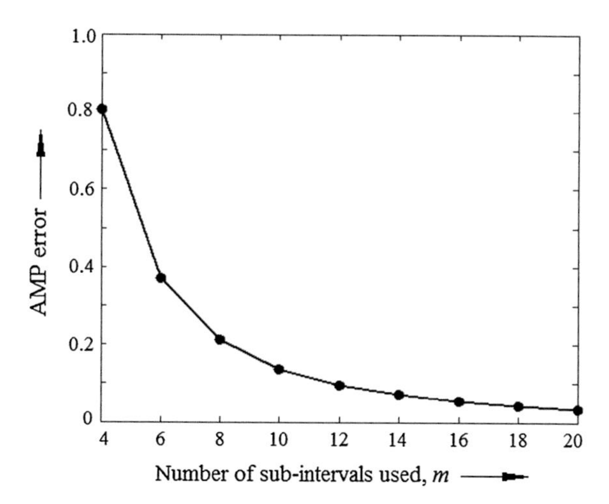

Fig. 14.4 AMP error of the samples computed via HF domain for different values of m for $T = 1$ s and $\varepsilon = 10^{-5}$ for Example 14.1

Comparing the SHF vectors of output $y(t)$ with Eq. (14.9), we get

$$\begin{bmatrix} y_0 & y_1 & y_2 & y_3 \end{bmatrix} = \frac{h}{6}\begin{bmatrix} g_0 & g_1 & g_2 & g_3 \end{bmatrix} \times \begin{bmatrix} 0 & (2e_1+e_0) & (2e_2+e_1) & (2e_3+e_2) \\ 0 & (e_1+2e_0) & (e_2+4e_1+e_0) & (e_3+4e_2+e_1) \\ 0 & 0 & (e_1+2e_0) & (e_2+4e_1+e_0) \\ 0 & 0 & 0 & (e_1+2e_0) \end{bmatrix} \tag{14.10}$$

Proceeding in the same way as for Eqs. (14.6) and (14.10) can be written as

$$\begin{aligned}
\begin{bmatrix} y_0 & y_1 & y_2 & y_3 \end{bmatrix} &= \begin{bmatrix} g_0 & g_1 & g_2 & g_3 \end{bmatrix} \begin{bmatrix} \varepsilon & \frac{h}{6}E_0 & \frac{h}{6}E_1 & \frac{h}{6}E_2 \\ 0 & \frac{h}{6}E_4 & \frac{h}{6}E_5 & \frac{h}{6}E_6 \\ 0 & 0 & \frac{h}{6}E_4 & \frac{h}{6}E_5 \\ 0 & 0 & 0 & \frac{h}{6}E_4 \end{bmatrix} \\
\text{or, } \begin{bmatrix} g_0 & g_1 & g_2 & g_3 \end{bmatrix} &= \begin{bmatrix} y_0 & y_1 & y_2 & y_3 \end{bmatrix} \begin{bmatrix} \varepsilon & \frac{h}{6}E_0 & \frac{h}{6}E_1 & \frac{h}{6}E_2 \\ 0 & \frac{h}{6}E_4 & \frac{h}{6}E_5 & \frac{h}{6}E_6 \\ 0 & 0 & \frac{h}{6}E_4 & \frac{h}{6}E_5 \\ 0 & 0 & 0 & \frac{h}{6}E_4 \end{bmatrix}^{-1}
\end{aligned} \tag{14.11}$$

As before, the computation is carried out for a typical values of ε, i.e., $\varepsilon = 10^{-5}$. We equate the last elements of the TF parts of Eqs. (11.13) and (14.9) to get

$$\begin{aligned}
y_4 - y_3 = \frac{h}{6}[&g_0(2e_4 - e_3 - e_2) + g_1(e_4 + 3e_3 - 3e_2 - e_1) + g_2(e_3 + 3e_2 - 3e_1 - e_0) \\
&+ g_3(e_2 + 3e_1 - e_0) + g_4(e_1 + 2e_0)]
\end{aligned}$$

Solving for coefficient g_4, we have

$$g_4 = \frac{[y_4 - y_3] - \frac{h}{6}[g_0(E_3 - E_2) + g_1(E_7 - E_6) + g_2(E_6 - E_5) + g_3(E_5 - E_4)]}{\frac{h}{6}E_4}$$

For an analysis involving m terms, the generalized expression for the last coefficient of the plant is

$$g_m = \frac{[y_m - y_{(m-1)}] - \frac{h}{6}\left[g_0\left\{E_{(m-1)} - E_{(m-2)}\right\} + \sum_{i=1}^{m-1} g_i\left\{E_{(2m-i)} - E_{(2m-i-1)}\right\}\right]}{\frac{h}{6}E_m} \tag{14.12}$$

From Eqs. (14.11) and (14.12), we can calculate all the coefficient of the plant impulse response

$$\mathbf{G} = \begin{bmatrix} g_0 & g_1 & g_2 & g_3 & g_4 \end{bmatrix} \tag{14.13}$$

If we intend to compute further samples of **G**, we need to know the respective HF coefficients of the error signal and the system response. This can easily be done by using (14.12).

14.1.2.1 Numerical Examples

An input $r(t)$ is applied to a causal SISO system, shown in Fig. 11.8, at $t = 0$. If the impulse response of the plant is $g(t)$, we obtain the output $y(t)$ simply by convolution of $e(t)$ and $g(t)$. Knowing $r(t)$, $h(t)$ and $y(t)$, we can employ Eqs. (14.11) and (14.12) for computing the samples of the plant $g(t)$ so that the result is obtained in HF domain.

Example 14.2 (vide Appendix B, Program no. 37) Consider the closed loop system of Fig. 11.8 with input $r_2(t) = u(t)$, feedback $h(t) = u(t)$ and output $y_2(t) = \frac{2}{\sqrt{3}}\exp\left(-\frac{t}{2}\right)\sin\left(\frac{\sqrt{3}}{2}\right)t$. The plant $g_2(t) = \exp(-t)$ is identified using HF domain via deconvolution and compared with the direct expansion of the plant.

Table 14.3 Samples of the plant of the system via direct expansion and samples obtained from HF domain identification for Example 14.2 for $T = 1$ s, $m = 4$ and $\varepsilon = 10^{-5}$ (vide Appendix B, Program no. 37)

t (s)	Samples obtained from direct expansion of $g_2(t)$ g_{2d}	Samples of $g_2(t)$ obtained via HF domain analysis using Eqs. (14.11) and (14.12) g_{2c}	% Error $\varepsilon = \frac{g_{2d}-g_{2c}}{g_{2d}} \times 100$
0	1.00000000	1.00000000	0.00000000
$\frac{1}{4}$	0.77880078	0.77656217	0.28744362
$\frac{2}{4}$	0.60653066	0.60289480	0.59945116
$\frac{3}{4}$	0.47236655	0.46770601	0.98663752
$\frac{4}{4}$	0.36787944	0.36267668	1.41425719

Table 14.4 Samples of the plant of the system via direct expansion and samples obtained from HF domain identification for Example 14.2 for $T = 1$ s, $m = 10$ and $\varepsilon = 10^{-5}$ (vide Appendix B, Program no. 37)

t (s)	Samples obtained from direct expansion of $g_2(t)$ g_{2d}	Samples of $g_2(t)$ obtained via HF domain analysis using Eqs. (14.11) and (14.12) g_{2c}	% Error $\varepsilon = \frac{g_{2d}-g_{2c}}{g_{2d}} \times 100$
0	1.00000000	1.00000000	0.00000000
$\frac{1}{10}$	0.90483742	0.90468094	0.01729318
$\frac{2}{10}$	0.81873075	0.81844307	0.03513718
$\frac{3}{10}$	0.74081822	0.74041517	0.05440636
$\frac{4}{10}$	0.67032005	0.66982167	0.07434830
$\frac{5}{10}$	0.60653066	0.60594820	0.09603091
$\frac{6}{10}$	0.54881164	0.54816110	0.11853513
$\frac{7}{10}$	0.49658530	0.49587437	0.14316533
$\frac{8}{10}$	0.44932896	0.44857050	0.16879892
$\frac{9}{10}$	0.40656966	0.40576860	0.19702903
$\frac{10}{10}$	0.36787944	0.36704625	0.22648450

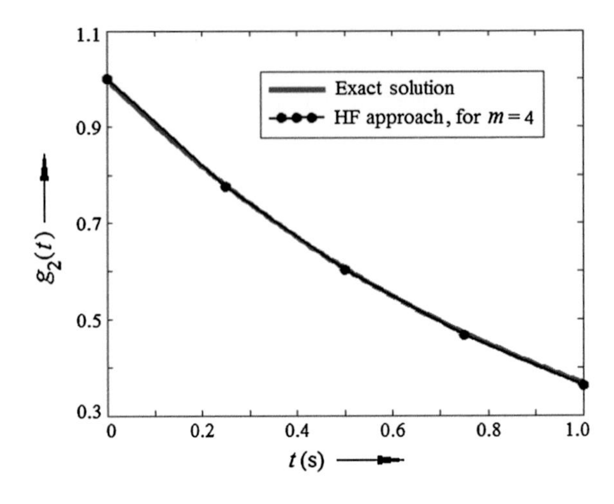

Fig. 14.5 Samples of the closed loop system of Example 14.2 identified using HF domain deconvolution along with the exact curve for $m = 4$ and $T = 1$ s (vide Appendix B, Program no. 37)

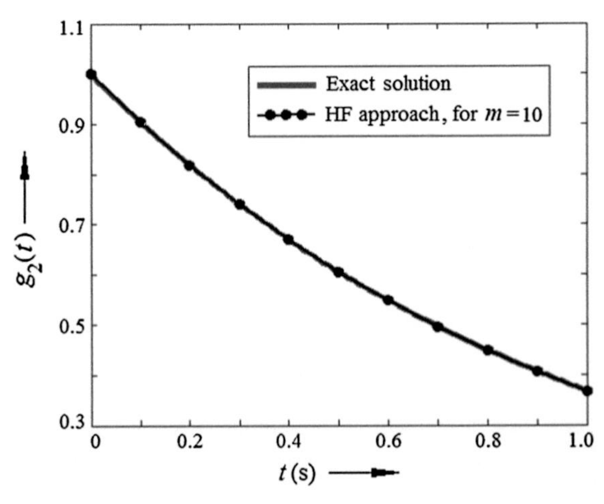

Fig. 14.6 Samples of the closed loop system of Example 14.2 identified using HF domain deconvolution for $m = 10$ and $T = 1$ s (vide Appendix B, Program no. 37)

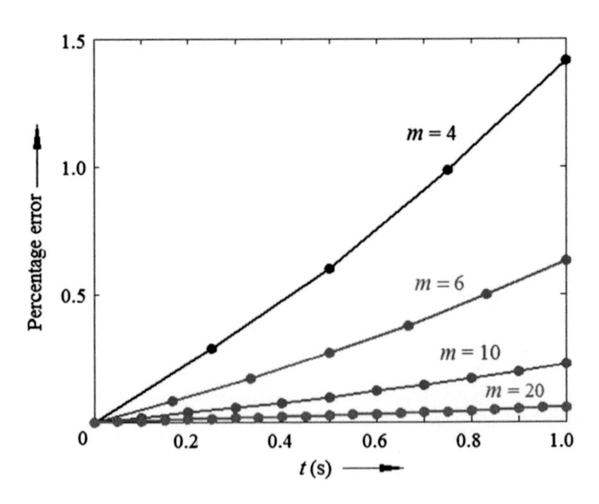

Fig. 14.7 Percentage error at different sample points, computed via HF domain, for different values of m with $T = 1$ s and $\varepsilon = 10^{-5}$ for Example 14.2. It is observed that percentage error decreases with increasing m

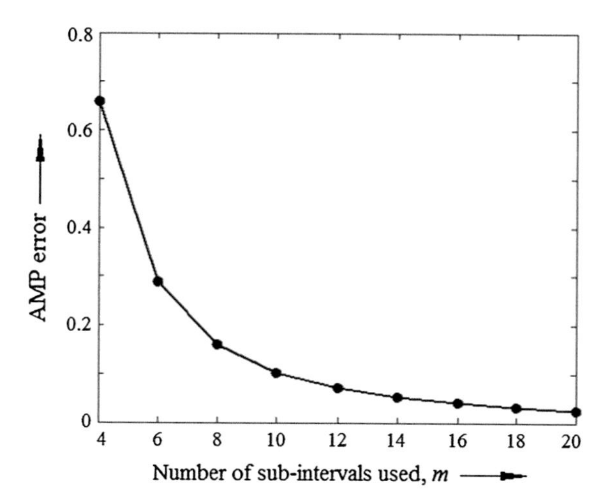

Fig. 14.8 AMP error of the samples for different values of m for $T = 1$ s and $\varepsilon = 10^{-5}$ for Example 14.2

Here Tables 14.3 and 14.4 present the quantitative results in HF domain along with the percentage errors. These facts are graphically compared with the exact impulse response of the plant in Figs. 14.5 and 14.6, for $m = 4$ and $m = 10$. Figures 14.7 and 14.8 show the characteristics of percentage errors for various values of m.

14.2 Conclusion

In this chapter, we have studied identification of linear control systems, open loop as well as closed loop, using the hybrid function platform employing the principle of 'deconvolution'.

As a foundation, convolution of basic component functions of the HF set was computed in Chap. 7. These sub-results were further used to determine the convolution of two time functions.

Applying the deconvolution concept, an open loop system has been identified for $m = 4$ and 10. Percentage errors at different sample points have been computed and presented in Tables 14.1 and 14.2. Figures 14.3 and 14.4 show that, with increasing m, the error is reduced.

Similarly, a closed loop system was identified successfully for $m = 4$ and $m = 10$. The percentage errors in identification, at different sampling instants, are tabulated in Tables 14.3 and 14.4, and for better clarity, they have been translated graphically in Figs. 14.7 and 14.8.

In case of system analysis or identification, block pulse domain approach showed oscillations in many instances indicating the onset of numerical instability.

We have noted such cases of instability in references [1, 4, 6–8]. But HF based technique achieves the objective without any numerical instability, even with a small number of sub-intervals m.

References

1. Deb, A., Sarkar, G., Sen, S.K.: Linearly pulse-width modulated block pulse functions and their application to linear SISO feedback control system identification. Proc. IEE, Part D, Control Theor. Appl. **142**(1), 44–50 (1995)
2. Unbehauen, H., Rao, G.P.: Identification of continuous systems. North-Holland, Amsterdam (1987)
3. Ljung, L.: System identification: theory for the user. Prentice-Hall Inc., New Jersey (1985)
4. Kwong, C.P., Chen, C.F.: Linear feedback system identification via block pulse functions. Int. J. Syst. Sci. **12**(5), 635–642 (1981)
5. Biswas, A.: Analysis and synthesis of continuous control systems using a set of orthogonal hybrid functions. Ph. D. dissertation, University of Calcutta (2015)
6. Jiang, J.H., Schaufelberger, W.: Block pulse functions and their application in control system, LNCIS, vol. 179. Springer, Berlin (1992)
7. Deb, A., Ghosh, S.: Power electronic systems, walsh analysis with MATLAB®. CRC Press, Boca Raton (2014)
8. Deb, A., Sarkar, G., Biswas, A., Mandal, P.: Numerical instability of deconvolution operation via block pulse functions. J. Franklin Inst. **345**(4), 319–327 (2008)

Chapter 15
System Identification: Parameter Estimation of Transfer Function

Abstract In this chapter, parameter estimation of the transfer function of a linear system has been done employing many non-sinusoidal orthogonal function sets, e.g., block pulse functions, non-optimal block pulse functions, triangular functions, hybrid functions and sample-and-hold functions. A comparative study of the parameters estimated by different methods are made with focus on estimated errors. One numerical example has been studied extensively and ten figures and fourteen tables have been presented as illustration.

A typical problem considered in system identification [1–4] is the design of estimators trying to recover a discrete time linear time invariant (LTI) system based on noise corrupted output sequence resulting from a known input sequence. The unknown components, usually the plant under control, are assumed to be described satisfactorily by its respective models.

In this chapter, as the name suggests, the orthogonal hybrid function (HF) set, a combination of sample-and-hold functions (SHF) and triangular functions (TF), is employed to identify an unknowing plant using method of deconvolution.

15.1 Transfer Function Identifications

The estimation method consists of finding a rough estimate of the impulse response from the sampled input and output data. The impulse response estimate is then transformed to a two dimensional time-frequency mapping [5]. The mapping provides a clear graphical method for distinguishing the noise from the system dynamics. The information believed to correspond to noise is discarded and a cleaner estimate of the impulse response is obtained from the remaining information. The new impulse response estimate is then used to obtain the transfer function estimate [6].

There are many transfer function estimation techniques available, given data limitations, but these may yield poor results. One such method is the Empirical Transfer Function Estimate (ETFE), which estimates the transfer function by taking

A. Deb et al., *Analysis and Identification of Time-Invariant Systems, Time-Varying Systems, and Multi-Delay Systems using Orthogonal Hybrid Functions*,
Studies in Systems, Decision and Control 46, DOI 10.1007/978-3-319-26684-8_15

the ratios of the Fourier transforms of the output $y(t)$ and the input $u(t)$. The estimate is given by

$$\hat{\mathbf{G}}(\omega) = \frac{\mathcal{F}\{y(t)\}}{\mathcal{F}\{u(t)\}} \quad (15.1)$$

If the data set is noisy, the resulting estimate is also noisy. Unfortunately, taking more data points does not help. The variance does not decrease as the number of data points increase, because there is no feature of information compression. There are as many independent estimates as there are data points [4].

Parametric estimation methods are another class of system identification techniques. The motivation behind these methods is to find an estimate or model of the system in terms of a small number (compared to the number of measurements) of numerical values or parameters. A linear system is typically represented by

$$y(t) = \mathbf{G}(s)u(t) + \mathbf{H}(s)e(t) \quad (15.2)$$

where $e(t)$ is the disturbance, $\mathbf{G}(s)$ is the transfer function from input to output, $\mathbf{H}(s)$ is the transfer function from disturbance to output, and s is the shift operator ($s = e^{j\omega}$) used when dealing with discrete systems. The most generalized model structure is

$$\mathbf{A}(s)y(t) = \frac{\mathbf{B}(s)}{\mathbf{F}(s)}u(t) + \frac{\mathbf{C}(s)}{\mathbf{D}(s)}e(t) \quad (15.3)$$

where $\mathbf{A}(s)$, $\mathbf{B}(s)$, $\mathbf{C}(s)$, $\mathbf{D}(s)$, $\mathbf{F}(s)$ are all parameter polynomials to be estimated.

15.2 Pade Approximation

The Pade approximant [7] of a given power series is a rational function of numerator degree L and denominator degree M whose power series agrees with the given one up to degree $L + M$ inclusively. A collection of Pade approximants formed by using a suitable set of values of L and M often provides a means of obtaining information about the function outside its circle of convergence, and of more rapidly evaluating the function within its circle of convergence. Applications of these ideas in physics, chemistry, electrical engineering, and other areas have led to a large number of generalizations of Pade approximants that are tailor-made for specific applications. Applications to statistical mechanics and critical phenomena are extensively covered, and there are newly extended sections devoted to circuit design, matrix Pade approximation, computational methods, and integral and algebraic approximants.

Approximants derived by expanding a function as a ratio of two power series and determining both the numerator and denominator coefficients. Pade approximations are usually superior to Taylor series when functions contain poles, because the use of rational functions allows them to be well-represented.

The relation between the Taylor series expansion and the function is given classically by the statement that if the series converges absolutely to an infinity differentiable function, then the series defines the function uniquely and the function uniquely defines the series.

The Pade approximants are a particular type of rational approximation. The L, M Pade approximant is denoted by

$$\left[\frac{L}{M}\right] = \frac{\mathrm{P}_L(x)}{\mathrm{Q}_M(x)} = f(x) \tag{15.4}$$

where $\mathrm{P}_L(x)$ is a polynomial of degree less than or equal to L, and $\mathbf{Q}_M(x)$ is a polynomial of degree less than or equal to M. Sometimes, when the function f being approximated is not clear from the context, the function name is appended as a subscript $[L/M]_f$. The formal power series

$$f(x) = \sum_{j=0}^{\infty} f_j x^j \tag{15.5}$$

determines the coefficients by the equation

$$f(x) - \frac{\mathrm{P}_L(x)}{\mathrm{Q}_M(x)} = O\left(x^{L+M+1}\right). \tag{15.6}$$

This is the classical definition. The Baker [8] definition adds the condition

$$\mathbf{Q}_M(0) = 1. \tag{15.7}$$

$$\text{Let, } f(x) = c_0 + c_1 x + c_2 x^2 + \cdots \tag{15.8}$$

Let the Pade approximation (that is the reduced model) be

$$f(x) = \frac{\mathbf{P}_L(x)}{\mathrm{Q}_M(x)} = \frac{a_0 + a_1 x + a_2 x^2 + \cdots + a_L x^L}{b_0 + b_1 x + b_2 x^2 + \cdots + b_{M-1} x^{M-1} + x^M} \tag{15.9}$$

Hence, in accordance with the definition, Eqs. (15.8) and (15.9) yield the following set of linear equations:

$$
\left.\begin{aligned}
a_0 &= b_0c_0 \\
a_1 &= b_0c_1 + b_1c_0 \\
a_2 &= b_0c_2 + b_1c_1 + b_2c_0 \\
&\vdots \\
a_m &= b_0c_m + b_1c_{m-1} + \cdots + b_mc_0 \\
0 &= b_0c_{m+1} + b_1c_m + \cdots + b_{m+1}c_0 \\
&\vdots \\
0 &= b_0c_{m+n} + b_1c_{m+n-1} + \cdots + b_{n-1}c_{m+1} + c_m
\end{aligned}\right\} \qquad (15.10)
$$

which serve to determine the coefficients of Eq. (15.10) uniquely.

Baker et al. [8] introduced the concept of Pade approximation about more than one point. They suggested that the Pade approximation be required to exactly satisfy conditions at other points (they imposed the value of the function at infinity on the Pade approximate) as well as at the origin. The required modifications in the linear Eqs. (15.10) are very simple. The equation which makes the last power series coefficient of the function and its approximant equal is replaced by one that makes the Pade approximant equal to a given value at infinity.

15.3 Parameter Estimation of the Transfer Function of a Linear System

Now we employ the set of hybrid functions to estimate the parameters of a linear system from its impulse response [6] data.

Let $\mathbf{h}(t)$ be the $p \times r$ impulse response matrix of a linear time-invariant multivariable system where $\mathbf{h}(t)$ is specified graphically or analytically. Also, let $\mathbf{H}(s) = \mathcal{L}[\mathbf{h}(t)]$ be the transfer function matrix of the system.

Now consider the rational function matrix $\mathbf{G}(s)$ of the form

$$
\mathbf{G}(s) = \frac{\mathbf{B_p}\, s^p + \mathbf{B_{p-1}}\, s^{p-1} + \cdots + \mathbf{B_1}\, s + \mathbf{B_0}}{s^q + \mathbf{a_{q-1}}\, s^{q-1} + \cdots + \mathbf{a_1}\, s + \mathbf{a_0}}, \quad p < q \qquad (15.11)
$$

Expansions of $\mathbf{H}(s)$ and $\mathbf{G}(s)$ in power series are as follows:

$$
\mathbf{H}(s) = \sum_{k=0}^{\infty} \mathrm{H_k}\, s^k \qquad (15.12)
$$

$$
\mathbf{G}(s) = \sum_{k=0}^{\infty} \mathrm{G_k}\, s^k \qquad (15.13)
$$

The problem now may be stated as

Given $\mathbf{h}(t)$, the impulse response, find the parameters $a_0,\ a_1,\ \ldots,\ a_{n-1}$ and the elements of the matrices $\mathbf{B}_0, \mathbf{B}_1, \cdots \mathbf{B}_p$ of $\mathbf{G}(s)$ such that $\mathbf{G}(s)$ matches $\mathbf{H}(s)$ in the Pade sense [7], i.e., such that the power series coefficient matrices of $\mathbf{H}(s)$ and $\mathbf{G}(s)$ match up to the power $(p + q)$.

Thus,

$\mathbf{H}_k = \mathbf{G}_k$, where, $k = 0, 1, \cdots, p+q$.

To solve the problem we first determine the power series coefficient matrices $\mathbf{H}_k$ of $\mathbf{H}(s)$. We use the following power series coefficient formula

$$\mathbf{H}(s) = \mathbf{H_0}s^0 + \mathbf{H_1}s^1 + \mathbf{H_2}s^2 + \mathbf{H_3}s^3 + \cdots + \mathbf{H}_k s^k + \cdots \tag{15.14}$$

Now, Eq. (7.14) is differentiated with respect to s to obtain

$$\frac{\partial \mathbf{H}(s)}{\partial s} = \mathbf{H_1} + 2\mathbf{H_2}\, s^1 + 3\mathbf{H_3}\, s^2 + \cdots + k\, \mathbf{H}_k\, s^{(\mathrm{k}-1)} + \cdots \tag{15.15}$$

Putting, $s = 0$ in Eq. (15.15), we get

$$\left.\frac{\partial \mathbf{H}(s)}{\partial s}\right|_{s=0} = \mathbf{H_1}$$

Similarly, we get

$$\left.\frac{\partial^k \mathbf{H}(s)}{\partial s^k}\right|_{s=0} = k!\ \mathbf{H}_k \tag{15.16}$$

where, $k = 0,\ 1,\ \cdots,\ (p+q)$.

Again, the transfer function $\mathbf{H}(s)$ can be written as

$$\mathbf{H}(s) = \int_0^{\infty} \mathbf{h}(t) \exp(-st) \mathrm{d}t \tag{15.17}$$

Now, differentiating Eq. (15.17) with respect to s, we get

$$\begin{aligned}
\frac{\partial \mathbf{H}(s)}{\partial s} &= (-1)\int_0^\infty t\,\mathbf{h}(t)\ \exp(-st)\,\mathrm{d}t \\
&\vdots \\
&\vdots \\
\frac{\partial^k \mathbf{H}(s)}{\partial s^k} &= (-1)^k \int_0^\infty t^k\,\mathbf{h}(t)\ \exp(-st)\,\mathrm{d}t
\end{aligned} \tag{15.18}$$

Putting $s = 0$ in Eq. (15.18), we get

$$\left.\frac{\partial^k \mathbf{H}(s)}{\partial s^k}\right|_{s=0} = (-1)^k \int_0^\infty t^k\,\mathbf{h}(t)\,\mathrm{d}t \tag{15.19}$$

Assuming the system to be asymptotically stable, we consider a large positive number λ to be the upper limit of the integral in (15.19) such that it converges to a finite value [9]. Thus, using Eqs. (15.16) and (15.19), we get

$$\mathbf{H}_k = \frac{(-1)^k}{k!}\int_0^\lambda t^k\,\mathbf{h}(t)\,\mathrm{d}t \tag{15.20}$$

Let, $t = \lambda\tau$. Then $\mathrm{d}t = \lambda\,\mathrm{d}\tau$, for $t = 0$, $\tau = 0$ and $t = \lambda$, $\tau = 1$. Thus Eq. (15.20) becomes

$$\begin{aligned}
\mathbf{H}_k &= \frac{(-1)^k}{k!}\lambda\int_0^1 \tau^k\lambda^k\,\mathbf{h}(\lambda\tau)\,\mathrm{d}\tau \\
\text{or,}\quad \mathbf{H}_k &= (-1)^k\lambda^{(k+1)}\int_0^1 \frac{\tau^k}{k!}\,\mathbf{h}(\lambda\tau)\,\mathrm{d}\tau
\end{aligned} \tag{15.21}$$

15.3.1 Using Block Pulse Functions

In block pulse function (BPF) domain [10], referring to Sect. 2.1, $\mathbf{h}(\lambda\tau)$ can be written as

$$\mathbf{h}(\lambda\tau) = \sum_{i=0}^{(m-1)} \mathbf{D1}_{(l-1)i}\psi_i(\tau), \quad 0 \leq \tau < 1 \quad i = 0, 1, 2, \ldots, (m-1) \tag{15.22}$$

where, $\mathbf{D1}_{(l-1)i} = \begin{bmatrix} d1_{0i} & d1_{1i} & d1_{2i} \cdots d1_{(l-1)i} \end{bmatrix}^{\mathrm{T}}$
and $d1_{(l-1)i}$'s are the expansion coefficients in block pulse domain.
and $\mathbf{D1}_{(l-1)\,i} = \frac{1}{h} \int_{ih}^{(i+1)h} \mathbf{h}(\lambda\tau)\,\psi_i(\tau)\,\mathrm{d}\tau$
where, l is the number of input variables of the system.
Again, let

$$\mathbf{u}_0^{\mathrm{T}}\boldsymbol{\Psi}(\tau) = \sum_{i=0}^{m-1} \psi_i(\tau) = 1 \tag{15.23}$$

where, $\mathbf{u}_0^{\mathrm{T}} = \underbrace{[1 \quad 1 \quad \cdots \quad 1 \quad 1]}_{m\ \text{columns}}$

Now, we know

$$\underbrace{\int_0^{\tau}\int_0^{\tau}\int_0^{\tau}\cdots\int_0^{\tau}}_{k\ \text{times}} (\mathrm{d}\tau)^k = \frac{(\tau)^k}{k!} \underbrace{\int_0^{\tau}\int_0^{\tau}\int_0^{\tau}\cdots\int_0^{\tau}}_{k\ \text{times}} \mathbf{u}_0^{\mathrm{T}}\boldsymbol{\Psi}(\tau)(\mathrm{d}\tau)^k = \mathbf{u}_0^{\mathrm{T}}\,\mathbf{P}^k\boldsymbol{\Psi}(\tau) \tag{15.24}$$

where, $\mathbf{P}$ is the operational matrix for integration in block pulse function domain [10].

Then, from Eqs. (15.21)–(15.24), we can write

$$\mathbf{H}_k = (-1)^k\,\lambda^{(k+1)}\,\mathbf{u}_0^{\mathrm{T}}\,\mathbf{P}^k \sum_{i=0}^{m-1} \left[\int_0^1 \psi_i(\tau)\boldsymbol{\Psi}(\tau)\,\mathrm{d}\tau\right] \mathbf{D1}_{(l-1)i} \tag{15.25}$$

Due to orthogonal property,

$$\int_0^1 \psi_i(\tau)\boldsymbol{\Psi}(\tau)\mathrm{d}\tau = \mathbf{e}_i = \underbrace{[0 \quad \cdots \quad 0 \quad \overset{(i+1)^{\text{th}}\ \text{position}}{\overset{\downarrow}{h}} \quad 0 \quad \cdots \quad 0]}_{m\ \text{columns}}^{\mathrm{T}}$$

So, Eq. (15.25) becomes,

$$\mathbf{H}_k = (-1)^k\lambda^{(k+1)}\mathbf{u}_0^{\mathrm{T}}\,\mathbf{P}^k \sum_{i=0}^{m-1} \mathbf{e}_i\,\mathbf{D1}_{(l-1)i} \tag{15.26}$$

For instance, if we take $m = 4$, then

$$\sum_{i=0}^{4-1} \mathbf{e}_i\,\mathbf{D1}_{(l-1)i} = \begin{bmatrix} h \\ 0 \\ 0 \\ 0 \end{bmatrix} \mathbf{D1}_{(l-1)0} + \begin{bmatrix} 0 \\ h \\ 0 \\ 0 \end{bmatrix} \mathbf{D1}_{(l-1)1} + \begin{bmatrix} 0 \\ 0 \\ h \\ 0 \end{bmatrix} \mathbf{D1}_{(l-1)2} + \begin{bmatrix} h \\ 0 \\ 0 \\ 0 \end{bmatrix} \mathbf{D1}_{(l-1)3}$$

Thus, Eq. (15.26) becomes

$$\mathbf{H}_k = (-1)^k \lambda^{(k+1)}\,\mathbf{u}_0^{\mathrm{T}}\,\mathbf{P}^k \left\{ \begin{bmatrix} h \\ 0 \\ 0 \\ 0 \end{bmatrix} \mathbf{D1}_{(l-1)0} + \begin{bmatrix} 0 \\ h \\ 0 \\ 0 \end{bmatrix} \mathbf{D1}_{(l-1)1} + \begin{bmatrix} 0 \\ 0 \\ h \\ 0 \end{bmatrix} \mathbf{D1}_{(l-1)2} + \begin{bmatrix} 0 \\ 0 \\ 0 \\ h \end{bmatrix} \mathbf{D1}_{(l-1)3} \right\}$$

$$\text{or,}\quad \mathbf{H}_k = \mathbf{a} \sum_{i=0}^{3} \mathbf{e}_i\,\mathbf{D1}_{(l-1)i} \tag{15.27}$$

where, $\mathbf{a} \triangleq (-1)^k \lambda^{(k+1)}\,\mathbf{u}_0^{\mathrm{T}}\,\mathbf{P}^k$

Now, we consider a SISO system having a transfer function of the following form:

$$\frac{a_0 + a_1 s}{b_0 + b_1 s + s^2} = \mathbf{H_0} + \mathbf{H_1} s + \mathbf{H_2} s^2 + \mathbf{H_3} s^3 \tag{15.28}$$

From Eq. (15.28), equating the coefficients of like powers of s, we get

$$a_0 = b_0 \mathbf{H_0} \tag{15.29}$$

$$a_1 = b_0 \mathbf{H_1} + b_1 \mathbf{H_0} \tag{15.30}$$

$$b_0 \mathbf{H_2} + b_1 \mathbf{H_1} = -\mathbf{H_0} \tag{15.31}$$

$$b_0 \mathbf{H_3} + b_1 \mathbf{H_2} = -\mathbf{H_1} \tag{15.32}$$

Solving from (15.29)–(15.32), the a_i's and b_i's may be obtained as

$$\left.\begin{aligned}
a_0 &= \frac{1}{\Delta}\begin{vmatrix} 0 & 0 & -\mathbf{H}_0 & 0 \\ 0 & 1 & -\mathbf{H}_1 & -\mathbf{H}_0 \\ -\mathbf{H}_0 & 0 & \mathbf{H}_2 & \mathbf{H}_1 \\ -\mathbf{H}_1 & 0 & \mathbf{H}_3 & \mathbf{H}_2 \end{vmatrix} \\
a_1 &= \frac{1}{\Delta}\begin{vmatrix} 1 & 0 & -\mathbf{H}_0 & 0 \\ 0 & 0 & -\mathbf{H}_1 & -\mathbf{H}_0 \\ 0 & -\mathbf{H}_0 & \mathbf{H}_2 & \mathbf{H}_1 \\ 0 & -\mathbf{H}_1 & \mathbf{H}_3 & \mathbf{H}_2 \end{vmatrix} \\
b_0 &= \frac{1}{\Delta}\begin{vmatrix} 1 & 0 & 0 & 0 \\ 0 & 1 & 0 & -\mathbf{H}_0 \\ 0 & 0 & -\mathbf{H}_0 & \mathbf{H}_1 \\ 0 & 0 & -\mathbf{H}_1 & \mathbf{H}_2 \end{vmatrix} \\
b_1 &= \frac{1}{\Delta}\begin{vmatrix} 1 & 0 & -\mathbf{H}_0 & 0 \\ 0 & 1 & -\mathbf{H}_1 & 0 \\ 0 & 0 & \mathbf{H}_2 & -\mathbf{H}_0 \\ 0 & 0 & \mathbf{H}_3 & -\mathbf{H}_1 \end{vmatrix} \\
\text{where, } \Delta &= \begin{vmatrix} 1 & 0 & -\mathbf{H}_0 & 0 \\ 0 & 1 & -\mathbf{H}_1 & -\mathbf{H}_0 \\ 0 & 0 & \mathbf{H}_2 & \mathbf{H}_1 \\ 0 & 0 & \mathbf{H}_3 & \mathbf{H}_2 \end{vmatrix}
\end{aligned}\right\} \tag{15.33}$$

15.3.1.1 Numerical Example

Example 15.1 Consider the function

$$g(t) = \exp(-t) - \exp(-2t) \tag{15.34}$$

We express it in s-domain in the form as in (15.28). That is

$$\mathbf{H}(s) = \frac{1}{s^2 + 3s + 2} = \frac{a_1 s + a_0}{s^2 + b_1 s + b_0} \tag{15.35}$$

Then using Eqs. (15.27) and (15.33), the unknown parameters may be solved for $m = 64$ and $\lambda = 1$, $\lambda = 6$ and $\lambda = 12$ as shown in Table 15.1. Also, percentage errors for such estimation for $\lambda = 12$ is presented in Table 15.2.

Table 15.1 Estimated parameters of the transfer function of Example 15.1 for m = 64 and For three different values of λ, in block pulse function domain

Parameters	Actual	λ = 1	λ = 6	λ = 12
a_0	1	2.54176555	0.76514514	0.97307956
a_1	0	- 0.25154866	0.05825146	0.01235613
b_0	2	12.72230064	1.55297667	1.94618304
b_1	3	6.11098678	2.40403634	2.94373198

Table 15.2 Percentage errors for parameter estimation of the transfer function of Example 15.1 for m = 64 and λ = 12, in BPF domain

Parameters	Actual values c_a	BPF domain values c_h	% Error $\varepsilon = \frac{(c_a - c_h)}{c_a} \times 100$
a_0	1	0.97307956	2.69204374
a_1	0	0.01235613	–
b_0	2	1.94618304	2.69084797
b_1	3	2.94373198	1.87560054

15.3.2 Using Non-optimal Block Pulse Functions (NOBPF)

Referring to Sect. 1.2.11, if we use a set of non-optimal block pulse functions $\mathbf{\Psi}'(\tau)$ [11] instead of the traditional block pulse functions, the results obtained are slightly different.

In non-optimal block pulse function domain $\mathbf{h}(\lambda\tau)$ can be written as

$$\mathbf{h}(\lambda\tau) = \sum_{i=0}^{m-1} \mathbf{D2}_{(l-1)i}\, \psi'_i(\tau), \quad 0 \le \tau < 1 \tag{7.36}$$

where, each element, $d2_{0i}$ (say), of $\mathbf{D2}_{(l-1)i}$ is the average of two consecutive samples, c_i and $c_{(i+1)}$ (say), of each component of the function $\mathbf{h}(\lambda\tau)$. That is

$$d2_{0i} = \frac{\left[c_i + c_{(i+1)}\right]}{2}$$

As before

$$\mathbf{u}_0^{\mathrm{T}}\, \mathbf{\Psi}'(\tau) = \sum_{i=0}^{(m-1)} \psi'_i(\tau) = 1$$

Using the above relation in Eq. (15.24), we have, as per Eq. (15.25)

$$\mathbf{H}_k = (-1)^k \lambda^{(k+1)} \mathbf{u}_0^{\mathrm{T}} \mathbf{P}^k \sum_{i=0}^{m-1} \left[\int_0^1 \psi_i'(\tau) \Psi'(\tau)\, \mathrm{d}\tau \right] \mathbf{D2}_{(l-1)i} \tag{15.37}$$

where, $\mathbf{P}$ is the operational matrix for integration in non-optimal block pulse function domain.

It is to be noted that for both the traditional block pulse function based derivation and non-optimal block pulse function based derivation $\mathbf{P}$ remains unaltered.

So, Eq. (15.37) becomes,

$$\mathbf{H}_k = (-1)^k \lambda^{(k+1)} \mathbf{u}_0^{\mathrm{T}} \mathbf{P}^k \sum_{i=0}^{m-1} \mathbf{e}_i \mathbf{D2}_{(l-1)i} \tag{15.38}$$

which is similar to Eq. (15.26).

If we take $m = 4$, then, following Sect. 15.3.1, we get

$$\mathbf{H}_k = (-1)^k \lambda^{(k+1)} \mathbf{u}_0^{\mathrm{T}} \mathbf{P}^k \left\{ \begin{bmatrix} h \\ 0 \\ 0 \\ 0 \end{bmatrix} \mathbf{D2}_{(l-1)0} + \begin{bmatrix} 0 \\ h \\ 0 \\ 0 \end{bmatrix} \mathbf{D2}_{(l-1)1} + \begin{bmatrix} 0 \\ 0 \\ h \\ 0 \end{bmatrix} \mathbf{D2}_{(l-1)2} + \begin{bmatrix} 0 \\ 0 \\ 0 \\ h \end{bmatrix} \mathbf{D2}_{(l-1)3} \right\}$$

$$\text{or,}\quad \mathbf{H}_k = \mathbf{b} \sum_{i=0}^{3} \mathbf{e}_i \mathbf{D2}_{(l-1)i} \tag{15.39}$$

where, $\mathbf{b} \triangleq (-1)^k \lambda^{(k+1)} \mathbf{u}_0^{\mathrm{T}} \mathbf{P}^k$

We use Eqs. (15.39) to solve for the parameters in (15.35) in NOBPF domain.

For $m = 64$ and $\lambda = 1$, $\lambda = 6$ and $\lambda = 12$ the results are shown in Table 15.3. Also, percentage errors for such estimation for $\lambda = 12$ is presented in Table 15.4.

Table 15.3 Estimated parameters of the transfer function of Example 15.1 for $m = 64$ and for three different values of λ, in non-optimal block pulse function domain

Parameters	Actual	$\lambda = 1$	$\lambda = 6$	$\lambda = 12$
a_0	1	2.54147603	0.77203411	0.96755155
a_1	0	−0.25160253	0.05617966	0.00373086
b_0	2	12.72227298	1.55404255	1.94648567
b_1	3	6.11095662	2.40494375	2.94403456

Table 15.4 Percentage errors for parameter estimation of the transfer function of Example 15.1 for m = 64 and λ = 12, in non-optimal block pulse function domain

Parameters	Actual values c_a	NOBPF domain values c_h	% Error $\varepsilon = \frac{(c_a - c_h)}{c_a} \times 100$
a_0	1	0.96755155	3.24484513
a_1	0	0.00373086	–
b_0	2	1.94648567	2.67571651
b_1	3	2.94403456	1.86551468

15.3.3 *Using Triangular Functions (TF)*

In the triangular function domain [12], referring to Sect. 2.3, $\mathbf{h}(\lambda\tau)$ is given by

$$\mathbf{h}(\lambda\tau) = \sum_{i=0}^{m-1} \left[\mathbf{D31}_{(l-1)i} T1_i(\tau) + \mathbf{D32}_{(l-1)i} T2_i(\tau)\right], \quad 0 \le \tau < 1 \tag{15.40}$$

where, each element, $d31_{0i}$ (say), of $\mathbf{D31}_{(l-1)i}$ and $d32_{0i}$ (say), of $\mathbf{D32}_{(l-1)i}$ are i- th and $(i$ + 1)th samples of each component of the function $\mathbf{h}(\lambda\tau)$.

For triangular functions, the unit step function is represented as

$$u(t) = \mathbf{u_0}^{\mathrm{T}}[\mathbf{T1}(\tau) + \mathbf{T2}(\tau)]$$

To determine the formula for repeated integration in triangular function domain, we proceed from the basic relation for first integration which is

$$\int \mathbf{T1}\,\mathrm{d}t = \mathbf{P1}\,\mathbf{T1} + \mathbf{P2}\,\mathbf{T2} = \int \mathbf{T2}\,\mathrm{d}t \tag{15.41}$$

where, **P1** and **P2** are operational matrices for integration related to **T1** and **T2** [12] respectively. Also **P1** and **P2** are related to **P**, the operational matrix for integration in BPF domain, by the following equation

$$\mathbf{P} = \mathbf{P1} + \mathbf{P2} \tag{15.42}$$

Using relations (15.41) and (15.42), second integration of **T1** is

$$\iint \mathbf{T1}\,\mathrm{d}t = \int (\mathbf{P1}\,\mathbf{T1} + \mathbf{P2}\,\mathbf{T2})\,\mathrm{d}t = \mathbf{P1}\int \mathbf{T1}\,\mathrm{d}t + \mathbf{P2}\int \mathbf{T2}\,\mathrm{d}t = (\mathbf{P1} + \mathbf{P2})\int \mathbf{T1}\,\mathrm{d}t$$
$$= \mathbf{P}\int \mathbf{T1}\,\mathrm{d}t$$

Similarly, third integration of **T1** is given by

$$\iiint \mathbf{T1}dt = \mathbf{P}^2 \int \mathbf{T1}\,\mathrm{d}t$$

Thus, keeping in mind relations (15.41) and (15.42) above, k-times integration of **T1** and **T2** yields

$$\begin{aligned}\underbrace{\int\int\int\cdots\int}_{k\,\text{times}} \mathbf{T1}\,\mathrm{d}t = \underbrace{\int\int\int\cdots\int}_{k\,\text{times}} \mathbf{T2}\,\mathrm{d}t &= \ \mathbf{P}^{k-1}\int\mathbf{T1}\,\mathrm{d}t\ \mathbf{P}^{k-1}\int\mathbf{T2}\,\mathrm{d}t \\ &= \mathbf{P}^{k-1}(\mathbf{P1}\,\mathbf{T1}+\mathbf{P2}\,\mathbf{T2})\end{aligned} \tag{15.43}$$

Hence, using (15.43) and from Eq. (15.21) we have

$$\begin{aligned}\mathbf{H}_k &= (-1)^k\,\lambda^{(k+1)}\,2\mathbf{u}_\mathbf{0}^{\mathrm{T}}\,\mathbf{P}^{(k-1)} \\ &\quad\times\int_0^1 [\mathbf{P1}\,\mathbf{T1}(\tau)+\mathbf{P2}\,\mathbf{T2}(\tau)]\left[\sum_{i=0}^{m-1}\mathbf{D31}_{(l-1)i}T1_i(\tau)+\sum_{i=0}^{m-1}\mathbf{D32}_{(l-1)i}T2_i(\tau)\right]\mathrm{d}\tau \\ &= (-1)^k\,\lambda^{(k+1)}\,2\mathbf{u}_\mathbf{0}^{\mathrm{T}}\,\mathbf{P}^{(k-1)}\ \times\sum_{i=0}^{m-1}\int_0^1\big[\mathbf{P1}\,\mathbf{D31}_{(l-1)i}T1_i(\tau)\mathbf{T1}(\tau)+\ \mathbf{P2}\,\mathbf{D31}_{(l-1)i}T1_i(\tau)\mathbf{T2}(\tau) \\ &\quad+\ \mathbf{P1}\,\mathbf{D32}_{(l-1)i}T2_i(\tau)\mathbf{T1}(\tau)+\ \mathbf{P2}\,\mathbf{D32}_{(l-1)i}T2_i(\tau)\mathbf{T2}(\tau)\big]\mathrm{d}\tau \\ &= (-1)^k\,\lambda^{(k+1)}\,2\mathbf{u}_\mathbf{0}^{\mathrm{T}}\,\mathbf{P}^{(k-1)}\times\left\{\sum_{i=0}^{m-1}\mathbf{P1}\,\mathbf{D31}_{(l-1)i}\int_0^1[T1_i(\tau)\mathbf{T1}(\tau)\,\mathrm{d}\tau\right. \\ &\quad+\sum_{i=0}^{m-1}\mathbf{P2}\,\mathbf{D31}_{(l-1)i}0\int_0^1 T1_i(\tau)\mathbf{T2}(\tau)\,\mathrm{d}\tau+\sum_{i=0}^{m-1}\mathbf{P1}\,\mathbf{D32}_{(l-1)i}\int_0^1 T2_i(\tau)\mathbf{T1}(\tau)\,\mathrm{d}\tau \\ &\quad\left.+\sum_{i=0}^{m-1}\mathbf{P2}\,\mathbf{D32}_{(l-1)i}\int_0^1 T2_i(\tau)\mathbf{T2}(\tau)\,\mathrm{d}\tau\right\}\end{aligned} \tag{15.44}$$

Due to orthogonal property

$$\int_0^1 T1_i(\tau)\mathbf{T1}(\tau)\,\mathrm{d}\tau = \mathbf{e1}_i = \underbrace{[0\ \ \cdots\ \ 0\ \ \overset{(i+1)^{\text{th}}\text{ position}}{\overset{\downarrow}{\tfrac{h}{3}}}\ \ 0\ \ \cdots\ \ 0]}_{m\,\text{columns}}^{\mathrm{T}} = \int_0^1 T2_i(\tau)\mathbf{T2}(\tau)\,\mathrm{d}\tau \tag{15.45}$$

$$\int_0^1 T1_i(\tau)\mathbf{T2}(\tau)\,\mathrm{d}\tau = \mathbf{e2}_i = \underbrace{[0 \quad \cdots \quad 0 \quad \overset{(i+1)^{\text{th}}\text{ position}}{\overset{\downarrow}{\tfrac{h}{6}}} \quad 0 \quad \cdots \quad 0]}_{m\text{ columns}}^{\mathrm{T}} = \int_0^1 T2_i(\tau)\mathbf{T1}(\tau)\,\mathrm{d}\tau \tag{15.46}$$

So Eq. (15.44) becomes

$$\mathbf{H}_k = (-1)^k \lambda^{(k+1)} 2\mathbf{u}_0^{\mathrm{T}}\mathbf{P}^{(k-1)} \times \left\{ \mathbf{P1}\sum_{i=0}^{m-1} \mathbf{e1}_i\mathbf{D31}_{(l-1)i} + \mathbf{P2}\sum_{i=0}^{m-1} \mathbf{e2}_i\mathbf{D31}_{(l-1)i} + \mathbf{P1}\sum_{i=0}^{m-1} \mathbf{e2}_i\mathbf{D32}_{(l-1)i} + \mathbf{P2}\sum_{i=0}^{m-1} \mathbf{e1}_i\mathbf{D32}_{(l-1)i} \right\}$$

For $m = 4$, following the earlier procedure, we get

$$\begin{aligned}
\mathbf{H}_k = {} & (-1)^k \lambda^{(k+1)} 2\,\mathbf{u}_0^{\mathrm{T}}\mathbf{P}^{(k-1)} \\
& \times \left[\left\{ \mathbf{P1}\begin{bmatrix} \frac{h}{3} \\ 0 \\ 0 \\ 0 \end{bmatrix}\mathbf{D31}_{(l-1)0} + \mathbf{P1}\begin{bmatrix} 0 \\ \frac{h}{3} \\ 0 \\ 0 \end{bmatrix}\mathbf{D31}_{(l-1)1} + \mathbf{P1}\begin{bmatrix} 0 \\ 0 \\ \frac{h}{3} \\ 0 \end{bmatrix}\mathbf{D31}_{(l-1)2} + \mathbf{P1}\begin{bmatrix} 0 \\ 0 \\ 0 \\ \frac{h}{3} \end{bmatrix}\mathbf{D31}_{(l-1)3} \right\} \right. \\
& + \left\{ \mathbf{P2}\begin{bmatrix} \frac{h}{6} \\ 0 \\ 0 \\ 0 \end{bmatrix}\mathbf{D31}_{(l-1)0} + \mathbf{P2}\begin{bmatrix} 0 \\ \frac{h}{6} \\ 0 \\ 0 \end{bmatrix}\mathbf{D31}_{(l-1)1} + \mathbf{P2}\begin{bmatrix} 0 \\ 0 \\ \frac{h}{6} \\ 0 \end{bmatrix}\mathbf{D31}_{(l-1)2} + \mathbf{P2}\begin{bmatrix} 0 \\ 0 \\ 0 \\ \frac{h}{6} \end{bmatrix}\mathbf{D31}_{(l-1)3} \right\} \\
& + \left\{ \mathbf{P1}\begin{bmatrix} \frac{h}{6} \\ 0 \\ 0 \\ 0 \end{bmatrix}\mathbf{D32}_{(l-1)0} + \mathbf{P1}\begin{bmatrix} 0 \\ \frac{h}{6} \\ 0 \\ 0 \end{bmatrix}\mathbf{D32}_{(l-1)1} + \mathbf{P1}\begin{bmatrix} 0 \\ 0 \\ \frac{h}{6} \\ 0 \end{bmatrix}\mathbf{D32}_{(l-1)2} + \mathbf{P1}\begin{bmatrix} 0 \\ 0 \\ 0 \\ \frac{h}{6} \end{bmatrix}\mathbf{D32}_{(l-1)3} \right\} \\
& \left. + \left\{ \mathbf{P2}\begin{bmatrix} \frac{h}{3} \\ 0 \\ 0 \\ 0 \end{bmatrix}\mathbf{D32}_{(l-1)0} + \mathbf{P2}\begin{bmatrix} 0 \\ \frac{h}{3} \\ 0 \\ 0 \end{bmatrix}\mathbf{D32}_{(l-1)1} + \mathbf{P2}\begin{bmatrix} 0 \\ 0 \\ \frac{h}{3} \\ 0 \end{bmatrix}\mathbf{D32}_{(l-1)2} + \mathbf{P2}\begin{bmatrix} 0 \\ 0 \\ 0 \\ \frac{h}{3} \end{bmatrix}\mathbf{D32}_{(l-1)3} \right\} \right]
\end{aligned}$$

$$\text{or,}\quad \mathbf{H}_k = \mathbf{c}\sum_{i=0}^{3}\left[\mathbf{c}_{1i}\mathbf{D31}_{(l-1)i} + \mathbf{c}_{2i}\mathbf{D31}_{(l-1)i} + \mathbf{c}_{3i}\mathbf{D32}_{(l-1)i} + \mathbf{c}_{4i}\mathbf{D32}_{(l-1)i}\right] \tag{15.47}$$

where, $\mathbf{c} \triangleq (-1)^k \lambda^{(k+1)} 2\,\mathbf{u}_0^{\mathrm{T}}\mathbf{P}^{(k-1)}$
and $\mathbf{c}_{1i} \triangleq \mathbf{P1e1}_i$, $\mathbf{c}_{2i} \triangleq \mathbf{P2e2}_i$, $\mathbf{c}_{3i} \triangleq \mathbf{P1e2}_i$, $\mathbf{c}_{4i} \triangleq \mathbf{P2e1}_i$

Table 15.5 Estimated parameters of the transfer function of Example 15.1 for m = 64 and for three different values of λ, in triangular function domain

Parameters	Actual	$\lambda = 1$	$\lambda = 6$	$\lambda = 12$
a_0	1	2.54176778	0.77261257	0.97296430
a_1	0	−0.25152488	0.05825568	0.01226366
b_0	2	12.72231195	1.55292352	1.94613618
b_1	3	6.11098640	2.40394316	2.94367405

Table 15.6 Percentage errors for parameter estimation of the transfer function of Example 15.1 for m = 64 and λ = 12, in triangular function domain

Parameters	Actual values c_a	TF domain values c_h	% Error, $\varepsilon = \frac{(c_a - c_h)}{c_a} \times 100$
a_0	1	0.97296430	2.70357003
a_1	0	0.01226366	–
b_0	2	1.94613618	2.69319093
b_1	3	2.94367405	1.87753173

Following the procedure used for Sect. 7.3.1, we determine the values of a_i's and b_i's in TF domain.

Tables 15.5 and 15.6 present the results of TF domain based estimation.

15.3.4 Using Hybrid Functions (HF)

In hybrid function domain [13], referring to Sect. 2.4, $\mathbf{h}(\lambda\tau)$ can be written as

$$\mathbf{h}(\lambda\tau) = \sum_{i=0}^{m-1} \mathbf{D41}_{(l-1)i} S_i(\tau) + \sum_{i=0}^{m-1} \mathbf{D42}_{(l-1)i} T_i(\tau), \quad 0 \leq \tau < 1 \tag{15.48}$$

where, $\mathbf{D41}_{(l-1)i}$'s are the samples and $\mathbf{D42}_{(l-1)i}$'s are difference between two consecutive samples of $\mathbf{h}(\lambda\tau)$ e.g. $(\mathbf{D41}_{(i+1)} - \mathbf{D41}_i)$.

As before, following (7.23), in hybrid function domain, we write

$$\mathbf{u}_{01}^{\mathrm{T}} \mathbf{S}(\tau) + \mathbf{u}_{02}^{\mathrm{T}} \mathbf{T}(\tau) = \mathbf{u}_{01}^{\mathrm{T}} \sum_{i=0}^{m-1} S_i(\tau) + \mathbf{u}_{02}^{\mathrm{T}} \sum_{i=0}^{m-1} T_i(\tau) = 1$$

where, $\mathbf{u}_{01}^{\mathrm{T}} = \underbrace{[1 \;\; 1 \;\; \cdots \;\; 1 \;\; 1]}_{m \text{ columns}}$ and $\mathbf{u}_{02}^{\mathrm{T}} = \underbrace{[0 \;\; 0 \;\; \cdots \;\; 0 \;\; 0]}_{m \text{ columns}}$

To determine the formula for repeated integration in hybrid function domain, we proceed from the relation for first integration of sample-and-hold functions.

From Eq. (4.9), we have

$$\int \mathbf{S}_{(m)} dt = h\sum_{i=1}^{m-1} Q^{i}_{(m)} S_{(m)} + h\,\mathbf{I}_{(m)}\,\mathbf{T}_{(m)} = \mathbf{P1ss}_{(m)}\mathbf{S}_{(m)} + \mathbf{P1st}_{(m)}\mathbf{T}_{(m)} \quad (15.49)$$

Comparing Eqs. (4.9) and (4.18), we have

$$\int \mathbf{T}_{(m)}\mathrm{d}t = \frac{1}{2}\int \mathbf{S}_{(m)}\mathrm{d}t \quad (15.50)$$

Dropping the subscript m for convenience and using (15.49) and (15.50), the second integration of $\mathbf{S}$ is given by

$$\begin{aligned}\iint \mathbf{S}\,\mathrm{d}t &= \int (\mathbf{P1ss}\,\mathbf{S} + h\,\mathbf{I}\,\mathbf{T})\mathrm{d}t = \mathbf{P1ss}\int \mathbf{S}\,\mathrm{d}t + h\int \mathbf{T}\,\mathrm{d}t = \mathbf{P1ss}\int \mathbf{S}\,\mathrm{d}t + \frac{1}{2}h\mathbf{I}\int \mathbf{S}\,\mathrm{d}t \\ &= (\mathbf{P1ss} + \frac{1}{2}h\mathbf{I})\int \mathbf{S}\,\mathrm{d}t\end{aligned}$$

The third integration of $\mathbf{S}$ is

$$\iiint \mathbf{S}\,\mathrm{d}t = (\mathbf{P1ss} + \frac{1}{2}h\mathbf{I})\int\int \mathbf{S}\,\mathrm{d}t = \left(\mathbf{P1ss} + \frac{1}{2}h\mathbf{I}\right)^{2}\int \mathbf{S}\,\mathrm{d}t$$

Repeating this procedure, we obtain

$$\begin{aligned}\underbrace{\iiint\cdots\int}_{k\,\text{times}} \mathrm{S}\,\mathrm{d}t &= \left(\mathbf{P1ss} + \frac{1}{2}h\mathbf{I}\right)^{(k-1)}(\mathbf{P1ss}\,\mathbf{S} + h\mathbf{I}\,\mathbf{T}) \\ &= \left(\mathbf{P1ss} + \frac{1}{2}h\mathbf{I}\right)^{(k-1)}(\mathbf{P1ss}\,\mathbf{S} + h\,\mathbf{T})\end{aligned}$$

Then, proceeding as before, we have

$$
\begin{aligned}
\mathbf{H}^k &= (-1)^k \lambda^{(k+1)} \mathbf{u}_{01}^{\mathrm{T}} \int_0^1 \frac{\tau^k}{k!} \mathbf{h}(\lambda\tau)\, \mathrm{d}\tau \\
&= (-1)^k \lambda^{(k-1)} \mathbf{u}_{01}^{\mathrm{T}} \int_0^1 \left[\left(\mathbf{P1ss} + \tfrac{h}{2}\mathbf{I} \right)^{(k-1)} (\mathbf{P1ss}\,\mathbf{S} + h\,\mathbf{I}\,\mathbf{T}) \left[\sum_{i=0}^{m-1} \mathbf{D41}_{(l-1)i} S_i(\tau) + \sum_{i=0}^{m-1} \mathbf{D42}_{(l-1)i} T_i(\tau) \right] \right] \mathrm{d}\tau \\
&= (-1)^k \lambda^{(k+1)} \mathbf{u}_{01}^{\mathrm{T}} \left(\mathbf{P1ss} + \tfrac{h}{2}\mathbf{I} \right)^{(k-1)} \times \sum_{i=0}^{m-1} \int_0^1 \big[\mathbf{P1ss}\,\mathbf{D41}_{(l-1)i} S_i(\tau)\mathbf{S}(\tau) + h\,\mathbf{D41}_{(l-1)i} S_i(\tau)\mathbf{T}(\tau) \\
&\quad + \mathbf{P1ss}\,\mathbf{D42}_{(l-1)i} T_i(\tau)\mathbf{S}(\tau) + h\,\mathbf{D42}_{(l-1)i} T_i(\tau)\mathbf{T}(\tau) \big] \mathrm{d}\tau \\
&= (-1)^k \lambda^{(k+1)} \mathbf{u}_{01}^{\mathrm{T}} \left(\mathbf{P1ss} + \tfrac{h}{2}\mathbf{I} \right)^{(k-1)} \\
&\quad \times \left\{ \sum_{i=0}^{m-1} \mathbf{P1ss}\,\mathbf{D41}_{(l-1)i} \int_0^1 S_i(\tau)\mathbf{S}(\tau)\, \mathrm{d}\tau + \sum_{i=0}^{m-1} h\,\mathbf{D41}_{(l-1)i} \int_0^1 S_i(\tau)\mathbf{T}(\tau)\, \mathrm{d}\tau \right. \\
&\quad \left. + \sum_{i=0}^{m-1} \mathbf{P1ss}\,\mathbf{D42}_{(l-1)i} \int_0^1 T_i(\tau)\mathbf{S}(\tau)\, \mathrm{d}\tau + \sum_{i=0}^{m-1} h\,\mathbf{D42}_{(l-1)i} \int_0^1 T_i(\tau)\mathbf{T}(\tau)\, \mathrm{d}\tau \right\}
\end{aligned} \tag{15.51}
$$

Due to orthogonal property, we have the three following relations

$$
\int_0^1 S_i(\tau)\,\mathbf{S}(\tau)\,\mathrm{d}\tau = \mathbf{e}_i = \underbrace{[0 \;\; \cdots \;\; 0 \;\; \overset{(i+1)^{\mathrm{th}}\text{ position}}{\overset{\downarrow}{h}} \;\; 0 \;\; \cdots \;\; 0]}_{m\text{ columns}}{}^{\mathrm{T}}
$$

$$
\int_0^1 S_i(\tau)\,\mathbf{T}(\tau)\,\mathrm{d}\tau = \int_0^1 T_i(\tau)\,\mathbf{S}(\tau)\,\mathrm{d}\tau = \mathbf{e3}_i = \underbrace{[0 \;\; \cdots \;\; 0 \;\; \overset{(i+1)^{\mathrm{th}}\text{ position}}{\overset{\downarrow}{\tfrac{h}{2}}} \;\; 0 \;\; \cdots \;\; 0]}_{m\text{ columns}}{}^{\mathrm{T}}
$$

$$
\int_0^1 T1_i(\tau)\,\mathbf{T}(\tau)\,\mathrm{d}\tau = \mathbf{e1}_i = \underbrace{[0 \;\; \cdots \;\; 0 \;\; \overset{(i+1)^{\mathrm{th}}\text{ position}}{\overset{\downarrow}{\tfrac{h}{3}}} \;\; 0 \;\; \cdots \;\; 0]}_{m\text{ columns}}{}^{\mathrm{T}}
$$

Using the above relations in Eq. (15.51), we get

$$
\begin{aligned}
\mathbf{H}_k = (-1)^k \lambda^{(k+1)} \mathbf{u}_{01}^{\mathrm{T}} \left(\mathbf{P1ss} + \tfrac{h}{2}\mathbf{I} \right)^{(k-1)} \times \left\{ \mathbf{P1ss} \sum_{i=0}^{m-1} \mathbf{e}_i \mathbf{D41}_{(l-1)i} + h \sum_{i=0}^{m-1} \mathbf{e3}_i \mathbf{D41}_{(l-1)i} \right. \\
\left. + \mathbf{P1ss} \sum_{i=0}^{m-1} \mathbf{e3}_i \mathbf{D42}_{(l-1)i} + h \sum_{i=0}^{m-1} \mathbf{e1}_i \mathbf{D42}_{(l-1)i} \right\}
\end{aligned} \tag{15.52}
$$

If we consider $m = 4$, then

$$\mathbf{H}_k = (-1)^k \lambda^{(k+1)} \mathbf{u}_{01}^{\mathrm{T}} \left(\mathbf{P1ss} + \frac{h}{2}\mathbf{I}\right)^{(k-1)}$$
$$\times \left[\left\{ \mathbf{P1ss} \begin{bmatrix} h \\ 0 \\ 0 \\ 0 \end{bmatrix} \mathbf{D41}_{(l-1)0} + \mathbf{P1ss} \begin{bmatrix} 0 \\ h \\ 0 \\ 0 \end{bmatrix} \mathbf{D41}_{(l-1)1} + \mathbf{P1ss} \begin{bmatrix} 0 \\ 0 \\ h \\ 0 \end{bmatrix} \mathbf{D41}_{(l-1)2} + \mathbf{P1ss} \begin{bmatrix} 0 \\ 0 \\ 0 \\ h \end{bmatrix} \mathbf{D41}_{(l-1)3} \right\} \right.$$
$$+ h \left\{ \begin{bmatrix} \frac{h}{2} \\ 0 \\ 0 \\ 0 \end{bmatrix} \mathbf{D41}_{(l-1)0} + \begin{bmatrix} 0 \\ \frac{h}{2} \\ 0 \\ 0 \end{bmatrix} \mathbf{D41}_{(l-1)1} + \begin{bmatrix} 0 \\ 0 \\ \frac{h}{2} \\ 0 \end{bmatrix} \mathbf{D41}_{(l-1)2} + \begin{bmatrix} 0 \\ 0 \\ 0 \\ \frac{h}{2} \end{bmatrix} \mathbf{D41}_{(l-1)3} \right\}$$
$$+ \left\{ \mathbf{P1ss} \begin{bmatrix} \frac{h}{2} \\ 0 \\ 0 \\ 0 \end{bmatrix} \mathbf{D42}_{(l-1)0} + \mathbf{P1ss} \begin{bmatrix} 0 \\ \frac{h}{2} \\ 0 \\ 0 \end{bmatrix} \mathbf{D42}_{(l-1)1} + \mathbf{P1ss} \begin{bmatrix} 0 \\ 0 \\ \frac{h}{2} \\ 0 \end{bmatrix} \mathbf{D42}_{(l-1)2} + \mathbf{P1ss} \begin{bmatrix} 0 \\ 0 \\ 0 \\ \frac{h}{2} \end{bmatrix} \mathbf{D42}_{(l-1)3} \right\}$$
$$\left. + h \left\{ \begin{bmatrix} \frac{h}{3} \\ 0 \\ 0 \\ 0 \end{bmatrix} \mathbf{D42}_{(l-1)0} + \begin{bmatrix} 0 \\ \frac{h}{3} \\ 0 \\ 0 \end{bmatrix} \mathbf{D42}_{(l-1)1} + \begin{bmatrix} 0 \\ 0 \\ \frac{h}{3} \\ 0 \end{bmatrix} \mathbf{D42}_{(l-1)2} + \begin{bmatrix} 0 \\ 0 \\ 0 \\ \frac{h}{3} \end{bmatrix} \mathbf{D42}_{(l-1)3} \right\} \right]$$
$$\text{or,} \quad \mathbf{H}_k = \mathbf{d} \sum_{i=0}^{3} \left[\mathbf{d}_{1i} \mathbf{D41}_{(l-1)i} + \mathbf{d}_{2i} \mathbf{D41}_{(l-1)i} + \mathbf{d}_{3i} \mathbf{D42}_{(l-1)i} + \mathbf{d}_{4i} \mathbf{D32}_{(l-1)i} \right] \tag{15.53}$$

where, $\mathbf{d} = (-1)^k \lambda^{(k+1)} \mathbf{u}_{01}^{\mathrm{T}} (\mathbf{P1ss} + \frac{h}{2}\mathbf{I})^{(k-1)}$
and $\mathbf{d}_{1i} \triangleq \mathbf{P1ss}\ \mathbf{e}_i$, $\mathbf{d}_{2i} \triangleq h\,\mathbf{e3}_i$, $\mathbf{d}_{3i} \triangleq \mathbf{P1ss}\ \mathbf{e3}_i$, $\mathbf{d}_{4i} \triangleq h\,\mathbf{e1}_i$

Using Eq. (15.53) and following the earlier procedure, the values of a_i's and b_i's determined via HF domain are tabulated in Tables 15.7 and 15.8.

Table 15.7 Estimated parameters of the transfer function of Example 15.1 for m = 64 and for three different values of λ, in hybrid function domain

Parameters	Actual	λ = 1	λ = 6	λ = 12
a_0	1	2.54431980	0.77367044	0.99396497
a_1	0	−0.23665257	0.06699767	0.04655428
b_0	2	12.72939710	1.54597031	1.94350586
b_1	3	6.11083725	2.39578631	2.94075968

Table 15.8 Percentage errors for parameter estimation of the transfer function of Example 15.1 for m = 64 and λ = 12, in hybrid function domain

Parameters	Actual values c_a	HF domain values c_h	% Error, $\varepsilon = \frac{(c_a - c_h)}{c_a} \times 100$
a_0	1	0.99396497	0.60350298
a_1	0	0.04655428	–
b_0	2	1.94350586	2.82470706
b_1	3	2.94075968	1.97467736

15.3.5 Solution in SHF Domain from the HF Domain Solution

To find out the solution in sample-and-hold function domain [14], we simply discard the triangular function component of the hybrid function part in Eq. (15.48) for $\mathbf{h}(\lambda\tau)$ and write

$$\mathbf{h}(\lambda\tau) = \sum_{i=0}^{m-1} \mathbf{D41}_{(l-1)i} S_i(\tau), \quad 0 \leq \tau < 1$$

where, $\mathbf{D41}_{(l-1)i}$'s are the samples of $\mathbf{h}(\lambda\tau)$ with the sampling period h.

In this case also,

$$\mathbf{u}_{01}^{\mathrm{T}} = \underbrace{[1 \quad 1 \quad \cdots \quad 1 \quad 1]}_{m\,\text{columns}}$$

Then, from (15.51), we write

$$\mathbf{H}_k = (-1)^k \lambda^{(k+1)} \mathbf{u}_{01}^{\mathrm{T}} \left(\mathbf{P1ss} + \frac{h}{2}\mathbf{I}\right)^{(k-1)} \sum_{i=0}^{m-1} \int_0^1 \left[\mathbf{P1ss}\,\mathbf{D41}_{(l-1)i}\, \mathrm{S}_i(\tau)\,\mathbf{S}(\tau)\right] \mathrm{d}\tau$$

Due to orthogonal property,

$$\int_0^1 S_i(\tau)\,\mathbf{S}(\tau)\,d\tau = \mathbf{e}_i = \underbrace{[0 \quad \cdots \quad 0 \quad \overset{(i+1)^{\text{th}}\ \text{position}}{\overset{\downarrow}{h}} \quad 0 \quad \cdots \quad 0]^{\mathrm{T}}}_{m\,\text{columns}}$$

Equation (15.51) now becomes,

$$\mathbf{H}_k = (-1)^k \lambda^{(k+1)} \mathbf{u}_{01}^{\mathrm{T}} \left(\mathbf{P1ss} + \frac{h}{2}\mathbf{I}\right)^{(k-1)} \sum_{i=0}^{m-1} \left\{\mathbf{P1ss}\,\mathbf{e}_i\,\mathbf{D41}_{(l-1)i}\right\}$$

If we take $m = 4$. Then

Table 15.9 Estimated parameters of the transfer function of Example 15.1 for m = 64 and for three different values of λ, in sample-and-hold function domain

Parameters	Actual	$\lambda = 1$	$\lambda = 6$	$\lambda = 12$
a_0	1	2.54442701	0.77194125	0.98292877
a_1	0	−0.22952649	0.06307111	0.02956877
b_0	2	12.73277718	1.54703132	1.94374640
b_1	3	6.11061106	2.39575881	2.94090502

Table 15.10 Percentage errors for parameter estimation of the transfer function of Example 15.1 for m = 64 and λ = 12, in sample-and-hold function domain

Parameters	Actual values c_a	SHF domain values c_h	% Error, $\varepsilon = \frac{(c_a - c_h)}{c_a} \times 100$
a_0	1	0.98292877	1.70712349
a_1	0	0.02956877	–
b_0	2	1.94374640	2.81267981
b_1	3	2.94090502	1.96983257

$$\mathbf{H}_k = (-1)^k \lambda^{(k+1)} \mathbf{u}_{01}^{\mathrm{T}} \left(\mathbf{P1ss} + \frac{h}{2}\mathbf{I}\right)^{(k-1)}$$

$$\times \mathbf{P3} \left\{ \begin{bmatrix} h \\ 0 \\ 0 \\ 0 \end{bmatrix} \mathbf{D41}_{(l-1)0} + \begin{bmatrix} 0 \\ h \\ 0 \\ 0 \end{bmatrix} \mathbf{D41}_{(l-1)1} + \begin{bmatrix} 0 \\ 0 \\ h \\ 0 \end{bmatrix} \mathbf{D41}_{(l-1)2} + \begin{bmatrix} 0 \\ 0 \\ 0 \\ h \end{bmatrix} \mathbf{D41}_{(l-1)3} \right\}$$

$$\text{or,} \quad \mathbf{H}_k = \mathbf{r} \sum_{i=0}^{3} \mathbf{e}_i \mathbf{D41}_{(l-1)i} \tag{15.54}$$

where, $\mathbf{r} \triangleq (-1)^k \lambda^{(k+1)} \mathbf{u}_{01}^{\mathrm{T}} (\mathbf{P1ss} + \frac{h}{2}\mathbf{I})^{(k-1)} \mathbf{P1ss}$

Using Eq. (15.54) the values of a_i 's and b_i's computed via SHF domain are tabulated in Tables 15.9 and 15.10.

15.4 Comparative Study of the Parameters of the Transfer Function Identified via Different Approaches [15]

The estimated values of the parameters for the transfer function, for λ = 1, 6 and 12 derived via five types of orthogonal basis functions i.e. BPF, NOBPF, TF, HF and SHF for Example 15.1 are shown in Tables 15.11, 15.12 and 15.13 respectively.

The comparative study of the error estimates for the system under study for λ = 12 via block pulse function, non-optimal block pulse function, triangular function, hybrid function and sample-and-hold function domains, are shown in Table 15.14.

Table 15.11 Comparative study of the parameters of the transfer function of Example 15.1 under investigation in different function domains for $\lambda = 1$

Parameters	Actual values	BPF	NOBPF	TF	HF	SHF
a_0	1	2.54176555	2.54147603	2.54176778	2.54431980	2.54442701
a_1	0	−0.25154866	−0.25160253	−0.25152488	−0.23665257	−0.22952649
b_0	2	12.72230064	12.72227298	12.72231195	12.72939710	12.73277718
b_1	3	6.11098678	6.11095662	6.11098640	6.11083725	6.11061106

Table 15.12 Comparative study of the parameters of the transfer function of Example 15.1 under investigation in different function domains for $\lambda = 6$

Para-meters	Actual values	BPF	NOBPF	TF	HF	SHF
a_0	1	0.76514514	0.77203411	0.77261257	0.77367044	0.77194125
a_1	0	0.05825146	0.05617966	0.05825568	0.06699767	0.06307111
b_0	2	1.55297667	1.55404255	1.55292352	1.54597031	1.54703132
b_1	3	2.40403634	2.40494375	2.40394316	2.39578631	2.39575881

Table 15.13 Comparative study of the parameters of the transfer function of Example 15.1 under investigation in different function domains for $\lambda = 12$

Para-meters	Actual values	BPF	NOBPF	TF	HF	SHF
a_0	1	0.97307956	0.96755155	0.97296430	0.99396497	0.98292877
a_1	0	0.01235613	0.00373086	0.01226366	0.04655428	0.02956877
b_0	2	1.94618304	1.94648567	1.94613618	1.94350586	1.94374640
b_1	3	2.94373198	2.94403456	2.94367405	2.94075968	2.94090502

Table 15.14 Comparative study of error estimates of the parameters for the system of Example 15.1 under study for $\lambda = 12$ via different function domains

Parameters	% Error in BPF	% Error in NOBPF	% Error in TF	% Error in HF	% Error in SHF
a_0	2.69204374	3.24484513	2.70357003	0.60350298	1.70712349
a_1	–	–	–	–	–
b_0	2.69084797	2.67571651	2.69319093	2.82470706	2.81267981
b_1	1.87560054	1.86551468	1.87753173	1.97467736	1.96983257

15.5 Comparison of Errors for BPF, NOBPF, TF, HF and SHF Domain Approaches [15]

Figures 15.1a–d show the estimated values of the parameters, for $\lambda = 1$, derived via five kinds of orthogonal function basis. Figure 15.2 shows estimated values of two parameters a_0 and b_1 for $\lambda = 6$ to compute the results obtained via five different basis functions, while Fig. 15.3 presents all the parameter for $\lambda = 12$.

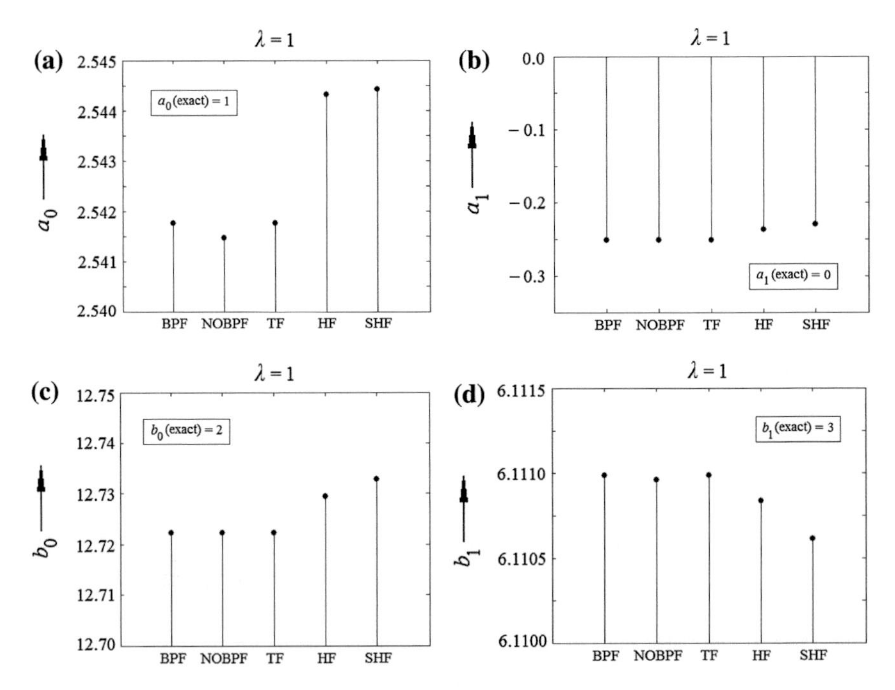

Fig. 15.1 Magnified view of the estimated parameters (a_0, a_1, b_0 and b_1) as per Tables 15.1, 15.3, 15.5, 15.7 and 15.9 for $\lambda = 1$ (vide Appendix B, Program no. 38)

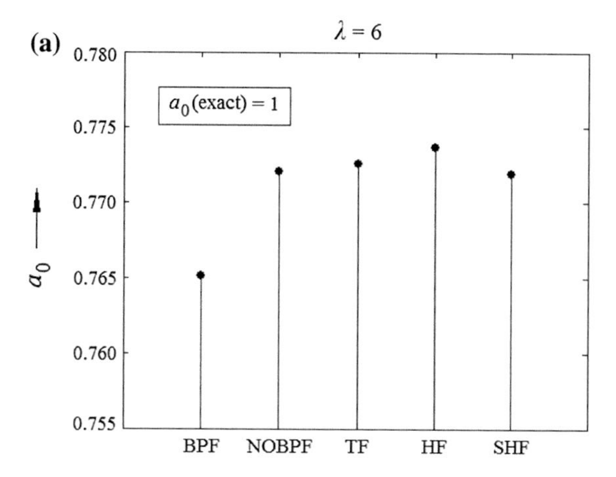

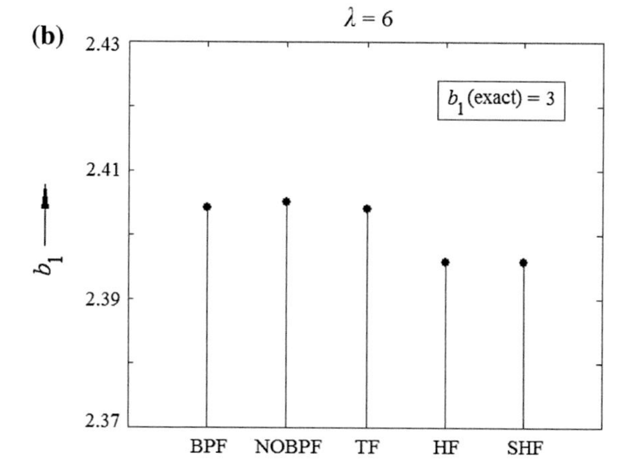

Fig. 15.2 Magnified view of two estimated parameters (a_0 and b_1) as per Tables 15.1, 15.3, 15.5, 15.7 and 15.9 for $\lambda = 6$ (vide Appendix B, Program no. 38)

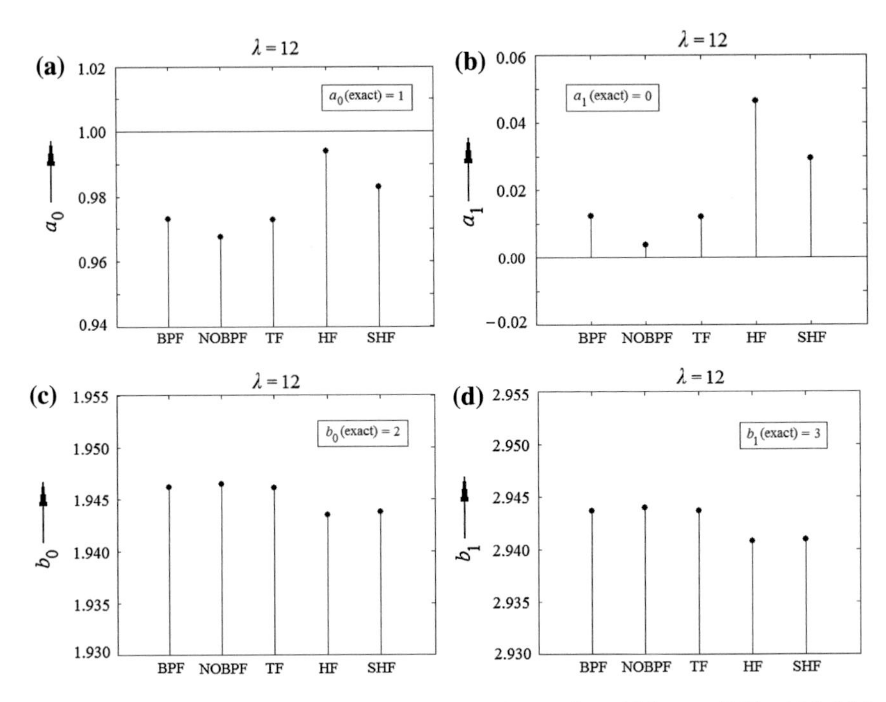

Fig. 15.3 Magnified view of the estimated parameters (a_0, a_1, b_0 and b_1) as per Tables 15.1, 15.3, 15.5, 15.7 and 15.9 for $\lambda = 12$ (vide Appendix B, Program no. 38)

15.6 Conclusion

The problem of parameter estimation of a transfer function of a multivariable system has been treated to determine the solution via orthogonal functions using a generalized algorithm. The derived algorithm is employed to solve for the parameters of the transfer function of a system via five different orthogonal sets e.g., (i) block pulse functions (BPF), (ii) non-optimal block pulse functions (NOBPF), (iii) triangular functions (TF), (iv) hybrid functions (HF) and (v) sample-and-hold functions (SHF).

Many tables are presented to compare the accuracies of different methods. Different curves are plotted are to indicate the values of the four parameters, a_0, a_1, b_0 and b_1 of a second order transfer function, computed via five different methods. From Tables 15.11, 15.12 and 15.13, we conclude that the parameter a_0 of the transfer function is closer to the actual value in HF domain.

It is noted that none of the presented methods proves itself absolutely superior to others, but from Table 15.14, it is observed that the minimum percentage error is obtained for a_0 for HF domain based computation.

Finally, an advantage of HF based analysis is, the sample-and-hold function based result may easily be obtained by simply dropping the triangular part solution of the hybrid function based result. This advantage may prove much significant for function approximation as well as for control system analysis.

References

1. Eykhoff, P.: System Identification: Parameter and State Estimation. Wiley, London (1974)
2. Chen, C.T.: Linear System Theory and Design. Holt Rinehart and Winston, Holt-Saunders, Japan (1984)
3. Unbehauen, H., Rao, G.P.: Identification of Continuous Systems. North-Holland, Amsterdam (1987)
4. Ljung, L.: System Identification: Theory for the User. Prentice-Hall Inc., New Jersey (1985)
5. Pintelon, R., Guillaume, P., Rolain, Y., Schoukens, J., Van Hamme, H.: Parametric identification of transfer functions in the frequency domain: a survey. IEEE Trans. Autom. Control **39**(11), 2245–2259 (1994)
6. Paraskevopoulos, P.N., Varoufakis, S.J.: Transfer function determination from impulse response via Walsh functions. Int. J. Circuit Theor. Appl. **8**, 85–89 (1980)
7. Baker Jr, G.A., Graves-Morris, P.: Padé Approximants. Cambridge University Press, New York (1996)
8. Baker Jr., G.A.: Padé Approximants in Theoretical Physics, pp. 27–38. Academic Press, New York (1975)
9. Zakian, V.: Simplification of linear time-invariant systems by moment approximants. Int. J. Control **18**, 455–460 (1973)
10. Jiang, J.H., Schaufelberger, W.: Block Pulse Functions and their Applications in Control System, LNCIS, vol. 179. Springer, Berlin (1992)
11. Deb, A., Sarkar, G., Mandal, P., Biswas, A., Sengupta, A.: Optimal block pulse function (OBPF) versus non-optimal block pulse function (NOBPF). In: Proceedings of International Conference of IEE (PEITSICON) 2005, pp. 195–199. Kolkata (2005) (28–29th Jan)
12. Deb, Anish, Sarkar, Gautam, Sengupta, Anindita: Triangular Orthogonal Functions for the Analysis of Continuous Time Systems. Anthem Press, London (2011)
13. Deb, Anish, Sarkar, Gautam, Mandal, Priyaranjan, Biswas, Amitava, Ganguly, Anindita, Biswas, Debasish: Transfer function Identification from impulse response via a new set of orthogonal hybrid function (HF). Appl. Math. Comput. **218**(9), 4760–4787 (2012)
14. Deb, Anish, Sarkar, Gautam, Bhattacharjee, Manabrata, Sen, Sunit K.: A new set of piecewise constant orthogonal functions for the analysis of linear SISO systems with sample-and-hold. J. Franklin Instt. **335B**(2), 333–358 (1998)
15. Biswas, A.: Analysis and synthesis of continuous control systems using a set of orthogonal hybrid functions. Ph. D. Dissertation, University of Calcutta (2015)

Appendix A
Introduction to Linear Algebra

A matrix is a rectangular array of variables, mathematical expressions, or simply numbers. Commonly a matrix is written as

$$\mathbf{A} = \begin{bmatrix} a_{11} & a_{12} & \cdots & a_{1n} \\ a_{21} & a_{22} & \cdots & a_{2n} \\ \vdots & \vdots & \ddots & \vdots \\ a_{m1} & a_{m2} & \cdots & a_{mn} \end{bmatrix}. \tag{A.1}$$

The size of the matrix, with m rows and n columns, is called an m-by-n (or, $m \times n$) matrix, where, m and n are called its dimensions.

A matrix with one row [a $(1 \times n)$ matrix] is called a row vector, and a matrix with one column [an $(m \times 1)$ matrix] is called a column vector. Any isolated row or column of a matrix is a row or column vector, obtained by removing all other rows or columns respectively from the matrix.

Square Matrices

A square matrix is a matrix with $m = n$, i. e., the same number of rows and columns. An n-by-n matrix is known as a square matrix of order n. Any two square matrices of the same order can be added, subtracted, or multiplied.

For example, each of the following matrices is a square matrix of order 4, with four rows and four columns:

$$\mathbf{A} = \begin{bmatrix} 1 & 2 & 7 & 0 \\ 3 & 4 & 3 & 1 \\ 8 & 3 & 1 & 1 \\ 2 & 4 & 0 & 3 \end{bmatrix}, \quad \mathbf{B} = \begin{bmatrix} 3 & 1 & 0 & -1 \\ 2 & -3 & 4 & 2 \\ 1 & 0 & 9 & 1 \\ -3 & 2 & -1 & 0 \end{bmatrix}$$

Then,

A. Deb et al., *Analysis and Identification of Time-Invariant Systems, Time-Varying Systems, and Multi-Delay Systems using Orthogonal Hybrid Functions*,
Studies in Systems, Decision and Control 46, DOI 10.1007/978-3-319-26684-8

$$\mathbf{A} + \mathbf{B} = \begin{bmatrix} 4 & 3 & 7 & -1 \\ 5 & 1 & 7 & 3 \\ 9 & 3 & 10 & 2 \\ -1 & 6 & -1 & 3 \end{bmatrix} \quad \text{and} \quad \mathbf{A} - \mathbf{B} = \begin{bmatrix} -2 & 1 & 7 & 1 \\ 1 & 7 & -1 & -1 \\ 7 & 3 & -8 & 0 \\ 5 & 2 & 1 & 3 \end{bmatrix}$$

Determinant

The determinant [written as det(**A**) or |**A**|] of a square matrix **A** is a number encoding certain properties of the matrix. A matrix is invertible, if and only if, its determinant is nonzero.

The determinant of a 2-by-2 matrix is given by

$$\det\begin{pmatrix} a & b \\ c & d \end{pmatrix} = ad - bc \tag{A.2}$$

Properties

The determinant of a product of square matrices equals the product of their determinants: det(**AB**) = det(**A**) · det(**B**).

Adding a multiple of any row to another row, or a multiple of any column to another column, does not change the determinant.

Interchanging two rows or two columns makes the determinant to be multiplied by −1.

Using these operations, any matrix can be transformed to a lower (or, upper) triangular matrix, and for such matrices the determinant equals the product of the entries on the main diagonal.

Orthogonal Matrix

An orthogonal matrix is a square matrix with real entries whose columns and rows are orthogonal vectors, i.e., orthonormal vectors.

Equivalently, a matrix **A** is orthogonal if its transpose is equal to its inverse. That is

$$\mathbf{A}^{\mathrm{T}} = \mathbf{A}^{-1}$$

which implies

$$\mathbf{A}^{\mathrm{T}}\mathbf{A} = \mathbf{A}\,\mathbf{A}^{\mathrm{T}} = \mathrm{I},$$

where, **I** is the identity matrix.

An orthogonal matrix **A** is necessarily invertible with inverse $\mathbf{A}^{-1} = \mathbf{A}^{\mathrm{T}}$. The determinant of any orthogonal matrix is either +**I** or −**I**.

Trace of a Matrix

In (A.1), the entries $a_{i,i}$ form the main diagonal of the matrix **A**. The trace, tr(**A**), of the square matrix **A** is the sum of its diagonal entries. The trace of the product of two matrices is independent of the order of the factors **A** and **B**. That is

$$\mathrm{tr}(\mathbf{AB}) = \mathrm{tr}(\mathbf{BA}).$$

Also, the trace of a matrix is equal to that of its transpose, i.e., $\mathrm{tr}(\mathbf{A}) = \mathrm{tr}(\mathbf{A}^{\mathrm{T}})$.

Diagonal, Lower Triangular and Upper Triangular Matrices

If all entries of a matrix except those of the main diagonal are zero, the matrix is called a *diagonal matrix*. If only all entries above (or, below) the main diagonal are zero, it is called a *lower triangular matrix* (or, *upper triangular matrix*).

For example, a diagonal matrix of 3rd order is

$$\begin{bmatrix} d_{11} & 0 & 0 \\ 0 & d_{22} & 0 \\ 0 & 0 & d_{33} \end{bmatrix}$$

A lower triangular matrix of 3rd order is

$$\begin{bmatrix} l_{11} & 0 & 0 \\ l_{21} & l_{22} & 0 \\ l_{31} & l_{32} & l_{33} \end{bmatrix}$$

and an upper triangular matrix of similar order is

$$\begin{bmatrix} u_{11} & u_{12} & u_{13} \\ 0 & u_{22} & u_{23} \\ 0 & 0 & u_{33} \end{bmatrix}$$

Symmetric Matrix

A square matrix **A** that is equal to its transpose, i.e., $\mathbf{A} = \mathbf{A}^{\mathrm{T}}$, is a symmetric matrix. If instead, **A** is equal to the negative of its transpose, i.e., $\mathbf{A} = -\mathbf{A}^{\mathrm{T}}$, then **A** is a skew-symmetric matrix.

Singular Matrix

If the determinant of a square matrix **A** is equal to zero, it is called a *singular matrix* and its inverse does not exist. Examples of two singular matrices are

$$\begin{bmatrix} 4 & 2 \\ 6 & 3 \end{bmatrix} \quad \text{and} \quad \begin{bmatrix} 2 & -1 & \frac{2}{7} \\ -7 & 0 & 3 \\ -2 & -1 & 2 \end{bmatrix}$$

Identity Matrix or Unit Matrix

If **A** is a square matrix, then

$$\mathbf{AI} = \mathbf{IA} = \mathbf{A}$$

where, **I** is the identity matrix of the same order.

The identity matrix $\mathbf{I}_{(n)}$ of size n is the n-by-n matrix in which all the elements on the main diagonal are equal to 1 and all other elements are equal to 0. An identity matrix of order 3 is

$$\mathbf{I}_{(3)} = \begin{bmatrix} 1 & 0 & 0 \\ 0 & 1 & 0 \\ 0 & 0 & 1 \end{bmatrix}.$$

It is called identity matrix because multiplication with it leaves a matrix unchanged. If **A** is an $(m \times n)$ matrix then

$$\mathbf{A}_{(m\times n)}\mathbf{I}_{(n)} = \mathbf{I}_{(m)}\mathbf{A}_{(m\times n)}$$

Transpose of a Matrix

The transpose $\mathbf{A}^{\mathrm{T}}$ of a square matrix $\mathbf{A}$ can be obtained by reflecting the elements along its main diagonal. Repeating the process on the transposed matrix returns the elements to their original position.

The transpose of a matrix may be obtained by any one of the following equivalent actions:

(i) reflect $\mathbf{A}$ over its main diagonal to obtain $\mathbf{A}^{\mathrm{T}}$
(ii) write the rows of $\mathbf{A}$ as the columns of $\mathbf{A}^{\mathrm{T}}$
(iii) write the columns of $\mathbf{A}$ as the rows of $\mathbf{A}^{\mathrm{T}}$

Formally, the *i*th row, *j*th column element of $\mathbf{A}^{\mathrm{T}}$ is the *j*th row, *i*th column element of $\mathbf{A}$. That is

$$[\mathbf{A}^{\mathrm{T}}]_{ij} = \mathbf{A}_{ji}$$

If $\mathbf{A}$ is an $m \times n$ matrix then $\mathbf{A}^{\mathrm{T}}$ is an $(n \times m)$ matrix.

Properties

For matrices $\mathbf{A}$, $\mathbf{B}$ and scalar c we have the following properties of transpose:

(i) $(\mathbf{A}^{\mathrm{T}})^{\mathrm{T}} = \mathbf{A}$
(ii) $(\mathbf{A}+\mathbf{B})^{\mathrm{T}} = \mathbf{A}^{\mathrm{T}} + \mathbf{B}^{T}$
(iii) $(\mathbf{AB})^{\mathrm{T}} = \mathbf{B}^{\mathrm{T}}\mathbf{A}^{\mathrm{T}}$
Note that the order of the factors above reverses. From this one can deduce that a square matrix $\mathbf{A}$ is invertible if and only if $\mathbf{A}^{\mathrm{T}}$ is invertible, and in this case, we have $(\mathbf{A}^{-1})^{\mathrm{T}} = (\mathbf{A}^{\mathrm{T}})^{-1}$. By induction this result extends to the general case of multiple matrices, where we find that

$$(\mathbf{A}_1\mathbf{A}_2\ldots\mathbf{A}_{k-1}\mathbf{A}_k)^{\mathrm{T}} = \mathbf{A}_k^{\mathrm{T}}\mathbf{A}_{k-1}^{\mathrm{T}}\ldots\mathbf{A}_2^{\mathrm{T}}\mathbf{A}_1^{\mathrm{T}}.$$

(iv) $(c\mathbf{A})^{\mathrm{T}} = c\mathbf{A}^{\mathrm{T}}$
The transpose of a scalar is the same scalar.
(v) $\det(\mathbf{A}^{\mathrm{T}}) = \det(\mathbf{A})$
(vi) $\det(\mathbf{A}^{-1}) = \frac{1}{\det(\mathbf{A})}$

Matrix Multiplication

Matrix multiplication is a binary operation that takes a pair of matrices, and produces another matrix. This term normally refers to the matrix product.

Multiplication of two matrices is defined only if the number of columns of the left matrix is the same as the number of rows of the right matrix. If **A** is an m-by-n matrix and **B** is an n-by-p matrix, then their matrix product **AB** is the m-by-p matrix whose entries are given by dot product of the corresponding row of **A** and the corresponding column of **B**. That is

$$[\mathbf{AB}]_{i,j} = A_{i,1}B_{1,j} + A_{i,2}B_{2,j} + \cdots + A_{i,n}B_{n,j} = \sum_{r=1}^{n} A_{i,r}B_{r,j}$$

where $1 \leq i \leq m$ and $1 \leq j \leq p$.

Matrix multiplication satisfies the rules

(i) (**AB**)**C** = **A**(**BC**) (associativity),
(ii) (**A** + **B**)**C** = **AC** + **BC**
(iii) **C**(**A** + **B**) = **CA** + **CB** (left and right distributivity),
whenever the size of the matrices is such that the various products are defined.

The product **AB** may be defined without **BA** being defined, namely, if **A** and **B** are m-by-n and n-by-k matrices, respectively, and $m \neq k$.

Even if both products are defined, they need not be equal, i.e., generally one has

$$\mathbf{AB} \neq \mathbf{BA},$$

i.e., matrix multiplication is not commutative, in marked contrast to (rational, real, or complex) numbers whose product is independent of the order of the factors. An example of two matrices not commuting with each other is:

$$\begin{bmatrix} 5 & 2 \\ 3 & 3 \end{bmatrix} \begin{bmatrix} 1 & 0 \\ 0 & 0 \end{bmatrix} = \begin{bmatrix} 5 & 0 \\ 3 & 0 \end{bmatrix}$$

where as

$$\begin{bmatrix} 1 & 0 \\ 0 & 0 \end{bmatrix} \begin{bmatrix} 5 & 2 \\ 3 & 3 \end{bmatrix} = \begin{bmatrix} 5 & 2 \\ 0 & 0 \end{bmatrix}$$

Since det(**A**) and det(**B**) are just numbers and so commute, det(**AB**) = det(**A**)det(**B**) = det(**B**)det(**A**) = det(**BA**), even when **AB** ≠ **BA**.

A Few Properties of Matrix Multiplication

(i) Associative

$$\mathbf{A}(\mathbf{BC}) = (\mathbf{AB})\mathbf{C}$$

(ii) Distributive over matrix addition

$$\mathbf{A}(\mathbf{B}+\mathbf{C}) = \mathbf{AB}+\mathbf{AC}, (\mathbf{A}+\mathbf{B})\mathbf{C} = \mathbf{AC}+\mathbf{BC}$$

(iii) Scalar multiplication is compatible with matrix multiplication

$$\lambda(\mathbf{AB}) = (\lambda\mathbf{A})\mathbf{B} \quad \text{and} \quad (\mathbf{AB})\lambda = \mathbf{A}(\mathbf{B}\lambda)$$

where, λ is a scalar. If the entries of the matrices are real or complex numbers, then all four quantities are equal.

Inverse of a Matrix

If $\mathbf{A}$ is a square matrix, there may be an inverse matrix $\mathbf{A}^{-1} = \mathbf{B}$ such that

$$\mathbf{AB} = \mathbf{BA} = \mathbf{I}$$

If this property holds, then $\mathbf{A}$ is an invertible matrix. If not, $\mathbf{A}$ is a *singular* or *degenerate* matrix.

Analytic Solution of the Inverse

Inverse of a square non-singular matrix $\mathbf{A}$ may be computed from the transpose of a matrix $\mathbf{C}$ formed by the *cofactors* of $\mathbf{A}$. Thus, $\mathbf{C}^{\mathrm{T}}$ is known as the *adjoint matrix* of $\mathbf{A}$. The matrix $\mathbf{C}^{\mathrm{T}}$ is divided by the determinant of $\mathbf{A}$ to compute $\mathbf{A}^{-1}$. That is

$$\mathbf{A}^{-1} = \frac{\mathbf{C}^{\mathrm{T}}}{\det(\mathbf{A})} = \frac{1}{\det(\mathbf{A})}\begin{bmatrix} c_{11} & c_{21} & \cdots & c_{n1} \\ c_{12} & c_{22} & \cdots & c_{n2} \\ \vdots & \vdots & \ddots & \vdots \\ c_{1n} & c_{2n} & \cdots & c_{nn} \end{bmatrix} \quad (\text{A.3})$$

where, $\mathbf{C}$ is the matrix of cofactors, and $\mathbf{C}^{\mathrm{T}}$ denotes the transpose of $\mathbf{C}$.

Cofactors

Let a (3 × 3) matrix be given by

$$\mathbf{A} = \begin{bmatrix} a & b & c \\ d & e & f \\ g & h & k \end{bmatrix}$$

Then a (3 × 3) matrix **P** formed by the cofactors of **A** is

$$\mathbf{P} = \begin{bmatrix} A & B & C \\ D & E & F \\ G & H & K \end{bmatrix}$$

where,

$$\begin{array}{lll} A = (ek - fh), & B = (fg - dk), & C = (dh - eg), \\ D = (ch - bk), & E = (ak - cg), & F = (gb - ah), \\ G = (bf - ce), & H = (cd - af) \quad \text{and} & K = (ae - bd). \end{array}$$

Inversion of a 2 × 2 Matrix

The cofactor equation listed above yields the following result for a 2 × 2 matrix. Let the matrix to be inverted be

$$\mathbf{A} = \begin{bmatrix} a & b \\ c & d \end{bmatrix}.$$

Then

$$\mathbf{A}^{-1} = \begin{bmatrix} a & b \\ c & d \end{bmatrix}^{-1} = \frac{1}{\det(\mathbf{A})} \begin{bmatrix} d & -b \\ -c & a \end{bmatrix} = \frac{1}{(ad - bc)} \begin{bmatrix} d & -b \\ -c & a \end{bmatrix}.$$

using Eqs. (A.2) and (A.3).

Inversion of a 3 × 3 Matrix

Let the matrix to be inverted be

$$\mathbf{A} = \begin{bmatrix} a & b & c \\ d & e & f \\ g & h & k \end{bmatrix}.$$

Then, its inverse is given by

$$\mathbf{A}^{-1} = \begin{bmatrix} a & b & c \\ d & e & f \\ g & h & k \end{bmatrix}^{-1} = \frac{1}{\det(\mathbf{A})} \begin{bmatrix} A & B & C \\ D & E & F \\ G & H & K \end{bmatrix}^{\mathrm{T}} = \frac{1}{\det(\mathbf{A})} \begin{bmatrix} A & D & G \\ B & E & H \\ C & F & K \end{bmatrix} \quad \text{(A.4)}$$

where, A, B, C, D, E, F, G, H, K are the cofactors of the matrix $\mathbf{A}$.

The determinant of $\mathbf{A}$ can be computed as follows:

$$\det(\mathbf{A}) = a(ek - fh) - b(dk - fg) + c(dh - eg).$$

If the determinant is non-zero, the matrix is invertible and determination of the cofactors subsequently lead to the computation of the inverse of $\mathbf{A}$.

Similarity Transformation

Two n-by-n matrices $\mathbf{A}$ and $\mathbf{B}$ are called similar if

$$\mathbf{B} = \mathbf{P}^{-1}\mathbf{A}\mathbf{P} \quad \text{(A.5)}$$

for some invertible n-by-n matrix $\mathbf{P}$.

Similar matrices represent the same linear transformation under two different bases, with $\mathbf{P}$ being the change of basis matrix.

The determinant of the similarity transformation of a matrix is equal to the determinant of the original matrix $\mathbf{A}$.

$$\det(\mathbf{B}) = \det\left(\mathbf{P}^{-1}\mathbf{A}\mathbf{P}\right) = \det\left(\mathbf{P}^{-1}\right)\det(\mathbf{A})\det(\mathbf{P}) = \frac{\det(\mathbf{A})}{\det(\mathbf{P})}\det(\mathbf{P}) = \det(\mathbf{A}) \quad \text{(A.6)}$$

Also, the eigenvalues of the matrices **A** and **B** are also same. That is

$$\begin{aligned}\det(\mathbf{B} - \lambda\mathbf{I}) &= \det\left(\mathbf{P}^{-1}\mathbf{A}\mathbf{P} - \lambda\mathbf{I}\right) \\ &= \det\left(\mathbf{P}^{-1}\mathbf{A}\mathbf{P} - \mathbf{P}^{-1}\lambda\mathbf{I}\mathbf{P}\right) \\ &= \det\left(\mathbf{P}^{-1}(\mathbf{A} - \lambda\mathbf{I})\mathbf{P}\right) \\ &= \det\left(\mathbf{P}^{-1}\right)\det(\mathbf{A} - \lambda\mathbf{I})\det(\mathbf{P}) \\ &= \det(\mathbf{A} - \lambda\mathbf{I})\end{aligned} \tag{A.7}$$

where, λ is a scalar.

The eigenvalues of an $n \times n$ matrix **A** are the roots of the characteristic equation

$$|\lambda\mathbf{I} - \mathbf{A}| = 0$$

Hence, eigenvalues are also called the *characteristic roots*. Also, the eigenvalues are invariant under any linear transformation.

Appendix B
Selected MATLAB Programs

1. Program for adding two time functions in hybrid function domain (Chap. 2, Fig. 2.5, p. 36)

```
clc
clear all
format long


%%---------- Number of Sub-intervals and Total Time ---------%%

m = input('Enter the number of sub-intervals chosen:\n')
T = input('Enter the total time period:\n')
h = T/m;
t = 0:h:T;


%%--------------- Functions for Addition -----------------%%

syms x
f = 1-exp(-x);
g = exp(-x);


%%------------------- Function Samples -------------------%%

F = subs(f,t)  %Samples of first function f(t) upto final time T
G = subs(g,t)  %Samples of second function g(t) upto final time T



%%--------- Hybrid Function Based Representation ---------%%

for i=1:m
   F_SHF(i)=F(i);  % Sample-and-Hold Function Coefficients for f(t)
end

for i=1:m
   F_TF(i)=F(i+1)-F(i); % Triangular Function Coefficients for f(t)
end
```


A. Deb et al., *Analysis and Identification of Time-Invariant Systems, Time-Varying Systems, and Multi-Delay Systems using Orthogonal Hybrid Functions*,
Studies in Systems, Decision and Control 46, DOI 10.1007/978-3-319-26684-8

```
for i=1:m
   G_SHF(i)=G(i);  % Sample-and-Hold Function Coefficients for g(t)
end

for i=1:m
   G_TF(i)=G(i+1)-G(i); % Triangular Function Coefficients for g(t)
end


%%---------- Addition in Hybrid Function Domain ----------%%

A_SHF = F_SHF + G_SHF;     % SHF Part of Addition
A_TF = F_TF + G_TF;        % TF Part of Addition

A_m = A_SHF(m) + A_TF(m);  % m-th coefficient of Addition
A = [A_SHF  A_m];           % Samples for plotting Addition


%%------------------- Function Plotting -------------------%%

plot(t,F,'-^k','LineWidth',2,'MarkerFaceColor','k')
hold on
plot(t,G,'-ok','LineWidth',2,'MarkerFaceColor','k')
hold on
plot(t,A,'-.>k','LineWidth',2)
ylim([0 1.2])
```

2. Program for dividing two time functions in hybrid function domain (Chap. 2, Fig. 2.13, p. 47)

```
clc
clear all
format long

%%------------ Number of Sub-intervals and Total Time -----------%%

m = input('Enter the number of sub-intervals chosen:\n')
T = input('Enter the total time period:\n')
h = T/m;
t = 0:h:T;

%%------------------- Functions for Division ---------------------%%

syms x
y = 1-exp(-x);
r = exp(-x);

%%----------------------- Function Samples -----------------------%%

Y = subs(y,t)   % Samples of first function y(t) upto final time T
R = subs(r,t)   % Samples of second function r(t) upto final time T

%%------------- Hybrid Function Based Representation -------------%%
```

```
for i=1:(m+1)
   Y_SHF(i)=Y(i);  % Sample-and-Hold Function Coefficients for y(t)
end


for i=1:(m+1)
   R_SHF(i)=R(i);  % Sample-and-Hold Function Coefficients for r(t)
end


%%-------------- Division in Hybrid Function Domain --------------%%

D_SHF=Y_SHF./R_SHF;    % SHF Part of Multiplication

for i=1:m
   D_TF(i)=D_SHF(i+1)-D_SHF(i);     % TF Part of Multiplication
end


%%----------------------- Function Plotting -----------------------%%

plot(t,Y,'-^k','LineWidth',2,'MarkerFaceColor','k')
hold on
plot(t,R,'-ok','LineWidth',2,'MarkerFaceColor','k')
hold on
plot(t,D_SHF,'-.>k','LineWidth',2)
ylim([0 1.8])
```

3. Program for approximating a function $f(t) = \sin(\pi t)$ in BPF domain (Chap. 3, Fig. 3.2, p. 51)

```
clc
clear all
format long

%%------------ Number of Sub-intervals and Total Time -----------%%

m = input('Enter the number of sub-intervals chosen:\n')
T = input('Enter the total time period:\n')
h = T/m;
t = 0:h:T;

%%------------------ Functions for Approximated ------------------%%

syms x
f=sin(pi*x);    % the function to be approximated

j=0:0.01:T;
plot(j,sin(pi*j),'--k','LineWidth',2)  % plot of the exact function
hold on
```

```
%%------------------- BPF Based Representation -------------------%%

C=zeros(1,m);
for k=1:m
   C(k)=(m/T)*int(f,t(k),t(k+1));  % Calculating BPF Coefficients
end

Coeff=[C C(m)];   % For plotting the BPF Coefficients of the function


%%---------------------- Function Plotting ----------------------%%

stairs(t,Coeff,'-k','LineWidth',2)
ylim([0 1])
```

4. Program for approximating a function $f(t) = \sin(\pi t)$ in hybrid function domain (Chap. 3, Fig. 3.5, p. 54)

```
clc
clear all
format long

%%----------- Number of Sub-intervals and Total Time -----------%%

m = input('Enter the number of sub-intervals chosen:\n')
T = input('Enter the total time period:\n')
h = T/m;
t = 0:h:T;

%%---------------- Function to be Approximated ----------------%%

syms x
f=sin(pi*x);      % Function to be Approximated
F=subs(f,t)       % Samples of first function f(t) upto final time T


%%------------ Hybrid Function Based Representation -----------%%

for i=1:m
   F_SHF(i)=F(i);  % Sample-and-Hold Function Coefficients for f(t)
end

for i=1:m
   F_TF(i)=F(i+1)-F(i);  % Triangular Function Coefficients for f(t)
end


%%---------------------- Function Plotting ----------------------%%

j=0:0.01:T;
plot(j,sin(pi*j),'-.k','LineWidth',2)  % plot of the exact function
hold on
plot(t,F,'-ok','LineWidth',2,'MarkerFaceColor','k')
ylim([0 1])
```

5. Program for approximating a function $f(t) = t$ in hybrid function domain and BPF domain. (Chap. 3, Fig. 3.6, p. 55)

```
clc
clear all
format long

%%------------ Number of Sub-intervals and Total Time -----------%%

m = input('Enter the number of sub-intervals chosen:\n')
T = input('Enter the total time period:\n')
h = T/m;
t = 0:h:T;

%%----------------- Function to be Approximated -----------------%%

syms x
f=x;               % Function to be Approximated
F=subs(f,t)        % Samples of function f(t) upto final time T


%%------------------- BPF Based Representation -------------------%%

C=zeros(1,m);
for k=1:m
   C(k)=(m/T)*int(f,t(k),t(k+1));  % Calculating BPF Coefficients
end
Coeff=[C C(m)];                   % BPF Coefficients of the function


%%------------- Hybrid Function Based Representation -------------%%

for i=1:m
   F_SHF(i)=F(i);  % Sample-and=Hold Function Coefficients for f(t)
end

for i=1:m
   F_TF(i)=F(i+1)-F(i);  % Triangular Function Coefficients for f(t)
end

F_m = F_SHF(m)+F_TF(m);      % m-th coefficient of Subtraction
F = [F_SHF F_m];             % Final Samples for plotting

%%----------------------- Function Plotting ----------------------%%

stairs(t,Coeff,'-k','LineWidth',2)
hold on
plot(t,t,'--k','LineWidth',2)  % plot of the exact function
hold on
plot(t,F,'-ok','LineWidth',2,'MarkerFaceColor','k')
ylim([0 1])
```

6. Program for comparing HFc and HFm based approach (Chap. 3, Fig. 3.16, p. 63)

```
clc
clear all
format long

%%----------- Number of Sub-intervals and Total Time -----------%%

m = input('Enter the number of sub-intervals chosen:\n')
T1 = input('Enter the time at jump instant:\n')
T2 = input('Enter the time after jump instant:\n')
T=T1+T2;          % Total time period
h = T/m;
m_jump = T1/h;    % Number of sub-intervals up to jump instant
t = 0:h:T;

%%----------------- Function to be Approximated -----------------%%

syms t
f1 = t+0.2;        % Function before jump instant
f2 = t+1.2;        % Function after jump instant


%%----------------------- Exact Solution -----------------------%%

t1=0:0.001:T1;
t2=(T1+0.001):0.001:T;

f1t=subs(f1,t1);
f2t=subs(f2,t2);

te=[t1 t2];
fe=[f1t f2t];


%%----------- Hybrid Function Based Representation -----------%%

th1=0:h:(T1-h);
th2=T1:h:T;
th=[th1 th2];

F1=subs(f1,th1);
F2=subs(f2,th2);
cfsx=[F1 F2];    % total Sample-and-Hold function coefficients
for i=1:m
   Cfsx(i)=cfsx(i);    % First m number of SHF coefficients
end

for i=1:m
   Cftx(i)=cfsx(i+1)-cfsx(i);
end
Cftx(m_jump)=0;  % to be considered only in HFm based approximation
Cftx;
```

```
%%---------------------- Function Plotting ----------------------%%

tf=(T1-h):0.001:T1;
cf=[F1(m/T)*ones(1,length(tf)) 1.2];
Tf=[tf T1];

plot(te,fe,'k-','Linewidth',2)    % plot of the exact function
hold on
plot(th,cfsx,'ko','MarkerFaceColor','k','MarkerSize',7)
hold on
plot(Tf,cf,'k-')
xlim([-0.001 2.001])
```

7. Program for comparing MISEs using HFc and HFm based approaches (Chap. 3, Table 3.2, , p. 65); Fig. 3.17, p. 64)

```
clc
clear all
format long

%%------------ Number of Sub-intervals and Total Time -----------%%

m = input('Enter the number of sub-intervals chosen:\n')
T1 = input('Enter the time at jump instant:\n')
T2 = input('Enter the time after jump instant:\n')
T = T1+T2;        % Total time period
h = T/m;
m_jump = T1/h;   % Number of sub-intervals up to jump instant
t = 0:h:T;


%%------------------ Function to be Approximated -----------------%%

syms t
f1=t+0.2;
f2=t+1.2;

%%-------------- Hybrid Function Based Approximation -------------%%

th1=0:h:(T1-h);
th2=T1:h:T;
th=[th1 th2];

F1=subs(f1,th1);
F2=subs(f2,th2);
cfsx=[F1 F2];
Cfsx=cfsx(1:m)

for i=1:m
   Cftx(i)=cfsx(i+1)-cfsx(i); % Triangular function coefficients
end

Cftx(m_jump)=0; % to be considered only in HFm based approximation

%%--------------------- Calculation of MISE ---------------------%%
```

```
mise=0;
for i=1:(T1/h)
   ft=f1;
   Fhf=(Cftx(i)/h)*(t-(i-1)*h)+Cfsx(i);
   Fd=(ft-Fhf)^2;
   mise1(i)=int(Fd,t,((i-1)*h),(i*h));
   mise=mise1(i)+mise;
end
miseone=mise/T1;

mise=0;
for j=1:(T2/h)
ft=f2;
Fhf=(Cftx(j+(T1/h))/h)*((t-1)-(j-1)*h)+Cfsx(j+T1/h);
Fd=(ft-Fhf)^2;
mise2(j)=int(Fd,t,(T1+(j-1)*h),(T1+j*h));
   mise=mise2(j)+mise;
end
misetwo=mise/T2;

MISE=double(miseone+misetwo)
```

8. Program for approximating a function using Legendre Polynomial (Chap. 3, Fig. 3.20, p. 68)

```
clc
clear all
format long

%%----------- Number of Sub-intervals and Total Time -----------%%

m = input('Enter the number of sub-intervals chosen:\n')
T = input('Enter the total time interval:\n')
h = T/m;

%%---------------- Function to be Approximated ----------------%%

syms t
f=exp(t-1);        % Function


%%---------------------- Exact Solution ----------------------%%

te=0:0.001:T;
ft=subs(f,te);
plot(te,ft,'k-','LineWidth',2)
hold on

%%----------- Legendre Polnomial Based Approximation ----------%%

P0=1;                % Legendre polynomial of degree 0
P1=t;                       % Legendre polynomial of degree 1
P2=((3*t^2)-1)/2;           % Legendre polynomial of degree 1
P3=((5*t^3)-(3*t))/2;                % Legendre polynomial of degree 1
P4=((35*t^4)-(30*t^2)+3)/8;          % Legendre polynomial of degree 1
P5=((63*t^5)-(70*t^3)+(15*t))/8;     % Legendre polynomial of degree 1
P6=((231*t^6)-(315*t^4)+(105*t^2)-5)/16;
                                     % Legendre polynomial of degree 1

P=[P0 P1 P2 P3 P4 P5 P6];


Func=subs(P,(t-1));

for i=1:7
   c=double(int(P(i)*exp(t),-1,1));
Coeff(i)=((2*(i-1)+1)/2)*c;    % Legendre coefficient for i-th degree
end

F=sum(Coeff.*Func);

%%-------------------- Function Plotting --------------------%%

th=0:h:T;
F_Legendre=double(subs(F,th));

plot(th,F_Legendre,'ok','MarkerFaceColor','k')
```

9. Program for calculating MISE of a function, approximated using Legendre Polynomial (Chap. 3, Table 3.4, p. 70)

```
clc
clear all
format long
%%----------------- Total Time Period considered-----------------%%

T = input('Enter the total time interval:\n')
Np = input('Enter the number of polynomials to be used:\n')

%%----------------- Function to be Approximated -----------------%%

syms t
f=exp(t-1);       % Function



%%------------ Legendre Polnomial Based Approximation ----------%%

P0=1;               % Legendre polynomial of degree 0
P1=t;                      % Legendre polynomial of degree 1
P2=((3*t^2)-1)/2;          % Legendre polynomial of degree 1
P3=((5*t^3)-(3*t))/2;               % Legendre polynomial of degree 1
P4=((35*t^4)-(30*t^2)+3)/8;         % Legendre polynomial of degree 1
P5=((63*t^5)-(70*t^3)+(15*t))/8;        % Legendre polynomial of degree 1
P6=((231*t^6)-(315*t^4)+(105*t^2)-5)/16;   % Legendre polynomial of degree 1

P=[P0 P1 P2 P3 P4 P5 P6];

Func=subs(P,(t-1));

for i=1:Np
   c=double(int(P(i)*exp(t),-1,1));
   Coeff(i)=((2*(i-1)+1)/2)*c;    % Legendre coefficient for i-th degree
end
Coeff
F=sum(Coeff.*Func(1:i));
Func(1:i)
F

%%-------------------- Calculation of MISE --------------------%%

for p=1:Np
   F_Legendre=sum(Coeff(1:p).*Func(1:p));
   Fd=(f-F_Legendre)^2;
   MISE=(1/T)*double(int(Fd,t,0,T));
end

MISE
```

10. Program for calculating MISE of a function, approximated in hybrid function domain (Chap. 3, Table 3.4, p. 70)

```
clc
clear all
format long
%%------------ Number of Sub-intervals and Total Time -----------%%

m = input('Enter the number of sub-intervals chosen:\n')
T = input('Enter the total time period: \n')
h = T/m;
th = 0:h:T;

%%----------------- Function to be Approximated -----------------%%

syms t
f=exp(t-1);

%%--------------- Hybrid Function Based Approximation -----------%%

cfsx=subs(f,th);
Cfsx=cfsx(1:m);   % Sample-and-Hold function coefficients

for i=1:m
   Cftx(i)=cfsx(i+1)-cfsx(i); % Triangular function coefficients
end

%%--------------------- Calculation of MISE ---------------------%%

for i=1:m
   Fhf=(Cftx(i)/h)*(t-(i-1)*h)+Cfsx(i);
   Fd=(f-Fhf)^2;
   mise(i)=(1/T)*double(int(Fd,t,((i-1)*h),(i*h)));
end

MISE=sum(mise)
```

11. Program for integrating a function $f(t) = \sin(\pi t)$ in hybrid function domain (Chap. 4, Fig. 4.8; Table 4.2, p. 100, 89)

```
clc
clear all
format long

%%------------ Number of Sub -intervals and Total Time -----------%%

m = input('Enter the number of sub  -intervals chosen: \n')
T = input('Enter the total time period:  \n')
h = T/m;
th = 0:h:T;

%%------------------ Function to be Integrated ------------------%%

syms x
f=sin(pi*x);      % Function to be Integrated
F=subs(f,th);     % Samples of the Function f(t)
fi=int(f,0,x);    % Function after Integration
Fi=subs(fi,th);   % Samples of integrated function fi(t)


%%----- HF Based Representation of function to be Integrated -----%%

for i=1:m
   F_SHF(i)=F(i);  % Sample-and-Hold Function Coefficients for f(t)
end

for i=1:m
   F_TF(i)=F(i+1)-F(i);  % Triangular Function Coefficients for f(t)
end

%%---------- HF domain Integration Operational Matrices ----------%%

ps=zeros(m,m);

for i=1:m
   for j=1:m
      if j-i>0
         Ps(i,j)=ps(i,j)+1;
      else
         Ps(i,j)=ps(i,j);
      end
   end
end

Pss=h*Ps;       % SHF part after integrating SHF components

Pst=h*eye(m);   % TF part after integrating SHF components

Pts=0.5*Pss;    % SHF part after integrating TF components

Ptt=0.5*Pst;    % TF part after integrating TF components
```

```
%%------- Integartion using Integration Operational Matrices -------%%

Cs = (F_SHF*Pss)+(F_TF*Pts);     % Sample-and-Hold Function Coefficients
                                         after integration in HF domain

Ct = (F_SHF*Pst)+(F_TF*Ptt);     % Triangular Function Coefficients
                                         after integration in HF domain

Cs_m = Cs(m)+Ct(m);              % m-th coefficient

Cs_plot=[Cs Cs_m];               % Samples for plotting the function
                                         after integration in HF domain

%%----------------------- Function Plotting -----------------------%%

plot(th,Fi,'r-','LineWidth',3)
hold on
plot(th,Cs_plot,'ok-','LineWidth',2,'MarkerFaceColor','k')
xlim([0 T])
```

12. Program for differentiating a function $f(t) = 1 - \exp(-t)$ in hybrid function domain (Chap. 4, Fig. 4.9, p. 105)

```
clc
clear all
format long

%%----------- Number of Sub-intervals and Total Time -----------%%

m = input('Enter the number of sub-intervals chosen: \n')
T = input('Enter the total time period: \n')
h = T/m;
th=0:h:T;


%%---------------- Function to be Differentiated ---------------%%

syms x
f=1-exp(-x);       % Function to be Differentiated
fd=diff(f);        % Function after Differentiation
F=subs(f,th);      % Samples of the Function f(t)
Fd=subs(fd,th);    % Samples of differentiated function fd(t)


%%----- HF Based Representation of Differentiated function -----%%

for i=1:m
   F_SHF(i)=F(i);  % Sample-and-Hold Function Coefficients for f(t)
end

for i=1:m
   F_TF(i)=F(i+1)-F(i);  % Triangular Function Coefficients for f(t)
end

%%-------------- HF domain Differentiation Matrices ------------%%
```

```
ds=(-1)*eye(m);

for i=1:m
   for j=1:m
      if (i-j)==1
         Ds(i,j)=ds(i,j)+1;
      else
         Ds(i,j)=ds(i,j);
      end
   end
end
Ds(m,m)=(F(m+1)-F(m))/F(m);

Ds=(1/h)*Ds;   % SHF part of Differentiation matrix

dt=(-1)*eye(m);

for i=1:m
   for j=1:m
      if (i-j)==1
         Dt(i,j)=dt(i,j)+1;
      else
         Dt(i,j)=dt(i,j);
      end
   end
end

F_(m+2)=subs(f,(T+h));
Dt(m,m)=((F_(m+2)-F(m+1))-(F(m+1)-F(m)))/(F(m+1)-F(m));

Dt=(1/h)*Dt;   % TF part of Differentiation matrix


%%---------- Differentiation using Operational Matrices ----------%%

Cs = (F_SHF*Ds);                % Sample-and-Hold Function Coefficients
                                    after differentiation in HF domain

Ct = (F_TF*Dt);                 % Triangular Function Coefficients
                                    after differentiation in HF domain


Cs_m=Cs(m)+Ct(m);

Cs_plot=[Cs Cs_m];              % Samples for plotting the function
                                    after differentiation in HF domain


%%---------------------- Function Plotting ----------------------%%

plot(th,F,'-^k','LineWidth',2,'MarkerFaceColor','k')
hold on
plot(th,Fd,'-ok','LineWidth',2,'MarkerFaceColor','k')
hold on
plot(th,Cs_plot,':>k','LineWidth',2)
```

13. Program for differentiating a function $f(t) = \sin(\pi t)/\pi$ in hybrid function domain (Chap. 4, Fig. 4.10, p. 106)

```
clc
clear all
format long

%%----------- Number of Sub-intervals and Total Time -----------%%

m = input('Enter the number of sub-intervals chosen:\n')
T = input('Enter the total time period:\n')
h = T/m;
th = 0:h:T;

%%---------------- Function to be Differentiated ---------------%%

syms x
f=sin(pi*x)/pi;  % Function to be Differentiated
fd=diff(f);      % Function after Differentiation
F=subs(f,th);    % Samples of the Function f(t)
Fd=subs(fd,th);  % Samples of differentiated function fd(t)


%%------ HF Based Representation of Differentiated function ------%%

for i=1:m
   F_SHF(i)=F(i);  % Sample-and-Hold Function Coefficients for f(t)
end

for i=1:m
   F_TF(i)=F(i+1)-F(i);  % Triangular Function Coefficients for f(t)
end


%%-------------- HF domain Differentiation Matrices --------------%%

ds=(-1)*eye(m);

for i=1:m
   for j=1:m
      if (i-j)==1
         Ds(i,j)=ds(i,j)+1;
      else
         Ds(i,j)=ds(i,j);
      end
   end
end
Ds(m,m)=(F(m+1)-F(m))/F(m);

Ds=(1/h)*Ds;   % SHF part of Differentiation matrix
dt=(-1)*eye(m);

for i=1:m
   for j=1:m
      if (i-j)==1
         Dt(i,j)=dt(i,j)+1;
      else
         Dt(i,j)=dt(i,j);
      end
```

```
    end
end
F_(m+2)=subs(f,(T+h));
Dt(m,m)=((F_(m+2)-F(m+1))-(F(m+1)-F(m)))/(F(m+1)-F(m));

Dt=(1/h)*Dt;    % TF part of Differentiation matrix


%%---------- Differentiation using Operational Matrices ----------%%

Cs=(F_SHF*Ds);                   % Sample-and-Hold Function Coefficients
                                       after differentiation in HF domain

Ct=(F_TF*Dt);                    % Triangular Function Coefficients
                                       after differentiation in HF domain

Cs_m=Cs(m)+Ct(m);

Cs_plot=[Cs Cs_m];               % Samples for plotting the function
                                       after differentiation in HF domain


%%----------------------- Function Plotting -----------------------%%

plot(th,F,'-^k','LineWidth',2,'MarkerFaceColor','k')
hold on
plot(th,Fd,'-ok','LineWidth',2,'MarkerFaceColor','k')
hold on
plot(th,Cs_plot,':>k','LineWidth',2)

z=zeros(length(th));
hold on
plot(th,z,'-k')
```

14. Program for calculating average of mod of percentage (AMP) error for *n*th repeated I-D operation of function $f(t) = \sin(\pi t)$ in hybrid function domain (Chap. 4, Fig. 4.12, p. 110)

```
function Per_Err=ID(n)

%%%%%%% Integration-Differentiation (I-D) operation %%%%%%%%
%%%%%%% For SINE Function
%Go to Command Window and type ID(required value of number of ID
%operations n) and ENTER

clc
clear all

%%---------- Defining the Function for I-D operation -----------%%

syms t
ft=sin(pi*t);


%%------------ Number of Sub-intervals and Total Time -----------%%

m = input('Enter the number of sub-intervals chosen:\n')
T = input('Enter the total time period:\n')
h = T/m;

t1 = 0:h:(T+n*h);
Cs = subs(ft,t1);     % SHF coefficients

Cid=zeros(1,(n+1));  %Create space for coefficients of ID operation

Cid(1)=1;
a=n;
for j=1:n
   c=a/factorial(j);
   Cid(j+1)=c;
   a=a*(n-j);
end
Cid;

Cs_n=zeros(1,m);
for k=1:m
   C=Cs(k:(n+k));
   Cs_n(k)=(1/2^n)*sum(C.*Cid);
end

%%-------------------- Calculation of AMP Error --------------------%%

Percentage_Error=(Cs(1:m)-Cs_n)*100./Cs(1:m);
AMP_Error=abs(sum(PE(2:m))/(m-1))
```

15. Program for plotting the effect of repeated I-D operation over a specific function $f(t) = \sin(\pi t)$ in hybrid function domain (Chap. 4, Fig. 4.15, p. 112)

```
clc
clear all

%%------------ Defining the Function for I-D opeartion ------------%%
syms t
ft=sin(pi*t);

%%------------- Number of Sub-intervals and Total Time ------------%%

m = input('Enter the number of sub-intervals chosen:\n');
T = input('Enter the total time period:\n');
h = T/m;

n = input('Enter the maximum number of ID operations required:\n');


%%--------------------- Repeated I-D Opeartion --------------------%%

for id=1:n
   t1=0:h:(T+id*h);
   Cs=subs(ft,t1);        % SHF component including m-th component
   Cs_MAT(1,:)=Cs(1:(m+1));

   Cid=zeros(1,(id+1));  %Create space for coefficients of ID operation

   Cid(1)=1;
   a=id;
      for j=1:id
         c=a/factorial(j);
         Cid(j+1)=c;
         a=a*(id-j);
      end

   Cs_n=zeros(1,m);

      for k=1:m
         C=Cs(k:(id+k));
         Cs_n(k)=(1/2^id)*sum(C.*Cid);
      end

   % For m-th SHF component

   t2=T:h:(T+id*h);
   Cm=Cs((m+1):(m+id+1));
   Cs_n(k+1)=(1/2^id)*sum(Cm.*Cid);


   Cs_MAT((id+1),:)=Cs_n(1:(m+1))
end

%%------------- For Plotting the Effect of I-D Opeartion -------------%%
tn=0:h:T;
z=zeros(length(tn));
plot(tn,Cs_MAT(1,:),'r-',tn,Cs_MAT(2,:),'b',tn,Cs_MAT(3,:)
       ,'g',tn,Cs_MAT(4,:),'k-',tn,Cs_MAT(11,:),'m-','LineWidth',2)

hold on
plot(tn,z,'k-')
ylim([-1.2 1.2])
```

16. Program for finding second order integration of function *f* (*t*) = *t*, using HF domain one-shot integration operational matrices (Chap. 5, Example 5.1, p. 130)

```
clc
clear all
format long

%%------------ Number of Sub-intervals and Total Time -----------%%

m = input('Enter the number of sub-intervals chosen:\n')
T = input('Enter the total time period:\n')
h = T/m;
th = 0:h:T;

%%------------------ Function to be Integrated ------------------%%

syms x
f=x;               % Function to be Integrated
F=subs(f,th);      % Samples of the Function f(t)


%%----------------------- Exact Solution -----------------------%%

fi=int(int(f));
Fi=subs(fi,th)

for j=1:m
   Fi_SHF(j)=Fi(j);     % Sample-and-Hold function part (Equation 5.28)
end

for j=1:m
   Fi_TF(j)=Fi(j+1)-Fi(j);  % Triangular Function part (Equation 5.28)
end


%%----- HF Based Representation of function to be Integrated -----%%

for i=1:m
   F_SHF(i)=F(i);  % Sample-and-Hold Function Coefficients for f(t)
end

for i=1:m
   F_TF(i)=F(i+1)-F(i);  % Triangular Function Coefficients for f(t)
end


%%------- Using One-Shot Integration Operational Matrices --------%%
%%--------------- Formation of P2SS matrices ----------------%%
n=2;     %For Second order Integration

%%%%Formation of P2SS matrices %%%%%%
p=zeros(m,m);

for i=1:m
   for j=1:m
      if j-i>0
          P2ss(i,j)=(j-i)^n-(j-i-1)^n;
       else
```

```
            P2ss(i,j)=p(i,j);
        end
    end
end
P2SS=(h^n/factorial(n))*P2ss;


%%%%Formation of P2ST matrices %%%%%%

p=eye(m);

for i=1:m
    for j=1:m
        if j-i>0
            P2st(i,j)=((j-i+1)^n-(j-i)^n)-((j-i)^n-(j-i-1)^n);
        else
            P2st(i,j)=p(i,j);
        end
    end
end
P2ST=(h^n/factorial(n))*P2st;


%%%%Formation of P2TS matrices %%%%%%

p=zeros(m,m);
for i=1:m
    for j=1:m
        if j-i>0
            P2ts(i,j)=(j-i)^(n+1)-(j-i-1)^(n+1)-(n+1)*(j-i-1)^n;
        else
            P2ts(i,j)=p(i,j);
        end
    end
end
P2TS=(h^n/factorial(n+1))*P2ts;


%%%%Formation of P2TT matrices %%%%%%

p=eye(m);
for i=1:m
    for j=1:m
        if j-i>0
                P2tt(i,j)=((j-i+1)^(n+1)-(j-i)^(n+1))-((j-i)^(n+1)
                                  -(j-i-1)^(n+1))-(n+1)*((j-i)^n-(j-i-1)^n);
        else
                P2tt(i,j)=p(i,j);
        end
    end
end
P2TT=(h^n/factorial(n+1))*P2tt;

CS_oneshot=(F_SHF*P2SS+F_TF*P2TS)          %SHF Coefficients after
                                           double integration in HF domain

CT_oneshot=(F_SHF*P2ST+F_TF*P2TT);         %TF Coefficients after
                                           double integration in HF domain
```

17. Program for finding deviation indices δ_R (using repeated integration) and δ_O (using one-shot integration operational matrices) for second order integration of a typical functions $f(t) = \exp(-t)$ (Chap. 5, Table 5.1, p. 131)

```
clc
clear all
format long

%%------------ Number of Sub-intervals and Total Time -----------%%

m = input('Enter the number of sub-intervals chosen:\n');
T = input('Enter the total time period:\n');
h = T/m;
th = 0:h:T;

%%---------------- Function under Consideration -----------------%%

syms x
f=exp(-x);

F_samples=subs(f,th);  % Samples of the Function f(t)


%%----- HF Based Representation of function to be Integrated -----%%

for i=1:m
   Cfsx(i)=F_samples(i);  % SHF Coefficients for f(t)
end

for i=1:m

   Cftx(i)=F_samples(i+1)-F_samples(i);  % TF Coefficients for f(t)
end


%%------- Exact Samples after double integration of function ------%%

FT=x+exp(-x)-1;

CFsx=subs(FT,th);

CFsx_Exact=CFsx;      % (m+1) number of exact samples

for i=1:m
   CFtx(i)=CFsx(i+1)-CFsx(i);  % TF coefficients
end
CFsx=CFsx(1:m)        % SHF coefficients


%%----------- HF domain Integration Operational Matrices ----------%%

p1s=zeros(m,m);

for i=1:m
   for j=1:m
      if j-i>0
         P1s(i,j)=p1s(i,j)+1;
      else
         P1s(i,j)=p1s(i,j);
```

```
      end
   end
end

P1SS=h*P1s;       % SHF part after integrating SHF components

P1ST=h*eye(m);    % TF part after integrating SHF components

P1TS=0.5*P1SS;    % SHF part after integrating TF components

P1TT=0.5*P1ST;    % TF part after integrating TF components


%%------- Integration of Function through Repeated use of -------%%
%%------- First Order Integration Operational Matrices --------%%

n=2;   %For Second order Integration

Cs=Cfsx;
Ct=Cftx;
for j=1:n
   Cs1=(Cs*P1SS)+(Ct*P1TS);
   Ct1=(Cs*P1ST)+(Ct*P1TT);
   Cs=Cs1;
   Ct=Ct1;
end

Cs_m1=Cs(m)+Ct(m);         % (m+1)-th sample
CS_repeated=[Cs Cs_m1];    % SHF samples after Second order Integration

%%------- Using One-Shot Integration Operational Matrices -------%%
%%--------------- Formation of P2SS matrices ----------------%%

n=2;     %For Second order Integration

p=zeros(m,m);
for i=1:m
   for j=1:m
      if j-i>0
        P2ss(i,j)=(j-i)^n-(j-i-1)^n;
      else
        P2ss(i,j)=p(i,j);
      end
   end
end
P2SS=(h^n/factorial(n))*P2ss;


%%%%Formation of P2ST matrices %%%%%%

p=eye(m);
for i=1:m
   for j=1:m
      if j-i>0
         P2st(i,j)=((j-i+1)^n-(j-i)^n)-((j-i)^n-(j-i-1)^n);
```

```
        else
           P2st(i,j)=p(i,j);
        end
    end
end
P2ST=(h^n/factorial(n))*P2st;

%%%%Formation of P2TS matrices %%%%%%

p=zeros(m,m);
for i=1:m
    for j=1:m
        if j-i>0
           P2ts(i,j)=(j-i)^(n+1)-(j-i-1)^(n+1)-(n+1)*(j-i-1)^n;
        else
           P2ts(i,j)=p(i,j);
        end
    end
end
P2TS=(h^n/factorial(n+1))*P2ts;

%%%%Formation of P2TT matrices %%%%%%

p=eye(m);
for i=1:m
    for j=1:m
        if j-i>0
           P2tt(i,j)=((j-i+1)^(n+1)-(j-i)^(n+1))-((j-i)^(n+1)
                                 -(j-i-1)^(n+1))-(n+1)*((j-i)^n-(j-i-1)^n);
        else
           P2tt(i,j)=p(i,j);
        end
    end
end
P2TT=(h^n/factorial(n+1))*P2tt;

CS=(Cfsx*P2SS+Cftx*P2TS);              %SHF Coefficients after
                                           double integration in HF domain

CT=(Cfsx*P2ST+Cftx*P2TT);              %TF Coefficients after
                                           double integration in HF domain

CS_m1=CS(m)+CT(m);          % (m+1)th sample
CS_oneshot=[CS CS_m1];      % SHF samples after Second order Integration

diff_repeated=CFsx_Exact-CS_repeated;
            % differences between exact samples and samples obtained
            % via repeated integration

diff_OneShot=CFsx_Exact-CS_oneshot;
            % differences between exact samples and samples obtained
            % using oneshot integration matrices

Deviation_Index_repeated=sum(abs(diff_repeated))/(m+1)

Deviation_Index_OneShot=sum(abs(diff_OneShot))/(m+1)
```

18. Program for obtaining the recursive solution of first order differential equation of Examples 6.1 and 6.2 (Chap. 6, Figs. 6.1, 6.2 and 6.3; Table 6.1, p. 143, 144, 148–150)

```
clc
clear all
format long

%%------------ Number of Sub-intervals and Total Time -----------%%

m = input('Enter the number of sub-intervals chosen:\n');
T = input('Enter the total time period:\n');
h = T/m;
th = 0:h:T;

%%-------------- Solution of Differential Equation -------------%%

sol=dsolve('Dy + 0.5*y =1.25','y(0)=0','t')
t1=0:0.01:T;

F=subs(sol,t1);

plot(t1,F,'k-','Linewidth',2)


%%---------------- Direct Expansion of Solution ----------------%%

C=subs(sol,th);

Cs=C(1:m)                    % Sample-and-Hold function coefficients


for i=1:m
   Ct(i)=C(i+1)-C(i);        % Triangular function coefficients
end


 %%--------- Solution in HF domain, using equation (6.7) --------%%

c1(1)=0;   % given initial value

a=0.5;  b=1.25;  % constants as per equations (6.1) and (6.8)

for i=1:m
   c1(i+1)=(b*h)+(1-a*h)*c1(i);  % recursive solution
end

Cs_HF=c1(1:m);    % SHF coefficients obtained using equation (6.7)

for i=1:m
   Ct_HF(i)=c1(i+1)-c1(i); %TF coefficients obtained using equation(6.7)
end

hold on
plot(th,c1,'ko','MarkerEdgeColor','k','MarkerFaceColor','k','MarkerSize',7)

ylim([-0.2 1.2])
```

```
%%--------- Solution in HF domain, using equation (6.23) --------%%

c2(1)=0;    % given initial value

f=2/(2+a*h);
for i=1:m
   c2(i+1)=(b*f*h)+(1-a*f*h)*c2(i);  % recursive solution
end

Cs_HF=c2(1:m)                % Sample-and-Hold function coefficients

for i=1:m
   Ct_HF(i)=c2(i+1)-c2(i);    % Triangular function coefficients
end

hold on
plot(th,c2,'ko','MarkerEdgeColor','k','MarkerFaceColor','k','MarkerSize',7)

ylim([0 1.2])

%%----------- Calculation of Sample wise Percentage Error -----------%%

Percentage_Error=(Cs-Cs_HF)./Cs*100
```

19. Program for obtaining the solution of first order differential equation of Example 6.2 using Runge-Kutta method (Chap. 6, Table 6.2, p. 151)

Function File

```
function v=f(t,y)
v=1.25-0.5*y;
```

M-file for obtaining solution via Runge-Kutta method

```
clc
clear all
format long

h = 1/12;    % step size
t = 0;       % initial time
y = 0;       % initial condition

for i=1:12
   k1=h*v(t,y);
   k2=h*v(t+h/2,y+k1/2);
   k3=h*v(t+h/2,y+k2/2);
   k4=h*v(t+h,y+k3);
   y=y+(k1+2*k2+2*k3+k4)/6;
   t=t+h;
   y
end
```

20. Program for obtaining the solution of second order differential equation of Examples 6.3 and 6.4, in Hybrid Function domain (Chap. 6, Tables 6.3 and 6.4; Figs. 6.5 and 6.6, pp. 153-158)

```
clc
clear all
format long

%%------------ Number of Sub-intervals and Total Time -----------%%

m = input('Enter the number of sub-intervals chosen:\n');
T = input('Enter the total time period:\n');
h = T/m;
th = 0:h:T;

%%-------------- Solution of Differential Equation -------------%%

% sol=dsolve('D2y + 3*Dy + 2*y = 2','Dy(0)=-1','y(0)=1','t')
sol=dsolve('D2y + 100*y = 0','Dy(0)=0','y(0)=2','t')


Exact_samples=subs(sol,th)

tt=0:0.01:T;

F=subs(sol,t1);

plot(tt,F,'k-','Linewidth',2)

%%---------------- Direct Expansion of Solution ----------------%%

C=subs(sol,th);

Cs=C(1:m);    % SHF components

for i=1:m
   Ct(i)=C(i+1)-C(i);    % TF components
end

%---- via repreated use of first order integration matrices ----%

% a=3; b=2; d=2;
% k1=1; k2=-1;

a=0; b=100; d=0;
k1=2; k2=0;

r2=(a*k1)+k2;
r3=k1;
U=ones(1,m);
```

```
I=eye(m);  % Identity matrix

%-------- Operational matrix for Integration in BPF domain, P --------%

P=zeros(m,m);
for i=1:m
   P(i,i)=h/2;
      for j=1:(i-1)
         P(j,i)=h;
      end
  end
  P;

  %%----------- HF domain Integration Operational Matrices ----------%%

  p1s=zeros(m,m);

  for i=1:m
     for j=1:m
        if j-i>0
           P1s(i,j)=p1s(i,j)+1;
        else
           P1s(i,j)=p1s(i,j);
        end
     end
  end

  P1SS=h*P1s;       % SHF part after integrating SHF components

  P1ST=h*eye(m);    % TF part after integrating SHF components

  P1TS=0.5*P1SS;    % SHF part after integrating TF components

  P1TT=0.5*P1ST;    % TF part after integrating TF components

  L=(2*d*P)+(2*r2*I);
  Q=-((a*I)+(b*P));

  Cs_repeated=(2/h)*U*(L+(2*r3*Q))*inv((2/h)*I-(4/h)*P1TS*Q-Q)*P1TS+(r3*U)

  Ct_repeated=U*(L+(2*r3*Q))*inv((2/h)*I-(4/h)*P1TS*Q-Q);

  Cs_R=[Cs_repeated  (Cs_repeated(m)+Ct_repeated(m))]


  %%------- Using One-Shot Integration Operational Matrices --------%%
  %%------------ Formation of P2SS matrix -------------%%

  n=2;      %For Second order Integration
  %%----------------Formation of P2SS matrix ---------------%%
  p=zeros(m,m);
  for i=1:m
     for j=1:m
        if j-i>0
           P2ss(i,j)=(j-i)^n-(j-i-1)^n;
```

```
        else
           P2ss(i,j)=p(i,j);
        end
     end
  end

  P2SS=(h^2/factorial(2))*P2ss;

%%----------------Formation of P2ST matrix ---------------%%

p=eye(m);

for i=1:m
   for j=1:m
      if j-i>0
         P2st(i,j)=((j-i+1)^n-(j-i)^n)-((j-i)^n-(j-i-1)^n);
      else
         P2st(i,j)=p(i,j);
      end
   end
end

P2ST=(h^2/factorial(2))*P2st;


%%----------------Formation of P2TS matrix ---------------%%

p=zeros(m,m);

for i=1:m
   for j=1:m
      if j-i>0
         P2ts(i,j)=(j-i)^(n+1)-(j-i-1)^(n+1)-(n+1)*(j-i-1)^n;
      else
         P2ts(i,j)=p(i,j);
      end
   end
end

P2TS=(h^2/factorial(3))*P2ts;

%%----------------Formation of P2TT matrix ---------------%%

p=eye(m);
for i=1:m
   for j=1:m
      if j-i>0
         P2tt(i,j)=((j-i+1)^(n+1)-(j-i)^(n+1))-((j-i)^(n+1)
                                    -(j-i-1)^(n+1))-(n+1)*((j-i)^n-(j-i-1)^n);
      else
         P2tt(i,j)=p(i,j);
      end
   end
end
```

```
P2TT=(h^n/factorial(n+1))*P2tt;


X=I+(a*P1SS)+(b*P2SS);
Y=0.5*((a*P1SS)+(b*P2SS));

W=(a*P1ST)+(b*P2ST);
Z=I+0.5*((a*P1ST)+(b*P2ST));

M1=(Y*inv(X))-(Z*inv(W));
M2=U*((d*P2SS)+(r2*P1SS)+(r3*I))*inv(X)-U*((d*P2ST)+(r2*P1ST))*inv(W);

Ct_oneshot=M2*inv(M1);

M3=M2*inv(M1)*Z*inv(W);
M4=U*((d*P2ST)+(r2*P1ST))*inv(W);

Cs_oneshot=M4-M3

Cs_O=[Cs_oneshot (Cs_oneshot(m)+Ct_oneshot(m))]    % For plotting the
                                                           samples


Percentage_Error_repeated=(C-Cs_R)./C*100


Percentage_Error_OneShot=(C-Cs_O)./C*100


%%------ Plotting -----%%

hold on
plot(th,Cs_R,'k<','MarkerEdgeColor','k','MarkerFaceColor','k'
                                                ,'MarkerSize',7)

figure
plot(t1,F,'k-','Linewidth',2)
hold on
plot(th,Cs_O,'ko','MarkerEdgeColor','k','MarkerFaceColor','k'
                                                ,'MarkerSize',7)
```

21. Program for obtaining the convolution of two time functions of Example 7.1, in Hybrid Function domain (Chap. 7, Table 7.1; Fig. 7.11, p. 181, 182 and 183)

```
clc
clear all
format long

%%------------ Number of Sub-intervals and Total Time -----------%%

m = input('Enter the number of sub-intervals chosen:\n');
T = input('Enter the total time period:\n');
h = T/m;
th = 0:h:(T+h);

%%-------------- Exact Solution  -------------%%

R=ones(1,(m+2));
R_SH=R(1:(m+1)); % SHF coefficients of r(t)

for i=1:(m+1)
   R_T(i)=R(i+1)-R(i); % TF coefficients of r(t)
end

syms t
gt=exp(-0.5*t)*(2*cos(2*t)-0.5*sin(2*t));
G=subs(gt,th);
G_SH=G(1:(m+1));    % Sample-and-Hold function coefficients of g(t)

for i=1:(m+1)
   G_T(i)=G(i+1)-G(i); % TF coefficients of g(t)
end

yt=exp(-0.5*t)*sin(2*t);

Y_direct=subs(yt,th(1:(m+1)))
plot(th(1:(m+1)),Y_direct,'k-','LineWidth',2)

%%-------------- Convolution of a time function -------------%%

%%----------- Formation of R1 matrix ----------%%

R1=zeros((m+1),(m+1));
for i=1:(m+1)
   for j=1:(m+1)
      if j-i>0
        R1(i,j)=2*R_SH(j)+R_SH(j-1);
      end
   end
end

G1=G_SH; % to be pre-multiplied with R1
%%----------- Formation of R2 matrix ----------%%

R2=zeros((m+1),(m+1));
for i=1:(m+1)
   for j=1:(m+1)
      if j-i>0
        R2(i,j)=R_SH(j)+2*R_SH(j-1);
      end
```

```
    end
end

G2=G(2:(m+2)); % to be pre-multiplied with R2

Y_HF=(h/6)*((G1*R1)+(G2*R2))

hold on
plot(th(1:(m+1)),Y_HF,'ko','LineWidth',2)
```

22. Program for analyzing a non-homogeneous system of Example 8.1, Hybrid Function domain (Chap. 8, Table 8.1; Fig. 8.1, pp. 197-198)

```
clc
clear all
format long

%%---------------- For step input -------------------%%

C_SH_input=1;            % SHF part of input represented in HF domain

C_Triangular_input=0;    % TF part of input represented in HF domain

C_input=C_SH_input+0.5*C_Triangular_input     % Samples in HF domain


%%--------------------- System ----------------------%%

A=[0 1; -2 -3];          % System matrix
B=[0;1];                 % Input matrix
x0=[0;0.5];              % Initial condition

m = input('Enter the number of sub-intervals chosen:\n');
T = input('Enter the total time period:\n');
h = T/m;


%%---------------- Exact Solution of States -------------------%%

syms s
Us=laplace(C_input,s);
I=eye(length(A));        % Identity matrix
x=inv(s*I-A)*(x0+B*Us);
xt=ilaplace(x)
t=0:0.01:T;
x_time=subs(xt,t);
xt1=x_time(1,:);         % Exact solution of state x1
xt2=x_time(2,:);         % Exact solution of state x2


%%---------- Solution in Hybrid Function domain ------------%%

xi=x0;           % Initial values of the states

for i=1:m
   Cs(:,i)=inv((2/h)*I-A)*[((2/h)*I+A)*xi+2*B*C_input];
   xi=Cs(:,i);
end
```

```
HF_coefficients=[x0 Cs]

HF_State1=HF_coefficients(1,:);
HF_State2=HF_coefficients(2,:);

th=0:h:T;

x_t=subs(xt,th);

x1_t=x_t(1,:)    % Direct expansion of state x1 in HF domain
x2_t=x_t(2,:)    % Direct expansion of state x2 in HF domain

plot(t,xt1,'k-',th,HF_State1,'ko','MarkerFaceColor','k','Linewidth',2)
hold on
plot(t,xt2,'k-',th,HF_State2,'ko','MarkerFaceColor','k','Linewidth',2)
xlabel('Time (in Sec)')
ylabel('States X_1, X_2')
```

23. Program for analyzing output of a non-homogeneous system of Example 8.2, Hybrid Function domain (Chap. 8, Table 8.2; Fig. 8.3, p. 201 and 201)

```
clc
clear all
format long

%%---------------- For step input -------------------%%

C_SH_input=1;            % SHF part of input represented in HF domain

C_Triangular_input=0;    % TF part of input represented in HF domain

C_input=C_SH_input+0.5*C_Triangular_input     % Samples in HF domain

%%--------------------- System ----------------------%%

A=[0 1; -2 -3];          % System matrix
B=[0;1];                 % Input matrix
C=[1 0];                 % Output matrix
D=0;                     % Direct transmission matrix

x0=[0;0.5];              % Initial condition

m = input('Enter the number of sub-intervals chosen:\n');
T = input('Enter the total time period:\n');
h = T/m;

%%---------------- Exact Solution of States ------------------%%

syms s
Us=laplace(C_input,s);
```

```
I=eye(length(A));          % Identity matrix

x=inv(s*I-A)*(x0+B*Us);
xt=ilaplace(x)
t=0:0.01:T;
x_time=subs(xt,t);
xt1=x_time(1,:);           % Exact solution of state x1
xt2=x_time(2,:);           % Exact solution of state x2
yt=C*x_time;               % Exact solution of output y


%%----------- Solution in Hybrid Function domain ------------%%

xi=x0;       % Initial values of the states


for i=1:m
    Cs(:,i)=inv((2/h)*I-A)*[((2/h)*I+A)*xi+2*B*C_input];
    xi=Cs(:,i);
end

HF_coefficients=[x0 Cs]

HF_State1=HF_coefficients(1,:);
HF_State2=HF_coefficients(2,:);

%%----------- Output in Hybrid Function domain ------------%%

th=0:h:T;

Xt=subs(xt,th);
Yt=C*Xt
y_h=C*[HF_State1; HF_State2]+D*C_SH_input*ones(1,m+1)

plot(t,yt,'k-','Linewidth',2)
hold on
plot(th,y_h,'ko','MarkerFaceColor','k')
```

24. Program for analyzing a homogeneous system of Example 8.3, Hybrid Function domain (Chap. 8, Table 8.3; Fig. 8.4, pp. 202-203)

```
clc
clear all
format long

%%---------------------- System ----------------------%%

A=[0 1; -2 -3];            % System matrix
B=[0;1];                   % Input matrix
C=[1 0];                   % Output matrix
D=0;                       % Direct transmission matrix

x0=[0;1];                  % Initial condition

%%----------- Number of Sub-intervals and Total Time -----------%%

m = input('Enter the number of sub-intervals chosen:\n');
T = input('Enter the total time period:\n');
h = T/m;


%%----------------- Exact Solution of States ------------------%%

I=eye(length(A));          % Identity matrix

x=inv(s*I-A)*x0;
xt=ilaplace(x)
t=0:0.01:T;
x_time=subs(xt,t);
xt1=x_time(1,:);           % Exact solution of state x1
xt2=x_time(2,:);           % Exact solution of state x2


%%---------- Solution in Hybrid Function domain ------------%%

xi=x0;      % Initial values of the states

for i=1:m
   Cs(:,i)=inv((2/h)*I-A)*((2/h)*I+A)*xi;
   xi=Cs(:,i);
end

HF_coefficients=[x0 Cs]
HF_State1=HF_coefficients(1,:);
HF_State2=HF_coefficients(2,:);

th=0:h:T;

x_t=subs(xt,th);

x1_t=x_t(1,:)    % Direct expansion of state x1 in HF domain
x2_t=x_t(2,:)    % Direct expansion of state x12 in HF domain


plot(t,xt1,'k-',th,HF_State1,'ko','MarkerFaceColor','k','Linewidth',2)
hold on
plot(t,xt2,'k-',th,HF_State2,'ko','MarkerFaceColor','k','Linewidth',2)
```

25. Program for analyzing a non-homogeneous system with jump discontinuity at input of Example 8.8, both in HFc and HFm based approaches (Chap. 8, Table 8.16; Fig. 8.16, p. 215, 218, 216)

```
clc
clear all
format long

%%---------------------- System ----------------------%%

A=[0 1; -2 -3];          % System matrix
B=[0;1];                 % Input matrix

x0=[0;0.5];              % Initial condition

%----------- Number of Sub-intervals and Total Time -----------%%

m = input('Enter the number of sub-intervals chosen:\n');
T = input('Enter the total time period:\n');
h = T/m;

a=2*h;       % Instant of jump in input

%%-------------------- Exact Plot of States ---------------------%%

t1=0:0.001:a;               % Interval before occurring jump
t2=(a+0.001):0.001:T;       % Interval after occurring jump
tt=[t1 t2];                 % Total interval

syms t

x1a=0.5-0.5*exp(-t);
X1a=subs(x1a,t1);           % State x1 before jump

x1b=1-(0.5*exp(-t))-exp(-(t-a))+(0.5*exp(-2*(t-a)));
X1b=subs(x1b,t2);           % State x1 after jump
X1=[X1a X1b];

x2a=0.5*exp(-t);
X2a=subs(x2a,t1);           % State x2 before jump

x2b=(0.5*exp(-t))+exp(-(t-a))-exp(-2*(t-a));
X2b=subs(x2b,t2);           % State x2 after jump

X2=[X2a X2b];


plot(tt,X1,'k-','LineWidth',2)
hold on
plot(tt,X2,'k--','LineWidth',2)


%%----------- Solution in Hybrid Function domain ------------%%

th1=0:h:(a-h);
th2=a:h:T;
th=[th1 th2];

% Approximation of input in Hybrid Function domain (convensional)

U_SH=ones(1,length(th))+[zeros(1,length(th1)) ones(1,length(th2))];
                                                    % SHF coefficients
```

```
U_TF=zeros(1,m);

for k=1:m
   U_TF(k)=U_SH(k+1)-U_SH(k);
End

U_TF; % Triangular function coefficients

U=U_SH(1:m)+0.5*U_TF        % Input samples

I=eye(length(A));

xi=x0;

for i=1:m
    Cs(:,i)=inv((2/h)*I-A)*[((2/h)*I+A)*xi+2*B*U(i)];
    xi=Cs(:,i);
end

HF_coefficients_HFc=[x0 Cs];

HF_State1_HFc=HF_coefficients_HFc(1,:);
HF_State2_HFc=HF_coefficients_HFc(2,:);


plot(tt,X1,'k-',th,HF_State1_HFc,'ko','LineWidth',2)
hold on
plot(tt,X2,'k-',th,HF_State2_HFc,'ko','LineWidth',2)
hold on

% Approximation of input in Hybrid Function domain (modified)

U_SH=ones(1,length(th))+[zeros(1,length(th1)) ones(1,length(th2))];
                                                        % SHF coefficients

U_TF=zeros(1,m);            % Triangular function coefficients

U=U_SH(1:m)+0.5*U_TF        % Input samples


I=eye(length(A));

xi=x0;

for i=1:m
    Cs(:,i)=inv((2/h)*I-A)*[((2/h)*I+A)*xi+2*B*U(i)];
    xi=Cs(:,i);
end

HF_coefficients_HFm=[x0 Cs];

HF_State1_HFm=HF_coefficients_HFm(1,:);
HF_State2_HFm=HF_coefficients_HFm(2,:);

plot(tt,X1,'k-',th,HF_State1_HFm,'ko','MarkerFaceColor','k'
                                                        ,'LineWidth',2)
hold on
plot(tt,X2,'k-',th,HF_State2_HFm,'ko','MarkerFaceColor','k'
                                                        ,'LineWidth',2)
xlabel('Time (in Sec)')
ylabel('States X_1, X_2')
```

26. Program for analyzing a time-varying non-homogeneous system of Example 9.1 (Chap. 9, Table 9.1; Fig. 9.1, p. 230, 231)

```
clc
clear all
format long

x0=input('Enter the initial values:\n')  % Initial values of the states

m = input('Enter the number of sub-intervals chosen:\n');
T = input('Enter the total time period:\n');
h = T/m;

B=input('Enter the input vector:\n')     % Input matrix
%%-------------------- Exact Plot of States ---------------------%%

t=0:0.01:T;

syms t1

X1=1+t1;
xt1=subs(X1,t);          % Exact values of state x1

X2=1+(t1^2/2)+(t1^3/3);
xt2=subs(X2,t);          % Exact values of state x2


%%----------- Solution in Hybrid Function domain ------------%%

tt=0:h:T;

syms t2

A=[0 0;t2 0];            % System matrix

s=size(A,1);
x=zeros(s,(m+1));

x(:,1)=x0;               % States x1 and x2 starts with initial values

I=eye(s);                % Identity matrix of dimension s


% Solution of States in HF domain
for k=1:m;
   for e=1:s;
      for f=1:s;
          a(e,f)=subs(A(e,f),tt(k+1));
      end
   end
   b=(2/h)*I-a;

   for e1=1:s;
      for f1=1:s;
          a1(e1,f1)=subs(A(e1,f1),tt(k));
      end
   end
```

```
    b1=(2/h)*I+a1;

  x(:,(k+1))=inv(b)*((b1)*x(:,k)+2*B);
  end


  x1=x(1,:);
  x2=x(2,:);

  plot(tt,x1,'ko','LineWidth',2,'MarkerFaceColor','k')
  hold on
  plot(t,xt1,'k-','LineWidth',2)
  hold on
  plot(tt,x2,'ko','LineWidth',2,'MarkerFaceColor','k')
  hold on
  plot(t,xt2,'k-','LineWidth',2)

  xlabel('Time t(s)')
  ylim([0 2.5])
```

27. Program for analyzing a time-varying homogeneous system of Example 9.5 (Chap. 9, Table 9.5; Fig. 9.6, pp. 236-238)

```
clc
clear all
format long

x0=input('Enter the initial values:\n')  % Initial values of the states

m = input('Enter the number of sub-intervals chosen:\n');
T = input('Enter the total time period:\n');
h = T/m;

B=input('Enter the input vector:\n')      % Input matrix

%%-------------------- Exact Plot of States ---------------------%%

t=0:0.01:T;

syms tt

Xt1=(cos(1-cos(tt))+2*sin(1-cos(tt)))*exp(sin(tt));
xt1=subs(Xt1,t);                          % Exact values of state x1

Xt2=(-sin(1-cos(tt))+2*cos(1-cos(tt)))*exp(sin(tt));
xt2=subs(Xt2,t);                          % Exact values of state x2

%%----------- Solution in Hybrid Function domain ------------%%

t1=0:h:T;

syms t2

A=[cos(t2) sin(t2);-sin(t2) cos(t2)];              % System matrix
```

```
s=size(A,1);
x=zeros(s,(m+1));

x(:,1)=x0;             % States x1 and x2 starts with initial values

I=eye(s);              % Identity matrix of dimension s
% Solution of States in HF domain
for k=1:m;
   for e=1:s;
      for f=1:s;
          a(e,f)=subs(A(e,f),t1(k+1));
      end
   end
   b=(2/h)*I-a;

   for e1=1:s;
      for f1=1:s;
          a1(e1,f1)=subs(A(e1,f1),t1(k));
      end
   end

b1=(2/h)*I+a1;
x(:,(k+1))=inv(b)*((b1)*x(:,k)+2*B);
end

x1=x(1,:);
x2=x(2,:);

plot(t1,x1,'ko','LineWidth',2,'MarkerFaceColor','k')
hold on
plot(t,xt1,'k-','LineWidth',2)
hold on
plot(t1,x2,'ko','LineWidth',2,'MarkerFaceColor','k')
hold on
plot(t,xt2,'k-','LineWidth',2)

xlabel('Time t(s)')
ylim([0 4.5])
```

28. Program for analyzing a time-delay non-homogeneous system of Example 10.5 (Chap. 10, Table 10.1; Fig. 10.6, pp. 260-262)

```
clc
clear all
format long

m = input('Enter the number of sub-intervals chosen:\n');
T = input('Enter the total time period:\n');
h = T/m;

d1=m/2;    % Number of intervals up to delay time
%%---------- Information regarding the system -----------%%

A=input('The System Matrix without delay:\n');

A1=input('The System Matrix with delay:\n');

B=input('The Input Matrix without delay:\n');


%%-------------------- Exact Plot of States --------------------%%

syms t

t1=0:.01:(1-0.01);
f1=1-(1.1*t)+(0.525*t^2);    % Solution for time 0 s to 1 s
f1t=subs(f1,t1);

t2=1:0.01:T;
f2=-0.25+(1.575*t)-(1.075*t^2)+(0.175*t^3);  % Solution for 1 s to 2 s
f2t=subs(f2,t2);

te=[t1 t2];
fe=[f1t f2t];
plot(te,fe,'k-','Linewidth',2)
hold on


%%---------- Solution in Hybrid Function domain -----------%%

I=eye(length(A1));  %Identity matrix

x=[1 zeros(1,m)];
xi=ones(1,m);         % Initial value of state in HF domain

t1=0:h:(1-h);
t2=1:h:T;

%%----- Input represented in HF domain ------%%

u=zeros(1,(m+1));
u1=-2.1+(1.05*t);            % Solution for time 0 s to 1 s
u1t=subs(u1,t1);

u2=-1.05;
u2t=u2*ones(1,length(t2));   % Solution for time 1 s to 2 s

U=[u1t u2t];
```

```
Csu=U(1:m);                    % SHF coefficients of input

for j=1:m
   Ctu(j)=U(j+1)-U(j);         % TF coefficients of input
end

Cu=Csu+(0.5*Ctu)


for i=1:d1
   if (i+d1)<=m
      x(i+1)=inv((2/h)*I-A)*(((2/h)+A)*x(i)+(2*A1*xi(i+d1))+2*B*Cu(i));
   else
      x(i+1)=inv((2/h)*I-A)*(((2/h)+A)*x(i)+2*B*Cu(i));
   end
end

for i=(d1+1):m
   if (i+d1)<=m
      x(i+1)=inv((2/h)*I-A)*(((2/h)+A)*x(i)+(A1*x(i-d1))+(A1*x(i+1
                                 -d1))+(2*A1*xi(i+d1))+2*B*Cu(i));
   else
      x(i+1)=inv((2/h)*I-A)*(((2/h)+A)*x(i)+(A1*x(i-d1))+(A1*x(i+1
                                                -d1))+2*B*Cu(i));
   end
end


th=0:h:T;
plot(th,x,'ko','MarkerFaceColor','k','MarkerSize',8)
hold on
plot(te,fe,'k-','Linewidth',2)
```

29. Program for analyzing a time invariant open loop system of Example 11.2 (Chap. 11, Table 11.3 and 11.4; Figs. 11.6 and 11.7, pp. 274-276)

```
clc
clear all
format long

%%----------- Number of Sub-intervals and Total Time -----------%%

m=input('Enter the number of sub-intervals chosen:\n');
T=input('Enter the total time period:\n');
h=T/m;
th=0:h:(T+h);

%%-------------------- Exact Solution ----------------------%%
R=ones(1,(m+2));
R_SH=R(1:(m+1));         % SHF coefficients of r(t)

for i=1:(m+1)
   R_T(i)=R(i+1)-R(i);   % TF coefficients of r(t)
end
```

```
syms t
gt=2*exp(-2*t)*(cos(2*t)-sin(2*t));
G=subs(gt,th);
G_SH=G(1:(m+1));          % SHF coefficients of g(t)

for i=1:(m+1)
   G_T(i)=G(i+1)-G(i);    % TF coefficients of g(t)
end

yt=exp(-2*t)*sin(2*t);
Y=subs(yt,th(1:(m+1)))

te=0:0.01:T;
Ye=subs(yt,te);

plot(te,Ye,'k-','LineWidth',2)
hold on


%%-------------- Convolution of a time function -------------%%

%%----------- Formation of R1 matrix ----------%%

R1=zeros((m+1),(m+1));
for i=1:(m+1)
   for j=1:(m+1)
      if j-i>0
         R1(i,j)=2*R_SH(j)+R_SH(j-1);
      end
   end
end

G1=G_SH;      % to be pre-multiplied with R1


%%----------- Formation of R1 matrix ----------%%

R2=zeros((m+1),(m+1));
for i=1:(m+1)
   for j=1:(m+1)
      if j-i>0
         R2(i,j)=R_SH(j)+2*R_SH(j-1);
      end
   end
end

G2=G(2:(m+2)); % to be pre-multiplied with R2

Y_HF=(h/6)*((G1*R1)+(G2*R2))


plot(th(1:(m+1)),Y_HF,'ko','MarkerFaceColor','k','LineWidth',2)
Percentage_Error=(Y-Y_HF)./Y*100

AMP_Error=sum((abs(Percentage_Error(2:(m+1)))))/m
```

30. Program for analyzing a time invariant closed loop system of Example 11.3 (Chap. 11, Table 11.5; Figs. 11.10, 11.11, and 11.12, pp. 283-286)

```
clc
clear all
format long

%%------------ Number of Sub-intervals and Total Time -----------%%

m=input('Enter the number of sub-intervals chosen:\n');
T=input('Enter the total time period:\n');
hh=T/m;
th=0:hh:T;


%%------------------ HF domain representation of ----------------%%
%%------------- input, output, plant and feedback signal ------------%%

r=ones(1,(m+1));   %% Input

syms t
g_t=2*exp(-4*t);           % Plant Impulse Response
g=subs(g_t,th);

h=4*ones(1,length(th));    % Feedback Gain

yt=exp(-2*t)*sin(2*t);     % Output
Y=subs(yt,th)

te=0:0.01:T;
Ye=subs(yt,te);           % Exact values of output

plot(te,Ye,'m-','LineWidth',2)
hold on


%%---------- Convolution of a time function in HF domain ----------%%

for i=1:m
   G(i)=2*g(i+1)+g(i);
end

for i=(m+1)
   G(i)=g(2)+2*g(1);
end

for i=(m+2):(2*m)
   G(i)=g(i-m+1)+(4*g(i-m))+g(i-m-1);
end
G;


for i=1:m
   H(i)=2*h(i+1)+h(i);
end

for i=(m+1)
   H(i)=h(2)+2*h(1);
end
```

```
for i=(m+2):(2*m)
   H(i)=h(i-m+1)+(4*h(i-m))+h(i-m-1);
end
H;

Denominator=1+((hh^2)/36)*H(m+1)*G(m+1);
y(1)=0;    % First term of output in HF domain after convolution


for i=1:m
    y1(i)=r(1)*G(i);

    for p=2:(i+1)
       y_part2(p-1)=r(p)*G(m+i-p+2);
    end

    y2(i)=sum(y_part2);

    for p=2:(i+1)
       y_part3(p-1)=H(p-1)*G(m+i-p+2)*y(1);
    end

    y3(i)=sum(y_part3);

    if i>=2
      for p=2:i
         j=i-p+2;
         k=zeros(1,i);
            for p1=1:j
               k(p1)=H(m+p1)*G(m+j-p1+1);
            end
            K=sum(k);
            y_part4(p-1)=K*y(p);
      end
      y4(i)=sum(y_part4);
    else
    y4(i)=0;
    end

     y4

     y(i+1)=((hh/6)*(y1(i)+y2(i)-((hh/6)*y3(i)) -((hh/6)*y4(i))))
                                                            /Denominator;
 end

 Y_HF=y


 plot(th(1:(m+1)),y,'ko-','MarkerFaceColor','k','LineWidth',2)

 Percentage_Error=(Y-Y_HF)./Y*100

 AMP_Error=sum((abs(Percentage_Error(2:(m+1)))))/m
```

31. Program for identifying a non-homogeneous time invariant system of Example 12.1 (Chap. 12, Table 12.1; Fig. 12.1, pp. 291-293)

```
clc
clear all
format long

%%------------ Number of Sub-intervals and Total Time -----------%%

m = input('Enter the number of sub-intervals chosen:\n');
T = input('Enter the total time period:\n');
h = T/m;
th = 0:h:(T+h);

%%---------------------- System ----------------------%%

A=[0 1; -2 -3];          % System matrix
B=[0;1];                 % Input matrix

x0=[0;1];                % Initial condition

%%---------- Approximating the step input in HF domain -----------%%

n=length(B);

C_SH_input=ones(1,n);

C_Triangular_input=zeros(1,n);

C_input=C_SH_input+0.5*C_Triangular_input


%%---------- Taking samples from system states ------------%%

syms t
x_expression=[1/2 - 1/(2*exp(t)); 1/(2*exp(t))]

x_time=subs(x_expression,th);

n1=1:n; n2=2:(n+1);      % Required Dimensions


%%-------------------- Define Matrices ---------------------%%

Matrix_plus=x_time(:,n2)+x_time(:,n1);

Matrix_minus=x_time(:,n2)-x_time(:,n1);


%%------------- Identification of system matrix -------------%%

A=((2/h)*Matrix_minus-(2*B*C_input))*inv(Matrix_plus)
```

32. Program for identifying the output matrix of a time invariant system of Example 12.4 (Chap. 12, Fig. 12.4, p. 296, 296)

```
clc
clear all
format short

%%---------------------- System ----------------------%%

B=[0;0;1];                    % Input matrix

%%------------ Number of Sub-intervals and Total Time -----------%%

m = input('Enter the number of sub-intervals chosen:\n');
T = input('Enter the total time period:\n');
h = T/m;
th = 0:h:(T+h);

%%---------- Approximating the step input in HF domain -----------%%

n=length(B);

C_SH_input=ones(1,n);

C_Triangular_input=zeros(1,n);

C_input=C_SH_input+0.5*C_Triangular_input


%%----------- Taking samples from system states ------------%%
syms t

x_expression=[5/(2*exp(t))-5/(2*exp(2*t))+5/(6*exp(3*t))+1/6;

              5/exp(2*t)-5/(2*exp(t))-5/(2*exp(3*t));

              5/(2*exp(t))-10/exp(2*t)+15/(2*exp(3*t))];

output =5/exp(2*t)-5/(3*exp(3*t))+2/3;

x_time=subs(x_expression,th);

output_time=subs(output,t);

n1=1:n; n2=2:(n+1);       %Required Dimensions

%%-------------------- Define Matrices ---------------------%%

Matrix_x=x_time(:,1:n);
Matrix_y=y_time(:,1:n);

%%------------- Identification of output matrix -------------%%

C=Matrix_output*inv(Matrix_x)
```

33. Program for identifying the system matrix of a non-homogeneous time invariant system involving jump discontinuity at the input, of Example 12.6 (Chap. 12, Tables 12.8 and 12.9; Fig. 12.6, pp. 301-304)

```
clc
clear all
format long

%%---------------------- System ----------------------%%

A=[0 1; -2 -3];           % System matrix
B=[0;1];                  % Input matrix

x0=[0;0.5];               % Initial condition

%----------- Number of Sub-intervals and Total Time -----------%%

m = input('Enter the number of sub-intervals chosen:\n');
T = input('Enter the total time period:\n');
h = T/m;

a=0.5;      % Instant of jump in input

%%-------------------- Exact Plot of States --------------------%%

syms t

t1=0:h:(a-h);
t2=a:h:T;
tt=[t1 t2];


x1a= -(3/4)+(0.5*t)+(3/2)*exp(-t)-(3/4)*exp(-2*t);
X1a=subs(x1a,t1);

x1b=-(1/4)+(0.5*t)+(3/2)*exp(-t)-(3/4)*exp(-2*t)-exp(-(t-a))
                                                    +0.5*exp(-2*(t-a));
X1b=subs(x1b,t2);

X1=[X1a X1b];


x2a=0.5-(3/2)*exp(-t)+(3/2)*exp(-2*t);
X2a=subs(x2a,t1);

x2b=0.5-(3/2)*exp(-t)+(3/2)*exp(-2*t)+exp(-(t-a))-exp(-2*(t-a));
X2b=subs(x2b,t2);

X2=[X2a X2b];

X=[X1;X2]


% Approximation of input in HF domain

U_SH_Ramp=tt;
```

```
U_SH_Delayed_Step=[zeros(1,length(t1))  ones(1,length(t2))];

U_SH = U_SH_Ramp + U_SH_Delayed_Step

U_TF=zeros(1,m);
for k=1:m
   U_TF(k)=U_SH(k+1)-U_SH(k);
end


% U_TF;                                  % For HFc method

% U_TF(a/h)=0;                           % For HFm method

U_TF(a/h)=U_SH(a/h)-U_SH((a/h)-1);      % For combined HFc and HFm method

U=U_SH(1:m)+0.5*U_TF     % Samples of input


%%-------------------- Required Dimensions -------------------%%

n=length(B);

% n1=1:n; n2=2:(n+1);       % for samples before jump point

% n1=(a/h+2):(a/h+1+n); n2=(a/h+1+n):(a/h+2+n);
                                                   % for samples after jump

n1=(a/h-1):(a/h);  n2=(a/h):(a/h+1);    % for samples involving jump


%%--------------------- Define Matrices ---------------------%%

Matrix_plus=X(:,n2)+X(:,n1);

Matrix_minus=X(:,n2)-X(:,n1);


%%------------- Identification of system matrix -------------%%

A=((2/h)*Matrix_minus-(2*B*U(n1)))*inv(Matrix_plus)
```

34. Program for identifying the time varying element of system matrix of a non-homogeneous system, of Example 13.1 (Chap. 13, Table 13.1; Fig. 13.1, p. 309, 310)

```
clc
clear all
format long

%%--------------------- System ----------------------%%

B=[1;0];                % Input matrix

x0=[1;1];               % Initial condition


%----------- Number of Sub-intervals and Total Time -----------%%

m = input('Enter the number of sub-intervals chosen:\n');
T = input('Enter the total time period:\n');
h = T/m;


%%---------- Taking samples from system states ------------%%

t=0:h:T;
syms tt

Xt1=1+tt;
xt1=subs(Xt1,t)

Xt2=1+(tt^2/2)+(tt^3/3);
xt2=subs(Xt2,t)

%%------------ Identification of system matrix -------------%%

z=zeros(1,(m+1));

a11=z; a12=z; a21=z; a22=z;

a210=input('Enter the initial value of a21:\n')

a21(1)=a210;

for k=1:m
   a21(k+1)=(1/xt1(k+1))*(((2/h)-a22(k+1))*xt2(k+1)-
                        ((2/h)+a22(k))*xt2(k)-a21(k)*xt1(k)-(2*B(2)));
end
a21

plot(t,t,'-k','LineWidth',2)
hold on
plot(t,a21,'ko-','MarkerFaceColor','k')
```

35. Program for identifying the time varying elements of system matrix of a homogeneous system, of Example 13.2 (Chap. 13, Figs. 13.2 and 13.3, p. 311, 312)

```
clc
clear all
format short

%%----------------- Initial values of the states ----------------%%

x0=[1;1];               % Initial condition


%----------- Number of Sub-intervals and Total Time -----------%%

m = input('Enter the number of sub-intervals chosen:\n');
T = input('Enter the total time period:\n');
h = T/m;

%%----------- Taking samples from system states ------------%%

t=0:h:T;
syms tt

Xt1=(cos(1-cos(tt))+2*sin(1-cos(tt)))*exp(sin(tt));
xt1=subs(Xt1,t)

Xt2=(-sin(1-cos(tt))+2*cos(1-cos(tt)))*exp(sin(tt));
xt2=subs(Xt2,t)

%%------------------- Identification of a11 -------------------%%

a12=sin(t);
a11(1)=1;

for k=1:m
   a11(k+1)=(2/h)-(1/x1(k+1))
                 *(((2/h)+a11(k))*x1(k)+(a12(k)*x2(k))+(a12(k+1)*x2(k+1)));
end

plot(t,a11,'-ko','LineWidth',2,'MarkerFaceColor','k')


%%------------------- Identification of a21 -------------------%%

a22=cos(t);
a21(1)=0;

for k=1:m
    a21(k+1)=(1/x1(k+1))*(((2/h)-a22(k+1))
                           *x2(k+1)-((2/h)+a22(k))*x2(k)-(a21(k)*x1(k)));
end

plot(t,a21,'-ko','LineWidth',2,'MarkerFaceColor','k')
```

36. Program for identifying the impulse response of the plant, of Example 14.1 via method of deconvolution (Chap. 14, Figs. 14.1 and 14.2; Tables 14.1 and 14.2, pp. 322-324)

```
clc
clear all
format long

%------------ Number of Sub-intervals and Total Time -----------%%

m = input('Enter the number of sub-intervals chosen:\n');
T = input('Enter the total time period:\n');
h = T/m;


%%--------------- Exact Plot of impulse response ----------------%%

syms t

t_exact=0:0.01:T;

tt=0:h:T;

r=ones(1,length(tt));       % samples of input applied to system

yt=1-exp(-t);
y=subs(yt,tt);              % samples of system output

gt=exp(-t);
g_direct=subs(gt,tt);       % samples of impulse response
g_exact=subs(gt,t_exact);


%%------------ Formation of R coefficients -------------%%

for i=1:m
   R(i)=2*r(i+1)+r(i);
end

R(m+1)=r(2)+2*r(1);

for i=(m+2):(2*m)
   R(i)=r(i-m+1)+(4*r(i-m))+r(i-m-1);
end

for i=(2*m+1):(2*m+3)
   R(i)=r(i-2*m+2)+r(i-2*m+1)-2*r(i-2*m);
end


for i=1:m
   for j=1:m
      if i==1
         RR(i,:)=[0 R(1:(m-1))];
      elseif j-i>=0
         RR(i,j)=R(m+(j-i)+1);
```

```
      end
   end
end

RR=(h/6)*RR;

epsilon=0.01;
RR(1,1)=epsilon;

W=inv(RR);
g=[epsilon y(2:m)]*W;


for i=2:m
   g_mult_R(i)=g(i)*(R(2*m-i+2)-R(2*m-i+1));
end

g_mult_R;

g(m+1)=((y(m+1)-y(m))-((h/6)*(g(1)*(R(m)-R(m-1))+sum(g_mult_R))))
                                                   /((h/6)*R(m+1));

plot(t_exact,g_exact,'-m','LineWidth',3)
hold on
plot(tt,g,'-ko','LineWidth',2,'MarkerFaceColor','k'
                                           ,'MarkerEdgeColor','k')
```

37. Program for identifying the impulse response of the plant of closed loop system, of Example 14.2 via method of deconvolution (Chap. 14, Figs. 14.5 and 14.6; Tables 14.3 and 14.4, p. 328)

```
clc
clear all
format long

%----------- Number of Sub-intervals and Total Time -----------%%

m = input('Enter the number of sub-intervals chosen:\n');
T = input('Enter the total time period:\n');
h = T/m;


%%-------------- Exact Plot of impulse response ----------------%%

syms t

t_exact=0:0.01:T;

tt=0:h:T;

r=ones(1,length(tt));        % samples of input applied to system

h1=ones(1,length(tt));       % samples of feedback signal
```

```
yt=(2/sqrt(3))*exp(-t/2)*sin((sqrt(3)*t)/2);
y=subs(yt,tt)                               % samples of system output

gt=exp(-t);
g_direct=subs(gt,tt);          % samples of system output
g_exact=subs(gt,t_exact);


%%--------------- Formation of H coefficients -----------------%%

for i=1:m
   H(i)=2*h1(i+1)+h1(i);
end

H(m+1)=h1(2)+2*h1(1);

for i=(m+2):(2*m)
   H(i)=h1(i-m+1)+(4*h1(i-m))+h1(i-m-1);
end

for i=(2*m+1):(2*m+3)
   H(i)=h1(i-2*m+2)+h1(i-2*m+1)-2*h1(i-2*m);
end


for i=1:m
   for j=1:m
      if i==1
         HH(i,:)=[0 H(1:(m-1))];
      elseif j-i>=0
         HH(i,j)=H(m+(j-i)+1);
      end
   end
end
HH
HH=(h/6)*HH
b=y(1:m)*HH


for j=(m+1)
   bm_part1=H(m)*y(1);
   for k=2:(m+1)
      bm_part2(k-1)=H(2*m-k+2)*y(k);
   end
   bm_2=sum(bm_part2);
   bm=bm_part1+bm_2;
end

bm=bm*(h/6);
b=[b bm]

e=r-b
```

```
    %%---------------- Formation of E coefficients ----------------%%

    for i=1:m
       E(i)=2*e(i+1)+e(i);
    end

    E(m+1)=e(2)+2*e(1);
    for i=(m+2):(2*m)
       E(i)=e(i-m+1)+(4*e(i-m))+e(i-m-1);
    end

    for i=(2*m+1):(2*m+3)
       E(i)=e(i-2*m+2)+e(i-2*m+1)-2*e(i-2*m);
    end

for i=1:m
   for j=1:m
      if i==1
        EE(i,:)=[0 E(1:(m-1))];
      elseif j-i>=0
        EE(i,j)=E(m+(j-i)+1);
      end
   end
end


EE
EE=(h/6)*EE;

epsilon=0.0001;
EE(1,1)=epsilon;
E_inverse=inv(EE);


g=[epsilon y(2:m)]*E_inverse

plot(t_exact,g_exact,'-m','LineWidth',3)
hold on

for i=2:m
   E_mult_G(i)=g(i)*(E(2*m-i+2)-E(2*m-i+1));
end

g(m+1)=((y(m+1)-y(m))-((h/6)*(g(1)*(E(m)-E(m-1))+sum(E_mult_G))))
                                                    /((h/6)*E(m+1));

plot(tt,g,'-ko','LineWidth',2,'MarkerFaceColor','k'
                                         ,'MarkerEdgeColor','k')
```

38. Program for drawing figures showing identified parameters of the transfer function of a plant from impulse response data, of Example 15.1 (Chap. 15, Figs. 15.1, 15.2 and 15.3, p. 339 and 325-354)

```
clc
format long

% BPF   NOBPF   TF   HF  SHF
x=[1 2 3 4 5];

%%------------ L=1, a0=1 -------------%%
color [ 0.267 0.647 0.106 ]
y=[2.54176555024836  2.54147603465520  2.54176778302942
                                 2.54431980212434  2.54442700797934 ];
stem(x,y,'ko:')
Title('\lambda = 1')
ylabel('a_o')
xlim([0 6])
ylim([2.54 2.545])


%%------------ L=1, a1=0 -------------%%
% y=[-0.25154866105028  -0.25160253204068   -0.25152488460955
                              -0.23665257425908   -0.22952648857943];
% stem(x,y,'ko:')
% text(1, 0.05, 'a_1 (exact) = 0')
% Title('\lambda = 1')
% ylabel('a_1')
% xlim([0 6])
% ylim([0 -0.35])


%%------------ L=1, b0=2 -------------%%
% y=[12.72230063919302  12.72227298081633   12.72231194744942
                              12.72939709991139   12.73277717740051];
% stem(x,y,'ko:')
% text(1, 12.7325, 'b_o (exact) = 2')
% Title('\lambda = 1')
% ylabel('b_o')
% xlim([0 6])
% ylim([12.7 12.75])


%%------------ L=1, b1=3 -------------%%
% y=[6.11098678009208   6.11095661720580    6.11098640031485
                              6.11083725422649    6.11061105677642];
% stem(x,y,'ko:')
% text(1, 6.1113, 'b_1 (exact) = 3')
% Title('\lambda = 1')
% ylabel('b_1')
% xlim([0 6])
% ylim([6.11 6.1115])


%%------------ L=6, a0=1 -------------%%
% y=[0.76514513692653   0.77203410891583    0.77261257243453
                              0.77367044331709    0.77194124707928];
```

```
% stem(x,y,'ko:')
% Title('\lambda = 6')
% ylabel('a_0')
% xlim([0 6])

%% ylim([6.11 6.1115])
% text(0.5, 0.7735, 'a_0 (exact) = 1')
% ylim([0.755 0.78])


%%------------ L=6, a1=0 -------------%%
% y=[0.05825145602514   0.05617965889910    0.05825567565826
                                0.06699766762652    0.06307111143826];
% stem(x,y,'bo:')
% Title('\lambda = 6')
% ylabel('a_1')
% xlim([0 6])
%% ylim([6.11 6.1115])
% text(1, 0.065, 'a_1 (exact) = 0')


%%------------ L=6, b0=2 -------------%%
% y=[1.55297667130353   1.55404255356324    1.55292351884170
                                1.54597030721091    1.54703131628115];
% stem(x,y,'ko:')
% Title('\lambda = 6')
% ylabel('b_o')
% xlim([0 6])
% ylim([2.37 2.43])
% text(1, 1.56, 'b_o (exact) = 2')

%%------------ L=6, b1=3 -------------%%
% y=[2.40403634491437   2.40494375238060    2.40394316332675
                                2.39578631301634    2.39575881018459];
% stem(x,y,'bo:')
% text(1, 2.407, 'b_1 (exact) = 3')
% Title('\lambda = 6')
% ylabel('b_1')
% xlim([0 6])


%%------------ L=12, a0=1 -------------%%
% y=[0.97307956262148   0.96755154868810    0.97296429969120
                                0.99396497018149    0.98292876505362];
% stem(x,y,'ko:')
% Title('\lambda = 12')
% ylabel('a_o')
% xlim([0 6])
% ylim([0.94 1.02])
% text(1, 1.015, 'a_o (exact) = 1')


%%------------ L=12, a1=0 -------------%%
% y=[0.01235612920984   0.00373085885751    0.01226365668376
                                0.04655427747571    0.02956876833446];
% stem(x,y,'ko:')
% Title('\lambda = 12')
% ylabel('a_1')
% xlim([0 6])
% ylim([-0.02 0.06])
% text(1, 1.94725, 'a_1 (exact) = 0')
```

```
%%------------ L=12, b0=2 -------------%%
% y=[1.94618304069512   1.94648566973870    1.94613618137098
                                  1.94350585887879    1.94374640375645];
% stem(x,y,'ko:')
% Title('\lambda = 12')
% ylabel('b_o')
% xlim([0 6])
% ylim([1.93 1.955])
% text(1, 1.948, 'b_o (exact) = 2')


%%------------ L=12, b1=3 -------------%%
% y=[2.94373198375025   2.94403455964053    2.94367404814152
                                  2.94075967907800    2.94090502275678];
% stem(x,y,'ko:')
% Title('\lambda = 12')
% ylabel('b_1')
% xlim([0 6])
% ylim([2.93 2.955])
% text(1, 2.9483, 'b_1 (exact) = 3')
```

Index

A. Deb et al., *Analysis and Identification of Time-Invariant Systems, Time-Varying Systems, and Multi-Delay Systems using Orthogonal Hybrid Functions*, Studies in Systems, Decision and Control 46, DOI 10.1007/978-3-319-26684-8

Printed in the United States
By Bookmasters